Essential Precalculus

HOUGHTON MIFFLIN COMPANY   BOSTON
Dallas   Geneva, Illinois   Hopewell, New Jersey   Palo Alto   London

Essential Precalculus

Doris S. Stockton • *University of Massachusetts, Amherst*

Acknowledgments

The author is grateful to the following people for their invaluable help: Natalie F. Noble, for typing manuscript after manuscript with prompt precision and pleasant good humor; Teresa Farnum and Veronica Gold, for working exercises and tests and making professional comments; and William Swyter, Montgomery College; C. Kenneth Bradshaw, California State University—San Jose; Jerry Amburgey, College of Du Page; Marc Glucksman, El Camino College; Stanley Lukawecki, Clemson University; Douglas Hall, Michigan State University; Mark Hale, University of Florida; and H. D. Perry, Texas A&M University, for reviewing the manuscript.

COVER AND PART OPENING ILLUSTRATIONS BY JOHN C. HILL

Library of Congress Catalog Card Number: 77—75647

ISBN: 0—395—25417—5

Contents

Part 3 Precalculus Trigonometry

Part 4 Precalculus College Algebra

Appendixes

Answers to Quizzes and Tests

Odd-Numbered Exercise Answers; Selected Graphs and Proofs

Index

Preface

The purpose of this text is to provide a preparation for calculus with the material arranged so that it can be used for a wide variety of courses. It is designed for students who need a rapid review of manipulative algebra, or need the elementary functions, analytic geometry, college algebra, or trigonometry essential for calculus. The topics are arranged so that trigonometry can be omitted, and the text can be used as a specific preparation for a calculus course that does not require trigonometry as well as for a standard calculus course. Exercises and examples indicate ways in which the material is used in calculus.

The topics appear in four major parts of eight to eleven sections each. The four parts are

Part 1 Algebraic Review A rapid review of algebraic manipulations and definitions used in calculus.

Part 2 Elementary Functions and Analytic Geometry An introduction to the function concept and analytic geometry including graphs of the conic sections.

Part 3 Precalculus Trigonometry An introduction to trigonometry covering those topics considered absolutely essential for standard calculus.

Part 4 Precalculus College Algebra Topics from college algebra that are especially useful in calculus.

Pedagogical features include references in the margins to pertinent exercises and a final test for each of the four major parts. An informal statement of objectives is given at the beginning of each section, and a check list of objectives for the student, followed by a self-scoring section quiz, appears at the end of each section. The format is usable in either self-paced courses or standard lecture courses.

Numerous exercises are provided for each section. Illustrative examples are given in the exercise sets prior to most new types of exercises as well as being given in the body of the text. Groups of exercises utilize notation employed in the most widely used calculus texts. Emphasis is placed on manipulations and techniques that usually cause calculus students to stumble.

Properties of the real numbers are listed in an appendix rather than in the text proper. Proofs are very seldom given.

Tables of logarithms and trigonometric functions are included. These tables are used some in the text along with appropriate references to the comparative efficiency and accuracy of hand calculators. Computations are made both with tables and with a calculator.

A placement test, a diagnostic pretest keyed to Part 1, and additional final tests, as well as all exercise and test answers that do not appear in the text, are available in an instructor's manual.

Students not necessarily planning to take calculus may feel more comfortable with the version of this text designed for them. That version is *Essential Algebra and Trigonometry*, by this author and published by Houghton Mifflin Company.

Algebraic topics reviewed rapidly in the two versions of this text for the precalculus or college algebra and trigonometry student are treated in detail for the beginning algebra student in Doris S. Stockton, *Essential Algebra* (Glenview, Illinois: Scott, Foresman and Company, 1973). The author is grateful to Scott, Foresman and Company for permitting her to use bits of that material.

It is always a pleasure to thank my sons, Fred and Tom, and my husband, Fred D. Stockton, for their cheerful encouragement and unique understanding. It is likewise a privilege to salute my parents, Mary and Roland Skillman, and their fifty-seven plus years of marriage.

D.S.S.

Note To The Student

This book is designed to prepare you for calculus. It is arranged in four major parts subdivided into sections. At the end of each section there is a set of exercises and a check list of objectives for you. There are illustrative examples throughout each section and within each exercise set. References in the margin refer to pertinent exercises.

There is a self-scoring quiz at the end of each section and a final test at the end of each major part. Answers to most odd-numbered exercises, including many graphs, are given at the end of the book. For your convenience, the pages containing the answers to the section quizzes, as well as the pages containing the quizzes and your check lists, are tinted in a second color.

One way to use this book is as follows.

1. With pencil or pen in hand, study the text and work through the illustrative examples until you encounter a reference to exercises in the margin.

2. Try some of the recommended exercises; check your answers in the back of the book.

3. When you reach the end of a section, do more exercises, selecting them at random from the various types.

4. Read the check list; go back and review any areas in which you feel insecure.

5. Take and score the section quiz. If necessary, review, and then correct your answers.

6. When you reach the end of a major part, take the final test.

I want this book to be as helpful and as useful as possible. Please feel free to address comments or criticism to me in care of the publisher.

My best wishes to you.

D.S.S.

Essential Precalculus

Part 1 Algebraic Review

Calculus students are expected to be able to perform algebraic manipulations with ease so that they can direct their attention to the beautiful ideas and applications of calculus without being distracted or hampered by inability to do algebra. Part 1 of this text consists of a rapid review of algebraic manipulations and definitions utilized in calculus.

1.1 Common Sets of Numbers

The purpose of this section is to review the most commonly used sets of numbers.

The numbers usually used in counting are called the **counting numbers**, the **natural numbers**, or the **positive integers**. The set of all the positive integers may be written

$$\{1, 2, 3, 4, \ldots\}. \qquad \text{The positive integers}$$

The three dots within the braces indicate that all the rest of the positive integers are also included in the set. The numeral for a positive integer might be prefixed with a plus sign for emphasis. The set $\{1, 2, 3, 4, \ldots\}$ might be written

$$\{+1, +2, +3, +4, \ldots\}.$$

The set consisting of the positive integers together with zero, 0, is called the set of **whole numbers** and may be written

$$\{0, 1, 2, 3, 4, \ldots\}. \qquad \text{The whole numbers}$$

Assiciated with each positive integer there is a number called its *negative.* The **negative** of a positive integer is designated by writing the numeral for the positive integer with a negative or minus sign in front of it. For example, the negative of 3 is written -3 and is read ''negative three.'' Although this is also sometimes read ''minus three,'' it is advisable to reserve the use of the word *minus* for cases in which the symbol $-$ denotes the operation subtraction rather than the negative of a number. The negatives of all the positive integers constitute the set of **negative integers**.

$$\{-1, -2, -3, -4, \ldots\} \qquad \text{The negative integers}$$

The set consisting of all the positive integers, zero, and all the negative integers is called the set of **integers**.

$$\{\ldots, -4, -3, -2, -1, 0, 1, 2, 3, 4, \ldots\} \qquad \text{The integers}$$

The integers may be conveniently represented by points on a line called a *number line.* Some integers are shown on a number line in

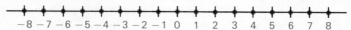

FIGURE 1.1 Integers on a number line

Figure 1.1. It is understood that the line continues indefinitely both to the left and to the right so that for each integer there is exactly one point on the number line.

The set of integers evenly divisible by the integer 2 is called the set of *even integers.* The set of integers other than the even integers is called the set of *odd integers.* Note that 0 is an even integer (0 divided by 2 is 0).

$$\{\ldots, -6, -4, -2, 0, 2, 4, 6, \ldots\} \quad \text{The even integers}$$
$$\{\ldots, -5, -3, -1, 1, 3, 5, \ldots\} \quad \text{The odd integers}$$

Any even integer can be considered to be of the form $2k$ where k is an integer. As indicated in the next display, the entire set of even integers can be generated by computing $2k$ as k successively takes on the values of all the integers:

$$k: \quad \ldots, -4, -3, -2, -1, 0, 1, 2, 3, 4, \ldots$$
$$2k: \quad \ldots, -8, -6, -4, -2, 0, 2, 4, 6, 8, \ldots$$

Any odd integer can be considered to be of the form $2k + 1$ or $2k - 1$ where k is an integer.

The entire set of odd integers can be generated by computing $2k + 1$ or $2k - 1$ as k successively takes on the values of all the integers.

$$k: \quad \ldots, -4, -3, -2, -1, \quad 0, 1, 2, 3, 4, \ldots$$
$$2k + 1: \quad \ldots, -7, -5, -3, -1, \quad 1, 3, 5, 7, 9, \ldots$$
$$2k - 1: \quad \ldots, -9, -7, -5, -3, -1, 1, 3, 5, 7, \ldots$$

A **rational number** is any number that can be expressed in the form m/n where m and n are integers and n is not zero. The symbols

$$\tfrac{2}{3}, \tfrac{17}{5}, \tfrac{-3}{7}, \tfrac{0}{6}, \tfrac{0}{1}, \tfrac{5}{1}, \text{ and } \tfrac{-8}{1}$$

represent rational numbers.

The word **fraction** is commonly used for any symbol of the form P/D where P and D are mathematical expressions. The P is called the **numerator** and the D is called the **denominator**. In this sense of the word *fraction,* any rational number can be written as a fraction whose numerator and denominator are integers and whose denominator is not zero.

Any rational number or any fraction can be written in more than one form. Two fractions P/D and R/S with nonzero denominators are said to be **equal** if the product *PS* equals the product *DR*. This definition leads to the so-called fundamental principle of fractions:

$$\frac{PN}{DN} = \frac{P}{D}, \qquad \text{for nonzero } N \text{ and } D.$$

$$\frac{5xy}{13y} = \frac{5x}{13}, \qquad \text{if } y \neq 0.$$

We can consider the integers to be rational numbers by identifying each integer with a rational number whose denominator is 1. For example, identify the integer 3 with the rational number 3/1; identify −7 with −7/1; identify 0 with 0/1.

For any integer *n*, let $n = n/1$.

We can represent any rational number as either a terminating or a repeating decimal by formally dividing its denominator into its numerator. For example,

$$\tfrac{3}{4} = 0.75 \qquad \tfrac{1}{3} = 0.333\ldots$$

$$\tfrac{-9}{2} = -4.5 \qquad \tfrac{-14}{11} = -1.272727\ldots$$

(In a repeating decimal, it is customary to designate the shortest complete repeating pattern by drawing a horizontal bar over it so that 0.333... could also be written $0.\overline{3}$, and −1.272727... could also be written $-1.\overline{27}$.) Likewise, each terminating or repeating decimal represents a rational number. For example, $1.5 = \tfrac{3}{2}$, $2.0 = 2$, and $-0.666\ldots = \tfrac{-2}{3}$. Thus we can say that the set of all rational numbers is the same as the set of all terminating or repeating decimals.

Each rational number is represented by exactly one point on a number line. Some rational numbers are shown on a number line in Figure 1.2.

FIGURE 1.2   Rational numbers on a number line

There do exist points on the number line that do not represent rational numbers. These points represent numbers called *irrational numbers*. An **irrational number** is not a rational number; it cannot be expressed in the form *m/n* where *m* and *n* are integers and *n* is not zero. An irrational number cannot be represented by a terminating or repeating decimal. Instead, any irrational number can be represented by a nonterminating, nonrepeating decimal. Likewise, any nonterminating, nonrepeating decimal represents an irrational number. We can say that the set of all nonterminating, nonrepeating decimals is the same as the set of all irrational numbers.

One example of an irrational number is the positive number whose square (whose product with itself) is equal to the integer 2. Algebraically, this irrational number is the positive solution of the equation $x \cdot x = 2$. It is denoted by the symbol $\sqrt{2}$. Another example of an irrational number is the number that first occurred in classical geometry and is commonly denoted by the Greek letter π. It is the ratio of the circumference to the diameter of any circle. Some other irrational numbers are $-\sqrt{2}$, $\sqrt{3}$, $-\sqrt{3}$, $\sqrt{5}$, and $-\sqrt{5}$. It can be shown that the sum or difference of any rational number and any irrational number is itself an irrational number. Likewise, the product or quotient of any nonzero rational number with any irrational number is itself irrational. Thus numbers such as $\frac{4}{5} + \sqrt{2}$, $7 - \sqrt{3}$, 2π, $3\sqrt{2}$, $\sqrt{2}/2$, and $2/\sqrt{3}$ are also irrational numbers.

Any irrational number can be approximated by a rational number. For example, a nine-decimal-place approximation to $\sqrt{2}$ is found on a hand calculator to be the rational number 1.414213562. Thus $\sqrt{2}$ is approximately equal to the three-place rational number 1.414.* Using the symbol $\simeq$ for "approximately equal to,"

$$\sqrt{2} \simeq 1.414.$$

Likewise, a fourteen-decimal-place approximation to π is the rational number 3.14159265358979. Thus π is approximately equal to the five-place rational number 3.14159.

$$\pi \simeq 3.14159$$

Note that the rational number $\frac{22}{7}$ is not equal to π. In fact, $\frac{22}{7}$ only agrees with π as far as the first two decimal places.

$$\frac{22}{7} = 3.\overline{142857}$$

An irrational number that is frequently encountered as a logarithm base in calculus and applications is denoted by the letter e. We have not developed sufficient vocabulary to define the number e precisely. Suffice it to say that e, as shown on a hand calculator, is approximately equal to the rational number 2.7182818284.

$$e \simeq 2.7182818284$$

The irrational numbers together with the rational numbers make up a set called the **real numbers**. Each real number is represented by exactly one point on a number line. Conversely, each point on a number line represents exactly one real number. Some real numbers are shown on the

---

*For one method of computing $\sqrt{2}$ to as many decimal places as you please, see Doris S. Stockton, *Essential Mathematics*, Scott, Foresman and Company, Glenview, Illinois, 1972, pp. 318–322.

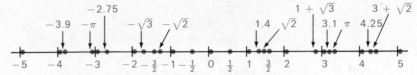

FIGURE 1.3   Real numbers on a number line

number line in Figure 1.3. Useful properties of the real numbers are listed in Appendix A.

Of any two real numbers, the one further to the right on a number line is said to be the greater. The set of all real numbers greater than 0 is called the set of **positive real numbers.** The set of all real numbers less than 0 is called the set of **negative real numbers.**

The symbol > is used for "is greater than," and the symbol < is used for "is less than." The symbols ≥ and ≤ are used for "is greater than or equal to" and "is less than or equal to," respectively. For example,

$$\pi > 3,$$
$$\pi < \tfrac{22}{7},$$

and
$$\sqrt{2} < \tfrac{3}{2}.$$

Also, if x is positive, then

$$x > 0,$$

and if x is nonnegative, then

$$x \geq 0.$$

It is sometimes convenient to denote sets of real numbers symbolically by using "set-builder notation." In this notation, the symbols

$$\{x \mid x \text{ is so and so}\}$$

are read

"the set of all x such that x is so and so,"

and unless otherwise specified, it is understood that x is a real number. Thus the set of all positive real numbers could be written $\{x \mid x > 0\}$, and these symbols could be read "the set of all x such that x is greater than zero." The set of all even integers could be written $\{x \mid x = 2k$ where k is an integer$\}$, and the set of all odd integers could be written $\{x \mid x = 2k + 1$ where k is an integer$\}$ or could be written $\{x \mid x = 2k - 1$ where k is an integer$\}$.

The graph of any set of real numbers is the set of all their corresponding points on a number line. The graphs of some common sets of real numbers

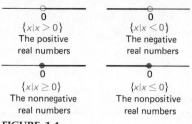

$$\{x|x>0\}$$
The positive
real numbers

$$\{x|x<0\}$$
The negative
real numbers

$$\{x|x\geq0\}$$
The nonnegative
real numbers

$$\{x|x\leq0\}$$
The nonpositive
real numbers

FIGURE 1.4

are shown in Figure 1.4. Heavy dots are used to indicate endpoints that are to be included in the graph, and circles are used to indicate excluded endpoints.

Exercise 1.1, 61–68

One commonly encountered set of real numbers is a set of positive integers called *primes*. A **prime** is any integer greater than the integer 1 which is evenly divisible by no positive integer other than itself and 1. The first five primes are 2, 3, 5, 7, and 11. Some primes are represented by dots on a number line in Figure 1.5.

0   2 3   5   7   11 13   17   19

FIGURE 1.5 Primes on a number line

We have noted that associated with each positive integer p there is a negative integer called its negative and denoted by $-p$. In general, associated with each real number x there is a real number called its negative and denoted by $-x$. The symbol $-x$ is read "negative x." The numbers x and $-x$ are represented by points on a number line on opposite sides of the zero point and equally distant from it. If x is positive, $-x$ is negative. If x is negative, $-x$ is positive. Also, the negative of $-x$ is x itself.

$$-(-x) = x$$
$$-(-7) = 7$$

The number of units on a number line between the point representing any real number and the point representing 0 is called the *absolute value* of the real number. For example, the absolute value of 3 is 3, and the absolute value of -3 is also 3, because 3 and -3 are each represented by a point located 3 units from the point representing 0. The absolute value of any real number x is denoted by the symbol $|x|$ and is formally defined as follows:

$$|x| = x, \qquad \text{if } x \text{ is positive or zero,}$$
$$|x| = -x, \qquad \text{if } x \text{ is negative.}$$

$$|-2| = -(-2) = 2$$

For example,

$$|3| = 3,$$
$$|0| = 0,$$

Exercise 1.1, 69–85 and
$$|-3| = -(-3) = 3.$$

We have noted that each real number is represented by exactly one point on a number line and that each point on a number line represents exactly one real number. There do exist numbers other than real numbers that arise quite naturally in algebra. Examples of such numbers are the solutions of the equation $x \cdot x = -4$. We know that no real number x is a solution of $x \cdot x = -4$ because the product of any real number with itself is not negative. (A real number is either positive, negative, or zero. A positive times a positive is positive, a negative times a negative is positive, and zero times zero is zero.) The solutions of the equation $x \cdot x = -4$ are numbers called *imaginary numbers*. If we define a new number denoted by the letter i and satisfying the requirement that $i \times i = -1$, then the solutions of $x \cdot x = -4$ are the so-called imaginary numbers $2i$ and $-2i$. This can be checked by substitution in the equation $x \cdot x = -4$.

$$x \cdot x = -4 \qquad\qquad x \cdot x = -4$$
$$(2i)(2i) = 4(i \times i) \qquad (-2i)(-2i) = +4(i \times i)$$
$$= 4(-1) \qquad\qquad = 4(-1)$$
$$= -4 \qquad\qquad = -4$$

$2i$ is a solution $\qquad -2i$ is a solution

In general, denoting $i \times i$ by the symbol i^2, an **imaginary number** is any number that can be written in the form $a + bi$ where b is a *nonzero* real number, a is any real number, and i is a number such that $i^2 = -1$.

$$a + bi \qquad\qquad \text{Imaginary number}$$

Real ⌐ ⌐ Nonzero real

$$i^2 = -1$$

For example, $3 + 2i$, $3 + (-2)i = 3 - 2i$, $0 + 2i = 2i$, and $0 + (-2)i = -2i$ are imaginary numbers.

Any real number or any imaginary number is a special kind of a more general class of numbers called *complex numbers*. A **complex number** is any number that can be written in the form $a + bi$ where a and b are real numbers (with b not necessarily nonzero) and $i^2 = -1$. The a is called the *real part*, and the b is called the *imaginary part* of the complex number $a + bi$.

PART ONE | ALGEBRAIC REVIEW

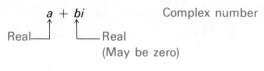

$$a + bi \qquad \text{Complex number}$$

Real $\llcorner$ $\llcorner$ Real
(May be zero)

$$i^2 = -1$$

For example,

$$\tfrac{3}{5} - 7i, \qquad \sqrt{2} + 6i, \qquad -7 + \pi i, \quad \text{and} \quad 8 + 0i$$

are complex numbers. Any real number is a special complex number in which the imaginary part equals zero. Any imaginary number is a special complex number in which the imaginary part does *not* equal zero. The numbers

$$7 = 7 + 0i, \qquad \tfrac{-3}{2} = \tfrac{-3}{2} + 0i, \qquad \text{and} \qquad \sqrt{2} = \sqrt{2} + 0i$$

are real numbers and complex numbers. The numbers

$$2i = 0 + 2i, \qquad 3 + 5i, \quad \text{and} \quad 3 - 5i$$

are imaginary numbers and complex numbers.

Note that we may say that

$$0 = 0 + 0i$$

is a real number and a complex number, and that

$$i = 0 + (1)i$$

is an imaginary number and a complex number. It is true that

$$0 \times i = 0$$

and $\qquad\qquad\qquad\qquad 1 \times i = i.$

In Sections 1.4 and 1.6 it will be shown that sums, differences, products, and quotients of complex numbers can be written in the form $a + bi$ where a and b are real numbers.

A diagram showing relationships we have noted among common sets of numbers appears in Figure 1.6 (see page 10).

EXAMPLE 1

Let A be the set

$$\{-7, 0, 7, \tfrac{2}{3}, \pi, \sqrt{2}, -\sqrt{2}, e, 3i, 2 + 5i, \tfrac{22}{7}\}.$$

List the integers, list the rational numbers, list the irrational numbers, and list the imaginary numbers in A.

Solution

The integers in A are -7, 0, and 7.

The rational numbers in A are -7, 0, 7, $\tfrac{2}{3}$, and $\tfrac{22}{7}$.

The irrational numbers in A are π, $\sqrt{2}$, $-\sqrt{2}$, and e.

Exercise 1.1, 1–15 The imaginary numbers in A are $3i$ and $2 + 5i$.

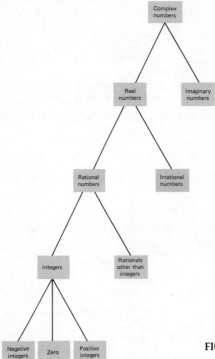

FIGURE 1.6 Relationships among common sets of numbers

EXAMPLE 2

List each of the following sets to which the number 3 belongs:

positive integers   real numbers
whole numbers   positive real numbers
negative integers   nonnegative real numbers
integers   negative real numbers
even integers   nonpositive real numbers
odd integers   primes
rational numbers   imaginary numbers
irrational numbers   complex numbers

Solution

Positive integers, whole numbers, integers, odd integers, rational numbers, real numbers, positive real numbers, nonnegative real numbers, primes, and complex numbers.

Exercise 1.1, 16–45

EXAMPLE 3

Classify each of the following as a terminating, repeating, or non-terminating and nonrepeating decimal, and state which kind of number, rational or irrational, it represents:

(a) $3.\overline{25}$ (b) 3.25 (c) $3.25225222522225\ldots$

(a) $3.\overline{25}$ is a repeating decimal. The pattern 25 repeats indefinitely. It represents a rational number.

(b) 3.25 is a terminating decimal. There are no nonzero digits beyond the second decimal place. It represents a rational number.

(c) 3.25225222522225 . . . is a nonterminating and nonrepeating decimal. After each digit 5 there appears one more digit 2 than appeared after the preceding 5, and this continues indefinitely. It represents an irrational number.

EXAMPLE 4

State which kind of decimal—terminating, repeating, or nonterminating and nonrepeating—represents each of the following:
(a) $\frac{3}{8}$ (b) $\sqrt{3}$ (c) $\frac{5}{3}$

Solution

(a) Formally dividing 8 into 3.000, we get 0.375. Thus $\frac{3}{8}$ is represented by a terminating decimal.

(b) $\sqrt{3}$ is an irrational number. It is represented by a nonterminating and nonrepeating decimal. (On a calculator, $\sqrt{3} \simeq 1.732050808$.)

(c) Formally dividing 3 into 5.000, we get 1.666 . . . $= 1.\overline{6}$. Thus $\frac{5}{3}$ is represented by a repeating decimal.

Exercise 1.1, 46–60

EXERCISE 1.1 COMMON SETS OF NUMBERS

▶ Let A be the set
$\{-5,\ -1,\ 0,\ 1,\ 4,\ \frac{3}{4},\ \pi,\ 7,\ 11,\ \sqrt{3},\ e,\ -3i,\ 4+3i,\ \frac{22}{7},\ -1.5,\ 1.\overline{5},$
$1.020020002 \ldots\}$.

In each of the following, list the specified members of A.

EXAMPLE

Nonnegative rational numbers

Solution

$0,\ 1,\ 4,\ \frac{3}{4},\ 7,\ 11,\ \frac{22}{7},\ 1.\overline{5}$

1. Nonnegative integers
2. Whole numbers
3. Negative integers
4. Integers
5. Even integers
6. Odd integers
7. Rational numbers
8. Irrational numbers
9. Real numbers
10. Nonpositive real numbers
11. Nonnegative real numbers
12. Primes
13. Imaginary numbers
14. Complex numbers
15. Nonpositive rational numbers

▶ List each of the following sets to which the given number belongs:

positive integers   real numbers
whole numbers   positive real numbers
negative integers   nonnegative real numbers
integers   negative real numbers
even integers   nonpositive real numbers
odd integers   primes
rational numbers   imaginary numbers
irrational numbers complex numbers

EXAMPLE

$\sqrt{2}$

Solution

Irrational numbers, real numbers, positive real numbers, nonnegative real numbers, and complex numbers.

16. 7   **17.** $\frac{2}{3}$ **18.** 0   **19.** $\sqrt{3}$

20. 125   **21.** -3 **22.** $\frac{-4}{5}$ **23.** $\frac{22}{7}$

24. -4 **25.** 17π **26.** 23   **27.** -23

28. $-\sqrt{2}$ **29.** $\sqrt{5}$ **30.** π **31.** e

32. $0 + 0i$ **33.** $2 - 3i$ **34.** $5i$ **35.** i

36. $3 + \sqrt{2}$ **37.** $-\sqrt{3}$ **38.** $2 + \sqrt{3}$ **39.** $-4\sqrt{2}$

40. $7 + 9i$ **41.** $\sqrt{2} + 3i$ **42.** $\sqrt{2}i$ **43.** $-3i$

44. $-\pi$ **45.** $\frac{8}{3}$

▶ Classify each of the following as a terminating, repeating, or nonterminating and nonrepeating decimal, and state which kind of number, rational or irrational, it represents.

EXAMPLE

$8.\overline{35}$

Solution

Repeating decimal; rational number

46. 1.25   **47.** $1.\overline{25}$ **48.** 4.37

49. $4.\overline{37}$ **50.** $4.373373337\ldots$ **51.** 23.5

52. $6.\overline{41}$ **53.** 1.0001   **54.** $1.\overline{01}$

55. $1.001000100001\ldots$ **56.** 1.414   **57.** 3.14159

58. 2.71828   **59.** 1.57   **60.** $5.\overline{9}$

▶ Use set-builder notation to denote each of the following sets, and make a graph of each set on a number line.

(a) All real numbers greater than or equal to 3

(b) All real numbers greater than 3 and less than 7

(c) All real numbers less than 3 or greater than 7

Solutions

(a) $\{x \mid x \geq 3\}$

0 1 2 3

(b) $\{x \mid 3 < x < 7\}$

0 1 2 3 4 5 6 7

(c) $\{x \mid x < 3 \text{ or } x > 7\}$

0 1 2 3 4 5 6 7

61. All real numbers greater than 2

62. All real numbers less than 2

63. All real numbers greater than 2 and less than 5

64. All real numbers less than 2 or greater than 5

65. All real numbers less than or equal to 2 and greater than or equal to -2

66. All integers less than or equal to 2 and greater than or equal to -2

67. All nonnegative integers less than or equal to 9

68. All nonpositive integers greater than -5

69. All integers with absolute value less than 5

70. All real numbers with absolute value less than 5

71. All real numbers with absolute value greater than or equal to 5

72. All real numbers with absolute value equal to $\frac{3}{2}$

73. All nonnegative integers with absolute value less than 7

74. All nonpositive integers with absolute value less than 7 and greater than 2

75. All real numbers greater than or equal to -3

▶ Find the value of each of the following.

EXAMPLE

$-|2 - 7|$

Solution

$-|2 - 7| = -|-5| = -(5) = -5$

76. $|9 - 11|$ **77.** $|9 - 9|$ **78.** $-|3 - 7|$

79. $-|-8|$ **80.** $-(-8)$ **81.** $-|-1 - 1|$

82. $-|1 - 1|$ **83.** $|-8| - |2|$ **84.** $|-8| + |-2|$

85. $(-8 - 2)$

CHECK LIST FOR THE STUDENT

If you have learned the material in Section 1.1, you should be able to do the following:

1. Write an expression representing each of the following sets: the positive integers, the whole numbers, the negative integers, the integers, the even integers, and the odd integers.

2. Complete the statement: A rational number is any number that can be expressed in the form m/n where m and n are _____ and n is _____.

3. State the definition of a fraction, and state the definition of the numerator and denominator of a fraction.

4. Complete the statement: $PN/DN = P/D$ for _____ N and D.

5. Complete the statement: The set of all rational numbers is the same as the set of all _____ or _____ decimals.

6. Complete the statement: The set of all irrational numbers is the same as the set of all _____, _____ decimals.

7. Give three-decimal-place approximations to $\sqrt{2}$, π, and e.

8. Use set-builder notation and write an expression for each of the following: the positive real numbers, the negative real numbers, the nonnegative real numbers, the nonpositive real numbers.

9. State the definition of a prime.

10. Complete the statement: If x is negative, then $-x$ is _____; if x is positive, then $-x$ is _____.

11. State the definition of the absolute value of any real number x.

12. State a definition of an imaginary number.

13. State a definition of a complex number; define the real part and the imaginary part of a complex number.

14. Show the relationships among the following sets on a diagram (see Figure 1.6): negative integers, zero, positive integers, integers, rational numbers other than integers, rational numbers, irrational numbers, real numbers, imaginary numbers, complex numbers.

SECTION QUIZ 1.1

The names of some common sets of numbers are: negative integers, positive integers, integers, rational numbers, irrational numbers, real numbers, imaginary numbers, and complex numbers. To which of these sets do the numbers in each of the following sets belong?

1. (a) $\{1, 2, 3, 4, \ldots\}$
   (b) $\{0, 1, 2, 3, 4, \ldots\}$

2. (a) $\{-1, -2, -3, -4, \ldots\}$
   (b) $\{\ldots, -4, -3, -2, -1, 0, 1, 2, 3, 4, \ldots\}$

3. (a) $\{2, 4, 6, 8, \ldots\}$
   (b) $\{\frac{2}{3}, \frac{3}{4}, \frac{4}{3}, 5, \frac{-7}{8}, 2.5, 0\}$

4. (a) $\{\sqrt{2}, \sqrt{3}, \pi)$
   (b) $\{2 + 3i, 5i, 2 - 3i\}$

Let A be the set

$$\{-3,\ -2,\ 0,\ 2,\ \tfrac{3}{5},\ 2.3,\ 2.\overline{3},\ 9,\ \sqrt{3},\ \pi,\ \tfrac{22}{7},\ 3.14,\ \tfrac{-2}{3},\ 1.732\}.$$

In each of the following, list the specified members of A.

5. All the rational numbers of A

6. All the irrational numbers of A

7. Classify each of the following as rational or irrational, and state which kind of decimal — terminating, repeating, or nonterminating and nonrepeating — represents it.
   (a) $\sqrt{2}$ (b) $\tfrac{3}{7}$ (c) $\tfrac{3}{5}$

Use set-builder notation to denote each of the following sets, and make a graph of each on a number line.

8. (a) All nonnegative real numbers
   (b) All negative real numbers

9. (a) All nonpositive real numbers
   (b) All positive real numbers

10. Find the value of $-|-7| + |-3|$.

1.2 Integral Exponents and Roots

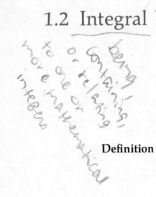

The purpose of this section is to review the definitions of integral exponents and roots of real numbers and to review the laws of exponents.

Our first main objective is to define a^m for real number a and integer m so that a^m is a real number. We will accomplish this by first defining a^k for positive integral k and then defining a^0 and a^{-k} for nonzero a. We will then define roots of real numbers and review the laws of exponents.

Definition If k is a positive integer and a is a real number, a^k represents the product of k a's.

$$a^5 = a \cdot a \cdot a \cdot a \cdot a$$

Five a's

$$a^2 = a \cdot a$$
$$a^1 = a$$

The number a^k is sometimes called the **kth power of a**. The a is called the **base,** and the k is called the **exponent** of a^k.

Note that if k is positive, $0^k = 0$.

Definition If a is any nonzero real number, $a^0 = 1$.

$$5^0 = 1 \qquad (\sqrt{2})^0 = 1$$
$$(\tfrac{-2}{3})^0 = 1 \qquad \underline{0^0 \text{ is not defined}}$$

Definition If k is a positive integer and a is any nonzero real number, $a^{-k} = 1/a^k$.

$$0^{-k} \text{ is not defined}$$
$$a^{-3} = 1/a^3, \qquad \text{if } a \text{ is not zero}$$
$$a^{-1} = 1/a, \qquad \text{if } a \text{ is not zero}$$

$$7^{-2} = 1/7^2 = \tfrac{1}{49} \qquad (\tfrac{3}{4})^{-2} = \frac{1}{(\tfrac{3}{4})^2} = \frac{1}{\tfrac{9}{16}} = \tfrac{16}{9}$$
$$7^{-1} = \tfrac{1}{7} \qquad (\tfrac{3}{4})^{-1} = \frac{1}{(\tfrac{3}{4})} = \tfrac{4}{3}$$

Definition If n is a positive integer and b is any real number, **an nth root of b** is a number whose nth power equals b.

For example, a fourth root of 81 is 3 because $3^4 = 81$. Another fourth root of 81 is -3 because $(-3)^4 = 81$.

If n is odd, there is exactly one real nth root of any real number b. It is designated by the symbol $\sqrt[n]{b}$. If $n = 3$, the nth root is called a **cube root.**

$$n \text{ odd} \rightarrow \text{Exactly one real } n\text{th root of } b, \ \sqrt[n]{b}$$
$$\sqrt[3]{8} = 2 \qquad \text{The only real cube root of 8}$$
$$\sqrt[3]{-8} = -2 \qquad \text{The only real cube root of } -8$$

If n is even and b is positive, there are exactly two real nth roots of b. One of these roots is positive, one is negative, and they have the same absolute value. The *positive root* is designated by the symbol $\sqrt[n]{b}$. If $n = 2$, an nth root is called a **square root**. It is customary to write $\sqrt{b}$ for $\sqrt[2]{b}$ so that the positive square root of positive number b is written $\sqrt{b}$.

n even, and

b positive $\rightarrow$ Exactly two real nth roots of b, $\sqrt[n]{b}$ and $-\sqrt[n]{b}$

$\sqrt{25} = 5$ The positive square root of 25

$-\sqrt{25} = -5$ The negative square root of 25

$\sqrt[4]{81} = 3$ The positive fourth root of 81

$-\sqrt[4]{81} = -3$ The negative fourth root of 81

If n is even and b is negative, there is no real nth root of b. In this case the nth roots of b are imaginary numbers and $\sqrt[n]{b}$ does not represent a real number.

n even, and

b negative $\rightarrow$ No real nth root of b

$\sqrt[n]{b}$ is not a real number.

$\sqrt[4]{-81}$ is not a real number.

$\sqrt{-9}$ is not a real number.

If n is 2 and b is negative, say $b = -p$ where p is positive, then $\sqrt{b} = \sqrt{-p}$ is defined to be the imaginary number $\sqrt{p}\,i$ where $i^2 = -1$.

$$\sqrt{-p} = \sqrt{p}\,i \quad \text{for } p \text{ positive.}$$
$$\sqrt{-2} = \sqrt{2}\,i$$
$$\sqrt{-9} = \sqrt{9}\,i$$
$$= 3i$$

If n is even or odd, $\sqrt[n]{0} = 0$.

$$\sqrt{0} = 0 \qquad \sqrt[3]{0} = 0$$

In the symbol $\sqrt[n]{b}$, n is called the **index**. The symbol $\sqrt[n]{b}$ itself is called a **radical**, and the b is called its **radicand**.

Our symbolism has been defined so that for any real number x, $\sqrt{x^2}$ is nonnegative. Thus if x is itself negative, $\sqrt{x^2}$ must be $-x$. If x is positive or zero, $\sqrt{x^2} = x$. In other words, for any real number x

$$\sqrt{x^2} = |x|.$$
$$\sqrt{2^2} = |2| = 2$$
$$\sqrt{(-2)^2} = |-2| = -(-2) = 2$$
$$\sqrt{0^2} = 0$$

Likewise, if n is odd, $\sqrt[n]{x^n} = x$. If n is even, the symbol $\sqrt[n]{x^n}$ is defined to be nonnegative, so if n is even, $\sqrt[n]{x^n} = |x|$.

$$\sqrt[n]{x^n} = x, \qquad \text{if } n \text{ is odd}$$
$$\sqrt[n]{x^n} = |x|, \qquad \text{if } n \text{ is even.}$$

$$\sqrt[3]{5^3} = 5, \qquad \sqrt[3]{(-5)^3} = -5, \qquad \sqrt[3]{0^3} = 0$$
$$\sqrt[4]{5^4} = 5, \qquad \sqrt[4]{(-5)^4} = |-5| = 5, \qquad \sqrt[4]{0^4} = 0$$

EXAMPLE 1

Find the value of each of the following.

(a) $\left(\frac{2}{3}\right)^2$ (b) $\left(\frac{2}{3}\right)^0$ (c) $\left(\frac{2}{3}\right)^{-2}$ (d) $\left(\frac{2}{3}\right)^{-1}$

(e) 0^2 (f) $\sqrt[3]{64}$ (g) $\sqrt[3]{-64}$ (h) $\sqrt{64}$

(i) $\sqrt{3^2}$ (j) $\sqrt{(-3)^2}$ (k) $\sqrt[3]{4^3}$ (l) $\sqrt[3]{(-4)^3}$

(m) $\sqrt[6]{5^6}$ (n) $\sqrt[6]{(-5)^6}$

Solution

(a) $\left(\frac{2}{3}\right)^2 = \frac{2}{3} \times \frac{2}{3} = \frac{4}{9}$ (b) $\left(\frac{2}{3}\right)^0 = 1$

(c) $\left(\frac{2}{3}\right)^{-2} = \dfrac{1}{\left(\frac{2}{3}\right)^2} = \dfrac{1}{\frac{4}{9}} = \frac{9}{4}$ (d) $\left(\frac{2}{3}\right)^{-1} = \dfrac{1}{\left(\frac{2}{3}\right)} = \frac{3}{2}$

(e) $0^2 = 0 \times 0 = 0$ (f) $\sqrt[3]{64} = 4$

(g) $\sqrt[3]{-64} = -4$ (h) $\sqrt{64} = 8$

(i) $\sqrt{3^2} = 3$ (j) $\sqrt{(-3)^2} = |-3| = 3$

(k) $\sqrt[3]{4^3} = 4$ (l) $\sqrt[3]{(-4)^3} = -4$

Exercise 1.2, 1–75 (m) $\sqrt[6]{5^6} = 5$ (n) $\sqrt[6]{(-5)^6} = |-5| = 5$

If a^r and a^s are real numbers with real-number base a and integral exponents r and s, then the given definitions can be used to show that the following so-called *exponent laws* are true. Since we shall note in the next section that these laws are also true for rational-number exponents r and s, we shall state them here for that more general case.

Exponent Laws If a^r, a^s, a^{rs}, and b^r are real numbers with a and b real and r and s rational, then

Law 1 $a^r a^s = a^{r+s}$

Law 2 $\dfrac{a^r}{a^s} = a^{r-s}, \quad a \neq 0$

Law 3 $(a^r)^s = a^{rs}$

Law 4 $(ab)^r = a^r b^r$

Law 5 $\left(\dfrac{a}{b}\right)^r = \dfrac{a^r}{b^r}, \qquad b \neq 0$

In Examples 2–6, let x and y represent real numbers.

EXAMPLE 2   (a) $x^3 x^4 = x^{3+4}$ Law 1: $a^r a^s = a^{r+s}$

                 $= x^7$

             (b) $10^3 \cdot 10^{-4} = 10^{3+(-4)}$ Law 1: $a^r a^s = a^{r+s}$

                       $= 10^{-1}$

                       $= \frac{1}{10}$

             (c) $2^{-3} \cdot 2^{-4} = 2^{-3+(-4)}$ Law 1: $a^r a^s = a^{r+s}$

                       $= 2^{-7}$

                       $= 1/2^7$

             (d) If $x \neq 0$,

                 $x \cdot x^{-1} = x^{1+(-1)}$ Law 1: $a^r a^s = a^{r+s}$

                       $= x^0$

                       $= 1$

EXAMPLE 3   (a) If $x \neq 0$,

                 $\dfrac{x^5}{x^3} = x^{5-3}$ Law 2: $\dfrac{a^r}{a^s} = a^{r-s}$,

                      $= x^2$ $a \neq 0$

             (b) If $x \neq 0$,

                 $\dfrac{x^3}{x^5} = x^{3-5}$ Law 2: $\dfrac{a^r}{a^s} = a^{r-s}$,

                      $= x^{-2}$ $a \neq 0$

                      $= 1/x^2$

             (c) If $x \neq 0$

                 $x^3/x^3 = x^{3-3}$ Law 2: $\dfrac{a^r}{a^s} = a^{r-s}$,

                      $= x^0$ $a \neq 0$

                      $= 1$

             (d) $\dfrac{10^{-3}}{10^{-2}} = 10^{-3-(-2)}$ Law 2: $\dfrac{a^r}{a^s} = a^{r-s}$,

                      $= 10^{-3+2}$ $a \neq 0$

                      $= 10^{-1}$

                      $= \frac{1}{10}$

EXAMPLE 4   (a) $(x^3)^2 = x^{3\times 2}$ Law 3: $(a^r)^s = a^{rs}$

                      $= x^6$

(b) If $x \neq 0$

$$(x^{-1})^2 = x^{(-1) \times (2)}$$

$$= x^{-2}$$

$$= \frac{1}{x^2}$$

Law 3: $(a^r)^s = a^{rs}$

(c) $(2^{-3})^{-2} = 2^{(-3) \times (-2)}$

$$= 2^6$$

$$= 64$$

Law 3: $(a^r)^s = a^{rs}$

EXAMPLE 5

(a) $(2x)^3 = 2^3 x^3$

$$= 8x^3$$

Law 4: $(ab)^r = a^r b^r$

(b) $(3\sqrt{2})^2 = 3^2 \times (\sqrt{2})^2$

$$= 9 \times 2$$

$$= 18$$

Law 4: $(ab)^r = a^r b^r$

(c) $(1.1 \times 10^{-3})^2 = (1.1)^2 \times (10^{-3})^2$

$$= 1.21 \times 10^{-6}$$

Law 4: $(ab)^r = a^r b^r$
Law 3: $(a^r)^s = a^{rs}$

EXAMPLE 6

(a) If $y \neq 0$

$$\left(\frac{x}{y}\right)^2 = \frac{x^2}{y^2}$$

Law 5: $\left(\dfrac{a}{b}\right)^r = \dfrac{a^r}{b^r}$,

$b \neq 0$

(b) $\left(\dfrac{\sqrt{2}}{2}\right)^2 = \dfrac{(\sqrt{2})^2}{2^2}$

$$= \tfrac{2}{4}$$

$$= \tfrac{1}{2}$$

Law 5: $\left(\dfrac{a}{b}\right)^r = \dfrac{a^r}{b^r}$,

$b \neq 0$

In the last part of Example 2 we saw that the product of nonzero x and x^{-1} is 1. The number $x^{-1} = 1/x$ is called the *reciprocal* of x. In general, each of two real numbers is called the **reciprocal** of the other provided that their product is 1.

EXAMPLE 7

Find the reciprocal of $\tfrac{2}{3}$.

Solution

The reciprocal of $\tfrac{2}{3}$ is $(\tfrac{2}{3})^{-1}$.

Exercise 1.2, 76–95

$$\left(\frac{2}{3}\right)^{-1} = \frac{1}{\frac{2}{3}} = \frac{3}{2}$$

We have asserted that the exponent laws are true for rational exponents and real-number bases. These laws are also true for integral exponents and bases that are powers of the imaginary number i where $i^2 = -1$. The powers of i are defined as are the powers of the nonzero real number a.

For k a positive integer, i^k represents the product of k i's, $i^{-k} = 1/i^k$, and $i^0 = 1$.

Note that even powers of i are either -1 or $+1$, and odd powers of i are either $-i$ or $+i$. For example, let us use the exponent laws and the definition of i to compute a few positive integral powers of i.

$$i^2 = -1 \qquad i^3 = i^2 \cdot i$$
$$= (-1) \cdot i \qquad i^2 = -1$$
$$= -i \qquad\qquad\qquad i^3 = -i$$

$$i^4 = (i^2)^2 \qquad i^5 = i^4 \cdot i$$
$$= (-1)^2 \qquad = 1 \cdot i \qquad i^4 = 1$$
$$= 1 \qquad\quad = i \qquad\qquad i^5 = i$$

$$i^6 = (i^2)^3 \qquad i^7 = i^6 \cdot i$$
$$= (-1)^3 \qquad = (-1) \cdot i \qquad i^6 = -1$$
$$= -1 \qquad\quad = -i \qquad\qquad i^7 = -i$$

Similarly, negative integral powers of i are either -1 or $+1$ or else are $-i$ or $+i$. For example:

$$i^{-2} = 1/i^2 \qquad i^{-1} = i^{-2} \cdot i$$
$$= \tfrac{1}{-1} \qquad\quad = (-1) \cdot i \qquad i^{-2} = -1$$

Exercise 1.2, 96–110

$$= -1 \qquad\qquad = -i \qquad\qquad i^{-1} = -i$$

The next two examples illustrate the use of integral exponents and the exponent laws in deriving basic calculus formulas. Example 8 suggests the definition of the irrational number e, and Example 9 demonstrates part of the derivation of the so-called power formula for integral exponent n. Once derived, the power formula is used frequently in calculus.

EXAMPLE 8

Find the values of $(1 + 1/n)^n$ for $n = 1, 2, 3$, and 4 to three decimal places, and fill in the blanks in the given table. Observe the three-place values of $(1 + 1/n)^n$ given for $n = 10, 100, 1000$, and 10,000. (These

n	$(1 + 1/n)^n$ to three places
1	___
2	___
3	___
4	___
10	2.594
100	2.705
1000	2.717
10,000	2.718

were obtained with a hand calculator.) What is the smallest one-place decimal that appears to be larger than any value of $(1 + 1/n)^n$?

Solution

Let $n = 1$. $(1 + \frac{1}{1})^1 = 2^1$

$$= 2$$

Let $n = 2$. $(1 + \frac{1}{2})^2 = (\frac{3}{2})^2$

$$= \frac{9}{4}$$

$$= 2.250$$

Let $n = 3$. $(1 + \frac{1}{3})^3 = (\frac{4}{3})^3$

$$= \frac{64}{27}$$

$$= 2.\overline{370}$$

Let $n = 4$. $(1 + \frac{1}{4})^4 = (\frac{5}{4})^4$

$$= \frac{625}{256}$$

$$= 2.441 \text{ to three places}$$

From the table we see that as n successively takes on larger and larger values, the values of $(1 + 1/n)^n$ gradually become larger. But as the values of $(1 + 1/n)^n$ become larger, they appear all to be less than the one-place decimal 2.8. The decimal 2.8 is the smallest one-place decimal that is larger than any value of $(1 + 1/n)^n$. Using techniques of the calculus, it can be shown that as n increases, the values of $(1 + 1/n)^n$ approach an irrational number less than 2.8. That irrational number is denoted by the letter e. As we said in Section 1.1, e is approximately equal to 2.7182818284. By making n large enough, we can make the value of $(1 + 1/n)^n$ be as close to the value of e as we please; and for all still larger values of n, the values of $(1 + 1/n)^n$ will be even closer to the value of e. We might say that e is the "limit" of $(1 + 1/n)^n$ as n increases indefinitely.

EXAMPLE 9

Let n be a positive integer, and let x be a positive real number. Show that if h were set equal to zero, the expression $(x + h)^{n-1} + (x + h)^{n-2}x + (x + h)^{n-3}x^2 + (x + h)^{n-4}x^3 + \cdots + (x + h)x^{n-2} + x^{n-1}$ would equal nx^{n-1}.

Solution

The three dots in the given expression represent missing terms that continue to follow the law of formation indicated by the first three terms. That is, in each term the power of $(x + h)$ is one less than in the preceding term, whereas the power of x is one more than in the preceding term.

Note that there are n terms. Indeed, counting left to right, the exponent on the x is one less than the number of the term. The term with x^2 is the third term, the term with x^3 is the fourth term, etc.; so the last term, the term with x^{n-1}, must be the nth term.

If $h = 0$, the given expression becomes:

$$x^{n-1} + x^{n-2}x + x^{n-3}x^2 + x^{n-4}x^3 + \cdots + xx^{n-2} + x^{n-1}$$

n terms

$$= x^{n-1} + x^{n-2+1} + x^{n-3+2} + x^{n-4+3} + \cdots + x^{1+n-2} + x^{n-1}$$

n terms

$$= x^{n-1} + x^{n-1} + x^{n-1} + x^{n-1} + \cdots + x^{n-1} + x^{n-1}$$

n terms

Exercise 1.2, 111–114 $= nx^{n-1}.$

EXERCISE 1.2 INTEGRAL EXPONENTS AND ROOTS

▶ Find the value of each of the following.

EXAMPLES

(a) $\sqrt[3]{\frac{-1}{8}}$ (b) $(\sqrt{3})^2$ (c) $(\sqrt{-3})^2$

Solutions

(a) $\sqrt[3]{\frac{-1}{8}} = -\frac{1}{2}$ (b) $(\sqrt{3})^2 = 3$

(c) $(\sqrt{-3})^2 = (\sqrt{3}i)^2$

$\qquad\qquad = (\sqrt{3})^2 i^2$

$\qquad\qquad = 3(-1)$

$\qquad\qquad = -3$

1. $(\frac{3}{5})^2$ 2. $(\frac{3}{5})^0$ 3. $(\frac{3}{5})^{-2}$ 4. $(\frac{3}{5})^{-1}$

5. 0^3 6. $\sqrt[3]{27}$ 7. $\sqrt[3]{-27}$ 8. $\sqrt{25}$

9. $\sqrt{\frac{1}{25}}$ 10. $\sqrt{17^2}$ 11. $\sqrt{(-17)^2}$ 12. $\sqrt[3]{8^3}$

13. $\sqrt[3]{(-8)^3}$ 14. $\sqrt[6]{2^6}$ 15. $\sqrt[6]{(-2)^6}$ 16. $(\frac{-7}{8})^2$

17. $(\sqrt{2})^2$ 18. $(\sqrt{2})^0$ 19. $(\sqrt{2})^{-2}$ 20. $(\frac{1}{10})^{-1}$

21. 0^{23} 22. $\sqrt[5]{32}$ 23. $\sqrt[5]{-32}$ 24. $\sqrt{16}$

25. $\sqrt{11^2}$ 26. $\sqrt{(-11)^2}$ 27. $\sqrt[5]{7^5}$ 28. $\sqrt[5]{(-7)^5}$

29. $\sqrt[4]{7^4}$ 30. $\sqrt[4]{(-7)^4}$ 31. $\sqrt{4^2}$ 32. $\sqrt{(-4)^2}$

33. $(\sqrt{-4})^2$ 34. $\sqrt{9^2}$ 35. $\sqrt{(-9)^2}$ 36. $(\sqrt{-9})^2$

37. $\sqrt{7^2}$ 38. $\sqrt{(-7)^2}$ 39. $(\sqrt{7})^2$ 40. $(\sqrt{-7})^2$

▶ In each of the following, if the expression is defined and is a real number, give its value. If the expression is not defined or is not real, say "not defined" or "not real."

EXAMPLES

(a) $\sqrt{0^3} = \sqrt{0} = 0$ (b) $3 \times 0 = 0$

(c) $3 + 0 = 3$ (d) $1/0^3 = 1/0$ is not defined

(e) $3^0 = 1$ (f) $(\sqrt{0})^0 = 0^0$ is not defined

(g) $\sqrt{-25}$ is not real   (h) $\sqrt[4]{-81}$ is not real

41. 0^3

42. $\sqrt[3]{0}$

43. 0^{-3}

44. 0^0

45. 0×0

46. $\sqrt{0}$

47. $(\sqrt{0})^0$

48. 1^0

49. 0^1

50. $\frac{1}{0}$

51. 0^{-1}

52. $\frac{0}{1}$

53. 1×0

54. $\frac{0}{0}$

55. $\frac{3}{0}$

56. $\frac{0}{3}$

57. $(0^2)^0$

58. $(3 \times 0)^2$

59. $(\frac{0}{1})^3$

60. $(\frac{0}{1})^4$

61. $\frac{-1}{0}$

62. 0^{-2}

63. $\frac{-2}{0}$

64. $\frac{0}{-2}$

65. $(-2) \times 0$

66. $\sqrt{-1}$

67. $\sqrt{(-36)^2}$

68. $\sqrt{-36}$

69. $\sqrt{-2}$

70. $\sqrt[4]{16}$

71. $\sqrt[4]{-16}$

72. $\sqrt{-81}$

73. $\sqrt[4]{1}$

74. $\sqrt[3]{1}$

75. $\sqrt[3]{-1}$

▶ Let x and y represent real numbers, and express each of the following in a simplified form containing no negative exponents. State any restrictions that must be placed on x and y so that the given expressions will be defined.

EXAMPLES

(a) $\dfrac{xy^{-2}}{x^2y}$ (b) $(3x)^0$ (c) $(\sqrt{x})^2$

Solutions

(a) $\dfrac{1}{xy^3}$, $x \neq 0, y \neq 0$

(b) $(3x)^0 = 1$, $x \neq 0$

(c) $(\sqrt{x})^2 = x$

76. $\dfrac{3x^3y^2}{7xy^3}$

77. $(x^3y^2)(x^5y^7)$

78. $\dfrac{x^3y^{-5}}{x^5y^2}$

79. $(\sqrt{2}xy)^2$

80. $(5x^2)^0$

81. $(\sqrt{2x})^2$

82. $\sqrt{4x^2}$

83. $x^{-3}y^{-3}$

84. $(x^3y^3)(x^{-3}y^{-3})$

85. $\dfrac{y^{-2}}{y^{-1}}$

86. $(x^3y^2)^{-1}$

87. $(x^2y^5)^2$

88. $(xy)(xy)^{-1}$

89. $\left(\dfrac{5x^3y^{-7}}{\sqrt{2}x^{-1}y^{-1}}\right)^2$

90. $\left(\dfrac{7x^{-5}y^{-9}}{\sqrt{3}x^2y^{-2}}\right)^{-1}$

91. $\left(\dfrac{x}{y}\right)^{-1}$

92. $\left(\dfrac{3x^2}{2y^3}\right)^{-1}$

93. $\left(\dfrac{\sqrt{2}x^{-1}}{y^2}\right)^2$

94. $\left(\dfrac{x}{y}\right)^{-2}$

95. $\left(\dfrac{x}{\sqrt{3}}\right)^2$

▶ Express each of the following as either -1 or $+1$ or as $-i$ or $+i$.

    96. i^8 **97.** i^9

    98. i^{-3} **99.** i^{-4}

    100. i^{10} **101.** i^{11}

    102. i^{-5} **103.** i^{-6}

    104. $i^6 \times i^2$ **105.** $i^{-3} \times i^7$

    106. $(i^{-1})^2$ **107.** i^{4k+2} where k is an integer

    108. i^{4k} where k is an integer   **109.** i^{4k+1} where k is an integer

    110. i^{4k+3} where k is an integer

▶ In each of the following let h be a positive integer, and let x be a positive real number. Show that if h were set equal to zero, the first expression would equal the second expression.

EXAMPLE

Solution

$(x + h)^3 + (x + h)^2 x + (x + h)x^2 + x^3; \; 4x^3$

Let $h = 0$. The first expression becomes

$$x^3 + x^2 x + xx^2 + x^3 = x^3 + x^3 + x^3 + x^3 = 4x^3.$$

    111. $(x + h)^2 + (x + h)x + x^2; \; 3x^2$

    112. $(x + h)^4 + (x + h)^3 x + (x + h)^2 x^2 + (x + h)x^3 + x^4; \; 5x^4$

    113. $(x + h)^5 + (x + h)^4 x + (x + h)^3 x^2 + (x + h)^2 x^3 + (x + h)x^4 + x^5; \; 6x^5$

    114. $(x + h)^6 + (x + h)^5 x + (x + h)^4 x^2 + (x + h)^3 x^3 + (x + h)^2 x^4 + (x + h)x^5 + x^6; \; 7x^6$

CHECK LIST FOR THE STUDENT

If you have learned the material in Section 1.2, you should be able to do the following:

1. State the meaning of a^k where a is a real number and k is a positive integer.

2. State the meaning of a^0 where a is any nonzero real number.

3. State the meaning of a^{-k} where a is any nonzero real number and k is a positive integer.

4. State the meaning of an nth root of b where n is a positive integer and b is any real number.

5. Tell how many real nth roots of b there are if n is an odd integer and b is any real number.

6. Tell how many real nth roots of b there are if n is an even integer and b is any positive real number.

7. State the meaning of $\sqrt{b}$ where b is any positive real number.

8. Tell how many real nth roots of b there are if n is an even integer and b is any negative real number.

9. Give the names for n and b in the symbol $\sqrt[n]{b}$.

10. Complete the statement: For any real number x, $\sqrt{x^2} = $ _____; $\sqrt[n]{x^n} = $ _____ if n is odd; $\sqrt[n]{x^n} = $ _____ if n is even.

11. State and use the five exponent laws.

12. State the definition of the reciprocal of nonzero x.

13. Express any given integral power of i as either -1, $+1$, $-i$, or $+i$.

14. Tell what irrational number the expression $(1 + 1/n)^n$ approaches as n takes on larger and larger values and increases indefinitely.

SECTION QUIZ 1.2

Find the value of each of the following.

1. (a) 5^3 (b) 5^1 (c) 5^{-3} (d) 5^0 (e) 0^5

2. (a) $\sqrt[3]{64}$ (b) $\sqrt[3]{-64}$ (c) $\sqrt{9}$ (d) $-\sqrt{9}$ (e) $\sqrt[3]{(-5)^3}$

3. (a) $(\sqrt{5})^2$ (b) $\sqrt{0}$ (c) $\sqrt{(3)^2}$ (d) $\sqrt{(-3)^2}$ (e) $(\frac{3}{5})^{-2}$

4. In each of the following, if the expression is defined and is a real number, give its value. If the expression is not defined or is not real, say "not defined" or "not real."

    (a) $5 + 0$ (b) 5×0 (c) $\frac{5}{0}$ (d) 0^0 (e) $\sqrt{-9}$

5. Let x and y represent real numbers, and give the integer that belongs in each of the following parentheses.

(a) $x^2 x^5 = x^{(\ \ )}$ (b) $\dfrac{x^5}{x^2} = x^{(\ \ )}$, $x \neq 0$ (c) $(x^2)^5 = x^{(\ \ )}$

(d) $(xy)^5 = x^{(\ \ )} y^{(\ \ )}$ (e) $\left(\dfrac{x}{y}\right)^5 = \dfrac{x^{(\ \ )}}{y^{(\ \ )}}$, $y \neq 0$

Let x and y represent real numbers, and express each of the following in a simplified form containing no negative exponents. State any restrictions that must be placed on x and y so that the given expressions will be defined.

6. $(x^3 y^{-2})(x^{-5} y^3)$ **7.** $\dfrac{xy^{-3}}{x^3 y^2}$ **8.** $\left(\dfrac{x^3}{y^2}\right)^{-1}$ **9.** $\left(\dfrac{3x^2 y^{-5}}{4x^{-1} y^{-1}}\right)^2$

10. Let $i^2 = -1$, and express each of the following as either -1, $+1$, $-i$, or $+i$:
(a) i^4 (b) i^3

1.3 Rational Exponents and Radicals

The purpose of this section is to review the definition of rational exponents, to continue to review the exponent laws, and to review properties of radicals.

Our primary objective is to extend the definition of a^r for real numbers a to include rational numbers r that are not necessarily integers and to state this definition so that a^r will be a real number and the exponent laws will hold true. To this end, let the rational number r be written in the form m/n where m and n are integers and n is positive. Then a^r is $a^{m/n}$, and we define

$$a^{m/n} = (\sqrt[n]{a})^m$$

where a and m are not both zero, a is nonzero if m is negative, and a is nonnegative if n is even. Note that a and m cannot both be zero because then $a^{m/n}$ would be 0^0, and 0^0 is not defined. Also, a must be nonzero if m is negative, because otherwise $(\sqrt[n]{a})^m$ would be of the form 0^{-k} where $-k$ is a negative integer, and such 0^{-k} is not defined. Finally, a must be nonnegative if n is even, because otherwise a^r might not be a real number. For example,

$$0^{0/2} \quad \text{is not defined}$$
$$0^{-3/2} \quad \text{is not defined}$$
$$(-16)^{3/2} \quad \text{is not defined.}$$

EXAMPLE 1

Find the values of $8^{2/3}$, $8^{-2/3}$, $(-8)^{2/3}$, $(16)^{3/2}$, and $0^{4/5}$.

Solution

$$8^{2/3} = (\sqrt[3]{8})^2 \qquad 8^{-2/3} = (\sqrt[3]{8})^{-2} \qquad (-8)^{2/3} = (\sqrt[3]{-8})^2$$
$$= 2^2 \qquad\qquad\quad = (2)^{-2} \qquad\qquad\quad = (-2)^2$$
$$= 4 \qquad\qquad\quad = \tfrac{1}{4} \qquad\qquad\qquad = 4$$

$$16^{3/2} = (\sqrt{16})^3 \qquad 0^{4/5} = (\sqrt[5]{0})^4$$
$$= 4^3 \qquad\qquad\quad = 0^4$$
$$= 64 \qquad\qquad\quad = 0$$

Note that $0^{m/n} = 0$ provided that m and n are both positive integers. Note also that $b^{1/n} = \sqrt[n]{b}$ provided that $\sqrt[n]{b}$ is a real number.

$$0^{2/3} = 0$$
$$25^{1/2} = \sqrt{25} = 5$$
$$64^{1/3} = \sqrt[3]{64} = 4$$

In defining $a^{m/n}$, we have placed restrictions on m, n, and a so that $\sqrt[n]{a}$ is a real number and $a^{m/n} = (\sqrt[n]{a})^m$ is a real number. In that case, $(\sqrt[n]{a})^m$ is equal to $\sqrt[n]{a^m}$ so that we can write

$$a^{m/n} = (\sqrt[n]{a})^m = \sqrt[n]{a^m}$$

provided that $\sqrt[n]{a}$ is a real number. For example:

$$8^{2/3} = (\sqrt[3]{8})^2 \quad \text{and} \quad 8^{2/3} = \sqrt[3]{8^2}$$
$$= 2^2 \qquad\qquad\qquad = \sqrt[3]{64}$$
$$= 4 \qquad\qquad\qquad = 4$$

$$4^{-3/2} = (\sqrt{4})^{-3} \quad \text{and} \quad 4^{-3/2} = \sqrt{4^{-3}}$$
$$= (2)^{-3} \qquad\qquad\qquad = \sqrt{1/4^3}$$
$$= 1/2^3 \qquad\qquad\qquad = \sqrt{\tfrac{1}{64}}$$
$$= \tfrac{1}{8} \qquad\qquad\qquad = \tfrac{1}{8}$$

The question sometimes arises of whether or not it is permissible to reduce a rational exponent to lowest terms. If $mk/nk = m/n$, can we say that $a^{mk/nk} = a^{m/n}$? It is true that $\frac{2}{6} = \frac{1}{3}$, but is it true that $a^{2/6} = a^{1/3}$? Not if a is negative, because $a^{2/6}$ has not been defined for negative a. If n is even and a is negative, $\sqrt[n]{a}$ is not a real number, and $a^{m/n}$ is not defined. If, however, $\sqrt[n]{a}$ and $\sqrt[nk]{a}$ are both real numbers so that $a^{m/n}$ and $a^{mk/nk}$ are both defined, then it is true that $a^{mk/nk} = a^{m/n}$ for any positive integer k. This says that

$$(64)^{2/6} = (64)^{1/3},$$

but it does not say that $(-64)^{2/6}$ is equal to $(-64)^{1/3}$, because $\sqrt[6]{-64}$ is not a real number and $(-64)^{2/6}$ has not been defined. If the base a of exponent m/n is nonnegative and $a^{m/n}$ is defined we can be sure that the value of $a^{m/n}$ does not depend on the form of the rational exponent. In particular, if a is positive,

$$a^{mk/nk} = a^{m/n}, \qquad m, n, \text{ and } k \text{ are integers},$$
$$n \text{ and } k \text{ are positive}.$$

This equation is also true if a is zero and both exponents are positive.

EXAMPLE 2   Find the value of $8^{14/21}$ and $(49)^{9/18}$.

Solution $8^{14/21} = 8^{2/3} = (\sqrt[3]{8})^2 = 2^2 = 4$

Exercise 1.3, 1–50 $(49)^{9/18} = (49)^{1/2} = \sqrt{49} = 7$

The given definitions can be used to show that the exponent laws stated in Section 1.2 are true for rational exponents. Use of these laws is demonstrated in the following examples.

EXAMPLE 3 $8^2 \times 8^{1/3} = 8^{2+(1/3)}$

$$= 8^{7/3}$$
$$= 128 \qquad\qquad \text{Law 1: } a^r a^s = a^{r+s}$$

EXAMPLE 4

$$\frac{8^{4/3}}{8^{1/3}} = 8^{(4/3)-(1/3)}$$

Law 2: $\dfrac{a^r}{a^s} = a^{r-s}, \quad a \neq 0$

$$= 8^{3/3}$$

$$= 8$$

EXAMPLE 5

$$(27^{1/3})^2 = 27^{2/3}$$

Law 3: $(a^r)^s = a^{rs}$

$$= 9$$

EXAMPLE 6

$$(3\sqrt{5})^2 = 3^2(\sqrt{5})^2$$

Law 4: $(ab)^r = a^r b^r$

$$= 9 \times 5$$

$$= 45$$

EXAMPLE 7

$$\left(\frac{4}{9}\right)^{3/2} = \frac{4^{3/2}}{9^{3/2}}$$

Law 5: $\left(\dfrac{a}{b}\right)^r = \dfrac{a^r}{b^r}, \quad b \neq 0$

Exercise 1.3, 51–60

$$= \tfrac{8}{27}$$

In beginning calculus, one often has occasion to use the exponent laws in performing algebraic simplifications such as those illustrated in the next three examples.

EXAMPLE 8

Express the following in the form ax^r where a is a real number written with no negative or fractional exponents, and r is a rational number. Assume that x is a positive real number.

$$\tfrac{1}{2}(3x^5)^{(1/2)-1}(15x^4)$$

Solution

$$\tfrac{1}{2}(3x^5)^{(1/2)-1}(15x^4) = \tfrac{15}{2}(3x^5)^{-1/2}x^4$$

$$= \tfrac{15}{2}(3)^{-1/2}x^{-5/2}x^4$$

$$= \tfrac{15}{2} \times \tfrac{1}{\sqrt{3}}x^{4-(5/2)}$$

$$= \frac{15}{2\sqrt{3}}x^{3/2}$$

EXAMPLE 9

Let m and n be integers with n not zero, and let x be a positive real number. Show that

$$m(x^{1/n})^{m-1} \cdot \frac{1}{n}x^{(1/n)-1} \quad \text{is equal to} \quad \frac{m}{n}x^{(m/n)-1}.$$

| Solution | $m(x^{1/n})^{m-1} \cdot \dfrac{1}{n}x^{(1/n)-1} = mx^{(m/n)-(1/n)} \cdot \dfrac{1}{n}x^{(1/n)-1}$ | Law 3: $(a^r)^s = a^{rs}$ |

$$m\left(x^{\frac{1}{n}}\right)^{m-1}$$
$$m\left(x^{\frac{1}{n} \cdot m-1}\right)$$
$$m\left(x^{\frac{m-1}{n}}\right)$$
$$m\,x^{\frac{m}{n}-\frac{1}{n}}$$

$$= \frac{m}{n}x^{(m/n)-(1/n)}x^{(1/n)-1}$$

$$= \frac{m}{n}x^{(m/n)-(1/n)+(1/n)-1} \qquad \text{Law 1: } a^r a^s = a^{r+s}$$

$$= \frac{m}{n}x^{(m/n)-1}$$

This last example represents part of a proof of the "power formula for rational exponents." That formula is used very frequently in differential calculus. Calculus students need to be able to perform manipulations such as these with great facility. The next example demonstrates part of the manipulations involved in deriving a special case of the power formula. Such work should be routinely easy for beginning calculus students.

EXAMPLE 10 — Express the following in the form ax^r where a is a real number, r is a rational number, and x is a positive real number.

$$\frac{1}{x^{2/3} + x^{1/3}x^{1/3} + x^{2/3}}$$

Solution

$$\frac{1}{x^{2/3} + x^{1/3}x^{1/3} + x^{2/3}} = \frac{1}{x^{2/3} + x^{2/3} + x^{2/3}} \qquad \text{Law 1: } a^r a^s = a^{rs}$$

$$= \frac{1}{3x^{2/3}}$$

$$= (3x^{2/3})^{-1}$$

$$= 3^{-1}x^{-2/3} \qquad \text{Law 4: } (ab)^r = a^r b^r$$

Exercise 1.3, 61–70 ax^r $= \tfrac{1}{3}x^{-2/3}$

Since a radical such as $\sqrt[n]{b}$ can be expressed in exponential form, radicals inherit properties from the exponent laws. Two of the most commonly used of these properties follow.

Properties of Radicals — If $\sqrt[n]{a}$ and $\sqrt[n]{b}$ are real numbers, then:

1. $\sqrt[n]{ab} = \sqrt[n]{a}\,\sqrt[n]{b}$ (Derived from Law 4: $(ab)^r = a^r b^r$)

2. $\sqrt[n]{\dfrac{a}{b}} = \dfrac{\sqrt[n]{a}}{\sqrt[n]{b}}, \ b \neq 0$ (Derived from Law 5: $\left(\dfrac{a}{b}\right)^r = \dfrac{a^r}{b^r}, \ b \neq 0$)

EXAMPLE 11 — Use the properties of radicals to find the values of $\sqrt[3]{64 \times 1000}$ and $\sqrt[3]{\frac{-27}{8}}$.

Solution

$$\sqrt[3]{64 \times 1000} = \sqrt[3]{64} \times \sqrt[3]{1000} \qquad \text{Property 1: } \sqrt[n]{ab} = \sqrt[n]{a}\,\sqrt[n]{b}$$
$$= 4 \times 10$$
$$= 40$$

$$\sqrt[3]{\frac{-27}{8}} = \frac{\sqrt[3]{-27}}{\sqrt[3]{8}} \qquad \text{Property 2: } \sqrt[n]{\frac{a}{b}} = \frac{\sqrt[n]{a}}{\sqrt[n]{b}},\ b \neq 0$$

Exercise 1.3, 71–80

$$= \frac{-3}{2}$$

In the following examples assume that the variables are all positive so that all the radicals represent real numbers.

EXAMPLE 12

Write $\sqrt[3]{5x^3y^6z^{10}}$ in a form whose radicand contains no factor with an exponent greater than or equal to the index 3.

Solution

$$\sqrt[3]{5x^3y^6z^{10}} = \sqrt[3]{5x^3y^6z^9 \cdot z}$$
$$= \sqrt[3]{5z}\,\sqrt[3]{x^3y^6z^9}$$
$$= \sqrt[3]{5z}\,\sqrt[3]{x^3}\,\sqrt[3]{y^6}\,\sqrt[3]{z^9}$$
$$= \sqrt[3]{5z} \cdot x \cdot y^2 \cdot z^3$$
$$= \sqrt[3]{5z}\,xy^2z^3$$

Some radicals can be written in an equivalent form having a smaller index. For example,

$$\sqrt[6]{x^2} = x^{2/6} = x^{1/3} = \sqrt[3]{x}.$$

The next example demonstrates this with a more complicated radicand.

EXAMPLE 13

Write $\sqrt[10]{49x^4y^6}$ in a form that cannot be rewritten with a smaller index.

Solution

$$\sqrt[10]{49x^4y^6} = \sqrt[10]{(7x^2y^3)^2}$$
$$= (7x^2y^3)^{2/10}$$
$$= (7x^2y^3)^{1/5}$$
$$= \sqrt[5]{7x^2y^3}$$

If a radical has a fractional radicand, it can be written otherwise. For example, $\sqrt{2/3}$ can be written $\sqrt{6}/3$ with no fraction in the radicand.

$$\sqrt{\frac{2}{3}} = \sqrt{\frac{6}{9}}$$

$$= \frac{\sqrt{6}}{\sqrt{9}} \qquad \text{Property 2: } \sqrt[n]{\frac{a}{b}} = \frac{\sqrt[n]{a}}{\sqrt[n]{b}},\ b \neq 0$$

$$= \frac{\sqrt{6}}{3}$$

EXAMPLE 14

Assume that the variables are positive and write $\sqrt[3]{7x/2y}$ in a form having no fraction in the radicand.

Solution

$$\sqrt[3]{\frac{7x}{2y}} = \sqrt[3]{\frac{(7x)(2y)^2}{(2y)(2y)^2}}$$

(handwritten) $\dfrac{7x \cdot 4y^2}{2y \cdot 4y^2} = \dfrac{28xy^2}{8y^3} = \dfrac{28xy^2}{(2y)^3}$

$$= \sqrt[3]{\frac{28xy^2}{(2y)^3}}$$

$$= \frac{\sqrt[3]{28xy^2}}{\sqrt[3]{(2y)^3}} \qquad \text{Property 2: } \sqrt[n]{\frac{a}{b}} = \frac{\sqrt[n]{a}}{\sqrt[n]{b}}, \ b \neq 0$$

$$= \frac{\sqrt[3]{28xy^2}}{2y}$$

To simplify a radical is to write it in a form

1. That cannot be rewritten with a smaller index, and
2. Whose radicand contains no factor with an exponent greater than or equal to the index.

EXAMPLE 15   Assume that the variables are all positive and simplify

$$\sqrt[4]{\frac{18^2 x^6 y^8}{25z^2}}.$$

Solution

Step 1   If possible, rewrite the given radical with a smaller index.

$$\sqrt[4]{\frac{18^2 x^6 y^8}{25z^2}} = \sqrt[4]{\left(\frac{18x^3 y^4}{5z}\right)^2}$$

$$= \left(\frac{18x^3 y^4}{5z}\right)^{2/4}$$

$$= \left(\frac{18x^3 y^4}{5z}\right)^{1/2}$$

$$= \sqrt{\frac{18x^3 y^4}{5z}}$$

Step 2   If possible, remove factors from the radicand.

$$\sqrt[4]{\frac{18^2 x^6 y^8}{25z^2}} = \sqrt{\frac{18x^3 y^4}{5z}}$$

$$= \sqrt{\frac{9 \cdot 2 \cdot x^2 \cdot x \cdot y^4}{5z}}$$

$$= \sqrt{9x^2 y^4}\sqrt{\frac{2x}{5z}}$$

Exercise 1.3, 81–100

$$= 3xy^2 \sqrt{\frac{2x}{5z}}$$

EXERCISE 1.3 RATIONAL EXPONENTS AND RADICALS

▶ Express each of the following in two radical forms.

EXAMPLES   (a) $7^{3/5} = \sqrt[5]{7^3} = (\sqrt[5]{7})^3$ (b) $7^{-3/5} = \sqrt[5]{7^{-3}} = (\sqrt[5]{7})^{-3}$

1. $8^{4/5}$ 2. $8^{-4/5}$ 3. $5^{2/3}$ 4. $5^{-2/3}$ 5. $\pi^{-1/2}$

6. $\pi^{3/5}$ 7. $2^{3/2}$ 8. $2^{-3/2}$ 9. $6^{7/8}$ 10. $6^{-7/8}$

▶ Express each of the following in exponential form.

EXAMPLE $\sqrt{10^3} = 10^{3/2}$

11. $\sqrt[3]{7^2}$ 12. $(\sqrt[3]{23})^2$ 13. $\sqrt{3}$ 14. $\sqrt[3]{2}$

15. $\sqrt{7^3}$ 16. $(\sqrt[3]{7})^2$ 17. $\sqrt{5}$ 18. $\sqrt[3]{5^2}$

19. $(\sqrt{2})^3$ 20. $\sqrt[3]{2^2}$

▶ Find the value of each of the following.

EXAMPLE $25^{3/2} = \sqrt{25^3} = (\sqrt{25})^3 = 5^3 = 125$

21. $49^{3/2}$ 22. $27^{2/3}$ 23. $(-27)^{2/3}$

24. $0^{2/3}$ 25. $27^{-2/3}$ 26. $(-27)^{-2/3}$

27. $16^{-3/2}$ 28. $100^{1/2}$ 29. $(0.01)^{-1/2}$

30. $64^{1/3}$ 31. $\left(\frac{8}{125}\right)^{1/3}$ 32. $(1000)^{2/3}$

33. $(0.001)^{-2/3}$ 34. $\left(\frac{9}{100}\right)^{1/2}$ 35. $(-32)^{1/5}$

36. $(32)^{-3/5}$ 37. $(144)^{-1/2}$ 38. $(\sqrt{11})^2$

39. $(\sqrt{9})^2$ 40. $\left(\frac{1}{36}\right)^{3/2}$ 41. $(0.09)^{1/2}$

42. $(0.0004)^{1/2}$ 43. $(0.008)^{1/3}$ 44. $(0.0016)^{3/4}$

45. $(0.00001)^{1/5}$ 46. $8^{-6/9}$ 47. $4^{4/8}$

48. $\left(\frac{1}{81}\right)^{6/4}$ 49. $(32)^{8/10}$ 50. $\left(\frac{8}{27}\right)^{2/6}$

▶ Using the exponent laws, express each of the following in a form containing no radicals or negative exponents. Assume all variables are positive.

EXAMPLE
$$(x^{3/2}y^{-3})(x^{1/2}y) = x^{3/2+1/2}y^{-3+1}$$
$$= x^2 y^{-2}$$
$$= \frac{x^2}{y^2}$$

51. $(x^{1/2}y^{3/2})(x^{-3/2}y^2)$

52. $(a^{3/2}b^{-1})(a^3b^{-3})$

53. $\dfrac{x^{1/2}y^{3/2}}{x^{-3/2}y^2}$

54. $\dfrac{a^{3/2}b^{-1}}{a^3b^{-3}}$

55. $\left(\dfrac{x^{5/2}}{y^{-1/2}}\right)^4$

56. $\left(\dfrac{x^{3/2}}{y^{1/2}}\right)^{-1/2}$

57. $(x^{-1}y^{-1})^{-1/2}(x^2y^3)^{1/2}$

58. $\left(\dfrac{a^3b^{-1/2}}{ab}\right)^{2/3}$

59. $(b^3a^{2/3})^{1/2}(ab)^2$

60. $\left(\dfrac{x^{-2/3}y^{2/3}}{x^{1/3}y}\right)^3$

▶ Express each of the following in the form ax^r where a is a real number written with no negative or fractional exponents, and r is a rational number. Assume that x is a positive real number.

EXAMPLE

$$\tfrac{1}{2}(5x^3)^{(1/2)-1}(15x^2) = \tfrac{1}{2}(5x^3)^{-1/2}(15x^2)$$

$$= \frac{15}{2\sqrt{5}}x^{-3/2}x^2$$

$$= \frac{15}{2\sqrt{5}}x^{1/2}$$

61. $\tfrac{1}{2}(7x^3)^{(1/2)-1}(21x^2)$

62. $\tfrac{1}{3}(5x^2)^{(1/3)-1}(10x)$

63. $\tfrac{1}{3}(2x^4)^{(1/3)-1}(8x^3)$

64. $(\tfrac{2}{3})(3)x^{(2/3)-1}$

65. $(\tfrac{-1}{2})(7x^3)^{(-1/2)-1}(21x^2)$

66. $(\tfrac{-1}{3})(2x^5)^{(-1/3)-1}(10x^4)$

67. $(\tfrac{-1}{2})(6x^5)^{(-1/2)-1}(30x^4)$

68. $\tfrac{1}{2}(3x^{2/3})^{-1/2}(3)(\tfrac{2}{3})x^{-1/3}$

69. $\tfrac{1}{2}(2^{1/2})(x^{1/3})^{(1/2)-1}(\tfrac{1}{3})x^{-2/3}$

70. $\dfrac{1}{x^{3/4} + x^{2/4} \cdot x^{1/4} + x^{1/4} \cdot x^{2/4} + x^{3/4}}$

▶ Use the properties of radicals to find the value of each of the following.

EXAMPLES

(a) $\sqrt[3]{125 \times 8} = \sqrt[3]{125} \times \sqrt[3]{8} = 5 \times 2 = 10$

(b) $\sqrt[3]{\dfrac{-1000}{8}} = \dfrac{\sqrt[3]{-1000}}{\sqrt[3]{8}} = \dfrac{-10}{2} = -5$

71. $\sqrt[3]{125 \times 1000}$

72. $\sqrt[3]{\tfrac{-27}{1000}}$

73. $\sqrt[5]{32 \times 243}$

74. $\sqrt[5]{\tfrac{243}{32}}$

75. $\sqrt{\tfrac{4}{8100}}$

76. $\sqrt{16 \times 2500}$

77. $\sqrt{16 \times 10^{10}}$

78. $\sqrt{\dfrac{1}{16 \times 10^{10}}}$

79. $\sqrt[3]{64 \times 125}$

80. $\sqrt[3]{\tfrac{-64}{125}}$

► Assume that all the variables are positive, and simplify each of the following.

EXAMPLE $\sqrt[6]{64x^8y^2} = \sqrt[6]{(8x^4y)^2} = \sqrt[3]{8x^4y} = 2x\sqrt[3]{xy}$

81. $\sqrt[6]{64x^6y^8}$ 82. $\sqrt[6]{64x^{15}y^9}$ 83. $\sqrt{81x^6y^3}$

84. $\sqrt[3]{54x^5y^2}$ 85. $\sqrt[3]{-16x^3y^7}$ 86. $\sqrt[4]{81x^3y^8z^7}$

87. $\sqrt[5]{64x^{10}y^{11}}$ 88. $\sqrt[6]{64x^2y^7}$ 89. $\sqrt[5]{-81x^9y^3z^{12}}$

90. $\sqrt{36x^2y^9}$

► Assume that all the variables are positive, and write each of the following in a form with a nonfractional radicand.

EXAMPLE $\sqrt{\dfrac{3}{2x}} = \sqrt{\dfrac{3(2x)}{4x^2}} = \dfrac{\sqrt{6x}}{2x}$

91. $\sqrt{\dfrac{2}{3x}}$ 92. $\sqrt{\dfrac{1}{3}}$ 93. $\sqrt[3]{\dfrac{5y}{2x}}$ 94. $\sqrt[3]{\dfrac{2}{x^2y}}$ 95. $\sqrt{\dfrac{1}{5}}$

96. $\sqrt[3]{\dfrac{7x^2}{2y}}$ 97. $\sqrt[4]{\dfrac{3x}{y^2}}$ 98. $\sqrt[5]{\dfrac{2xy}{y^3}}$ 99. $\sqrt[3]{\dfrac{1}{2}}$ 100. $\sqrt[3]{\dfrac{3x^2}{y^2}}$

CHECK LIST FOR THE STUDENT

If you have learned the material in Section 1.3, you should be able to do the following:

1. Complete the statement: Where m and n are integers and n is positive, we define $a^{m/n}$ to be _____ with the understanding that a and m are not both zero, that a is _____ if m is negative, and that a is _____ if n is even.

2. Write real number $a^{m/n}$ in two radical forms.

3. Write real number $(\sqrt[n]{a})^m$ and real number $\sqrt[n]{a^m}$ in exponential form.

4. Complete the statement: If $a^{m/n}$ and $a^{mk/nk}$ are both defined, then it is true that $a^{mk/nk} =$ _____ for positive integer k.

5. Use the exponent laws with rational exponents.

6. Complete the statements of the two main properties of radicals:

$$\sqrt[n]{ab} = \underline{\hspace{1.5cm}}, \qquad \sqrt[n]{\frac{a}{b}} = \underline{\hspace{1.5cm}}, \quad b \neq 0.$$

7. Use the two main properties of radicals.

8. Simplify radicals.

SECTION QUIZ 1.3

1. (a) Express $5^{3/4}$ in two radical forms.
   (b) Express $\sqrt[3]{5^2}$ in exponential form.

2. Find the value of each of the following:
   (a) $64^{2/3}$ (b) $64^{-2/3}$ (c) $(-64)^{2/3}$ (d) $64^{1/2}$ (e) $0^{2/3}$

3. Using the exponent laws, express each of the following in a form containing no radicals or negative exponents. Assume all variables are positive.

   (a) $(x^{2/3}y^{4/5})(x^{-4/3}y^{6/5})$ (b) $\left(\dfrac{x^{1/2}y^{-2}}{xy^2}\right)^{1/2}$

4. Use the properties of radicals to find the value of each of the following:
   (a) $\sqrt[3]{64 \times 27}$ (b) $\sqrt[3]{\frac{27}{-125}}$

5. Assume that all variables are positive, and tell what belongs in the blank parentheses in each of the following:
   (a) $\sqrt{8x^2y^5} = (\quad)\sqrt{2y}$
   (b) $\sqrt{5x/3y} = \sqrt{(\quad)}/3y$
   (c) $\sqrt[6]{64x^4y^8} = (\quad)\sqrt[3]{x^2y}$

1.4 Polynomials

The purpose of this section is to define polynomials, to review addition, subtraction, multiplication, and division of polynomials, and to show how to evaluate polynomials. Synthetic division and addition, subtraction, and multiplication of complex numbers are also covered.

Our first objective is to introduce some preliminary definitions so that we can use them in stating the definition of a polynomial.

A **term** is a combination of numbers, variables, or parenthetical expressions formed by multiplication or division. A single number, variable, or parenthetical expression is also called a term. For example,

$$5x^3, \quad -2x^2y, \quad -x^3, \quad (x + h)^2x^3, \quad \frac{3x^2y}{2z}, \quad \sqrt{2}, \quad x, \quad \text{and} \quad (x + h)$$

are all terms. A term containing no variables is called a **constant term.** The numbers $\sqrt{2}$ and -6 are constant terms.

The numerical factor of a term is called its **coefficient.** (In any product ab, the a and the b are called *factors*.) The coefficient of $5x^3$ is 5, the coefficient of $-2x^2y$ is -2, the coefficient of $-x^3 = (-1)x^3$ is -1, and the coefficient of $3x^2y/2z$ is $\frac{3}{2}$. If two terms are exactly alike except for their coefficients, they are said to be **like terms.** The terms

$$2x^2y \quad \text{and} \quad \tfrac{7}{5}x^2y$$

Exercise 1.4, 1–10 are like terms.

A **polynomial in one variable x** is a sum of terms of the form ax^n where a is the coefficient and n is a nonnegative integer. If $n = 0$, the term ax^n simply consists of the coefficient a because $x^0 = 1$. A polynomial of just one term is called a **monomial,** a polynomial of exactly two terms is called a **binomial,** and a polynomial of exactly three terms is called a **trinomial.**

A polynomial *over the reals* is a polynomial whose coefficients are all real numbers. A polynomial *over the integers* is one whose coefficients are all integers. The expression

$$\sqrt{2}x^3 + \tfrac{3}{4}x^2 - 2x + 7$$

is a polynomial over the reals. It is the sum of the terms $\sqrt{2}x^3, \tfrac{3}{4}x^2, -2x$, and 7. The expression

$$9x^3 - 3x^2 - 2x + 7$$

is a polynomial over the integers. It is the sum of the terms $9x^3, -3x^2, -2x$, and 7. Note that since any integer is a real number, a polynomial over the integers is also a polynomial over the reals.

A polynomial in two variables x and y is a sum of terms of the form ax^ny^m where a is the coefficient and n and m are nonnegative integers. The expression

$$3x^2y^3 - 2xy + 7x - 8y + 5$$

is a polynomial in x and y. It is the sum of the terms $3x^2y^3$, $-2xy$, $7xy^0$, $-8x^0y$, and $5x^0y^0$. In general a polynomial in more than one variable is a sum of terms having no variables in their denominators and having only nonnegative exponents on their variables. The expression

$$3x^2y^3z - 5x^2y + \tfrac{7}{2}yz - 3x$$

is a polynomial in x, y, and z. The expression

$$3x^2y^3/z - 5x^2y + \tfrac{7}{2}yz - 3x$$

is not a polynomial. There is a variable in the denominator of the first term. The expression

$$3x^2y^3z - 5x^{-2}y + \tfrac{7}{2}y^{1/2}z - 3x$$

Exercise 1.4, 11–40 is not a polynomial. There is a negative exponent on x in the second term, and there is a nonintegral exponent on y in the third term.

The **degree of a term** is the sum of the exponents of its variables. The degree of a term *in one particular variable* is the exponent of that variable. The degree of $7x^2y^3$ is 5. The degree of $7x^2y^3$ in x is 2 and in y is 3. The degree of the monomial $7x^2$ is 2, and the degree of the monomial $8 = 8x^0$ is 0. (Any nonzero real number can be considered a monomial of zero degree. The number zero itself is not assigned a degree.)

The **degree of a polynomial** is the degree of its term having the highest degree. The degree of

$$7x^2y^3 + 5x^4 - 2x$$

is 5, and the degree of

$$5x^3 - 2x^2 + 7$$

is 3. The degree of a polynomial in one particular variable is the highest exponent appearing on that variable anywhere in the polynomial. The degree of $7x^2y^3 + 5x^4 - 2x$ in the variable x is 4 and in the variable y is 3.

A polynomial of third degree in x technically has four terms, even though some of the terms may not appear because their coefficients are zero. For example,

$$5x^3 - 2x^2 + 7$$

technically has the four terms $5x^3$, $-2x^2$, $0x$, and 7. Likewise, a polynomial of fourth degree in x technically has five terms. The fourth-degree polynomial

$$7x^4 - 2x^3 + 9x - 8$$

has the five terms $7x^4$, $-2x^3$, $0x^2$, $9x$, and -8. In general, an nth-degree polynomial in x has $n + 1$ terms. If we denote the $n + 1$ coefficients by the symbols

$$a_0, a_1, a_2, a_3, \ldots, a_{n-1}, a_n$$

while acknowledging that any of these might be zero except for the coefficient of x^n, then an nth-degree polynomial in x may be written in the form

$$a_0 x^n + a_1 x^{n-1} + a_2 x^{n-2} + a_3 x^{n-3} + \cdots + a_{n-1} x + a_n$$

where $a_0 \neq 0$ and n is a nonnegative integer. Unless otherwise specified, it is understood that the variable x may only represent real numbers and that the coefficients are all real numbers, so that unless otherwise specified, by *polynomial* we mean *polynomial over the reals.*

Polynomials in x are sometimes represented by symbols such as $P(x)$, $Q(x)$, $R(x)$, etc. These are read "P of x," "Q of x," "R of x," etc. If $P(x)$ represents a polynomial, $P(a)$ is said to be the *value of the polynomial at $x = a$.* For example, if

$$P(x) = 5x^3 - 2x^2 - 4x + 1$$

then $P(-2)$ is the value of $P(x)$ at $x = -2$.

$$
\begin{aligned}
P(-2) &= 5(-2)^3 - 2(-2)^2 - 4(-2) + 1 \\
&= 5(-8) - 2(4) + 8 + 1 \\
&= -40 - 8 + 8 + 1 \\
&= -39
\end{aligned}
$$

To find $P(a)$, we merely go through the expression for $P(x)$ and replace each x by a.

$$P(a) = 5a^3 - 2a^2 - 4a + 1$$

In calculus we might have to find $P(a + h)$ or $P(x + h)$. To find $P(a + h)$, we go through the expression for $P(x)$ and replace each x by $a + h$.

$$P(x) = 5x^3 - 2x^2 - 4x + 1$$
$$P(a + h) = 5(a + h)^3 - 2(a + h)^2 - 4(a + h) + 1$$

To find $P(x + h)$, we replace each x in $P(x)$ by $x + h$.

$$P(x) = 5x^3 - 2x^2 - 4x + 1$$
$$P(x + h) = 5(x + h)^3 - 2(x + h)^2 - 4(x + h) + 1$$

EXAMPLE 1

If $P(x) = 2x^3 - 8x^2 + 7x$ and $Q(x) = 3x^2 - 4x$, find $P(2)$, $Q(2)$, $P(Q(2))$, and $Q(P(2))$.

PART ONE | ALGEBRAIC REVIEW

$$P(x) = 2x^3 - 8x^2 + 7x \qquad Q(x) = 3x^2 - 4x$$
$$P(2) = 2(2)^3 - 8(2)^2 + 7(2) \qquad Q(2) = 3(2)^2 - 4(2)$$
$$= 16 - 32 + 14 \qquad\qquad = 12 - 8$$
$$= -2 \qquad\qquad\qquad = 4$$

$P(Q(2)) = P(4)$ since $Q(2) = 4.$
$$P(4) = 2(4)^3 - 8(4)^2 + 7(4)$$
$$= 128 - 128 + 28$$
$$= 28$$

$Q(P(2)) = Q(-2)$ since $P(2) = -2.$
$$Q(-2) = 3(-2)^2 - 4(-2)$$
$$= 12 + 8$$

Exercise 1.4, 61–65

$$= 20$$

Like terms may be added or subtracted simply by adding or subtracting their coefficients and multiplying the result by the common variable part. For example,

$$3x^3 + 5x^3 = (3 + 5)x^3 = 8x^3$$

and $\qquad 2x^2y - 7x^2y = (2 - 7)x^2y = -5x^2y.$

Such operations are sometimes called _collecting like terms._ Collecting like terms is an application of the distributive property of the real numbers (Appendix A). To _simplify_ a polynomial is to collect its like terms.

EXAMPLE 2

Simplify the polynomial

$$3x^3 + 9 - 2x + 5x^3 - 7x + 8x^2 - 7 - 9x^2.$$

Solution

Since the commutative property of addition (Appendix A) assures us that we may rearrange the order in which the terms are added, we may write the given polynomial in the form

$$3x^3 + 5x^3 + 8x^2 - 9x^2 - 2x - 7x + 9 - 7.$$

Collecting like terms, we have

Exercise 1.4, 51–60

$$8x^3 - x^2 - 9x + 2.$$

Polynomials may be added as demonstrated in the next example.

EXAMPLE 3

Simplify the sum of

$$3x^3 - 2x^2 + 5x - 8 \quad \text{and} \quad 2x - 7x^2 + 6x^3 - 1.$$

Solution

The work may be arranged either horizontally or vertically. Arranging it horizontally, we simply collect like terms.

$$3x^3 - 2x^2 + 5x - 8 + 2x - 7x^2 + 6x^3 - 1$$
$$= 3x^3 + 6x^3 - 2x^2 - 7x^2 + 5x + 2x - 8 - 1$$
$$= 9x^3 - 9x^2 + 7x - 9$$

Arranging the work vertically, we place like terms in the same column and then add.

$$
\begin{array}{r}
3x^3 - 2x^2 + 5x - 8 \\
+\ 6x^3 - 7x^2 + 2x - 1 \\
\hline
9x^3 - 9x^2 + 7x - 9
\end{array}
$$

If A and B are algebraic expressions, $A - B$ is the same as $A + (-B)$. Thus to subtract B from A, we simply change the sign of B and add. For example, $(5x^3 - 9x^2 + 8) - (2x^3 - 6x^2 - 3x + 1)$ is the sum of $(5x^3 - 9x^2 + 8)$ and $-(2x^3 - 6x^2 - 3x + 1)$.

A minus sign in front of a parenthesis may be read (-1) *times* the entire expression within the parentheses. It follows from the distributive property of the real numbers (Appendix A) that to remove parentheses with a minus sign in front, we must change the sign of every term inside. On the other hand, if a parenthesis is preceded by no sign or by a plus sign, the parentheses may simply be removed without changing the expression inside.

$$-(2x^3 - 6x^2 - 3x + 1) = -2x^3 + 6x^2 + 3x - 1$$
$$(5x^3 - 9x^2 + 8) = 5x^3 - 9x^2 + 8$$
$$+(5x^3 - 9x^2 + 8) = 5x^3 - 9x^2 + 8$$

The work performed in subtracting polynomials may be arranged horizontally or vertically. This is shown in the next example.

EXAMPLE 4

Simplify $(5x^3 - 9x^2 + 8) - (2x^3 - 6x^2 - 3x + 1)$.

Solution

Arranging the work horizontally, we remove parentheses and collect like terms.

$$(5x^3 - 9x^2 + 8) - (2x^3 - 6x^2 - 3x + 1)$$
$$= 5x^3 - 9x^2 + 8 - 2x^3 + 6x^2 + 3x - 1$$
$$= 5x^3 - 2x^3 - 9x^2 + 6x^2 + 3x + 8 - 1$$
$$= 3x^3 - 3x^2 + 3x + 7$$

Arranging the work vertically, we place like terms in the same column, and then change signs and add.

$$
\begin{array}{r}
5x^3 - 9x^2 \quad\ + 8 \\
\mp\ 2x^3 \mp 6x^2 \mp 3x \pm 1 \\
\hline
3x^3 - 3x^2 + 3x + 7
\end{array}
$$

Exercise 1.4, 66–75

To multiply two polynomials, we take each term of one of them, multiply it by every term of the other one, and then add the individual products. This is illustrated in Example 5.

EXAMPLE 5 Simplify the product $(3x^2 + 4)(2x^3 - 5x^2 - 7)$.

Solution Arranging the work horizontally, we multiply each term of $(3x^2 + 4)$ by every term of $(2x^3 - 5x^2 - 7)$, add the products, and collect like terms.

$$(3x^2 + 4)(2x^3 - 5x^2 - 7)$$
$$= 3x^2(2x^3 - 5x^2 - 7) + 4(2x^3 - 5x^2 - 7)$$
$$= 6x^5 - 15x^4 - 21x^2 + 8x^3 - 20x^2 - 28$$
$$= 6x^5 - 15x^4 + 8x^3 - 21x^2 - 20x^2 - 28$$
$$= 6x^5 - 15x^4 + 8x^3 - 41x^2 - 28$$

Arranging the work vertically, we put like terms among the individual products in the same column, and then add.

$$2x^3 - 5x^2 - 7$$
$$3x^2 + 4$$
$$\overline{}$$
$$6x^5 - 15x^4 \qquad\quad - 21x^2 \longleftarrow \quad 3x^2(2x^3 - 5x^2 - 7)$$
$$\qquad\qquad\qquad 8x^3 - 20x^2 - 28 \longleftarrow \quad 4(2x^3 - 5x^2 - 7)$$
$$\overline{6x^5 - 15x^4 + 8x^3 - 41x^2 - 28} \longleftarrow \text{Sum of individual products}$$

Exercise 1.4, 76–110

EXAMPLE 6 If $P(x) = 3x^2 - 5x + 1$, express $P(a + h)$ as a polynomial in a and h.

Solution
$$P(a + h) = 3(a + h)^2 - 5(a + h) + 1$$
$$(a + h)^2 = (a + h)(a + h)$$
$$= a^2 + ah + ha + h^2$$
$$= a^2 + ah + ah + h^2$$
$$= a^2 + 2ah + h^2$$

$$P(a + h) = 3(a^2 + 2ah + h^2) - 5(a + h) + 1$$
$$= 3a^2 + 6ah + 3h^2 - 5a - 5h + 1$$

EXAMPLE 7 If $P(x) = 3x^2 - 5x + 1$, express $P(x + h) - P(x)$ as a polynomial in x and h.

Solution
$$P(x + h) = 3(x + h)^2 - 5(x + h) + 1$$
$$P(x + h) - P(x) = 3(x + h)^2 - 5(x + h) + 1 - (3x^2 - 5x + 1)$$
$$= 3(x^2 + 2xh + h^2) - 5x - 5h + 1 - 3x^2$$
$$+ 5x - 1$$
$$= 3x^2 + 6xh + 3h^2 - 5x - 5h + 1 - 3x^2$$
$$+ 5x - 1$$
$$= 3x^2 - 3x^2 - 5x + 5x + 1 - 1 + 6xh + 3h^2$$
$$- 5h$$

Exercise 1.4, 111–140

$$= 6xh + 3h^2 - 5h$$

If p and d are positive integers, there exist nonnegative integers q and r such that

$$\frac{p}{d} = q + \frac{r}{d}$$

where r is less than d. Otherwise stated,

$$p = qd + r.$$

$$\frac{17}{3} = 5 + \frac{2}{3} \quad \text{or} \quad 17 = 5 \times 3 + 2$$
$$\frac{12}{4} = 3 + \frac{0}{4} \quad \text{or} \quad 12 = 3 \times 4 + 0$$
$$\frac{4}{7} = 0 + \frac{4}{7} \quad \text{or} \quad 4 = 0 \times 7 + 4$$

A similar situation exists for polynomials. If $P(x)$ and $D(x)$ are polynomials with $D(x)$ not zero, there exist polynomials $Q(x)$ and $R(x)$ such that

$$\frac{P(x)}{D(x)} = Q(x) + \frac{R(x)}{D(x)}$$

where $R(x)$ is of lower degree than $D(x)$ or $R(x) = 0$. That is,

$$P(x) = Q(x) \cdot D(x) + R(x).$$

Here the numerator $P(x)$ is called the **dividend**, the denominator $D(x)$ is called the **divisor**, the $Q(x)$ is called the **quotient**, and the $R(x)$ is called the **remainder**. If the division comes out even, $R(x)$ is zero. To find the quotient and the remainder, we divide the polynomial $P(x)$ by the polynomial $D(x)$ in much the same manner as we perform long division of integers. To illustrate, we will divide $x - 3x^2 + 2x^3 - 2$ by $x^2 - 4x + 5$. Note that before the process is begun, the dividend and divisor are arranged in descending powers of x, and note that the process is repeated until it produces a remainder that is zero or is of lower degree than the divisor.

$$
\begin{array}{r}
2x + 5 \\
x^2 - 4x + 5 \overline{\smash{)}\, 2x^3 - 3x^2 + x - 2}
\end{array}
\qquad \text{Divide: } \frac{2x^3}{x^2} = 2x
$$

$2x(x^2 - 4x + 5) \longrightarrow \underline{2x^3 - 8x^2 + 10x}$ Multiply

$\qquad\qquad\qquad\qquad\qquad 5x^2 - 9x - 2$ Subtract

$\qquad\qquad\qquad\qquad\qquad\qquad$ Divide: $\frac{5x^2}{x^2} = 5$

$5(x^2 - 4x + 5) \longrightarrow \underline{5x^2 - 20x + 25}$ Multiply

$\qquad\qquad\qquad\qquad\qquad\quad 11x - 27$ Subtract

$$
\underset{\text{Dividend}}{} \quad \underset{\text{Quotient}}{} \quad \underset{\text{Remainder}}{}
$$

$$\frac{2x^3 - 3x^2 + x - 2}{x^2 - 4x + 5} = 2x + 5 + \frac{11x - 27}{x^2 - 4x + 5}$$

$$\underset{\text{Divisor}}{} \qquad\qquad\qquad\qquad\quad \underset{\text{Divisor}}{}$$

The student should check that

$$2x^3 - 3x^2 + x - 2 = (2x + 5)(x^2 - 4x + 5) + (11x - 27).$$
$$\text{Dividend} \quad = \text{(Quotient)(Divisor)} \quad + \quad \text{Remainder.}$$

EXAMPLE 8   Find the quotient and remainder.

$$\frac{x^2 - 2x + 5}{x - 3}$$

Solution

$$
\begin{array}{r}
x + 1 \\
x - 3 \overline{\smash{)}\, x^2 - 2x + 5} \\
\underline{x^2 - 3x} \\
\cdot\ x + 5 \\
\underline{x - 3} \\
8
\end{array}
$$

$$\frac{x^2 - 2x + 5}{x - 3} = x + 1 + \frac{8}{x - 3}$$

Check $(x + 1)(x - 3) + 8 = x^2 - 2x - 3 + 8$
$$= x^2 - 2x + 5$$

Exercise 1.4, 141–150   The quotient is $x + 1$. The remainder is 8.

In Example 8, the divisor is a polynomial of the form $x - c$ where c is the real number 3. Note that the remainder 8 happens to be the value of the dividend at $x = 3$. Indeed, if

$$P(x) = x^2 - 2x + 5,$$
then
$$P(3) = 3^2 - 2(3) + 5$$
$$= 9 - 6 + 5$$
$$= 8.$$

This is not just a coincidence. Whenever a polynomial $P(x)$ is divided by a polynomial of the form $x - c$, the remainder is equal to the value of $P(x)$ at $x = c$. This fact is known as the *remainder theorem*.

Remainder Theorem   If the polynomial $P(x)$ is divided by $x - c$, the remainder equals $P(c)$.

Proof   If $P(x)$ is divided by $x - c$, the remainder must be a number R because it must be zero or of lower degree than $x - c$. So if $Q(x)$ is the quotient,

$$P(x) = Q(x) \cdot (x - c) + R.$$
It follows that
$$P(c) = Q(c) \cdot (c - c) + R$$
$$= Q(c) \cdot 0 \quad\quad + R$$
$$= R.$$

Exercise 1.4, 151–160   That is, the remainder R equals $P(c)$.

Since there is a short method of dividing a polynomial by $x - c$, the remainder theorem can readily be used to evaluate a polynomial. This short method is called _synthetic division_ or _synthetic substitution_. It amounts to leaving out all unnecessary symbols from the usual long-division process. For example, consider the division of $x^4 - 7x^3 + 3x - 5$ by $x - 2$.

$$
\begin{array}{r}
x^3 - 5x^2 - 10x - 17 \\
x - 2\overline{\smash{)}\,x^4 - 7x^3 \qquad\quad + 3x - 5} \\
\underline{x^4 - 2x^3} \\
-5x^3 \qquad\quad + 3x - 5 \\
\underline{-5x^3 + 10x^2} \\
-10x^2 + 3x - 5 \\
\underline{-10x^2 + 20x} \\
-17x - 5 \\
\underline{-17x + 34} \\
-39
\end{array}
$$

If we examine this work, we see that we have done an unnecessary amount of writing. The colored numbers and the coefficients of x^3 and x^4 in the dividend and quotient are really the only essential symbols. For example, we know that the dividend is of fourth degree and that the quotient is going to be a third-degree polynomial in x, so the only symbols that we need in order to identify dividend and quotient are their coefficients. We know that the divisor is of the form $x - c$, so the only identifying symbol we need for the divisor is the number $c = 2$. If we eliminate almost all nonessential symbols, insert ones for the coefficients of x^3 and x^4, and take care to acknowledge any zero coefficients (in the dividend, the x^2 term is missing so its coefficient is zero), we are left with the following array:

$$
\begin{array}{r}
1 \qquad\qquad\qquad\qquad\qquad \\
-2\overline{\smash{)}\,1 - 7 \quad 0 \quad 3 - 5} \\
\underline{-2} \\
-5 \\
\underline{10} \\
-10 \\
\underline{20} \\
-17 \\
\underline{34} \\
-39
\end{array}
$$

The actual arithmetic computations that produced this array consisted of successive multiplications by -2 followed by subtractions. But if we work with 2 rather than -2, the signs of the products will be changed, and we can add rather than subtract. In that case, the array becomes

$$
\begin{array}{r}
1 \\
2\overline{)1 - 7 \quad\; 0 \quad\;\; 3 - 5} \\
\underline{2} \\
- 5 \\
\underline{- 10} \\
- 10 \\
- 20 \\
\underline{- 17} \\
- 34 \\
\underline{- 39}
\end{array}
$$

If we now compress this array to two lines, we have

$$
\begin{array}{r}
2\overline{)1 - 7 \quad\;\;\; 0 \quad\;\;\; 3 - 5} \\
\underline{2 - 10 - 20 - 34} \\
1 - 5 - 10 - 17 - 39
\end{array}
$$

The remainder is -39, and the numbers 1, -5, -10, and -17 are the coefficients of the quotient. Since the quotient is a third-degree polynomial, it must be

$$x^3 - 5x^2 - 10x - 17.$$

Now the work and the arithmetic computations that produced the coefficients 1, -5, -10, and -17 and the remainder -39 can be outlined as follows. To divide $x^4 - 7x^3 + 3x - 5$ by $x - 2$ using synthetic division:

Step 1 Write down the 2, and write down all the coefficients (including any zeros) of the dividend.

Step 2 Bring down the first coefficient and multiply it by 2. $\quad 1 \times 2 = 2$

Step 3 Add. $\quad -7 + 2 = -5$

Step 4 Multiply by 2. $\quad -5 \times 2 = -10$

Step 5 Add. $\quad 0 + (-10) = -10$

Step 6 Multiply by 2. $\quad -10 \times 2 = -20$

Step 7 Add. $\quad 3 + (-20) = -17$

Step 8 Multiply by 2. $\quad -17 \times 2 = -34$

Step 9 Add. $\quad -5 + (-34) = -39$

EXAMPLE 9

Use synthetic division to find the quotient and remainder.

$$\frac{3x^3 - 2x^2 - 8}{x + 5}$$

Solution

Here $x - c$ is $x + 5$, so $c = -5$. The quotient will be a second-degree polynomial.

$$
\begin{array}{r}
-5\overline{)\,3 \;-\; 2 \quad\;\; 0 \;-\; 8} \\
\underline{-\;15 \quad 85 \;-\; 425} \\
3 \;-\; 17 \quad 85 \;-\; 433
\end{array}
$$

The quotient is $3x^2 - 17x + 85$. The remainder is -433. (Note that the value of $3x^3 - 2x^2 - 8$ at $x = -5$ is indeed -433.)

The synthetic division process will work *only* for divisors of the form $x - c$ where c may be positive or negative. It will not work for divisors of any other form.

Exercise 1.4, 161–180

A complex number was defined in Section 1.1 as a number that can be written in the form $a + bi$ where a and b are real numbers and $i^2 = -1$. Keeping in mind that $i^2 = -1$, we may add, subtract, or multiply complex numbers as if they were polynomials. These operations are demonstrated in the following examples.

EXAMPLE 10

Express the sum $(3 + 2i) + (5 - 7i)$ in the form $a + bi$ where a and b are real numbers.

Solution

Arranging the work horizontally,

$$
\begin{aligned}
(3 + 2i) + (5 - 7i) &= 3 + 2i + 5 - 7i \\
&= (3 + 5) + (2i - 7i) \\
&= 8 - 5i.
\end{aligned}
$$

Arranging the work vertically,

$$
\begin{array}{r}
3 + 2i \\
+\; 5 - 7i \\
\hline
8 - 5i.
\end{array}
$$

EXAMPLE 11

Express the difference $(3 + 2i) - (5 - 7i)$ in the form $a + bi$ where a and b are real numbers.

Solution

Arranging the work horizontally,

$$
\begin{aligned}
(3 + 2i) - (5 - 7i) &= 3 + 2i - 5 + 7i \\
&= (3 - 5) + (2i + 7i) \\
&= -2 + 9i.
\end{aligned}
$$

Arranging the work vertically, we change signs and add.

$$3 + 2i$$
$$+ \ -5 + 7i$$
$$\overline{-2 + 9i}$$

EXAMPLE 12

Express the product $(3 + 2i)(5 - 7i)$ in the form $a + bi$ where a and b are real numbers.

Solution

Arranging the work horizontally, we multiply each term of $3 + 2i$ by every term of $5 - 7i$, add the products, and collect like terms.

$$(3 + 2i)(5 - 7i) = 15 - 21i + 10i - 14i^2$$
$$= 15 - 21i + 10i - 14(-1)$$
$$= (15 + 14) + (-21i + 10i)$$
$$= 29 - 11i$$

Arranging the work vertically,

$$5 - \ 7i$$
$$3 + \ 2i$$
$$\overline{15 - 21i}$$
$$+ \ 10i - 14i^2$$
$$\overline{15 - 11i - 14i^2} = 15 - 11i - 14(-1)$$
$$= 29 - 11i.$$

Exercise 1.4, 181–200 (In Section 1.6 we shall see how to express a fraction whose numerator and denominator are complex numbers as a single complex number.)

EXERCISE 1.4 POLYNOMIALS

▶ Identify like terms in each of the following.

EXAMPLE

$5x^3y,\ 7xy^3,\ -3x^3y,\ \frac{5}{3}y^3,\ \frac{9}{2}xy^3,\ 7x^3$

Solution

$5x^3y,\ -3x^3y$ are like terms.

$7xy^3,\ \frac{9}{2}xy^3$ are like terms.

1. $2x^2y,\ 3x,\ 7y,\ 5x^2y,\ -7x,\ 8y,\ 6x^2,\ 8xy$
2. $\sqrt{2}x,\ 7x^3,\ \sqrt{2}y,\ 7x^3y,\ -x^3,\ 9x,\ 9y,\ 7xy^3$
3. $x^5y^2,\ -3x^4y^3,\ 2x^2y^5,\ -x^5y^2,\ -x^2y^5,\ x^2,\ y^5,\ x^5$
4. $y^3,\ -2y^2x,\ \frac{2}{5}yx^2,\ \frac{7}{8}x^3,\ -9y^2x,\ \sqrt{2}y^3,\ x^3y^3,\ xy^2$
5. $-x^5y,\ 7x^4y^2,\ \frac{4}{5}x^5,\ 3y,\ -8x^2y^4,\ 7x^5y,\ -x^4y^2,\ 19y$
6. $3yz,\ 3xz,\ 3xy,\ 3x,\ 3y,\ 3z,\ 2yz,\ 2xz,\ 2xy,\ 2x$

7. x^3y, xy^3, x^2y^2, x^3y^3, $2x^3y$, $2xy^3$, $2xy$, xy

8. $13x^3y$, $5x^5yz$, $6x^6yz$, $-13x^3y$, $7x^7yz$, $8x^8yz$

9. $\sqrt{2}x^2y$, $3x^3y^7$, $\sqrt{2}x^2y^2$, $3x^7y^3$, $\sqrt{2}xy$, $3xy$, $3x^2y$

10. $5x$, $5x^2y$, $5y$, $5yz$, $5z$, $3x^2y$, $-yz$, $5xyz$

State whether or not each of the following is a polynomial. If it is a polynomial, state whether or not it is a polynomial over the integers.

EXAMPLES

(a) $5x^3 - 2x^{-4} + 7x - 8$
(b) $5x^3 - 2x^{3/2} + 7x - 8$
(c) $5x^3 - \frac{2}{3}x^4 + 7x - 8$
(d) $5x^3 - 2x^4 + 7x - 8$

Solutions

(a) Not a polynomial. (Not all exponents are nonnegative integers.)
(b) Not a polynomial. (Not all exponents are nonnegative integers.)
(c) A polynomial, but not over the integers.
(d) A polynomial over the integers.

11. $2x^3 - \frac{3}{2}x^2 + 7x - 8$ **12.** $2x^{3/2} - 3x^2 + 7x - 8$

13. $2x^3 - 5x^2 + 6x^{-1} - 8$ **14.** $\sqrt{2}x^2 - 3x + 7$

15. $x - 9$ **16.** $x^2 - 25$

17. $x^3 + 8$ **18.** $x^3 - 1$

19. $3x^2 - \sqrt{2}x^3 + 5x$ **20.** $3x^{-2} - \sqrt{2}x^3 + 5x$

21. $\frac{2}{3}x^2 - 2x^3 + 5x$ **22.** $3x^2 - 2x^3 + 5x$

23. $3x^{2/3} - 2x^3 + 5x$ **24.** $3x^{-1} - 2x^3 + 5x$

25. $3x + 7$ **26.** $-x^2 - 2x + 4$

27. $2x - \frac{3}{5}$ **28.** $x^2 - 2x^{-1} + 4$

29. $\sqrt{x} - 2$ **30.** $\sqrt{x} + 3$

▶ State whether or not each of the following is a polynomial.

EXAMPLES

(a) $2x^2y^3z^2 - 5xy^3 + \frac{7}{2}y^2$
(b) $2x^2y^3z - 5xy^{3/2} + \frac{7}{2}y^{-2}$
(c) $\dfrac{2x^2y^3}{z} - 5xy^3 + \frac{7}{2}y^2$

Solutions

(a) A polynomial.
(b) Not a polynomial. (Not all exponents are nonnegative integers.)
(c) Not a polynomial. (There is a variable in the denominator of $2x^2y^3/z$.)

31. $5x^3y^2 - 2xy^3 + 7x - 9$ **32.** $5x^{-3}y^2 - 2xy^3 + 7x - 9$

33. $5x^{3/2}y^2 - 2xy^3 + 7x - 9$ **34.** $\dfrac{5x^3}{y^2} - 2xy^3 + 7x - 9$

35. $x + y - z$ **36.** $x - y$

37. $x^2 - xy + y^2$ **38.** $2xy^3 - 5xy^{-1} + 1$

39. $2xy^3 - 5xy^{1/2} + 1$ **40.** $2xy^3 - \dfrac{5}{xy} + 1$

▶ State the degree of each of the following, and then state its degree in each variable.

EXAMPLE

$3x^2y^4 - 5x^3y^5 + 2xy^6$

Solution

Eighth-degree polynomial. In x, third degree. In y, sixth degree.

41. $2x^3y^5 - 7x^4y^6 + 3x^2y^7$ **42.** $3xy - 2x + 7y - 2$

43. $7xy^2 - 3x^3y - 8x^5$ **44.** $2x^3yz - 5x^4 + 3x^2z - 2yz^2$

45. $3x + 5y - 7$ **46.** $y - 3x - 6$

47. $y - 3x^2 + 2x - 5$ **48.** $x^2 + xy + y^2$

49. $x - y$ **50.** $6x^4yz - 7z^5 + 2x^3y^2 - 3zy^3$

▶ Simplify each of the following (collect like terms).

EXAMPLE

$5x^3 - 3x^2 + 7x^3 - 2x^2 + 4x$

Solution

$5x^3 + 7x^3 - 3x^2 - 2x^2 + 4x = 12x^3 - 5x^2 + 4x$

51. $6x^3 - 2x^4 + 5x^3 - 3x^4 - 7$

52. $3x^2y - 5xy + 7x^2y - 3xy$

53. $x^2 - 2xy + y^2 - 3xy + 7$

54. $x^2 + 2xh + h^2 + 3x + 3h - 5 - x^2 - 3x + 5$

55. $3x + 3h + 7 - 3x - 7$

56. $3x^2 + 6xh + 3h^2 - 2x - 2h + 1 - 3x^2 + 2x - 1$

57. $5x + 5h - 8 - 5x + 8$

58. $2x^2 + 4xh + 2h^2 - 3x - 3h + 4 - 2x^2 + 3x - 4$

59. $x^3 + 3x^2h + 3xh^2 + h^3 - x^3$

60. $5x^2 + 10xh + 5h^2 - 3x - 3h - 5x^2 + 3x$

EXAMPLE

If $P(x) = 5x^3$ and $Q(x) = 7x - 5$, find $P(Q(1))$.

Solution

$$Q(1) = 7(1) - 5 = 2$$
$$P(Q(1)) = P(2) = 5 \times (2)^3$$
$$= 5 \times (8) = 40$$

61. If $P(x) = 2x^3 - 5x^2$ and $Q(x) = x^4 - 3x$, find $P(1)$, $Q(1)$, $P(-1)$, $Q(-1)$, $P(Q(1))$, and $Q(P(1))$.

62. If $P(x) = 3x^2$ and $Q(x) = 6x + 4$, find $P(0)$, $Q(0)$, $P(1)$, $Q(1)$, $P(-1)$, $Q(-1)$, $P(Q(0))$, and $Q(P(0))$.

63. If $P(x) = x^2 - 3x$ and $Q(x) = 3x^2 - 2x$, find $P(3)$, $Q(3)$, $P(-3)$, $Q(-3)$, $P(Q(-3))$, and $Q(P(-3))$.

64. If $P(x) = 3x - 5$ and $Q(x) = x^4 - 6$, find $P(2)$, $Q(2)$, $P(-2)$, $Q(-2)$, $P(Q(2))$, and $Q(P(2))$.

65. If $P(x) = 3x^2 - 2x$ and $Q(x) = 2x + 7$, find $P(a)$, $Q(a)$, $P(Q(a))$, and $Q(P(a))$.

▶ Simplify each of the following.

66. $(3x^3 - 5x^2 + 7) + (2x^3 - 6x^2 + 8)$

67. $(3x^3 - 5x^2 + 7) - (2x^3 - 6x^2 + 8)$

68. $(6x^4 - 5x^3 + 2x^2 - 8) + (2x^4 - 3x^3 - 8x^2 + x - 16)$

69. $(6x^4 - 5x^3 + 2x^2 - 8) - (2x^4 - 3x^3 - 8x^2 + x - 16)$

70. $(x^2 - x + 1) + (x^2 - x + 1)$

71. $(x^2 - x + 1) - (x^2 - x + 1)$

72. $(3x^5 - 2x^4 + 3x^2 - 7) + (8x^5 - 7x^3 - 7x^4 - 3x^2 + 9)$

73. $(3x^5 - 2x^4 + 3x^2 - 7) - (8x^5 - 7x^3 - 7x^4 - 3x^2 + 9)$

74. $(2x^3 + 3x^2 - 5x - 9) + (-x^3 - 2x^2 + 7x - 8)$

75. $(2x^3 + 3x^2 - 5x - 9) - (-x^3 - 2x^2 + 7x - 8)$

▶ Simplify each of the following products.

76. $(2x^2 + 3)(5x^3 - 2x^2 + 4)$ 77. $(8x^2 - 2)(6x^3 - 5x^2 + 1)$

78. $(3x + 7)(x^2 - 5x + 6)$

79. $(5x^2 - 2x)(x^3 - 3x^2 + 2x + 1)$

80. $(x^2 - 4)(x^2 + 4)$ 81. $(3x + 7)(3x - 8)$

82. $(2x - 4)(5x + 6)$ 83. $(x - 7)(x + 4)$

84. $(4x - 3)(2x + 5)$ 85. $(x - 2)(x + 2)$

86. $(x - 2)(x^2 + 2x + 4)$ 87. $(x + 2)(x^2 - 2x + 4)$

88. $(5x - 1)(3x + 7)$ 89. $(x - y)(x^2 + xy + y^2)$

90. $(x^2 - y^2)(x^2 + y^2)$ 91. $(x + y)(x^2 - xy + y^2)$

92. $(x + 5)^2$ 93. $(x - 5)^2$

94. $(a + b)^2$

95. $(a - b)^2$

96. $(a - b)^3$

97. $(3 - \sqrt{2})(3 + \sqrt{2})$

98. $(2 - \sqrt{3})(2 + \sqrt{3})$

99. $(\sqrt{2} - \sqrt{3})(\sqrt{2} + \sqrt{3})$

100. $(\sqrt{5} - \sqrt{7})(\sqrt{5} + \sqrt{7})$

▶ Assume that all the radicands are positive, and write out each of the following products.

EXAMPLE

$(\sqrt{x} - 5)^2$

Solution

$x - 10\sqrt{x} + 25$

101. $(\sqrt{x} - 3)^2$

102. $(\sqrt{x} + 3)^2$

103. $(\sqrt{x} - \sqrt{y})^2$

104. $(\sqrt{x} - \sqrt{y})(\sqrt{x} + \sqrt{y})$

105. $(\sqrt{x - 2} + 2)^2$

106. $(\sqrt{x + 5} - 3)(\sqrt{x + 5} + 3)$

107. $(\sqrt{x - 3} + 2\sqrt{x})^2$

108. $(\sqrt{x - 3} - 2\sqrt{x})^2$

109. $(7 - \sqrt{x + 4})^2$

110. $(7 + \sqrt{x + 4})^2$

▶ In each of the following, express $P(a + h)$ as a polynomial in a and h and then express $P(x + h)$ as a polynomial in x and h.

EXAMPLE

$P(x) = 3x^2 - 2x + 7$

Solution

$P(a + h) = 3(a + h)^2 - 2(a + h) + 7$

$\qquad\qquad = 3a^2 + 6ah + 3h^2 - 2a - 2h + 7$

$P(x + h) = 3(x + h)^2 - 2(x + h) + 7$

$\qquad\qquad = 3x^2 + 6xh + 3h^2 - 2x - 2h + 7$

111. $P(x) = 2x^2 - 3x + 5$

112. $P(x) = 5x^2 - 7x + 6$

113. $P(x) = x^2 + 5x - 3$

114. $P(x) = 2x^3 - 5x^2 + 1$

115. $P(x) = x^3 - x^2 + 3x + 2$

116. $P(x) = 3x - 2$

117. $P(x) = 7x + 15$

118. $P(x) = 6x^2 - 3x$

119. $P(x) = 9x^2 + 1$

120. $P(x) = 8x^3 - 1$

▶ In each of the following, express $P(x + h) - P(x)$ as a polynomial in x and h.

EXAMPLE

$P(x) = 3x^2 - 4x + 5$

Solution

$\qquad P(x + h) = 3(x + h)^2 - 4(x + h) + 5$

$P(x + h) - P(x) = 3(x + h)^2 - 4(x + h) + 5 - (3x^2 - 4x + 5)$

$\qquad\qquad = 3x^2 + 6xh + 3h^2 - 4x - 4h + 5 - 3x^2 + 4x - 5$

$\qquad\qquad = 6xh + 3h^2 - 4h$

121. $P(x) = 3x^2 - 7x + 2$ **122.** $P(x) = -5x^2 + 3x + 1$

123. $P(x) = 2x^3 - x^2$ **124.** $P(x) = 6x - 5$

125. $P(x) = 4x^3$ **126.** $P(x) = 2x^2 - 5x + 3$

127. $P(x) = -x^2 + x - 1$ **128.** $P(x) = 7x - 4$

129. $P(x) = 2x^3 - 5x$ **130.** $P(x) = -3x^2 - 2x + 4$

▶   In each of the following, express $P(Q(x))$ as a polynomial.

EXAMPLE

$P(x) = 3x^2 + 4x$

$Q(x) = 2x + 5$

Solution

$P(Q(x)) = 3(2x + 5)^2 + 4(2x + 5)$

$\qquad\qquad = 12x^2 + 30x + 75 + 8x + 20$

$\qquad\qquad = 12x^2 + 38x + 95$

131. $P(x) = 5x^2 + 3x,\ Q(x) = 3x + 2$

132. $P(x) = 3x + 7,\ Q(x) = 3x^2 - 2x$

133. $P(x) = x^3,\ Q(x) = x + 2$

134. $P(x) = x^3,\ Q(x) = x^2$

135. $P(x) = x^2 - 3x,\ Q(x) = 3x^2 - 2$

136. $P(x) = 2x - 5,\ Q(x) = 6x^5 - 3x^4 + 2$

137. $P(x) = 3,\ Q(x) = 2x^2 + 5x$

138. $P(x) = 7,\ Q(x) = 8x^3 - 5x^2$

139. $P(x) = 2x^2 - 3x + 1,\ Q(x) = x + 1$

140. $P(x) = x^5 + x,\ Q(x) = 2x^3$

▶   In each of the following, find the quotient and the remainder.

141. $\dfrac{5x^3 - 2x^2 + x - 2}{x^2 - 3x + 6}$ **142.** $\dfrac{2x^3 - 5x^2 + 7}{x^2 - 2x}$

143. $\dfrac{4x^3 - 6x^2 + 2x - 1}{2x^2 + x + 3}$ **144.** $\dfrac{x^4 - 3x^2 + 2x - 1}{x^2 - 2x + 3}$

145. $\dfrac{2x^4 - x^3 + 3x^2 - 5}{x^3 - 5x^2 + 6}$ **146.** $\dfrac{x^3 - 2x + 8}{x - 2}$

147. $\dfrac{6x^2 + 17x - 45}{3x - 5}$ **148.** $\dfrac{10x^2 - 27x - 28}{2x - 7}$

149. $\dfrac{x^2 - 9}{x - 3}$ **150.** $\dfrac{12x^2 - 64x + 50}{6x - 5}$

▶ In each of the following, evaluate $P(c)$ by substitution, and then divide $P(x)$ by $x - c$ and find the remainder. Compare the remainder with $P(c)$.

EXAMPLE

$P(x) = 3x^2 - 2x - 3$, $c = 2$

Solution

$P(2) = 3(2)^2 - 2(2) - 3 = 12 - 4 - 3 = 5$

$$
\begin{array}{r}
3x + 4 \\
x - 2 \overline{)\,3x^2 - 2x - 3} \\
\underline{3x^2 - 6x} \\
4x - 3 \\
\underline{4x - 8} \\
5 \leftarrow \text{Remainder}
\end{array}
$$

151. $P(x) = 2x^2 - x - 9$, $c = 3$

152. $P(x) = 2x^2 + 11x + 20$, $c = -3$

153. $P(x) = 3x^3 - x^2 - 3x - 7$, $c = 1$

154. $P(x) = 6x^2 + 7x - 8$, $c = -2$

155. $P(x) = 2x^3 - x^2 - 7x - 4$, $c = 2$

156. $P(x) = x^5 - 3x^3 - x^2 + x^4 + 2x + 3$, $c = -1$

157. $P(x) = 5x^3 + 17x^2 - 12x - 10$, $c = -4$

158. $P(x) = x^4 - 2x^2 + 17x - 7x^3 - 23$, $c = 7$

159. $P(x) = 7x^3 + 33x^2 - 7x + 11$, $c = -5$

160. $P(x) = x^7 + 3x^4 - x^6 - 3x^3 - 2x^2 + 9x - 9$, $c = 1$

▶ In each of the following, use synthetic division to find the quotient and remainder.

161. $\dfrac{x^5 - 2x^4 + 3x^3 - x^2 + 2x + 1}{x - 2}$

162. $\dfrac{x^5 - 2x^4 + 3x^2 - 1}{x + 2}$

163. $\dfrac{3x^4 - 2x^3 + 7x - 4}{x + 3}$

164. $\dfrac{3x^4 - 2x^3 + 7x - 4}{x - 3}$

165. $\dfrac{5x^2 - 2x + 7}{x - 4}$

166. $\dfrac{5x^2 - 2x + 7}{x + 4}$

167. $\dfrac{2x^5 - 3x^3 + 5x^2 - 2}{x - 1}$

168. $\dfrac{2x^5 - 3x^3 + 5x^2 - 2}{x + 1}$

169. $\dfrac{5x^4 - 3x^2 - 4x + 6}{x - 7}$

170. $\dfrac{5x^4 - 3x^2 - 4x + 6}{x + 7}$

▶ In each of the following, use synthetic division to find $P(c)$.

EXAMPLE

$P(x) = 2x^3 - 5x^2 + 7$, $c = 3$

Solution

$$3\overline{)\begin{array}{cccc} 2 & -5 & 0 & 7 \\ & 6 & 3 & 9 \end{array}}$$
$$\begin{array}{cccc} 2 & 1 & 3 & 16 \end{array} \quad P(3) = 16$$

171. $P(x) = 5x^3 - 4x^2 - 2x + 8,\ c = 2$

172. $P(x) = 2x^4 - 5x^3 + 3x,\ c = 3$

173. $P(x) = x^5 - 4x^4 + 5x^2 - 11,\ c = 2$

174. $P(x) = 2x^3 - 5x^2 + 7x - 18,\ c = -3$

175. $P(x) = 8x^5 - 3x^3 + 2x - 1,\ c = -2$

176. $P(x) = 7x^4 - 2x^2 + 3x - 11,\ c = -1$

177. $P(x) = 6x^3 - 5x^2 + 3x - 12,\ c = 4$

178. $P(x) = 11x^5 - 3x^4 + 2x^3 - 8,\ c = 6$

179. $P(x) = x^3 - 18,\ c = -6$

180. $P(x) = 2x^6 - 5x^4 + 3x^3 - 2x^2 + 5x - 8,\ c = 2$

▶ Express each of the following in the form $a + bi$ where a and b are real numbers and $i^2 = -1$.

EXAMPLES

$(2 + 5i) + (6 - 3i) = 8 + 2i$

$(2 + 5i) - (6 - 3i) = -4 + 8i$

$(2 + 5i)(6 - 3i) = 12 + 30i - 6i - 15i^2$
$$= 27 + 24i$$

$(2 + 5i)^2 = 4 + 20i + 25i^2 = -21 + 20i$

181. $(3 - 7i) + (4 + 6i)$

182. $(3 - 7i) - (4 + 6i)$

183. $(3 - 7i)(4 + 6i)$

184. $(3 - 7i)^2$

185. $5(3 - 7i)$

186. $5 + (3 - 7i)$

187. $(3 - 7i) - 5$

188. $(2 + 3i)^2 - 2(2 + 3i) + 4$

189. $3(4 - i)^2 + 2(4 - i) + 6$

190. $(\sqrt{2} + 3i)^2 - 4(\sqrt{2} + 3i) + 1$

191. $(\sqrt{2} + \sqrt{3}i)(\sqrt{2} - \sqrt{3}i)$

192. $(5 - 2i)(5 + 2i)$

193. $(2 + \sqrt{3}i)^2 - 4(2 + \sqrt{3}i) + 7$

194. $(4 + 3i)(4 - 3i)$

195. $(\frac{3}{5} + i)(\frac{3}{5} - i)$

▶ In each of the following, find the value of the polynomial for each of the given values of x.

EXAMPLE

$x^2 - 4x + 13; \; x = 2 + 3i, \; x = 2 - 3i$

Solution

$$
\begin{aligned}
(2 + 3i)^2 - 4(2 + 3i) + 13 &= 4 + 12i + 9i^2 - 8 - 12i + 13 \\
&= 9 + 9i^2 + 12i - 12i \\
&= 9 - 9 \\
&= 0
\end{aligned}
$$

$$
\begin{aligned}
(2 - 3i)^2 - 4(2 - 3i) + 13 &= 4 - 12i + 9i^2 - 8 + 12i + 13 \\
&= 9 + 9i^2 - 12i + 12i \\
&= 0
\end{aligned}
$$

196. $x^2 - 8x + 41; \; x = 4 + 5i, \; x = 4 - 5i$

197. $x^2 - 10x + 74; \; x = 5 + 7i, \; x = 5 - 7i$

198. $x^2 + 4x + 31; \; x = 2 + 5i, \; x = 2 - 5i$

199. $x^2 - 2x + 7; \; x = 1 + 2i, \; x = 1 - 2i$

200. $x^2 - 6x + 10; \; x = 3 + i, \; x = 3 - i$

CHECK LIST FOR THE STUDENT

If you have learned the material in Section 1.4, you should be able to do the following:

1. Give examples of terms, constant terms, coefficients, factors, and like terms.
2. Define polynomial, monomial, binomial, and trinomial.
3. Define polynomial over the reals and polynomial over the integers.
4. Give an example of a polynomial in two variables x and y.
5. Define the degree of a term and the degree of a polynomial.
6. Write an expression for an nth-degree polynomial in x.
7. Use the notation $P(x)$ for a polynomial in x, and find $P(a)$ for some specific real number a and a given polynomial $P(x)$.
8. Simplify a polynomial.
9. Remove parentheses with either a plus sign or a minus sign in front.
10. Add, subtract, multiply, and divide polynomials.
11. Express $P(x + h) - P(x)$ as a polynomial in x and h.
12. State and verify the remainder theorem.
13. Use synthetic division or synthetic substitution.
14. Add, subtract, and multiply complex numbers.

SECTION QUIZ 1.4

1. State whether or not each of the following is a polynomial.
   (a) $3x^4 - 2x^{-3} + 7x^2 - 9$
   (b) $3x^4 - 2x^3 + 7x^{3/5} - 9$
   (c) $3x^4 - 5x^3 + 7x - 9$
   (d) $3x^4 - 5x^3 + 2x^2 + 7x - 9$
   (e) $3x^4 - 5x^3 + \frac{4}{5}x^2 + 7x - 9$

2. (a) State the degree of $5x^3 - 2x^4 + 3x^2 + 9x - 8$.
   (b) Simplify $2x^3 - 3x^2 - 7x^3 + x^2 - 3x + 7$ by collecting like terms.

3. Let $P(x) = x^3 - 5x^2 + 2x - 1$ and $Q(x) = x^2 + 3$.
   (a) Find $P(2)$.   (b) Find $Q(1)$.   (c) Find $P(Q(1))$.

Let $P(x) = 2x^3 + 3x^2 - 5x - 6$ and $Q(x) = x^2 - 4x - 7$.
4. Find $P(x) + Q(x)$.

5. Find $P(x) - Q(x)$.

6. Find the product $P(x) \times Q(x)$.

7. Find the quotient and remainder.

$$\frac{2x^3 + 3x^2 - 5x - 6}{x^2 - 4x - 7}$$

8. Let $P(x) = 3x^2 - 2x - 5$, and express $P(x + h)$ as a polynomial in x and h.

9. Use synthetic division to find the quotient and remainder.

$$\frac{2x^3 + 3x^2 - 5x - 6}{x - 1}$$

10. Express each of the following in the form $a + bi$ where a and b are real numbers and $i^2 = -1$.
    (a) $(2 + 5i) + (3 - 7i)$
    (b) $(2 + 5i) - (3 - 7i)$
    (c) $(2 + 5i) \times (3 - 7i)$

1.5 Factoring

The purpose of this section is to review factoring of polynomials with integral coefficients.

As we said in Section 1.4, in any product ab, the a and the b are called **factors.** Thus in the product 3×4, the 3 and the 4 are factors, and in the product $(x - 3)(x + 3)$, the $(x - 3)$ and the $(x + 3)$ are factors. To factor an integer or polynomial is to express it as a product of factors.

It can be shown that (except for the order of the factors) any positive integer that is not itself a prime can be factored into *primes* (Figure 1.5, Section 1.1) in exactly one way. For example, except for order, the only way to express 12 as a product of primes is to write

$$12 = 3 \times 2 \times 2$$
$$= 3 \times 2^2,$$

and the only way to express 72 as a product of primes is to write

$$72 = 2 \times 2 \times 2 \times 3 \times 3$$
$$= 2^3 \times 3^2.$$

Exercise 1.5, 1–10

A polynomial over the integers is said to be a **prime polynomial over the integers** if it is evenly divisible (the remainder is zero) by no polynomial with integral coefficients other than itself and $+1$ or -1. For example,

$$x - 7, \quad 2x + 3, \quad \text{and} \quad x^2 + 5$$

are prime polynomials over the integers.

It can be shown that (except for the order of the factors) any polynomial over the integers that is not itself a prime can be factored into prime factors with integral coefficients in exactly one way. For example, except for order, the only way to express $x^2 - 9$ as a product of such prime factors is to write

$$x^2 - 9 = (x - 3)(x + 3),$$

and the only way to express $3x^2 + 13x - 10$ as a product of prime polynomials with integral coefficients is to write

$$3x^2 + 13x - 10 = (3x - 2)(x + 5).$$

Expressing a polynomial over the integers as a product of prime polynomials with integral coefficients is sometimes called **factoring completely.** If a polynomial over the integers *can* be factored into prime polynomials with integral coefficients, then it is said to be **factorable.** Otherwise it is said to be **not factorable over the integers.**

In calculus, as in all of mathematics, some types of factorable polynomials occur more frequently than others. Before listing some of these more common types, let us first list the special products giving rise to them.

SPECIAL PRODUCTS

1. $a(b + c) = ab + ac$
2. $(a - b)(a + b) = a^2 - b^2$
3. $(a + b)^2 = a^2 + 2ab + b^2$

   $(a - b)^2 = a^2 - 2ab + b^2$ Square of a binomial
4. $(a + b)(a^2 - ab + b^2) = a^3 + b^3$

   $(a - b)(a^2 + ab + b^2) = a^3 - b^3$

Note particularly that the square of a binomial is the sum of *three* terms. We caution all students to avoid the very common error of overlooking the "middle term." In other words, $(a + b)^2$ is *not* equal to just $a^2 + b^2$, but $(a + b)^2$ is equal to a^2 plus $2ab$ plus b^2.

$$(a + b)^2 = a^2 + 2ab + b^2$$

Likewise, $(a - b)^2$ is *not* equal to $a^2 - b^2$, as so many students seem to believe; $(a - b)^2$ is equal to a^2 minus $2ab$ plus b^2.

$$(a - b)^2 = a^2 - 2ab + b^2$$

EXAMPLE 1   Write out the product $(x + 7)^2$ and the product $(x - 7)^2$.

Solution

$(x + 7)^2 = x^2 + 14x + 49$

$(a + b)^2 = a^2 + 2ab + b^2$

$(x - 7)^2 = x^2 - 14x + 49$

$(a - b)^2 = a^2 - 2ab + b^2$

The common types of factorable polynomials arising in our list of special products are factored in the following list. Students should verify each of the examples by multiplying out the factors.

Exercise 1.5, 11–70

COMMON TYPES OF FACTORABLE POLYNOMIALS

I. $ab + ac = a(b + c)$ Common monomial factor

    $15x - 6 = 3(5x - 2)$

    $15x^3 - 6x^2 = 3x^2(5x - 2)$

II. $a^2 - b^2 = (a - b)(a + b)$ Difference of two squares

    $9x^2 - 4 = (3x - 2)(3x + 2)$

III. $a^2 + 2ab + b^2 = (a + b)^2$ Perfect-square
$a^2 - 2ab + b^2 = (a - b)^2$ trinomials
$x^2 + 6x + 9 = (x + 3)^2$
$x^2 - 6x + 9 = (x - 3)^2$

IV. $a^3 + b^3 = (a + b)(a^2 - ab + b^2)$ Sum of two cubes
$a^3 - b^3 = (a - b)(a^2 + ab + b^2)$ Difference of two cubes
$x^3 + 125 = (x + 5)(x^2 - 5x + 25)$
$x^3 - 125 = (x - 5)(x^2 + 5x + 25)$

EXAMPLE 2

Factor each of the following completely.
(a) $5x^3 - 5x$ (b) $x^4 - y^4$
(c) $9x^2 + 6xy + y^2$ (d) $9x^2 - 6xy + y^2$
(e) $8x^3 + 27$ (f) $8x^3 - 27$

Solution

We are to express each of the given polynomials as a product of prime factors with integral coefficients. In each case, if there is a common monomial factor, we should remove that (factor that out) first.

(a) $5x^3 - 5x = 5x(x^2 - 1)$ Common monomial factor

       $= 5x(x - 1)(x + 1)$ Difference of two squares

(b) $x^4 - y^4 = (x^2 - y^2)(x^2 + y^2)$ Difference of two squares

       $= (x - y)(x + y)(x^2 + y^2)$ Difference of two squares

(c) $9x^2 + 6xy + y^2 = (3x + y)^2$ Perfect-square trinomial

(d) $9x^2 - 6xy + y^2 = (3x - y)^2$ Perfect-square trinomial

(e) $8x^3 + 27 = (2x + 3)(4x^2 - 6x + 9)$ Sum of two cubes

(f) $8x^3 - 27 = (2x - 3)(4x^2 + 6x + 9)$ Difference of two cubes

Type I of the common types we have listed arises from the distributive law (Appendix A). An extension of the distributive law is used in reverse to remove any monomial that is common to all the terms of any polynomial. For example, the terms of the polynomial

$$6x^5 - 14x^4 - 8x^3$$

all have the factor $2x^3$ in common, so we can factor out $2x^3$ (divide each term by $2x^3$) and write

$$6x^5 - 14x^4 - 8x^3 = 2x^3(3x^2 - 7x - 4).$$

EXAMPLE 3

Solution

Factor $3x^3 - 30x^2 + 75x$ completely.

$$3x^3 - 30x^2 + 75x = 3x(x^2 - 10x + 25)$$
$$= 3x(x - 5)^2$$

Type II, the difference of two squares, is easy to recognize and usually causes no problems.

Polynomials of Type III, perfect-square trinomials, are also relatively easy to recognize. Each is the sum of *three* terms; the first and last terms of each are perfect squares; the "middle term" of each is either plus or minus twice the product of the numbers whose squares constitute the first and last terms. Each perfect-square trinomial is the square of a binomial.

Perhaps it will help to emphasize that the square of a binomial, a perfect-square trinomial, is the sum of *three* terms, if we note the following.

Where k is a real number, the expression $x^2 + kx$ becomes a perfect-square trinomial if the number $(k/2)^2$ is added to it.

In particular, $x^2 + 6x$ becomes a perfect square if the number $(6/2)^2 = (3)^2$ is added to it.

$$x^2 + 6x + (3)^2 = (x + 3)^2$$

The process of determining what should be added to an expression like $x^2 + kx$ in order to make it become a perfect square trinomial is called *completing the square.* We will use it in Section 1.7.

EXAMPLE 4

Tell what number should be put in the parentheses in order to make $x^2 - 18x + (\ \ )$ become a perfect-square trinomial, and then factor that perfect-square trinomial.

Solution

The number that should go in the parentheses is the square of half the coefficient of x. Half the coefficient of x is $\frac{-18}{2} = -9$, and its square is $(\frac{-18}{2})^2 = (-9)^2 = 81$.

$$x^2 - 18x + (81) = (x - 9)^2$$

Exercise 1.5, 116–125 This can be checked by finding the product $(x - 9)^2$.

Type IV is another source of many common errors. Note that $a^3 + b^3$ is *not equal to* $(a + b)^3$. In fact,

$$a^3 + b^3 = (a + b)(a^2 - ab + b^2),$$

whereas
$$(a + b)^3 = (a + b)(a + b)^2$$
$$= (a + b)(a^2 + 2ab + b^2).$$

Similarly, note that $a^3 - b^3$ is *not equal to* $(a - b)^3$;

$$a^3 - b^3 = (a - b)(a^2 + ab + b^2),$$

whereas
$$(a - b)^3 = (a - b)(a - b)^2$$
$$= (a - b)(a^2 - 2ab + b^2).$$

EXAMPLE 5

(a) Factor $x^3 + 8$ completely, and factor $x^3 - 8$ completely.

(b) Write out the product $(x + 2)^3$ and the product $(x - 2)^3$.

Solution

(a) $x^3 + 8 = (x + 2)(x^2 - 2x + 4)$
$\quad x^3 - 8 = (x - 2)(x^2 + 2x + 4)$

(b) $(x + 2)^3 = (x + 2)(x + 2)^2$
$\qquad\qquad = (x + 2)(x^2 + 4x + 4)$
$\qquad\qquad = x^3 + 6x^2 + 12x + 8$
$\quad (x - 2)^3 = (x - 2)(x - 2)^2$
$\qquad\qquad = (x - 2)(x^2 - 4x + 4)$
$\qquad\qquad = x^3 - 6x^2 + 12x - 8$

In addition to the common types of factorable polynomials we have listed, factorable second-degree trinomials that are not perfect squares are frequently encountered. For example,

$$15x^2 + 2x - 8$$

is such a trinomial. Each such trinomial is the product of two prime binomial factors of first degree. For example, $15x^2 + 2x - 8$ is the product of $(3x - 2)$ and $(5x + 4)$. Any factorable trinomial of the form $ax^2 + bx + c$ where a, b, and c are integers is the product of two binomials of the form $(px + q)$ and $(rx + s)$ where p, q, r, and s are integers.

$$(px + q)(rx + s) = \underbrace{(pr)x^2}_{} + \underbrace{(ps)x + (qr)x}_{} + \underbrace{(qs)}_{}$$

Product of the first terms

Sum of the "cross products"

Product of the second terms

$$(3x - 2)(5x + 4) = 15x^2 + 12x - 10x + (-2)(4)$$

Where a, b, and c are integers, to factor $ax^2 + bx + c$ so that

$$ax^2 + bx + c = (px + q)(rx + s),$$

we systematically look for a pair of binomial factors $(px + q)$ and $(rx + s)$ with integral coefficients such that

1. The product of the first terms is ax^2,

$$pr = a$$

2. The product of the second terms is c,

$$qs = c$$

3. The sum of the "cross products" (the sum of the products of the first and last terms) is bx.

$$(ps) + (qr) = b$$

For example, to factor $15x^2 + 2x - 8$, we look for factors of the form $(px + q)(rx + s)$ with integral coefficients such that

1. $pr = 15$
2. $qs = -8$,
3. $ps + qr = 2$

We can do this by first listing all possibilities for integers p, q, r, and s such that $pr = 15$ and $qs = -8$ and then selecting a combination for which $ps + qr = 2$. Such a list would reveal that if $p = 3$, $q = -2$, $r = 5$, and $s = 4$, $ps + qr = 2$ so that

$$15x^2 + 2x - 8 = (3x - 2)(5x + 4).$$

EXAMPLE 6

Factor $10x^3 + 26x^2 - 12x$ completely.

Solution

The monomial factor $2x$ is common to all three terms so we factor $2x$ out first.

$$10x^3 + 26x^2 - 12x = 2x(5x^2 + 13x - 6)$$
$$= 2x(5x - 2)(x + 3)$$

This can be checked by multiplying out the factors.

EXAMPLE 7

Factor $x^2 + 3x + 7$ completely, if possible. If not possible, write "not factorable over the integers."

Solution

There is no monomial factor (other than 1) common to all three terms of $x^2 + 3x + 7$. If $x^2 + 3x + 7$ is factorable, there must exist integers p, q, r, and s such that $(px + q)(rx + s)$ are the factors. That is, there must exist integers p, q, r, and s such that

$$pr = 1, \qquad qs = 7,$$

and $ps + qr = 3$. The only possibilities for p, r, q, and s such that $pr = 1$ and $qs = 7$ are:

p	r	q	s	$ps + qr$
1	1	1	7	8
1	1	7	1	8
1	1	-1	-7	-8
1	1	-7	-1	-8
-1	-1	1	7	-8
-1	-1	7	1	-8
-1	-1	-1	-7	8
-1	-1	-7	-1	8

But in none of these possibilities is the sum of the cross products, $ps + qr$, equal to 3. It is not possible to factor $x^2 + 3x + 7$ into a product of polynomials with integral coefficients that are prime polynomials. The polynomial $x^2 + 3x + 7$ is itself a prime polynomial. We say that $x^2 + 3x + 7$ is *not factorable over the integers.*

Exercise 1.5, 71–115

We have seen that an extension of the distributive law can be used in reverse to factor any existing common *monomial* factors out of a polynomial. Sometimes the terms of a polynomial can be grouped in such a way that an extension of the distributive law can be used in reverse to factor out common *binomial* factors. For example, consider the polynomial

$$x^4 - 3x^3 - 2x^2 + 6x + 7x - 21.$$

We can group and factor the terms so that the new expression has the binomial factor $(x - 3)$ common to all the terms. We can then factor out that common binomial factor. Indeed,

$$x^4 - 3x^3 - 2x^2 + 6x + 7x - 21 = x^3(x - 3) - 2x(x - 3) + 7(x - 3)$$
$$= (x - 3)(x^3 - 2x + 7)$$

EXAMPLE 8

Group the terms of

$$2x^5 - 3x^4 - 10x^3 + 15x^2 + 18x - 27$$

in pairs and then factor out a common binomial factor.

Solution

$$2x^5 - 3x^4 - 10x^3 + 15x^2 + 18x - 27 = x^4(2x - 3) - 5x^2(2x - 3) + 9(2x - 3)$$

Exercise 1.5, 126–130

$$= (2x - 3)(x^4 - 5x^2 + 9)$$

Earlier in this section we saw that

$$a^2 - b^2 = (a - b)(a + b)$$

and

$$a^3 - b^3 = (a - b)(a^2 + ab + b^2).$$

By multiplying out the factors, it can also be shown that

$$a^4 - b^4 = (a - b)(a^3 + a^2b + ab^2 + b^3)$$

and

$$a^5 - b^5 = (a - b)(a^4 + a^3b + a^2b^2 + ab^3 + b^4).$$

In general, if n is any positive integer, the "formula" for factoring $(a - b)$ out of $a^n - b^n$ is

$$a^n - b^n = (a - b)(a^{n-1} + a^{n-2}b + a^{n-3}b^2 + \cdots + ab^{n-2} + b^{n-1}).$$

Note the pattern of the terms in the second factor. Reading left to right, the first term is a^{n-1}. Each term thereafter contains a raised to an exponent that is one less than the exponent of a in the preceding term. The next to the last term contains $a = a^1$, and the last term technically contains $a^{1-1} = a^0 = 1$ as a factor. The b's start appearing in the second term, and each term contains b raised to an exponent that is one more than the exponent of b in the preceding term. Note that the degree of every single term is $n - 1$. If we use this formula to factor $a^7 - b^7$, we get

$$a^7 - b^7 = (a - b)(a^6 + a^5b + a^4b^2 + a^3b^3 + a^2b^4 + ab^5 + b^6).$$

Note that the degree of every single term in the second factor is 6. The terms are formed by starting with $a^6 = a^6b^0$ and then successively dropping the exponent on the a by 1 while raising the exponent on the b by 1 until $b^6 = b^6a^0$ is reached. The factorization can be checked by multiplying out the factors.

Widely used beginning calculus texts base their derivations of the so-called "power formula" on the factorization of $a^n - b^n$. The following two examples utilize notation that is common to such texts.

EXAMPLE 9

Using the formula for factoring $(a - b)$ out of $a^n - b^n$, first factor $(x + h)^4 - x^4$ and then simplify the quotient

$$\frac{(x + h)^4 - x^4}{h}$$

where h represents a positive real number.

Solution

The expression $(x + h)^4 - x^4$ is of the form $a^4 - b^4$ where a is $(x + h)$ and b is x.

$$a^4 - b^4 = \quad (a - b)\,(a^3 \quad + a^2b \quad + \quad ab^2 + b^3)$$

$$(x + h)^4 - x^4 = (x + h - x)[(x + h)^3 + (x + h)^2x + (x + h)x^2 + x^3]$$

$$= \quad h \quad [(x + h)^3 + (x + h)^2x + (x + h)x^2 + x^3]$$

$$\frac{(x + h)^4 - x^4}{h} = \frac{h[(x + h)^3 + (x + h)^2x + (x + h)x^2 + x^3]}{h}$$

$$= (x + h)^3 + (x + h)^2x + (x + h)x^2 + x^3$$

Note: If h were successively assigned values such as $\frac{1}{2}, \frac{1}{4}, \frac{1}{8}, \frac{1}{16}, \frac{1}{32}, \frac{1}{64}, \cdots$ that successively "approach" zero, the simplified expression for

$$\frac{(x + h)^4 - x^4}{h}$$

would successively take on values that "approach"

$$(x + 0)^3 + (x + 0)^2x + (x + 0)x^2 + x^3 = x^3 + x^3 + x^3 + x^3$$

$$= 4x^3$$

In calculus, $4x^3$ is called the *limit* of the quotient

$$\frac{(x + h)^4 - x^4}{h}$$

as h approaches zero. It has various physical interpretations, such as the velocity with which a particle travels, and it is defined to be something called the "derivative of x^4 at x."

EXAMPLE 10

Where n is a positive integer, factor $(x + h)^n - x^n$.

Solution

The expression $(x + h)^n - x^n$ is of the form $a^n - b^n$ where a is $(x + h)$ and b is x.

$$a^n - b^n = (a - b)(a^{n-1} + a^{n-2}b + a^{n-3}b^2 + \cdots$$
$$+ ab^{n-2} + b^{n-1})$$
$$(x + h)^n - x^n = (x + h - x)[(x + h)^{n-1} + (x + h)^{n-2}x$$
$$+ (x + h)^{n-3}x^2 + \cdots + (x + h)x^{n-2} + x^{n-1}]$$
$$= h[(x + h)^{n-1} + (x + h)^{n-2}x + (x + h)^{n-3}x^2 + \cdots$$

Exercise 1.5, 131–135

$$+ (x + h)x^{n-2} + x^{n-1}]$$

EXERCISE 1.5 FACTORING

▶ Factor each of the following into a product of primes.

EXAMPLE

$180 = 5 \times 36 = 5 \times 6 \times 6 = 5 \times 2 \times 3 \times 2 \times 3 = 5 \times 2^2 \times 3^2$

1. 225	2. 1800	3. 1400	4. 1350	5. 338
6. 3388	7. 714	8. 539	9. 9020	10. 1287

▶ Write out each of the following products.

EXAMPLE

$(3x - 2)(3x + 2) = 9x^2 - 4$

11. $(a - b)(a + b)$ 12. $(a + b)^2$

13. $(a - b)^2$ 14. $(a + b)(a^2 - ab + b^2)$

15. $(a - b)(a^2 + ab + b^2)$ 16. $(a - b)(a^3 + a^2b + ab^2 + b^3)$

17. $(a - b)(a^4 + a^3b + a^2b^2 + ab^3 + b^4)$

18. $(x - 7)(x + 8)$ 19. $(x + 3)(x - 5)$

20. $(x - 2)(x + 9)$ 21. $(x + 6)(x - 4)$

22. $(x + 8)(x - 2)$ 23. $(3x - 7)(2x + 1)$

24. $(2x - 5)(3x + 8)$

25. $(6x - 4)(2x + 3)$

26. $(5x - 3)(3x - 2)$

27. $(7x - 2)(3x + 4)$

28. $2x^3(7x - 5)$

29. $3x^5(7x^2 - 2x + 9)$

30. $2x^7(3x^3 - 2x^2 + 5x - 1)$

31. $x(5x^4 - 4x^2 + 2x - 7)$

32. $-x(x^3 - x^2 + x - 1)$

33. $-5x^2(x^4 + 3x^2 - 2x - 9)$

34. $(2x - 3)(x^3 - 2x^2 + 7)$

35. $(x - 8)(3x^5 - 2x^3 + x^2 - 1)$

36. $(x - 7)(x^3 - 2x^2 + 3x - 1)$

37. $(3x + 7)(x^7 - 2x^5 + 6)$

38. $(3x - 4)(3x + 4)$

39. $(x - 2y)(x + 2y)$

40. $(3x + 4)^2$

41. $(3x - 4)^2$

42. $(x - 2y)^2$

43. $(x + 2y)^2$

44. $(x^2 - y^2)(x^2 + y^2)$

45. $(x^3 - y^3)(x^3 + y^3)$

46. $(x^2 - y^2)^2$

47. $(x^3 + y^3)^2$

48. $(2 - 3x)^2$

49. $(-x + 2)^2$

50. $(-x - 1)^2$

51. $(3x + 2)(3x - 2)$

52. $(3x + 2)^2$

53. $(x^3 - 2)(x^3 + 2)$

54. $(x^3 - 2)^2$

55. $(x + h)^2$

56. $(a - 2)(a^2 + 2a + 4)$

57. $(x - 1)^2$

58. $(3x + 7)^2$

59. $(3x + 7)(3x - 7)$

60. $(3x - 7)^2$

61. $(3x - 7)(3x + 1)$

62. $(x + 1)(x^2 - x + 1)$

63. $(x - 1)(x^2 + x + 1)$

64. $(3x - 7)(9x^2 + 21x + 49)$

65. $(3x + 7)(9x^2 - 21x + 49)$

66. $(x - 2y)(x^2 + 2xy + 4y^2)$

67. $(x + 2y)(x^2 - 2xy + 4y^2)$

68. $(x + 2)(x^2 - 2x + 4)$

69. $(2 - 3x)(8 + 12x + 18x^2 + 27x^3)$

70. $(2 - 3x)(16 + 24x + 36x^2 + 54x^3 + 81x^4)$

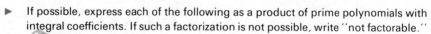

▶ If possible, express each of the following as a product of prime polynomials with integral coefficients. If such a factorization is not possible, write "not factorable."

71. $25x^2 - 16$

72. $25x^2 - 40x + 16$

73. $3x^2yz - 2x^5y^2 + x^3y^2z$

74. $8w^3 - 27z^3$

75. $8w^3 + 27z^3$

76. $x^2 - 5x - 36$

77. $9x^2 - 49y^2$

78. $9x^4 - 42x^3y + 49x^2y^2$

79. $27x^4y - 64xy^4$

80. $3x^2 - 19x + 20$

81. $x^8 - 2x^4y^4 + y^8$

82. $x^8 - y^8$

83. $x^9 - y^9$

84. $81x^2 - 100y^2$

85. $x^2 - 7x - 44$

86. $6x^2 - 13x - 5$

87. $64x^6 - y^6$

88. $x^3y^6 + x^6y^3$

89. $2x^2 - 3x - 20$

90. $x^2 + 5x + 3$

91. $36x^2 - 4y^2$

92. $36x^2 - 24xy + 4y^2$

93. $2a^4b^2 + 4a^3b^3 + 2a^2b^4$

94. $x^2 - 2x + 2$

95. $121a^2 - 169y^2$

96. $12x^3 - 34x^2 + 14x$

97. $9w^2 + 48wz + 64z^2$

98. $9w^2 - 64z^2$

99. $9w^2 + 64z^2$

100. $14x^2 + 33x - 5$

101. $27a^3b + 8b^4$

102. $15x^2 - 8x + 1$

103. $16x^3y^2 - 25xz^2$

104. $6x^4 + 19x^2 - 20x^2$

105. $x^4y - x^3y - 56x^2y$

106. $2a^2 - 2a - 40$

107. $x^2 + y^2$

108. $a^6 + 2a^3b^3 + b^6$

109. $6x^2 - 6x - 12$

110. $21a^2 - 13a + 2$

111. $x^4 - 16y^4$

112. $32a^5 - b^5$

113. $a^{10} - b^{10}$

114. $x^{11} - y^{11}$

115. $x^{12} - y^{12}$

▶ In each of the following, tell what number should be put in the parentheses in order to make the expression become a perfect-square trinomial, and then write that trinomial as the square of a binomial with rational coefficients.

EXAMPLE

$x^2 - 5x + (\quad)$

Solution

$(\frac{-5}{2})^2 = \frac{25}{4};\ x^2 - 5x + (\frac{25}{4}) = (x - \frac{5}{2})^2$

116. $x^2 + 5x + (\quad)$ **117.** $x^2 + 7x + (\quad)$ **118.** $x^2 - 3x + (\quad)$

119. $x^2 - 9x + (\quad)$ **120.** $x^2 + x + (\quad)$ **121.** $x^2 - 2x + (\quad)$

122. $x^2 - 4x + (\quad)$ **123.** $x^2 + 2x + (\quad)$ **124.** $x^2 + 4x + (\quad)$

125. If a and b are real numbers and $a \neq 0$, tell what should be put in the parentheses in order to make the expression

$$x^2 + \frac{b}{a}x + (\quad)$$

become a perfect-square trinomial, and then write that trinomial as the square of a binomial.

▶ In each of the following, group the terms in pairs and then factor out a common binomial factor.

$$x^4 - 7x^3 - 2x^2 + 14x + 6x - 42 = (x^4 - 7x^3) - (2x^2 - 14x)$$
$$+ (6x - 42)$$
$$= x^3(x - 7) - 2x(x - 7) + 6(x - 7)$$
$$= (x - 7)(x^3 - 2x + 6)$$

$$xy^5 - 2x + 3y^5 - 6 = (xy^5 + 3y^5) - (2x + 6)$$
$$= y^5(x + 3) - 2(x + 3)$$
$$= (x + 3)(y^5 - 2)$$

126. $x^4 - 5x^3 - 4x^2 + 20x + 8x - 40$

127. $xy^7 - 3x + 2y^7 - 6$

128. $a^5 - 3a^4b - ab^2 + 3b^3 + ab - 3b^2$

129. $3a^6 - 2a^4 - 3a^5 + 2a^3$

130. $x^3 + yx^2 - 2x^2y - 2xy^2 + 3x + 3y$

▶ In each of the following, first use the formula for factoring $(a - b)$ out of $a^n - b^n$ to factor $(x + h)^n - x^n$ for the given value of n, and then simplify the quotient

$$\frac{(x + h)^n - x^n}{h}$$

where h is a positive real number.

$n = 3$

$$(x + h)^3 - x^3 = (x + h - x)[(x + h)^2 + (x + h)x + x^2]$$
$$= \quad h \quad [(x + h)^2 + (x + h)x + x^2]$$
$$\frac{(x + h)^3 - x^3}{h} = (x + h)^2 + (x + h)x + x^2$$

131. $n = 2$ **132.** $n = 5$ **133.** $n = 6$ **134.** $n = 7$ **135.** $n = 8$

CHECK LIST FOR THE STUDENT

If you have learned the material in Section 1.5, you should be able to do the following:

1. Define prime polynomial over the integers.

2. Tell how many ways (except for the order of the factors) any polynomial over the integers that is not itself a prime can be factored into prime factors with integral coefficients.

3. Tell what is meant by factoring a polynomial over the integers completely.

4. Factor out a common monomial factor.

5. Factor the difference of two squares.

6. Factor perfect-square trinomials.

7. Tell what should be added to $x^2 + kx$ in order to make it become a perfect-square trinomial.

8. Factor the sum of two cubes.

9. Factor the difference of two cubes.

10. Factor any factorable second-degree trinomial over the integers.

11. Group terms and factor out a common binomial factor.

12. Factor $a^n - b^n$.

13. Factor $(x + h)^n - x^n$.

SECTION QUIZ 1.5

1. Write out each of the following products.
   (a) $(x - y)(x + y)$ (b) $(x + y)^2$
   (c) $(x - y)(x^2 + xy + y^2)$ (d) $(3x + 2)(x - 7)$
   (e) $-2x(3x^2 - x + 1)$

If possible, express each of the following as a product of prime polynomials with integral coefficients. If such a factorization is not possible, write "not factorable."

2. $9x^2 - 12x + 4$ 3. $9x^2 - 4$

4. $27x^3 + 8$ 5. $x^2 + 6x + 9$

6. $x^2 + 6x + 1$ 7. $9x^2 + 6x - 8$

8. $2a + 2b - 3ab - 3a^2$ 9. $2a^5b - 6a^3b^2 + 14a^2b$

10. Tell what number should be put in the parentheses in order to make the following expression become a perfect-square trinomial, and then factor that trinomial.

$$x^2 - 12x + (\quad)$$

1.6 Fractions; Rational Expressions

The purpose of this section is to review manipulations with fractions, particularly those fractions in which the numerator and denominator are polynomials. Some fractions with radicals in their denominators and fractions with imaginary-number denominators are also considered.

In Section 1.1, we saw that where P and nonzero D and N are mathematical expressions,

$$\frac{PN}{DN} = \frac{P}{D}.$$ Fundamental principle of fractions

It is likewise true that for nonzero D,

$$\frac{P}{D} = \frac{-P}{-D} = -\frac{P}{-D} = -\frac{-P}{D}.$$

For example,

$$\frac{3}{4} = \frac{-3}{-4} = -\frac{3}{-4} = -\frac{-3}{4}$$

$$\frac{x+2}{5x} = \frac{-(x+2)}{-5x} = -\frac{x+2}{-5x} = -\frac{-(x+2)}{5x}, \quad \text{if } x \neq 0.$$

Note that any fraction may be thought of as having three signs associated with it: the sign in front, the sign of the numerator, and the sign of the denominator. (If the sign in front is plus, it is not usually written.) We may change exactly two of these three signs without changing the value of the fraction.

EXAMPLE 1

State which sign belongs in the parentheses in each of the following, assuming that $x + y \neq 0$.

(a) $\dfrac{x-y}{x+y} = (\quad)\dfrac{y-x}{-x-y}$

(b) $\dfrac{x-y}{x+y} = (\quad)\dfrac{x-y}{-x-y}$

(c) $\dfrac{x-y}{x+y} = (\quad)\dfrac{y-x}{x+y}$

Solution

(a) $\dfrac{x-y}{x+y} = (\quad)\dfrac{y-x}{-x-y}$

Since $y - x = -(x - y)$ and $-x - y = -(x + y)$, the signs of both the numerator and the denominator have been changed. The sign in front stays the same.

$$\frac{x - y}{x + y} = (+) \frac{y - x}{-x - y} \qquad \frac{P}{D} = \frac{-P}{-D}$$

(b) $\dfrac{x - y}{x + y} = (\ ) \dfrac{x - y}{-x - y}$

Only the sign of the denominator has been changed, and since the numerator is unchanged, the sign in the front must be changed.

$$\frac{x - y}{x + y} = (-) \frac{x - y}{-x - y} \qquad \frac{P}{D} = - \frac{P}{-D}$$

(c) $\dfrac{x - y}{x + y} = (\ ) \dfrac{y - x}{x + y}$

Only the sign of the numerator has been changed, and since the denominator is unchanged, the sign in front must be changed.

Exercise 1.6, 1–15

$$\frac{x - y}{x + y} = (-) \frac{y - x}{x + y} \qquad \frac{P}{D} = - \frac{-P}{D}$$

If variables occur in a denominator, let it be understood that they are not permitted to take on values for which the denominator is zero. A fraction P/D may be considered to be an indicated division $P \div D$, and division by zero is not defined.

Numerical fractions—that is, fractions in which the numerator and denominator are real numbers—can sometimes be simplified or "reduced to lowest terms" by using the fundamental principle of fractions. A fraction in which the numerator and nonzero denominator are integers (that is, a fraction that is a rational number) is said to be in **lowest terms** if the numerator and denominator have no common integral factors other than $+1$ or -1. A fraction in which the numerator and nonzero denominator are polynomials with integral coefficients is said to be in lowest terms if the numerator and denominator have no common factors with integral coefficients other than $+1$ or -1. The fractions

$$\frac{6}{8} \quad \text{and} \quad \frac{x^2 - 9}{x^2 + 5x + 6}$$

are not in lowest terms but can be reduced to lowest terms by factoring their numerators and denominators and using the fundamental principle of fractions.

$$\frac{6}{8} = \frac{3 \times 2}{4 \times 2} = \frac{3}{4}$$

$$\frac{x^2 - 9}{x^2 + 5x + 6} = \frac{(x - 3)(x + 3)}{(x + 2)(x + 3)}$$

$$= \frac{x - 3}{x + 2}, \qquad \text{if } x + 3 \neq 0 \quad \text{and} \quad x + 2 \neq 0.$$

To reduce a fraction such as these to lowest terms, factor its numerator and denominator completely and then divide the numerator and denominator by their common factors. Dividing the numerator and denominator of a fraction by a common factor is sometimes called *canceling* that common factor.

EXAMPLE 2   Reduce each of the following to lowest terms, if possible.

(a) $\dfrac{(x - 3) \times 5(x - 2)}{(x - 3)(x + 7)}$ (b) $\dfrac{(x - 3) + 5(x - 2)}{(x - 3)(x + 7)}$

Solution   (a) $\dfrac{(x - 3) \times 5(x - 2)}{(x - 3)(x + 7)}$

The factor $(x - 3)$ is common to both numerator and denominator. Dividing both numerator and denominator by $(x - 3)$ and indicating this division by slash or "cancellation" marks, we have

$$\frac{(x - 3) \times 5(x - 2)}{(x - 3)(x + 7)} = \frac{\cancel{(x - 3)} \times 5(x - 2)}{\cancel{(x - 3)}(x + 7)}$$

$$= \frac{5(x - 2)}{x + 7}$$

$$= \frac{5x - 10}{x + 7}$$

(b) $\dfrac{(x - 3) + 5(x - 2)}{(x - 3)(x + 7)}$

Here $(x - 3)$ is not a factor of the numerator. The numerator and denominator do not have any common factors. The given fraction is already in lowest terms. Note that the term $(x - 3)$ cannot be canceled here. Students should beware of trying to cancel terms simply because they appear on the same page or in the same problem. In any fraction, only terms that are *factors* of *both* numerator and denominator can be canceled.

$$\frac{(x-3)+5(x-2)}{(x-3)(x+7)} = \frac{x-3+5x-10}{(x-3)(x+7)}$$

$$= \frac{6x-13}{(x-3)(x+7)}$$

Note that this result is *not* what we would get if we erroneously crossed out the $(x-3)$ simply because we saw it in two places in the same fraction.

Exercise 1.6, 16–40

The fundamental principle of fractions can also be used to express a fraction in an equivalent form with a larger or higher-degree denominator. For example, to express $\frac{3}{4}$ in a form with denominator 36, multiply numerator and denominator by 9.

$$\frac{3}{4} = \frac{3 \times 9}{4 \times 9} = \frac{27}{36}$$

To express $(x-3)/(x+2)$ in a form with denominator $(x+2)(x-7)$, multiply numerator and denominator by $(x-7)$.

$$\frac{x-3}{x+2} = \frac{(x-3)(x-7)}{(x+2)(x-7)} = \frac{x^2-10x+21}{x^2-5x-14}$$

Exercise 1.6, 41–50

We must perform such manipulations when it is necessary to express two or more fractions in forms having the same denominator.

A fraction in which the numerator and nonzero denominator are polynomials in one variable is called a **rational expression**. The fractions

$$\frac{x^2-9}{x^2+5x+6}, \quad \frac{x^2+5x+6}{1}, \quad \frac{3x^5}{7}, \quad \frac{\sqrt{2}x^3-5x^2+7}{\frac{3}{8}x^4-\sqrt{3}x+6}$$

are all rational expressions, even though the last one has irrational coefficients. The numerators and denominators are all polynomials (1, 7, and $3x^5$ are special polynomials of one term each) over the real numbers. Most of the rational expressions with which we shall work, however, have numerators and denominators that are polynomials with integral coefficients.

Fractions are multiplied or divided as follows:

$$\frac{A}{B} \times \frac{C}{D} = \frac{AC}{BD}, \qquad B \neq 0, \quad D \neq 0$$

$$\frac{A}{B} \div \frac{C}{D} = \frac{A}{B} \times \frac{D}{C} = \frac{AD}{BC}, \qquad B \neq 0, \quad D \neq 0, \quad C \neq 0$$

EXAMPLE 3

Find the product in each of the following.

(a) $\dfrac{3}{5} \times \dfrac{2}{7}$ (b) $\dfrac{x-2}{x+5} \times \dfrac{x+2}{x-3}$

Solution

(a) $\dfrac{3}{5} \times \dfrac{2}{7} = \dfrac{3 \times 2}{5 \times 7} = \dfrac{6}{35}$ (b) $\dfrac{x-2}{x+5} \times \dfrac{x+2}{x-3} = \dfrac{(x-2)(x+2)}{(x+5)(x-3)}$

$$= \dfrac{x^2 - 4}{x^2 + 2x - 15}$$

EXAMPLE 4

Perform the division in each of the following.

(a) $\dfrac{3}{5} \div \dfrac{8}{7}$ (b) $\dfrac{x-3}{x+2} \div \dfrac{x^2+5}{4x}$

Solution

(a) $\dfrac{3}{5} \div \dfrac{8}{7} = \dfrac{3}{5} \times \dfrac{7}{8}$ (b) $\dfrac{x-3}{x+2} \div \dfrac{x^2+5}{4x} = \dfrac{x-3}{x+2} \times \dfrac{4x}{x^2+5}$

$\qquad = \dfrac{21}{40}$ $\qquad\qquad = \dfrac{(x-3)(4x)}{(x+2)(x^2+5)}$

EXAMPLE 5

Divide $\sqrt{2}/2$ by $\frac{1}{3}$.

Solution

$$\dfrac{\sqrt{2}}{2} \div \dfrac{1}{3} = \dfrac{\sqrt{2}}{2} \times \dfrac{3}{1}$$

$$= \dfrac{3\sqrt{2}}{2}$$

Fractions with common nonzero denominators are added or subtracted as follows:

$$\dfrac{A}{B} + \dfrac{C}{B} = \dfrac{A+C}{B}, \quad B \neq 0$$

$$\dfrac{A}{B} - \dfrac{C}{B} = \dfrac{A-C}{B}, \quad B \neq 0$$

EXAMPLE 6

Find the sum in each of the following.

(a) $\dfrac{8}{13} + \dfrac{17}{13}$ (b) $\dfrac{x^2+2x+9}{x+5} + \dfrac{3x^2-8x+4}{x+5}$

Solution

(a) $\dfrac{8}{13} + \dfrac{17}{13} = \dfrac{8+17}{13} = \dfrac{25}{13}$

(b) $\dfrac{x^2+2x+9}{x+5} + \dfrac{3x^2-8x+4}{x+5}$

$$= \dfrac{x^2+2x+9+3x^2-8x+4}{x+5}$$

$$= \dfrac{4x^2-6x+13}{x+5}$$

EXAMPLE 7

Find the difference in each of the following.

(a) $\dfrac{8}{13} - \dfrac{17}{13}$ (b) $\dfrac{x^2 + 2x + 9}{x + 5} - \dfrac{3x^2 - 8x + 4}{x + 5}$

Solution

(a) $\dfrac{8}{13} - \dfrac{17}{13} = \dfrac{8 - 17}{13}$ (b) $\dfrac{x^2 + 2x + 9}{x + 5} - \dfrac{3x^2 - 8x + 4}{x + 5}$

$= \dfrac{-9}{13}$ $= \dfrac{x^2 + 2x + 9 - (3x^2 - 8x + 4)}{x + 5}$

$= \dfrac{x^2 + 2x + 9 - 3x^2 + 8x - 4}{x + 5}$

$= \dfrac{-2x^2 + 10x + 5}{x + 5}$

Fractions not having common denominators may be added or subtracted by first writing them in equivalent forms having common denominators.

EXAMPLE 8

Find the sum in each of the following.

(a) $\dfrac{5}{7} + \dfrac{3}{14}$ (b) $\dfrac{x^2 + 5}{(x + 2)(x - 7)} + \dfrac{x - 3}{x + 2}$

Solution

(a) $\dfrac{5}{7} + \dfrac{3}{14} = \dfrac{10}{14} + \dfrac{3}{14}$

$= \dfrac{10 + 3}{14}$

$= \dfrac{13}{14}$

(b) $\dfrac{x^2 + 5}{(x + 2)(x - 7)} + \dfrac{x - 3}{x + 2} = \dfrac{x^2 + 5}{(x + 2)(x - 7)} + \dfrac{(x - 3)(x - 7)}{(x + 2)(x - 7)}$

$= \dfrac{x^2 + 5 + (x - 3)(x - 7)}{(x + 2)(x - 7)}$

$= \dfrac{x^2 + 5 + (x^2 - 10x + 21)}{(x + 2)(x - 7)}$

$= \dfrac{x^2 + 5 + x^2 - 10x + 21}{(x + 2)(x - 7)}$

$= \dfrac{2x^2 - 10x + 26}{(x + 2)(x - 7)}$

When more than one operation is indicated in any one example, work from left to right, performing the multiplications and divisions first. If grouping symbols such as parentheses or brackets are involved, work from inside out and then from left to right.

EXAMPLE 9

Perform the indicated operations in each of the following.

(a) $\frac{2}{5} \times \frac{3}{7} + \frac{3}{10} \div \frac{7}{3}$ (b) $\left(\dfrac{3x}{x+2} + \dfrac{7}{x}\right) \times \dfrac{x^4}{3} \div \dfrac{x^3}{x+2}$

Solution

(a) $\frac{2}{5} \times \frac{3}{7} + \frac{3}{10} \div \frac{7}{3} = \frac{6}{35} + \frac{3}{10} \times \frac{3}{7}$

$$= \frac{6}{35} + \frac{9}{70}$$

$$= \frac{12}{70} + \frac{9}{70}$$

$$= \frac{21}{70}$$

$$= \frac{3}{10}$$

multiply both sides by what they need to make them equivalent

(b) $\left(\dfrac{3x}{x+2} + \dfrac{7}{x}\right) \times \dfrac{x^4}{3} \div \dfrac{x^3}{x+2}$

$$= \left(\frac{3x^2}{x(x+2)} + \frac{7(x+2)}{x(x+2)}\right) \times \frac{x^4}{3} \div \frac{x^3}{x+2}$$

$$= \frac{3x^2 + 7x + 14}{x(x+2)} \times \frac{x^4}{3} \div \frac{x^3}{x+2}$$

$$= \frac{x^4(3x^2 + 7x + 14)}{3x(x+2)} \times \frac{x+2}{x^3}$$

$$= \frac{x^4(3x^2 + 7x + 14)(x+2)}{3x^4(x+2)}$$

$$= \frac{3x^2 + 7x + 14}{3}$$

In expressing fractions in equivalent forms so that they will have common denominators, any common denominator is acceptable. Computations are usually simpler, however, if the smallest positive or lowest-degree common denominator is used. The **lowest common denominator (least common multiple)** of two or more positive integral denominators is the smallest positive integer that is evenly divisible by each of them. The **lowest common denominator (least common multiple)** of two or more denominators that are polynomials with integral coefficients is the lowest-degree polynomial with integral coefficients which is evenly divisible by each of them. For example, the lowest common denominator of $\frac{7}{24}$, $\frac{19}{30}$, and $\frac{13}{18}$ is 360, because 360 is the smallest positive integer which is evenly divisible by 24, by 18, and also by 30. The lowest common denominator of

$$\frac{7}{(x - 1)^3(x + 5)}, \quad \frac{19}{(x - 1)(x + 5)(x - 4)}, \quad \text{and} \quad \frac{13}{(x - 1)(x + 5)^2}$$

is $(x - 1)^3(x + 5)^2(x - 4)$, because this product is the lowest-degree polynomial over the integers which is evenly divisible by each of the given denominators.

To find the lowest common denominator (LCD) of a set of denominators that are integers or polynomials over the integers, factor each denominator completely and then form the product of the prime factors, taking each prime factor the largest number of times it appears in any individual factorization. For example, to find the LCD of $\frac{7}{24}$, $\frac{19}{30}$, and $\frac{13}{18}$, first factor 24, 30, and 18 completely, and then form the product of prime factors as follows:

$$24 = 2^3 \times 3 \qquad \text{LCD} = 2^3 \times 3^2 \times 5$$
$$30 = 2 \times 3 \times 5 \qquad \quad = 8 \times 9 \times 5$$
$$18 = 3^2 \times 2 \qquad \quad = 360$$

To find the LCD of

$$\frac{7}{(x - 1)^3(x + 5)}, \quad \frac{19}{(x - 1)(x + 5)(x - 4)}, \quad \text{and} \quad \frac{13}{(x - 1)(x + 5)^2},$$

first make sure each denominator is factored completely, and then form the product of prime factors as follows:

$$(x - 1)^3(x + 5) \qquad \text{is factored completely.}$$
$$(x - 1)(x + 5)(x - 4) \qquad \text{is factored completely.}$$
$$(x - 1)(x + 5)^2 \qquad \text{is factored completely.}$$
$$\text{LCD} = (x - 1)^3(x + 5)^2(x - 4)$$

Exercise 1.6, 51–60

EXAMPLE 10

Perform the indicated operations.

$$\frac{3}{x^2 - 4} + \frac{5x}{x^2 - 4x + 4} - \frac{9}{(x + 1)(x^2 + 2x + 1)}$$

Solution

First factor each denominator completely, and then find the LCD.

$$x^2 - 4 = (x - 2)(x + 2)$$
$$x^2 - 4x + 4 = (x - 2)^2$$
$$(x + 1)(x^2 + 2x + 1) = (x + 1)(x + 1)^2 = (x + 1)^3$$
$$\text{LCD} = (x - 2)^2(x + 2)(x + 1)^3$$

Next write each of the individual fractions in equivalent form having the LCD as denominator.

$$\frac{3}{x^2 - 4} = \frac{3}{(x - 2)(x + 2)} = \frac{3(x - 2)(x + 1)^3}{(x - 2)(x + 2)(x - 2)(x + 1)^3}$$

$$= \frac{3(x - 2)(x + 1)^3}{(x - 2)^2(x + 2)(x + 1)^3}$$

$$\frac{5x}{x^2 - 4x + 4} = \frac{5x}{(x - 2)^2} = \frac{5x(x + 2)(x + 1)^3}{(x - 2)^2(x + 2)(x + 1)^3}$$

$$\frac{9}{(x + 1)(x^2 + 2x + 1)} = \frac{9}{(x + 1)^3} = \frac{9(x - 2)^2(x + 2)}{(x + 1)^3(x - 2)^2(x + 2)}$$

$$= \frac{9(x - 2)^2(x + 2)}{(x - 2)^2(x + 2)(x + 1)^3}$$

Now perform the indicated operations.

$$\frac{3}{x^2 - 4} + \frac{5x}{x^2 - 4x + 4} - \frac{9}{(x + 1)(x^2 + 2x + 1)}$$

$$= \frac{3(x - 2)(x + 1)^3}{(x - 2)^2(x + 2)(x + 1)^3} + \frac{5x(x + 2)(x + 1)^3}{(x - 2)^2(x + 2)(x + 1)^3}$$

$$- \frac{9(x - 2)^2(x + 2)}{(x - 2)^2(x + 2)(x + 1)^3}$$

$$= \frac{3(x - 2)(x + 1)^3 + 5x(x + 2)(x + 1)^3 - 9(x - 2)^2(x + 2)}{(x - 2)^2(x + 2)(x + 1)^3}$$

We may stop here because we have not been instructed to express the numerator and denominator as polynomials.

EXAMPLE 11   Find the sum of $5/x$, $3/(x - 1)$, and $2/(x + 1)$; express its numerator and denominator as polynomials.

Solution $LCD = x(x - 1)(x + 1)$

$$= x(x^2 - 1)$$

$$\frac{5}{x} + \frac{3}{x - 1} + \frac{2}{x + 1} = \frac{5(x^2 - 1)}{x(x^2 - 1)} + \frac{3x(x + 1)}{x(x^2 - 1)} + \frac{2x(x - 1)}{x(x^2 - 1)}$$

$$= \frac{5(x^2 - 1) + 3x(x + 1) + 2x(x - 1)}{x(x^2 - 1)}$$

$$= \frac{5x^2 - 5 + 3x^2 + 3x + 2x^2 - 2x}{x^3 - x}$$

$$= \frac{10x^2 + x - 5}{x^3 - x}$$

Exercise 1.6, 61–100

We have now seen how to express the sum of rational expressions as a single rational expression. In performing certain operations in the calculus, it is necessary to reverse this procedure and decompose single rational expressions into sums of fractions with lower-degree denominators. For example, by actually performing the addition, we can verify that

$$\frac{9x^2 + 8x - 12}{x^3 - 4x}$$

can be decomposed into the sum of

$$\frac{3}{x}, \quad \frac{5}{x - 2}, \quad \text{and} \quad \frac{1}{x + 2}.$$

The fractions in such a decomposition are commonly called *partial fractions*. We shall work some with partial fractions in Section 1.10, and we shall see how to decompose certain rational expressions into sums of

Exercise 1.6, 101–110 partial fractions in Section 4.4.

Fractions in which the numerator and denominator contain fractions are called **complex fractions**. They can be simplified by multiplying numerator and denominator through by the LCD of all the fractions appearing in the numerator and denominator. This is illustrated in the next example.

EXAMPLE 12

Express the following as a fraction having no fractions in the numerator or denominator.

$$\frac{\dfrac{y}{3x^2} - \dfrac{1}{6x}}{\dfrac{y^2}{x} + \dfrac{5y}{12}}$$

Solution

The LCD of the fractions $y/3x^2$, $1/6x$, y^2/x, and $5y/12$ is $12x^2$. Multiplying numerator and denominator through by $12x^2$, we have

$$\frac{12x^2\left(\dfrac{y}{3x^2} - \dfrac{1}{6x}\right)}{12x^2\left(\dfrac{y^2}{x} + \dfrac{5y}{12}\right)} = \frac{12x^2\left(\dfrac{y}{3x^2}\right) - 12x^2\left(\dfrac{1}{6x}\right)}{12x^2\left(\dfrac{y^2}{x}\right) + 12x^2\left(\dfrac{5y}{12}\right)}$$

$$= \frac{4y - 2x}{12xy^2 + 5x^2y}.$$

Note that *every term* of the numerator and denominator must be multi-
Exercise 1.6, 111–115 plied by the LCD.

Some fractions with radicals in their denominators may be written in equivalent forms having rational denominators, as illustrated in the next example.

EXAMPLE 13   Express $7/(\sqrt{5} - \sqrt{2})$ in a form having no irrational number in the denominator.

Solution   We may use the fundamental principle of fractions and multiply numerator and denominator by $\sqrt{5} + \sqrt{2}$.

$$\frac{7}{\sqrt{5} - \sqrt{2}} = \frac{7(\sqrt{5} + \sqrt{2})}{(\sqrt{5} - \sqrt{2})(\sqrt{5} + \sqrt{2})}$$

$$= \frac{7\sqrt{5} + 7\sqrt{2}}{(\sqrt{5})^2 - (\sqrt{2})^2}$$

$$= \frac{7\sqrt{5} + 7\sqrt{2}}{5 - 2}$$

$$= \frac{7\sqrt{5} + 7\sqrt{2}}{3}$$

Any fraction of the form

$$\frac{c}{\sqrt{a} - \sqrt{b}} \quad \text{or} \quad \frac{c}{\sqrt{a} + \sqrt{b}}$$

where c and positive a and b are real numbers may be expressed in an equivalent form having no radical in the denominator by multiplying numerator and denominator by $\sqrt{a} + \sqrt{b}$ or by $\sqrt{a} - \sqrt{b}$, respectively. This process is sometimes called *rationalizing the denominator.*

$$\frac{c}{\sqrt{a} - \sqrt{b}} = \frac{c(\sqrt{a} + \sqrt{b})}{(\sqrt{a} - \sqrt{b})(\sqrt{a} + \sqrt{b})}$$

$$= \frac{c\sqrt{a} + c\sqrt{b}}{(\sqrt{a})^2 - (\sqrt{b})^2}$$

$$= \frac{c\sqrt{a} + c\sqrt{b}}{a - b}$$

$$\frac{c}{\sqrt{a} + \sqrt{b}} = \frac{c(\sqrt{a} - \sqrt{b})}{(\sqrt{a} + \sqrt{b})(\sqrt{a} - \sqrt{b})}$$

$$= \frac{c\sqrt{a} - c\sqrt{b}}{(\sqrt{a})^2 - (\sqrt{b})^2}$$

$$= \frac{c\sqrt{a} - c\sqrt{b}}{a - b}$$

Exercise 1.6, 116–120

As the next example illustrates, any fraction with an imaginary-number denominator may be written in an equivalent form having a real-number denominator.

EXAMPLE 14   Express

$$\frac{3 + 2i}{5 - 7i}$$

in a form having a real-number denominator, and express the fraction as a single complex number. $a + bi$

Solution   We may use the fundamental principle of fractions and multiply numerator and denominator by $5 + 7i$.

$$\frac{3 + 2i}{5 - 7i} = \frac{(3 + 2i)(5 + 7i)}{(5 - 7i)(5 + 7i)}$$

$$= \frac{15 + 21i + 10i + 14i^2}{25 - 49i^2}$$

$$= \frac{15 + 31i + 14(-1)}{25 - 49(-1)}$$

$$= \frac{15 + 31i - 14}{25 + 49}$$

$$= \frac{1 + 31i}{74}$$

Now use the definition of addition of fractions with common denominators.

$$\frac{3 + 2i}{5 - 7i} = \frac{1 + 31i}{74} = \frac{1}{74} + \frac{31}{74}i$$

$a + bi$

We have shown in the last example that the fraction $\dfrac{3 + 2i}{5 - 7i}$ is equal to the single complex number whose real part is $\frac{1}{74}$ and whose imaginary part is $\frac{31}{74}$. Any fraction in which the numerator and nonzero denominator are complex numbers can be written as a single complex number. If the denominator is an imaginary number, $a - bi$, multiply numerator and denominator by $a + bi$; if the denominator is an imaginary number, $a + bi$, multiply numerator and denominator by $a - bi$. Each of the two numbers $a - bi$ and $a + bi$ is called the **conjugate** of the other.

EXAMPLE 15   Express

$$\frac{2 + 5i}{4 + 7i}$$

as a single complex number.

Solution

We multiply numerator and denominator by the conjugate of $4 + 7i$.

$$\frac{2 + 5i}{4 + 7i} = \frac{(2 + 5i)(4 - 7i)}{(4 + 7i)(4 - 7i)}$$

$$= \frac{8 - 14i + 20i - 35i^2}{16 - 49i^2}$$

$$= \frac{8 + 6i + 35}{16 + 49}$$

$$= \frac{43 + 6i}{65}$$

$$= \frac{43}{65} + \frac{6}{65}i$$

Exercise 1.6, 121–125

EXERCISE 1.6 FRACTIONS; RATIONAL EXPRESSIONS

▶ State which sign belongs in the parentheses in each of the following.

EXAMPLES

(a) $\dfrac{x - y}{y - x} = (\ \ ) \dfrac{x - y}{x - y}$ (b) $\dfrac{x - y}{y - x} = (\ \ ) \dfrac{y - x}{x - y}$

Solutions

(a) Minus   (b) Plus

1. $\dfrac{c + d}{c - d} = (\ \ ) \dfrac{d + c}{d - c}$

2. $\dfrac{c + d}{c - d} = (\ \ ) \dfrac{-c - d}{c - d}$

3. $\dfrac{c + d}{c - d} = (\ \ ) \dfrac{-c - d}{d - c}$

4. $-\dfrac{x - y}{y - x} = (\ \ ) \dfrac{y - x}{x - y}$

5. $-\dfrac{x - y}{y - x} = (\ \ ) \dfrac{x - y}{x - y}$

6. $-\dfrac{x - y}{y - x} = (\ \ ) \dfrac{y - x}{y - x}$

7. $\dfrac{a^2 - 2ab + b^2}{a - b} = (\ \ ) \dfrac{-a^2 + 2ab - b^2}{b - a}$

8. $\dfrac{a^2 - 2ab + b^2}{a - b} = (\ \ ) \dfrac{a^2 - 2ab + b^2}{b - a}$

9. $\dfrac{a^2 - 2ab + b^2}{a - b} = (\ \ ) \dfrac{-a^2 + 2ab - b^2}{a - b}$

10. $\dfrac{(a - b)^2}{a + b} = (\ \ ) \dfrac{(b - a)^2}{a + b}$

11. $\dfrac{(a - b)^2}{a + b} = (\ \ ) \dfrac{(a - b)^2}{-a - b}$

12. $\dfrac{(a - b)^2}{a + b} = (\ \ ) \dfrac{(b - a)^2}{-a - b}$

13. $\dfrac{(x - y)(x + y)}{x(y - x)} = (\quad) \dfrac{(y - x)(x + y)}{-x(x - y)}$

14. $\dfrac{(x - y)(x + y)}{x(y - x)} = (\quad) \dfrac{(y - x)(-x - y)}{x(x - y)}$

15. $\dfrac{(x - y)(x + y)}{x(y - x)} = (\quad) \dfrac{(x - y)(x + y)}{-x(x - y)}$

▶ If possible, reduce each of the following to lowest terms. If the given fraction is already in lowest terms, simply write ''in lowest terms.''

EXAMPLES

(a) $\dfrac{75}{90} = \dfrac{5^2 \cdot 3}{5 \cdot 3^2 \cdot 2} = \dfrac{5}{3 \cdot 2} = \dfrac{5}{6}$

(b) $\dfrac{3(x - 1) + 2x(x - 1)}{(x - 1)(x + 8)} = \dfrac{(x - 1)(3 + 2x)}{(x - 1)(x + 8)} = \dfrac{3 + 2x}{x + 8}$

16. $\dfrac{36}{108}$ **17.** $\dfrac{72}{324}$ **18.** $\dfrac{45}{56}$ **19.** $\dfrac{392}{224}$

20. $\dfrac{1092}{735}$ **21.** $\dfrac{65}{24}$ **22.** $\dfrac{360}{900}$

23. $\dfrac{x^2 - 4}{x^2 + 5x - 6}$ **24.** $\dfrac{x^3 - 25x}{x^5 - 4x^4 - 5x^3}$

25. $\dfrac{x^2 - 5x - 6}{x^2 - 5x + 6}$ **26.** $\dfrac{3(x - 2)(x + 5)}{(x - 2)(x - 7)}$

27. $\dfrac{3(x - 2) + (x + 5)}{(x - 2)(x - 7)}$ **28.** $\dfrac{3(x - 2) + (x + 5)(x - 2)}{(x - 2)(x - 7)}$

29. $\dfrac{6(x + 1)(x - 9)}{(x - 9)(x + 4)}$ **30.** $\dfrac{6 + (x + 1)(x - 9)}{(x - 9)(x + 4)}$

31. $\dfrac{6(x - 9) + (x + 1)(x - 9)}{(x - 9)(x + 4)}$ **32.** $\dfrac{2x^3(x - 5) - x(x - 5)}{x^2(x - 5)}$

33. $\dfrac{2x^3 - x(x - 5)}{x^2(x - 5)}$ **34.** $\dfrac{2x^3(x - 5) - x(x - 5)}{x^2}$

35. $\dfrac{17(x^2 - y^2) + y^2}{17xy}$ **36.** $\dfrac{29x^2 - (x^2 + y^2)}{29x^2}$

37. $\dfrac{x^2 - 3x - 10}{x^2 - 5x - 14}$ **38.** $\dfrac{3x^2 - 5x - 2}{3x^2 - 7x + 2}$

39. $\dfrac{x^3 - 1}{x^2 - 2x + 1}$ **40.** $\dfrac{x^5y^2 + x^2y^5}{x^3y + 2x^2y^2 + xy^3}$

► Find the missing numerator in each of the following.

EXAMPLES

(a) $\dfrac{3}{7} = \dfrac{}{112}$ (b) $\dfrac{x - 2}{x + 5} = \dfrac{}{x^2 + 4x - 5}$

Solutions

(a) $112 \div 7 = 16, \quad 112 = 7 \times 16$

$$\dfrac{3}{7} = \dfrac{3 \times 16}{7 \times 16} = \dfrac{48}{112}$$

(b) $x^2 + 4x - 5 = (x + 5)(x - 1)$

$$\dfrac{x - 2}{x + 5} = \dfrac{(x - 2)(x - 1)}{(x + 5)(x - 1)}$$

$$= \dfrac{x^2 - 3x + 2}{x^2 + 4x - 5}$$

41. $\dfrac{2}{5} = \dfrac{}{275}$

42. $\dfrac{x - 3}{x + 7} = \dfrac{}{x^2 + 5x - 14}$

43. $\dfrac{3}{8} = \dfrac{}{200}$

44. $\dfrac{7}{x - 9} = \dfrac{}{x^2 - 81}$

45. $\dfrac{19}{x^2 + x - 6} = \dfrac{}{(x^2 - 4)(x + 3)}$

46. $\dfrac{7}{90} = \dfrac{}{15660}$

47. $\dfrac{-5}{3 - x} = \dfrac{}{x^2 - x - 6}$

48. $\dfrac{3x}{x^2 + x - 6} = \dfrac{}{(x^2 - 4)(x^2 - 9)}$

49. $\dfrac{1}{x^5(x - 2)} = \dfrac{}{x^{10} - 8x^7}$

50. $\dfrac{7x}{x^2 - 3x + 9} = \dfrac{}{x^3 + 27}$

► Find the LCD in each of the following.

EXAMPLES

(a) $\dfrac{3}{40}, \dfrac{7}{10}, \dfrac{9}{16}$

(b) $\dfrac{3}{x^2 - 16}, \dfrac{5x}{(x - 4)^2}, \dfrac{7x^4}{x^2 - x - 12}$

Solutions

(a) $\quad 40 = 2^3 \cdot 5$

$\quad 10 = 2 \cdot 5$

$\quad 16 = 2^4$

$\quad \text{LCD} = 2^4 \cdot 5 = 80$

(b) $\quad x^2 - 16 = (x - 4)(x + 4)$

$\quad (x - 4)^2 = (x - 4)^2$

$x^2 - x - 12 = (x - 4)(x + 3)$

$\quad \text{LCD} = (x - 4)^2(x + 4)(x + 3)$

51. $\frac{7}{12}, \frac{8}{30}, \frac{11}{36}, \frac{4}{15}$ **52.** $\frac{1}{40}, \frac{19}{315}, \frac{8}{210}, \frac{3}{380}$

53. $\dfrac{5}{63x^3}, \dfrac{-7}{27x^5}, \dfrac{4}{81x}$

54. $\dfrac{3x}{x-2}, \dfrac{2x^3}{(x-2)^3(x+1)}, \dfrac{5x^7}{(x-2)(x+1)^2}$

55. $\dfrac{-9}{x^2-2x+1}, \dfrac{5}{(x-1)^3}, \dfrac{7x^3}{x^3-1}$

56. $\dfrac{2x^7}{x^2-4x+4}, \dfrac{7x^3}{x^2-3x+2}, \dfrac{-8}{x^3-8}$

57. $\dfrac{5}{(x-2)^5(x+1)^3}, \dfrac{-6}{x^3(x-2)^7(x+1)}, \dfrac{3}{x(x-2)(x-1)}$

58. $\dfrac{1}{x}, \dfrac{1}{x+1}, \dfrac{1}{x-1}$

59. $\dfrac{1}{x^2-y^2}, \dfrac{1}{x+y}, \dfrac{1}{(x-y)^2}$

60. $\dfrac{1}{x^3}, \dfrac{-1}{y^3}, \dfrac{1}{x^3-y^3}$

▶ Perform the indicated operations in each of the following. If possible, reduce the answer to lowest terms.

EXAMPLES

(a) $\frac{7}{8} \times \frac{3}{5} = \frac{21}{40}$

(b) $\dfrac{3x}{x-1} \times \dfrac{7x^2}{x^2-1} = \dfrac{21x^3}{(x-1)(x^2-1)}$

(c) $\frac{7}{8} \div \frac{3}{5} = \frac{7}{8} \times \frac{5}{3} = \frac{35}{24}$

(d) $\dfrac{3x}{x-1} \div \dfrac{7x^2}{x^2-1} = \dfrac{3x}{x-1} \times \dfrac{x^2-1}{7x^2}$

$\qquad\qquad = \dfrac{3x\cancel{(x-1)}(x+1)}{\cancel{(x-1)} \cdot 7x^2}$

$\qquad\qquad = \dfrac{3x+3}{7x}$

(e) $\frac{7}{8} + \frac{3}{5} = \frac{35}{40} + \frac{24}{40}$

$\qquad = \frac{59}{40}$

(f) $\dfrac{3x}{x-1} + \dfrac{7x^2}{x^2-1} = \dfrac{3x(x+1)}{(x-1)(x+1)}$

$\qquad\qquad + \dfrac{7x^2}{(x-1)(x+1)}$

$\qquad\qquad = \dfrac{3x(x+1)+7x^2}{(x-1)(x+1)}$

$\qquad\qquad = \dfrac{10x^2+3x}{x^2-1}$

61. $\frac{5}{8} \times \frac{7}{11}$ **62.** $\frac{5}{8} \div \frac{7}{11}$ **63.** $\frac{5}{8} + \frac{7}{11}$ **64.** $\frac{5}{8} - \frac{7}{11}$

65. $\frac{8}{9} \div \frac{5}{27}$ **66.** $\frac{7}{36} + \frac{9}{48} - \frac{11}{60}$

67. $\frac{5}{42} - \frac{13}{45} + \frac{1}{105}$ **68.** $\frac{1}{20} + \frac{5}{21} - \frac{8}{45} + \frac{7}{30}$

69. $\frac{3}{8} \times \frac{2}{7} + \frac{6}{5} \div \frac{9}{4}$ **70.** $(\frac{3}{8} + \frac{2}{7}) \times \frac{6}{5} - \frac{9}{4}$

71. $(\frac{3}{8} \div \frac{2}{7}) \times \frac{6}{5} + \frac{9}{4}$ **72.** $[3 + (2 - \frac{4}{5})] \div \frac{7}{8}$

73. $(\frac{13}{14} - \frac{5}{21}) + \frac{11}{28} \div \frac{31}{32}$ **74.** $(\frac{23}{20} + \frac{7}{45}) \div [\frac{2}{3}(\frac{17}{12} - \frac{7}{30})]$

75. $\frac{13}{24} + \frac{13}{54} - \frac{5}{20} \times \frac{36}{13}$ **76.** $\frac{\sqrt{2}}{2} \times \frac{1}{2}$

77. $\frac{1}{\sqrt{3}} \times \frac{\sqrt{3}}{3}$ **78.** $\frac{\sqrt{3}}{2} \div \frac{1}{2}$

79. $\frac{1}{2} \div \frac{\sqrt{3}}{2}$ **80.** $1 \div \frac{\sqrt{3}}{3}$

81. $\frac{2x}{x - 3} \times \frac{5x^2}{x^2 - 9}$ **82.** $\frac{2x}{x - 3} \div \frac{5x^2}{x^2 - 9}$

83. $\frac{x^2 + 5x + 6}{x^2 + x - 6} \times \frac{(x + 1)^2}{x^2 - 9}$ **84.** $\frac{3xy^3}{25a^2b} \times \frac{10a^5b}{6x^3y}$

85. $\frac{2a^4b^4}{3x^2y^3} \div \frac{a^2b^2}{8xy^4}$ **86.** $\left(\frac{9x^2y^3}{5a^2b^3} \div \frac{xy}{7a^3b^2}\right) \times \frac{10ab^3}{3xy^4}$

87. $\left(\frac{4x^6y^3z}{6a^2b^2c^2} \times \frac{7a^3b^3c^3}{6x^2y^2z^2}\right) \div \frac{42x^2y^2z^2}{24a^2b^2c^2}$

88. $\frac{4x - 8}{x^5} \times \frac{x^3}{x - 2}$ **89.** $\frac{7x + 21}{x - 2} \times \frac{x^2 - 4}{x^2 - 9}$

90. $\frac{2x - 10}{x + 2} \div \frac{3x - 15}{x^2 - 4}$ **91.** $\frac{3x}{x^2 - 9} + \frac{5x^2}{x^2 - 9}$

92. $\frac{7x^3}{(x - 1)^3} - \frac{8x^2}{(x - 1)^3}$ **93.** $\frac{2x}{x - 3} + \frac{5x^2}{x^2 - 9}$

94. $\frac{8}{x - 5} + \frac{7}{x + 3}$ **95.** $\frac{2}{x - 2} + \frac{-7}{x^2 - 10x + 25}$

96. $\frac{3x}{(x - 2)(x + 1)} - \frac{2x^2}{x^2 + 2x + 1}$ **97.** $\frac{x + 1}{x - 1} + \frac{x - 2}{x + 2}$

98. $\frac{5}{x} - \frac{7}{x^2 - 1} + \frac{8}{x^4 - 2x^3 + x^2}$ **99.** $\left(\frac{7x}{x + 3} + \frac{3}{x}\right) \times \left(\frac{x^5}{5} \div \frac{x^4}{x + 3}\right)$

100. $\left(\frac{2x^3}{x - 2} + \frac{5}{x^2}\right) \div \left(\frac{2}{3} \times \frac{x - 2}{x^5}\right)$

▶ In each of the following, verify that the given rational expression has been properly decomposed into partial fractions.

EXAMPLE

$$\frac{14x^2 - 5x - 5}{x^3 - x} = \frac{5}{x} + \frac{2}{x - 1} + \frac{7}{x + 1}$$

Solution

$$\frac{5}{x} + \frac{2}{x - 1} + \frac{7}{x + 1} = \frac{5(x - 1)(x + 1) + 2x(x + 1) + 7x(x - 1)}{x(x - 1)(x + 1)}$$

$$= \frac{5x^2 - 5 + 2x^2 + 2x + 7x^2 - 7x}{x(x^2 - 1)}$$

$$= \frac{14x^2 - 5x - 5}{x^3 - x}$$

101. $\dfrac{17x - 1}{3x^2 + 5x - 2} = \dfrac{2}{3x - 1} + \dfrac{5}{x + 2}$

102. $\dfrac{16x^2 + x - 27}{(x + 2)(x^2 - 2x - 3)} = \dfrac{7}{x + 2} + \dfrac{3}{x + 1} + \dfrac{6}{x - 3}$

103. $\dfrac{11x^2 - 15x - 50}{x^3 - 25x} = \dfrac{3}{x - 5} + \dfrac{6}{x + 5} + \dfrac{2}{x}$

104. $\dfrac{3x + 2}{x^2 - 2x + 1} = \dfrac{3}{x - 1} + \dfrac{5}{(x - 1)^2}$

105. $\dfrac{7x + 22}{x^2 + 6x + 9} = \dfrac{7}{x + 3} + \dfrac{1}{(x + 3)^2}$

106. $\dfrac{x^2 + 7x + 15}{(x + 2)^3} = \dfrac{1}{x + 2} + \dfrac{3}{(x + 2)^2} + \dfrac{5}{(x + 2)^3}$

107. $\dfrac{4x^2 - 1}{x^3 - 1} = \dfrac{3x + 2}{x^2 + x + 1} + \dfrac{1}{x - 1}$

108. $\dfrac{2x^2 + 3x - 2}{x^3 - 8} = \dfrac{x + 3}{x^2 + 2x + 4} + \dfrac{1}{x - 2}$

109. $\dfrac{8x^2 + 49x + 63}{x^3 + 6x^2 + 9x} = \dfrac{7}{x} + \dfrac{1}{x + 3} + \dfrac{4}{(x + 3)^2}$

110. $\dfrac{2x^4 + x^3 + 4x^2 - 2x + 2}{(x^2 - x + 1)(x^3 + 1)} = \dfrac{x + 3}{x^2 - x + 1} + \dfrac{x - 2}{(x^2 - x + 1)^2}$

$$+ \dfrac{1}{x + 1}$$

▶ Express each of the following as a fraction having no fractions in the numerator or denominator.

EXAMPLE

$$\frac{\dfrac{4}{x^3} - \dfrac{2}{x}}{\dfrac{7}{x^5} - \dfrac{1}{3x^2}} = \frac{3x^5\left(\dfrac{4}{x^3} - \dfrac{2}{x}\right)}{3x^5\left(\dfrac{7}{x^5} - \dfrac{1}{3x^2}\right)} = \frac{12x^2 - 6x^4}{21 - x^3}$$

111. $\dfrac{\dfrac{5}{x^4} - \dfrac{3}{x^7}}{\dfrac{3}{x} + \dfrac{1}{2x^3}}$

112. $\dfrac{\dfrac{6}{x-3} + \dfrac{1}{x+3}}{x - \dfrac{4}{x-3}}$

113. $\dfrac{\dfrac{5}{x+2} - \dfrac{2}{x-2}}{\dfrac{3}{x+2} + \dfrac{1}{x}}$

114. $\dfrac{\dfrac{6}{x} + \dfrac{5}{x^2}}{\dfrac{7}{x^3} - \dfrac{3}{x^4}}$

115. $\dfrac{\dfrac{1}{x-1} - \dfrac{x}{x+1}}{\dfrac{x^2}{x^2-1} + 3}$

▶ Express each of the following in a form having no irrational number in the denominator.

EXAMPLE

$$\frac{5}{7 + \sqrt{3}} = \frac{5(7 - \sqrt{3})}{(7 + \sqrt{3})(7 - \sqrt{3})} = \frac{35 - 5\sqrt{3}}{49 - 3} = \frac{35 - 5\sqrt{3}}{46}$$

116. $\dfrac{4}{5 - \sqrt{2}}$

117. $\dfrac{4}{5 + \sqrt{2}}$

118. $\dfrac{2}{\sqrt{3} - \sqrt{5}}$

119. $\dfrac{2}{\sqrt{3} + \sqrt{5}}$

120. $\dfrac{-3}{4 - \sqrt{7}}$

▶ Express each of the following as a single complex number.

EXAMPLE

$$\frac{2 + 3i}{5 - 4i} = \frac{(2 + 3i)(5 + 4i)}{(5 - 4i)(5 + 4i)} = \frac{-2 + 23i}{25 + 16} = \frac{-2}{41} + \frac{23}{41}i$$

121. $\dfrac{2 + 3i}{5 + 4i}$

122. $\dfrac{3 + 7i}{2 - 4i}$

123. $\dfrac{3 + 7i}{2 + 4i}$

124. $\dfrac{5 + 8i}{7 + i}$

125. $\dfrac{7 + 9i}{i}$

CHECK LIST FOR THE STUDENT

If you have learned the material in Section 1.6, you should be able to do the following:

1. Write a fraction in three equivalent forms by changing signs.

2. Reduce to lowest terms a fraction whose numerator and nonzero denominator are integers or are polynomials with integral coefficients.

3. Express a fraction whose numerator and nonzero denominator are integers, or are polynomials with integral coefficients, in an equivalent form with a larger or higher-degree denominator.

4. Multiply, divide, add, and subtract fractions with nonzero denominators.

5. Perform more than one of the operations addition, subtraction, multiplication, and division in the same example.

6. Find the lowest common denominator (least common multiple) of two or more fractions whose numerators and nonzero denominators are integers or are polynomials with integral coefficients.

7. Express a complex fraction in a form having no fractions in the numerator or denominator.

8. Express any fraction of the form

$$\frac{c}{\sqrt{a} - \sqrt{b}} \quad \text{or} \quad \frac{c}{\sqrt{a} + \sqrt{b}},$$

where c and positive a and positive b are real numbers, in an equivalent form having no radical in the denominator.

9. Express any fraction in which the numerator and nonzero denominator are complex numbers as a single complex number.

SECTION QUIZ 1.6

1. Find what belongs in the parentheses in each of the following.

(a) $-\dfrac{x - y}{y - x} = (\quad) \dfrac{y - x}{x - y}$

(b) $\dfrac{a - 3}{a + 4} = \dfrac{(\qquad\qquad)}{a^2 + 3a - 4}$

2. Reduce the following to lowest terms.

$$\frac{2x^2 - 18}{4x^2 - 4x - 24}$$

3. Find the lowest common denominator.

$$\frac{4}{x^2 - 10x + 25}, \quad \frac{3x}{x^2 - 25}, \quad \frac{x^2}{(x - 5)^3}, \quad \frac{6x^4}{(x^2 + 10x + 25)(x + 1)}$$

Perform the indicated operations in each of the following. If possible, reduce the answer to lowest terms.

4. $\dfrac{5x}{x-2} \times \dfrac{x^2-4}{7x^3}$

5. $\dfrac{x^2-6x+9}{7(x^2-1)} \div \dfrac{x-3}{x-1}$

6. $\dfrac{5x^2}{x^2+2x-3} + \dfrac{2x}{x^2+5x+6}$

7. $\dfrac{3}{x-1} - \dfrac{2}{x^2+2x-3}$

8. $\left(\dfrac{3}{x} - \dfrac{2}{x^2}\right) \div \dfrac{1}{x^5}$

9. Express the following as a fraction having no fractions in the numerator or denominator.

$$\dfrac{\dfrac{3}{x^2} - \dfrac{5}{x^3}}{\dfrac{7}{x} - \dfrac{1}{x^4}}$$

10. (a) Express the following in a form having no irrational number in the denominator.

$$\dfrac{3}{\sqrt{5}-1}$$

(b) Express the following in the form $a + bi$ where a and b are real numbers and $i^2 = -1$.

$$\dfrac{3+5i}{2-7i}$$

1.7 Linear and Quadratic Equations

The purpose of this section is to review properties of equations, to review the solution of linear equations, to review the solution of quadratic equations by factoring and by the quadratic formula, and to review completing the square. A derivation of the quadratic formula is included at the end of the section.

An **equation** is a statement of equality. The expression on the left of the equals sign is called its *left side;* the expression on the right is called its *right side.* A **solution** or **root** of an equation in one variable is a number such that the equation becomes a true statement when the variable is replaced by that number. A solution of an equation is said to *satisfy* the equation. The set of all solutions of an equation is sometimes called the **solution set** of the equation. To *solve* an equation is to find all its solutions or to find its solution set. If an equation has no solutions, we say that its solution set is *empty* or that its solution set is the *empty set* or the *null* set. Two equations are said to be **equivalent** if they have exactly the same solutions, that is, if they have the same solution set. If two equations are equivalent, each is said to be an *equivalent form* of the other.

Some properties of equations are as follows:

Property 1 If the same number or polynomial in the variable(s) of an equation is added to or subtracted from both sides of an equation, the result is an equation with exactly the same solution(s) as the original equation. See Examples 1(a) and 1(b).

Property 2 If both sides of an equation are multiplied or divided by the same nonzero number, the result is an equation with exactly the same solution(s) as the original equation. See Example 1(c).

Property 3 If both sides of an equation in one variable are multiplied by a nonzero polynomial in that variable, the result is an equation such that any solution of the original equation is also a solution of the new equation, but the new equation may have additional solutions that are not solutions of the original equation. See Example 1(d).

Property 4 If both sides of an equation in one variable are divided by a nonzero polynomial in that variable, any solution of the new equation is also a solution of the original equation, but the original equation may have additional solutions that are not solutions of the new equation. Solutions of the original equation may have been "lost." See Example 1(e).

Property 5 If both sides of an equation in one variable are raised to the same positive integral power, any solution of the original equation is also a solution of the new equation, but the new equation may have additional solutions that are not solutions of the original equation. See Example 1(f).

Note: When both sides of an equation in one variable are multiplied by a polynomial in that variable or are raised to the same positive integral power, the solutions of the new equation must be considered candidates for solutions of the original equation, and each must be checked by substitution in the original equation to ascertain whether or not it actually satisfies the original equation. When both sides of an equation in one variable are divided by a polynomial in that variable, solutions of the original equation might be lost, in the sense that the new equation might not have as solutions all the solutions of the original equation.

Whenever Property 1 or Property 2 is applied, the resulting equation is equivalent to the original and is said to be the original equation written in an equivalent or another form. These and the other properties of equations on our list are illustrated in the following example.

EXAMPLE 1   In each of the following, state the relationship between the solutions of the original equation (Equation (1)) and the new equation (Equation (2)).

(a) (1) $x - 3 = 7$
    (2) $x = 7 + 3$

(b) (1) $10 = -x^2 + 7x$
    (2) $x^2 - 7x + 10 = 0$

(c) (1) $3x = 7$
    (2) $x = \frac{7}{3}$

(d) (1) $15 = 5x$
    (2) $15(x - 2) = 5x(x - 2)$

(e) (1) $3(x - 2)(x - 5) = 6(x - 5)$
    (2) $3(x - 2) = 6$

(f) (1) $2x + 1 = 3x$
    (2) $4x^2 + 4x + 1 = 9x^2$

Solution   (a) (1) $x - 3 = 7$
    (2) $x = 7 + 3$

The number 3 was added to both sides of Equation (1) to produce Equation (2). Equations (1) and (2) have exactly the same solution(s). Property 1

(b) (1) $10 = -x^2 + 7x$
    (2) $x^2 - 7x + 10 = 0$

The polynomial $x^2 - 7x$ was added to both sides of Equation (1) to produce Equation (2). Equations (1) and (2) have exactly the same solution(s). Property 1

(c) (1) $3x = 7$
    (2) $x = \frac{7}{3}$

Both sides of Equation (1) were divided by the number 3 to produce Equation (2). Equations (1) and (2) have exactly the same solution(s). Property 2

(d) (1) $15 = 5x$
    (2) $15(x - 2) = 5x(x - 2)$

Both sides of Equation (1) were multiplied by the polynomial $x - 2$ to produce Equation (2). Any solution of Equation (1) is also a solution of Equation (2), but Equation (2) may have additional solutions that are not solutions of Equation (1). Property 3

In this particular example, Equation (2) has one more solution than Equation (1). It can be verified by substitution that both 2 and 3 are solutions of Equation (2), but of these only 3 is a solution of Equation (1).

(e) (1) $3(x - 2)(x - 5) = 6(x - 5)$
    (2) $3(x - 2) = 6$

Both sides of Equation (1) were divided by the polynomial $x - 5$ to produce Equation (2). Any solution of Equation (2) is also a solution of Equation (1), but Equation (1) may have additional solutions that are not solutions of Equation (2). Property 4

In this example, Equation (1) has one more solution than Equation (2). It can be verified by substitution in Equation (1) that both 5 and 4 are solutions of it, but of these only 4 is a solution of Equation (2).

(f) (1) $2x + 1 = 3x$
    (2) $4x^2 + 4x + 1 = 9x^2$

Both sides of Equation (1) were squared to produce Equation (2). Any solution of Equation (1) is also a solution of Equation (2), but Equation (2) may have additional solutions that are not solutions of Equation (1). Property 5

In this particular example, Equation (2) has one more solution than Equation (1). It can be verified by substitution that both 1 and $-\frac{1}{5}$ are solutions of $4x^2 + 4x + 1 = 9x^2$, but of these only 1 is a solution of the original equation $2x + 1 = 3x$.

Exercise 1.7, 1–10

An **nth-degree polynomial equation** is an equation of the form $P = Q$ where P and Q are polynomials and the polynomial $P - Q$ is of degree n. For example,

$$3x - 2 = 7x + 5$$

is a first-degree polynomial equation, and

$$3x^2 + 13x = 10$$

is a second-degree polynomial equation.

A **linear** or **first-degree equation in one variable**, say x, is a first-degree polynomial equation in one variable that can be written in the form

$$ax + b = 0$$

where a and b are specific numbers and $a \neq 0$. Any linear equation in one variable has exactly one solution. The solution of $ax + b = 0$ is

$$x = \frac{-b}{a}.$$

As illustrated in the next example, the properties of equations can be used to find the solution of a linear equation in one variable.

EXAMPLE 2

Find the solution of $x - 2 = 4 - 2x$.

Solution

We shall use the properties of equations to isolate the terms involving x on one side of the equation and hence produce an equivalent equation whose solution is obvious.

$$x - 2 = 4 - 2x$$
$$x + 2x - 2 = 4 \qquad \text{Adding } 2x \text{ to both sides. Property 1}$$
$$3x = 6 \qquad \text{Adding 2 to both sides. Property 1}$$
$$x = 2 \qquad \text{Dividing both sides by 3. Property 2}$$

Check Show that the original equation becomes a true statement when x is replaced by 2.

$$x - 2 = 4 - 2x$$
$$2 - 2 = 4 - 2 \,(2)$$

Exercise 1.7, 11–30

$$0 = 4 - 4 \qquad \text{True. The solution is 2.}$$

A **quadratic** or **second-degree equation** is a second-degree polynomial equation in one variable, say x, that can be written in the "standard" form

$$ax^2 + bx + c = 0$$

where a, b, and c are specific real numbers and $a \neq 0$. Any quadratic equation has either

1. Exactly one real-number solution,
2. Exactly two distinct real-number solutions, or
3. Exactly two distinct imaginary-number solutions.

The solution(s) of $ax^2 + bx + c = 0$ is (are) given by the *quadratic formula*

$$x = \frac{-b \pm \sqrt{b^2 - 4ac}}{2a}. \qquad \text{Quadratic formula}$$

If $b^2 - 4ac = 0$, this formula gives the single real-number solution $x = -b/2a$, but if $b^2 - 4ac$ is not zero, the formula gives the two distinct solutions

$$x = \frac{-b + \sqrt{b^2 - 4ac}}{2a} \quad \text{and} \quad x = \frac{-b - \sqrt{b^2 - 4ac}}{2a}.$$

A derivation of the quadratic formula appears near the end of this section.

We see from the quadratic formula that the nature of the solutions of a quadratic equation depends on the value of the expression $b^2 - 4ac$. In particular,

1. If $b^2 - 4ac = 0$, $ax^2 + bx + c = 0$ has exactly one real solution.
2. If $b^2 - 4ac > 0$, $ax^2 + bx + c = 0$ has exactly two distinct real solutions.
3. If $b^2 - 4ac < 0$, $ax^2 + bx + c = 0$ has exactly two distinct imaginary solutions. (The two imaginary solutions are conjugates.)

EXAMPLE 3

In each of the following determine the nature of the solutions without actually finding them.
(a) $4x^2 = 12x - 9$ (b) $4x^2 - 4x = 3$ (c) $2x^2 - 2x + 5 = 0$

Solution

(a) To determine a, b, and c, we must first write the equation in standard form.

$$4x^2 - 12x + 9 = 0$$

Since $a = 4$, $b = -12$, and $c = 9$,

$$b^2 - 4ac = (-12)^2 - 4(4)(9)$$
$$= 144 - 144$$
$$= 0 \quad b^2 - 4ac \text{ is zero.}$$

There is exactly one real-number solution of $4x^2 = 12x - 9$.

(b) Written in standard form, the given equation is

$$4x^2 - 4x - 3 = 0.$$

Here $a = 4$, $b = -4$, and $c = -3$.

$$b^2 - 4ac = (-4)^2 - 4(4)(-3)$$
$$= 16 + 48$$
$$= 64 \quad b^2 - 4ac \text{ is positive.}$$

There are exactly two distinct real-number solutions of $4x^2 - 4x = 3$.

(c) The equation

$$2x^2 - 2x + 5 = 0$$

is already written in standard form. Here $a = 2$, $b = -2$, and $c = 5$.

$$b^2 - 4ac = (-2)^2 - 4(2)(5)$$
$$= 4 - 40$$
$$= -36 \qquad b^2 - 4ac \text{ is negative.}$$

There are exactly two distinct imaginary-number solutions of $2x^2 - 2x + 5 = 0$.

Exercise 1.7, 31–40

If the coefficients of a quadratic equation are integers, and if the left side of the standard form can be readily factored into first-degree polynomials with integral coefficients, probably the easiest way to find the solutions is to utilize the *zero factor law.*

Zero Factor Law   If the product of two factors is equal to zero, then one or the other or both of the factors is zero.

For example, if

$$(x - 3)(x + 7) = 0,$$

then $\qquad x - 3 = 0 \quad \text{or} \quad x + 7 = 0.$

If the product of more than two factors is equal to zero, an extension of the zero factor law is applicable. For example, if

$$(2x - 3)(x - 7)(x^2 - 6x + 3) = 0,$$

Exercise 1.7, 41–50 then $\quad 2x - 3 = 0, \quad \text{or} \quad x - 7 = 0, \quad \text{or} \quad x^2 - 6x + 3 = 0.$

Using the zero factor law to solve a quadratic equation by factoring, we do the following:

1. Make sure the equation is in standard form (0 is isolated on the right side).
2. Factor the left side.
3. Equate each factor to zero, and solve for the variable.

Check by substituting back in the original equation.

EXAMPLE 4   Solve $3x^2 + 13x = 10$ by factoring.

Solution   1. Write the equation in standard form.

$$3x^2 + 13x - 10 = 0$$

2. Factor the left side.

$$(3x - 2)(x + 5) = 0$$

3. Equate each factor to zero, and solve for the variable.

$$3x - 2 = 0 \qquad x + 5 = 0$$
$$x = \tfrac{2}{3} \qquad\quad x = -5$$

Check by showing that $\frac{2}{3}$ and -5 both satisfy the original equation.

$$3x^2 + 13x = 10$$

$$3\left(\tfrac{2}{3}\right)^2 + 13\left(\tfrac{2}{3}\right) = 10 \qquad\qquad 3(-5)^2 + 13(-5) = 10$$
$$3\left(\tfrac{4}{9}\right) + \tfrac{26}{3} \;\;\;= 10 \qquad\qquad\qquad 3(25) - 65 \;\;\;\;= 10$$
$$\tfrac{4}{3} + \tfrac{26}{3} \;\;\;\;= 10 \qquad\qquad\qquad\qquad 75 - 65 \;\;\;= 10 \quad \text{True}$$
$$\tfrac{30}{3} \;\;\;\;\;= 10 \quad \text{True}$$

The solutions of $3x^2 + 13x = 10$ are $\frac{2}{3}$ and -5.

EXAMPLE 5

Solution

Solve $4x^2 - 12x = -9$ by factoring.

1. Write the equation in standard form.

$$4x^2 - 12x + 9 = 0$$

2. Factor the left side.

$$(2x - 3)^2 = 0$$

3. Equate each factor to zero, and solve for the variable.

$$2x - 3 = 0 \qquad 2x - 3 = 0$$
$$x = \tfrac{3}{2}$$

Check

$$4x^2 - 12x \;\;\;= -9$$
$$4\left(\tfrac{3}{2}\right)^2 - 12\left(\tfrac{3}{2}\right) = -9$$
$$4\left(\tfrac{9}{4}\right) - 18 \;\;\;= -9$$
$$9 - 18 \;\;\;= -9 \quad \text{True}$$

Exercise 1.7, 51–60 The solution is $\frac{3}{2}$.

If a quadratic equation cannot easily be solved by factoring, it can always be solved by using the quadratic formula.

EXAMPLE 6

Solution

Solve $3x^2 + 13x = 10$ by using the quadratic formula.

$$3x^2 + 13x - 10 = 0$$
$$a = 3, \qquad b = 13, \qquad c = -10$$

$$x = \frac{-b \pm \sqrt{b^2 - 4ac}}{2a}$$

$$x = \frac{-13 \pm \sqrt{(13)^2 - 4(3)(-10)}}{6}$$

$$= \frac{-13 \pm \sqrt{169 + 120}}{6}$$

$$= \frac{-13 \pm \sqrt{289}}{6} = \frac{-13 \pm 17}{6}$$

PART ONE | ALGEBRAIC REVIEW

$$x = \frac{-13 + 17}{6} \quad \text{or} \quad x = \frac{-13 - 17}{6}$$

$$= \frac{4}{6} \qquad\qquad = \frac{-30}{6}$$

$$= \frac{2}{3} \qquad\qquad = -5$$

The solutions of $3x^2 + 13x = 10$ are $\frac{2}{3}$ and -5. These were checked in Example 4, where this same equation was solved by factoring.

EXAMPLE 7   Solve $2x^2 - 2x + 5 = 0$.

Solution   The equation $2x^2 - 2x + 5 = 0$ is already in standard form, and the left side cannot be factored into first-degree polynomials with integral coefficients. Here

$$b^2 - 4ac = (-2)^2 - 4(2)(5)$$
$$= -36.$$

There are two distinct imaginary-number solutions. They can be found by using the quadratic formula.

$$a = 2, \qquad b = -2, \qquad c = 5$$

$$x = \frac{-(-2) \pm \sqrt{(-2)^2 - 4(2)(5)}}{4}$$

$$= \frac{2 \pm \sqrt{-36}}{4}$$

$$= \frac{2 \pm 6i}{4}$$

$$x = \frac{2 + 6i}{4} \quad \text{or} \quad x = \frac{2 - 6i}{4}$$

$$= \frac{2}{4} + \frac{6}{4}i \qquad\qquad = \frac{2}{4} - \frac{6}{4}i$$

$$= \frac{1}{2} + \frac{3}{2}i \qquad\qquad = \frac{1}{2} - \frac{3}{2}i$$

Check
$$2x^2 - 2x + 5 = 0$$
$$2(\tfrac{1}{2} + \tfrac{3}{2}i)^2 - 2(\tfrac{1}{2} + \tfrac{3}{2}i) + 5 = 0$$
$$2(\tfrac{1}{4} + \tfrac{3}{2}i + \tfrac{9}{4}i^2) - (1 + 3i) + 5 = 0$$
$$\tfrac{1}{2} + 3i - \tfrac{9}{2} - 1 - 3i + 5 = 0$$
$$\tfrac{-8}{2} + 4 = 0$$
$$-4 + 4 = 0 \qquad \text{True}$$

The student should show by similar substitution that $\frac{1}{2} - \frac{3}{2}i$ also checks. The solutions of $2x^2 - 2x + 5 = 0$ are $\frac{1}{2} + \frac{3}{2}i$ and $\frac{1}{2} - \frac{3}{2}i$.

The quadratic formula is derived by employing the process called *completing the square* that was first introduced in Section 1.5. The technique itself provides an additional method for solving a quadratic equation. As a method of solution of a quadratic equation, it tends to be impractical, but completing the square is extremely useful in other contexts. For example, in some applications of the calculus, completing the square is needed to write equations in forms from which it is easy to identify various features of their graphs.

The process of completing the square hinges on the fact that where k is a specific real number, the expression $x^2 + kx$ becomes a perfect-square trinomial when the number $(k/2)^2$ is added to it. Solution of a quadratic by completing the square will be illustrated in the next example, and then the technique will be used to derive the quadratic formula. Other uses of completing the square will be illustrated in the exercises.

EXAMPLE 8

Solution

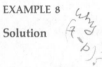

Solve $3x^2 + 13x - 10 = 0$ by completing the square.

The idea is to write the equation in the form $(x - p)^2 = q$ so that we can conclude that $x - p = +\sqrt{q}$ or $x - p = -\sqrt{q}$. To do this, we proceed as follows:

1. Divide the equation through by the coefficient of x^2.

$$x^2 + \tfrac{13}{3}x - \tfrac{10}{3} = 0$$

2. Write the equation in an equivalent form with the nonvariable term isolated on the right side.

$$x^2 + \tfrac{13}{3}x = \tfrac{10}{3}$$

3. To both sides of the equation add the square of half the coefficient of x. (This "completes the square" on the left side and makes the left side become a perfect-square trinomial.)
Half the coefficient of x is $\frac{1}{2} \times \frac{13}{3} = \frac{13}{6}$.
The square of half the coefficient of x is $(\frac{13}{6})^2$.

$$x^2 + \tfrac{13}{3}x + (\tfrac{13}{6})^2 = \tfrac{10}{3} + (\tfrac{13}{6})^2$$

4. Write the equation in the form $(x - p)^2 = q$.

$$(x + \tfrac{13}{6})^2 = \tfrac{10}{3} + \tfrac{169}{36}$$
$$(x + \tfrac{13}{6})^2 = \tfrac{120}{36} + \tfrac{169}{36}$$
$$= \tfrac{289}{36}$$

5. Conclude that $x - p = +\sqrt{q}$ or $x - p = -\sqrt{q}$.

$$x + \tfrac{13}{6} = +\sqrt{\tfrac{289}{36}} \quad \text{or} \quad x + \tfrac{13}{6} = -\sqrt{\tfrac{289}{36}}$$
$$= \tfrac{17}{6} \qquad\qquad\qquad = -\tfrac{17}{6}$$

PART ONE | ALGEBRAIC REVIEW

$$x = \frac{17}{6} - \frac{13}{6} \quad \text{or} \quad x = \frac{-17}{6} - \frac{13}{6}$$
$$= \frac{4}{6} \qquad\qquad\quad = \frac{-30}{6}$$
$$= \frac{2}{3} \qquad\qquad\quad = -5$$

The solutions of $3x^2 + 13x - 10 = 0$ are $\frac{2}{3}$ and -5. These same solutions were found by factoring in Example 4 and by the quadratic formula in Example 6. They were checked in Example 4.

Exercise 1.7, 71–80

The quadratic formula will now be derived by completing the square. Given the equation

$$ax^2 + bx + c = 0$$

where $a \neq 0$.

1. Divide through by the coefficient of x^2.

$$x^2 + \frac{b}{a}x + \frac{c}{a} = 0$$

2. Write the equation in an equivalent form with the nonvariable term isolated on the right side.

$$x^2 + \frac{b}{a}x = \frac{-c}{a}$$

3. To both sides of the equation add the square of

$$\frac{1}{2} \times \frac{b}{a} = \frac{b}{2a}.$$

$$x^2 + \frac{b}{a}x + \left(\frac{b}{2a}\right)^2 = \frac{-c}{a} + \left(\frac{b}{2a}\right)^2$$

4. Write the equation in the form $(x - p)^2 = q$.

$$\left(x + \frac{b}{2a}\right)^2 = \frac{-c}{a} + \frac{b^2}{4a^2}$$
$$= \frac{-4ac + b^2}{4a^2}$$
$$= \frac{b^2 - 4ac}{4a^2}$$

5. Conclude that $x - p = +\sqrt{q}$ or $x - p = -\sqrt{q}$.

$$x + \frac{b}{2a} = +\sqrt{\frac{b^2 - 4ac}{4a^2}} \quad \text{or} \quad x + \frac{b}{2a} = -\sqrt{\frac{b^2 - 4ac}{4a^2}}$$
$$= \frac{\sqrt{b^2 - 4ac}}{|2a|} \qquad\qquad = \frac{-\sqrt{b^2 - 4ac}}{|2a|}$$

Since $|2a|$ is either *plus* $2a$ or *minus* $2a$, $x + b/2a$ is either plus or minus $\sqrt{b^2 - 4ac}/2a$. In other words,

$$x + \frac{b}{2a} = \frac{\pm\sqrt{b^2 - 4ac}}{2a}$$

$$x = \frac{-b}{2a} \pm \frac{\sqrt{b^2 - 4ac}}{2a}$$

$$x = \frac{-b \pm \sqrt{b^2 - 4ac}}{2a}.$$

Additional exercises, Exercise 1.7, 81–105

EXERCISE 1.7 LINEAR AND QUADRATIC EQUATIONS

▶ In each of the following, state the relationship between the solutions of Equation (1) and Equation (2) without actually finding those solutions.

EXAMPLES

(a) (1) $x - 1 = 4$
    (2) $x^2 + 4x - 5 = 4x + 20$

(b) (1) $x^2 - 4 = 3x + 6$
    (2) $x - 2 = 3$

Solutions

(a) Both sides of (1) were multiplied by $(x + 5)$. Any solution of (1) is also a solution of (2), but (2) may have additional solutions. (The solutions of (2) are 5 and -5. The only solution of (1) is 5.)

(b) Both sides of (1) were divided by $(x + 2)$. Any solution of (2) is also a solution of (1), but (1) may have additional solutions. (The only solution of (2) is 5, but the solutions of (1) are 5 and -2.)

1. (1) $x - 2 = 5$
   (2) $x^2 + x - 6 = 5x + 15$

2. (1) $x^2 - 9 = 2x - 6$
   (2) $x + 3 = 2$

3. (1) $x - 2 = 5$
   (2) $x - 2 + x + 3 = 5 + x + 3$

4. (1) $7x^2 - 14x = 8$
   (2) $x^2 - 2x = \frac{8}{7}$

5. (1) $x + 3 = 2x$
   (2) $x^2 + 6x + 9 = 4x^2$

6. (1) $5x^2 + 2x = 11$
   (2) $x^2 + \frac{2}{5}x + \frac{1}{25} = \frac{11}{5} + \frac{1}{25}$

7. (1) $x + 4 = 7$
   (2) $x^2 + 3x - 4 = 7x - 7$

8. (1) $x^2 + x - 6 = 5x - 10$
   (2) $x + 3 = 5$

9. (1) $2x - 1 = 5x$
   (2) $4x^2 - 4x + 1 = 25x^2$

10. (1) $3x^2 + 6x = 9$
    (2) $x^2 + 2x + 1 = 4$

▶ Solve each of the following.

11. $3x - 7 = -2x + 8$

12. $5x + 3 = 4 - x$

13. $3(x - 2) = 2(x + 1)$

14. $0 = 8 - 2x$

15. $5x^2 - 3x + 7 = 8x + 5x^2$

16. $\frac{2}{3}x - \frac{4}{5} = x$

17. $x + \dfrac{2}{3} = \dfrac{x}{2}$ **18.** $2x + 16 - x^2 = 8 - x^2$

19. $\frac{2}{3}(x + 4) = \frac{1}{2}(x - 1)$ **20.** $5x - \frac{4}{5} = \frac{2}{5}x + \frac{3}{5}$

▶ Solve each of the following for the letter indicated.

EXAMPLE $l = a + (n - 1)d, \quad n \neq 1; \qquad$ solve for d.

Solution $l - a = (n - 1)d$

$\dfrac{l - a}{n - 1} = d$

21. $V = 2\pi rh, \quad r \neq 0; \qquad$ solve for h.

22. $A = \frac{1}{2}bh, \quad b \neq 0; \qquad$ solve for h.

23. $V = \frac{1}{3}\pi r^2 h, \quad r \neq 0; \qquad$ solve for h.

24. $r = \dfrac{1}{1 + i}, \quad r \neq 0; \qquad$ solve for i.

25. $A = P + Pni, \quad ni \neq -1; \qquad$ solve for P.

26. $S = \dfrac{a}{1 - r}, \quad r \neq 1, S \neq 0; \qquad$ solve for r.

▶ In differential calculus we sometimes encounter equations of first degree in the variable denoted by y'. Solve each of the following for y', and make note of any restrictions that must be put on the variables.

EXAMPLE $4x + 6xy + 3x^2 y' + 3y^2 y' = 0$

Solution $3x^2 y' + 3y^2 y' = -4x - 6xy$

$(3x^2 + 3y^2)y' = -4x - 6xy$

$y' = \dfrac{-4x - 6xy}{3x^2 + 3y^2}, \qquad 3x^2 + 3y^2 \neq 0$

27. $2x + xy' + y + 2y' = 0$ **28.** $6x + 7xy' + 7y + 6yy' = 0$

29. $2x + 2yy' = 0$ **30.** $6x + 5x^2 y' + 10xy - 8y^2 y' = 0$

▶ In each of the following, determine the nature of the solutions without actually finding them.

EXAMPLE $3x^2 - 2x = 5$

Solution $3x^2 - 2x - 5 = 0, \qquad b^2 - 4ac = 4 - 4(3)(-5) = 4 + 60 > 0$

There are two distinct real-number solutions.

31. $3x^2 = 11x - 8$ **32.** $2x^2 - 7x + 10 = 0$

33. $x^2 - 6x = -7$ **34.** $x^2 - 6x = -9$ **35.** $x^2 - 6x = -10$

36. $3x^2 = 2x - 1$ **37.** $4x^2 = -28x - 49$

38. $3x^2 = 20x + 7$ **39.** $3x^2 = 5x - 4$ **40.** $x^2 - 17 = 0$

▶ Use the zero factor law to solve each of the following.

EXAMPLES

Solutions

(a) $(x - 5)(x^2 - 16) = 0$ (b) $x(x + 1) = 6$

(a) $(x - 5)(x - 4)(x + 4) = 0$

$x - 5 = 0$ $x - 4 = 0$ $x + 4 = 0$

$x = 5$ $x = 4$ $x = -4$

(b) $x^2 + x - 6 = 0$

$(x - 2)(x + 3) = 0$

$x - 2 = 0$ $x + 3 = 0$

$x = 2$ $x = -3$

41. $(x - 8)(x + 4)(x - 7) = 0$

42. $(x^2 - 1)(x^2 - 4)(x^2 - 9) = 0$

43. $(x + 4)(x^2 + x - 6) = 0$ **44.** $x(x + 1) = 0$

45. $x(x + 1) = 12$ **46.** $x^3 - 5x^2 - 14x = 0$

47. $x(x - 3) = 0$ **48.** $x(x - 3) = 28$

49. $(x - 4)^2(x^2 - 4) = 0$ **50.** $x(x + 7)(x - 3)(x^2 - 9) = 0$

▶ Solve each of the following by factoring.

51. $x^2 - 7x - 18 = 0$ **52.** $x^2 + 5x - 24 = 0$

53. $x^2 - 2x = 35$ **54.** $x^2 = 12 - 4x$

55. $6x^2 + x - 1 = 0$ **56.** $15x^2 = 2 - 7x$

57. $56x^2 = x + 1$ **58.** $9x^3 = 4x^2$

59. $9x^2 + 30x = -25$ **60.** $x^3 = 4x^2 - 4x$

▶ Use the quadratic formula to solve each of the following.

61. $3x^2 + 5x - 2 = 0$ **62.** $x^2 = 2x - 10$

63. $x^2 = 2x + 10$ **64.** $4x^2 - 20x = -25$

65. $5x^2 = 16x - 20$ **66.** $x^2 + 9 = 0$

67. $x^2 - 9 = 0$ **68.** $5x^2 + 4x - 1 = 0$

69. $2x^2 + 2x + 13 = 0$ **70.** $8x^2 = 2x + 1$

Solve each of the following by completing the square.

EXAMPLE

$2x^2 - x - 3 = 0$

Solution

$$x^2 - \tfrac{1}{2}x - \tfrac{3}{2} = 0$$
$$x^2 - \tfrac{1}{2}x = \tfrac{3}{2}$$
$$x^2 - \tfrac{1}{2}x + (\tfrac{-1}{4})^2 = \tfrac{3}{2} + (\tfrac{-1}{4})^2$$
$$(x - \tfrac{1}{4})^2 = \tfrac{25}{16}$$
$$x - \tfrac{1}{4} = \tfrac{5}{4} \text{ or } x - \tfrac{1}{4} = \tfrac{-5}{4}$$
$$x = \tfrac{6}{4} \qquad\qquad x = \tfrac{-4}{4}$$
$$= \tfrac{3}{2} \qquad\qquad = -1$$

71. $6x^2 - 7x - 5 = 0$ 72. $2x^2 - 7x + 2 = 0$

73. $15x^2 + 2x - 1 = 0$ 74. $x^2 + 2x - 35 = 0$

75. $x^2 - 10x + 25 = 0$ 76. $10x^2 + 3x - 1 = 0$

77. $3x^2 - 22x + 7 = 0$ 78. $2x^2 + 11x + 12 = 0$

79. $2x^2 + 3x - 5 = 0$ 80. $9x^2 + 3x - 20 = 0$

Solve each of the following equations.

EXAMPLE

$x^4 - 13x^2 + 36 = 0$

Solution

This may be treated as a quadratic in x^2. Let $x^2 = y$. The equation becomes

$$y^2 - 13y + 36 = 0.$$

We can solve for y by factoring.

$$(y - 4)(y - 9) = 0$$
$$y = 4 \text{ or } y = 9$$

But $y = x^2$, so $x^2 = 4$ or $x^2 = 9$.

$$x = 2 \text{ or } -2 \quad \text{or} \quad x = 3 \text{ or } -3$$

81. $x^4 + 15x^2 - 16 = 0$ 82. $x^4 - 17x^2 + 16 = 0$

83. $x^4 + 5x^2 - 36 = 0$ 84. $x^6 - 9x^3 + 8 = 0$

85. $x^6 - 28x^3 + 27 = 0$

An equation of the form $Ax^2 + By^2 + Cx + Dy + E = 0$, where A, B, C, D, and E are all real numbers and A and B are not both zero, represents a graph called a conic section. In order to identify features of the graph it is frequently convenient to write the equation in a "standard" form. In each of the following, write the given equation in the indicated form, and state the values of h, k, r, p, a, or b where these letters represent real numbers and a, b, and r are positive. (For the student's information, the name of each conic section is given in parentheses. Conic sections are treated in detail in Part 2 of this text.)

EXAMPLE

$4x^2 + 9y^2 - 8x + 36y + 4 = 0$; $\dfrac{(x-h)^2}{a^2} + \dfrac{(y-k)^2}{b^2} = 1$ (Ellipse)

Solution

$4x^2 - 8x + 9y^2 + 36y = -4$

$4(x^2 - 2x) + 9(y^2 + 4y) = -4$

Completing the square in both parentheses by adding appropriate numbers to both sides of the equation, we have

$$4(x^2 - 2x + 1) + 9(y^2 + 4y + 4) = -4 + 4 + 36.$$

$$4(x-1)^2 + 9(y+2)^2 = 36$$

$$\frac{(x-1)^2}{9} + \frac{(y+2)^2}{4} = 1$$

$$h = 1, \quad k = -2, \quad a = 3, \quad b = 2$$

86. $x^2 + y^2 - 4x + 6y + 9 = 0$; $(x-h)^2 + (y-k)^2 = r^2$ (Circle)

87. $4x^2 - 9y^2 - 8x - 36y - 68 = 0$; $\dfrac{(x-h)^2}{a^2} - \dfrac{(y-k)^2}{b^2} = 1$
(Hyperbola)

88. $x^2 - 2x - 6y - 11 = 0$; $(x-h)^2 = 2p(y-k)$ (Parabola)

89. $y^2 - 8x - 6y + 1 = 0$; $(y-k)^2 = 2p(x-h)$ (Parabola)

90. $16x^2 + 25y^2 - 32x - 150y - 159 = 0$; $\dfrac{(x-h)^2}{a^2} + \dfrac{(y-k)^2}{b^2} = 1$
(Ellipse)

91. $-4x^2 + 9y^2 - 8x - 54y + 41 = 0$; $\dfrac{(y-k)^2}{a^2} - \dfrac{(x-h)^2}{b^2} = 1$
(Hyperbola)

92. $3x^2 + 3y^2 - 6x + 12y - 12 = 0$; $(x-h)^2 + (y-k)^2 = r^2$
(Circle)

93. $x^2 + 4y^2 + 10x - 32y + 85 = 0$; $\dfrac{(x-h)^2}{a^2} + \dfrac{(y-k)^2}{b^2} = 1$
(Ellipse)

94. $4x^2 - 16y^2 - 40x + 36 = 0$; $\dfrac{(x-h)^2}{a^2} - \dfrac{(y-k)^2}{b^2} = 1$
(Hyperbola)

95. $4x^2 + 16y^2 - 24x + 32y - 12 = 0$; $\dfrac{(x-h)^2}{a^2} + \dfrac{(y-k)^2}{b^2} = 1$

(Ellipse)

▶ In order to perform certain standard substitutions in integral calculus, it is sometimes necessary to write expressions of the type $\sqrt{ax^2 + bx + c}$ or $\sqrt{-ax^2 + bx + c}$, where b, c and positive a are real numbers, in forms having the sum or difference of two squares under a radical. Where h and positive a and k are real numbers, write each of the following in the form indicated and state the values of h and k.

EXAMPLE

$\sqrt{4x^2 - 24x + 20}$; $\sqrt{a}\,\sqrt{(x-h)^2 - k^2}$

Solution

$\sqrt{4x^2 - 24x + 20} = \sqrt{4}\,\sqrt{x^2 - 6x + 5}$
$\qquad\qquad\qquad\qquad = 2\sqrt{(x^2 - 6x \quad\;\;) + 5}$

Completing the square in the parentheses by adding and subtracting the appropriate number, we have

$\sqrt{4x^2 - 24x + 20} = 2\sqrt{(x^2 - 6x + 9) + 5 - 9}$
$\qquad\qquad\qquad\qquad = 2\sqrt{(x - 3)^2 - 4}$ $\qquad\qquad h = 3,\ k = 2$

96. $\sqrt{9x^2 - 36x - 45}$; $\sqrt{a}\,\sqrt{(x-h)^2 - k^2}$

97. $\sqrt{-9x^2 + 36x + 45}$; $\sqrt{a}\,\sqrt{k^2 - (x-h)^2}$

98. $\sqrt{4x^2 - 8x + 13}$; $\sqrt{a}\,\sqrt{(x-h)^2 + k^2}$

99. $\sqrt{4x^2 - 12x - 135}$; $\sqrt{a}\,\sqrt{(x-h)^2 - k^2}$

100. $\sqrt{-x^2 + 16x - 15}$; $\sqrt{a}\,\sqrt{k^2 - (x-h)^2}$

101. $\sqrt{x^2 + 10x + 34}$; $\sqrt{a}\,\sqrt{(x-h)^2 + k^2}$

102. $\sqrt{9x^2 - 72x + 135}$; $\sqrt{a}\,\sqrt{(x-h)^2 - k^2}$

103. $\sqrt{4x^2 - 20x - 11}$; $\sqrt{a}\,\sqrt{(x-h)^2 - k^2}$

104. $\sqrt{-4x^2 + 4x + 3}$; $\sqrt{a}\,\sqrt{k^2 - (x-h)^2}$

105. $\sqrt{4x^2 + 4x + 17}$; $\sqrt{a}\,\sqrt{(x-h)^2 + k^2}$

CHECK LIST FOR THE STUDENT

If you have learned the material in Section 1.7, you should be able to do the following:

1. Give the meaning of a solution or root of an equation in one variable.

2. Tell what is meant by equivalent equations and by an equivalent form of an equation.

3. Tell how the solutions of a given equation are related to the solutions of a new equation obtained from the given equation by doing each of the following:

(a) Adding or subtracting the same polynomial in its variable(s) to or from both sides,

(b) Multiplying or dividing both sides by the same nonzero number.

4. Tell how the solutions of a given equation in one variable are related to the solutions of a new equation obtained from the given equation by doing each of the following:

(a) Multiplying both sides by the same nonzero polynomial in the one variable,

(b) Dividing both sides by the same nonzero polynomial in the one variable,

(c) Raising both sides to the same positive integral power.

5. Give an example of a linear or first-degree equation in one variable.

6. Tell the number of solutions of any linear equation in one variable; solve any linear equation in one variable.

7. Give an example of a quadratic or second-degree equation in one variable.

8. List the three possibilities as far as the number and nature of the solutions of any quadratic equation are concerned.

9. State the quadratic formula for the solutions of $ax^2 + bx + c = 0$.

10. Tell how the number and nature of the solutions of $ax^2 + bx + c = 0$ are related to the value of $b^2 - 4ac$.

11. State and use the zero factor law.

12. Solve a quadratic equation by factoring.

13. Solve a quadratic equation using the quadratic formula.

14. Solve a quadratic equation by completing the square.

15. Complete squares and write certain equations or expressions in prescribed forms.

SECTION QUIZ 1.7

Solve each of the following equations.
1. $3x - 2 = -5x + 7$

2. $\frac{2}{5}(x - 1) = \frac{-1}{4}(x + 2)$

3. $x(x - 1)(x + 3) = 0$

4. Solve the equation

$$C = \frac{A}{A + B}$$

for B. (Assume that $A + B \neq 0$.)

In each of the following determine the number of distinct real-number solutions without actually finding them.

5. (a) $x^2 - 2x - 2 = 0$
   (b) $x^2 - 2x + 2 = 0$

6. (a) $x^2 - 2x + 1 = 0$
   (b) $x^2 - 2x = 0$

7. Solve each of the following by factoring.
   (a) $x^2 - 3x = 0$
   (b) $x^2 - 3x = 10$

8. Use the quadratic formula to solve the following equation.

$$9x^2 - 12x - 1 = 0$$

9. Solve the following equation by completing the square.

$$5x^2 - 9x - 2 = 0$$

10. Where h and positive a and k are real numbers, write $\sqrt{9x^2 - 18x - 27}$ in the form $\sqrt{a}\ \sqrt{(x - h)^2 - k^2}$, and state the values of h and k.

1.8 Radical, Fractional, and Absolute Value Equations

The purpose of this section is to consider some radical equations, fractional equations, and some equations involving absolute values.

A **radical equation** is an equation containing one or more radicals with one or more variables in their radicands. We shall consider radical equations in which the radicals are square roots. An example of such an equation in one variable is

$$\sqrt{2x - 4} - \sqrt{x - 1} = 1.$$

To solve such an equation in one variable, we do the following:

1. Isolate a radical on one side.

2. Square both sides.

3. Repeat this procedure until the equation contains no square roots of variable expressions.

4. Solve the latest equation for the variable.

5. Substitute in the original equation to determine which solutions of the latest equation are solutions of the original. (This step is necessary because the latest equation may have some solutions that are not solutions of the original equation.)

EXAMPLE 1

Solution

Solve the equation $\sqrt{2x - 4} - \sqrt{x - 1} = 1$.

1. Isolate a radical on one side.

$$\sqrt{2x - 4} = 1 + \sqrt{x - 1}$$

2. Square both sides.

$$2x - 4 = 1 + 2\sqrt{x - 1} + (x - 1)$$

3. Repeat this procedure: Isolate a radical and square both sides.

$$2x - x - 4 = 2\sqrt{x - 1}$$
$$x - 4 = 2\sqrt{x - 1}$$
$$x^2 - 8x + 16 = 4(x - 1)$$

4. Solve the latest equation.

$$x^2 - 8x + 16 = 4x - 4$$
$$x^2 - 12x + 20 = 0$$
$$(x - 2)(x - 10) = 0$$
$$x = 2 \quad \text{or} \quad x = 10$$

The numbers 2 and 10 are solutions of $x^2 - 12x + 20 = 0$; they are candidates for solutions of the original equation.

5. Substitute 2 and 10 for x in the original equation to determine whether or not they are solutions of it.

$$\sqrt{2x - 4} - \sqrt{x - 1} = 1$$
$$x = 2: \quad \sqrt{4 - 4} - \sqrt{2 - 1} = 1$$
$$0 - 1 \qquad = 1 \quad \text{False}$$

The number 2 is not a solution of the original equation.

$$x = 10: \quad \sqrt{20 - 4} - \sqrt{10 - 1} = 1$$
$$\sqrt{16} - \sqrt{9} \qquad = 1$$
$$4 - 3 \qquad = 1 \quad \text{True}$$

The number 10 is the only solution of the equation $\sqrt{2x - 4} - \sqrt{x - 1} = 1$.

Exercise 1.8, 1–10

In Part 2 of this text it is shown that certain equations containing square roots of expressions in two variables represent certain curves. For example, the equation

$$\sqrt{(x - 3)^2 + y^2} + \sqrt{(x + 3)^2 + y^2} = 10$$

represents a curve called an ellipse. In applications, it is frequently necessary to write such equations in standard form without radicals. This is illustrated in the next example.

EXAMPLE 2

Write the equation $\sqrt{(x - 3)^2 + y^2} + \sqrt{(x + 3)^2 + y^2} = 10$ in the form

$$\frac{x^2}{a^2} + \frac{y^2}{b^2} = 1$$

where a and b are positive real numbers. Give the values of a and b.

Solution

To write the equation in a form containing no radicals, we proceed as we did in Example 1.

1. Isolate a radical on one side.

2. Square both sides.

3. Repeat this procedure.

First we isolate a radical.

$$\sqrt{(x - 3)^2 + y^2} = 10 - \sqrt{(x + 3)^2 + y^2}$$

Then we square both sides.

$$(x - 3)^2 + y^2 = 100 - 20\sqrt{(x + 3)^2 + y^2} + (x + 3)^2 + y^2$$

Collecting like terms before repeating the procedure will make the algebraic manipulations much less cumbersome.

$$\cancel{x^2} - 6x + \cancel{9} + \cancel{y^2} = 100 - 20\sqrt{(x+3)^2 + y^2} + \cancel{x^2} + 6x + \cancel{9} + \cancel{y^2}$$

The slashes indicate that $x^2 + 9 + y^2$ was subtracted from both sides of the equation. Also subtracting $6x$ from both sides, we have:

$$-12x = 100 - 20\sqrt{(x+3)^2 + y^2}$$

Dividing both sides through by the common factor -4 further simplifies the equation.

$$3x = -25 + 5\sqrt{(x+3)^2 + y^2}$$

Now we begin to repeat the procedure by isolating the term containing the radical.

$$3x + 25 = 5\sqrt{(x+3)^2 + y^2}$$

Squaring both sides eliminates the remaining radical.

$$(3x + 25)^2 = 25[(x+3)^2 + y^2]$$

Since our aim is to write this in the form

$$\frac{x^2}{a^2} + \frac{y^2}{b^2} = 1,$$

we multiply out the indicated products and collect like terms.

$$9x^2 + 150x + 625 = 25[x^2 + 6x + 9 + y^2]$$
$$9x^2 + 150x + 625 = 25x^2 + 150x + 225 + 25y^2$$
$$-16x^2 - 25y^2 = 225 - 625$$
$$-16x^2 - 25y^2 = -400$$

To put this in the required form, we divide both sides through by -400.

$$\frac{16x^2}{400} + \frac{25y^2}{400} = 1$$

Since $400 = 16 \times 25$, we finally obtain the required form.

$$\frac{x^2}{25} + \frac{y^2}{16} = 1$$

Comparing this with

$$\frac{x^2}{a^2} + \frac{y^2}{b^2} = 1,$$

we see that $a^2 = 25$ and $b^2 = 16$. Since a and b are positive,

$$a = 5 \quad \text{and} \quad b = 4.$$

Exercise 1.8, 11–25

A **fractional equation** is an equation containing at least one fraction with a variable expression for denominator. An example of a fractional equation in one variable is

$$\frac{x-4}{x-2} - \frac{1}{x} = \frac{-4}{x(x-2)}.$$

To solve such an equation we do the following:

1. Multiply both sides through by the LCD of all the fractions.

2. Solve the new equation.

3. Substitute in the original equation to determine which solutions of the new equation are also solutions of the original. (The new equation may have some solutions that are not solutions of the original equation.)

EXAMPLE 3

Solve the equation

$$\frac{x-4}{x-2} - \frac{1}{x} = \frac{-4}{x(x-2)}.$$

Solution

1. Multiply both sides through by $x(x-2)$, the LCD of all the fractions.

$$\frac{x(x-2)(x-4)}{x-2} - \frac{x(x-2)}{x} = \frac{x(x-2)(-4)}{x(x-2)}$$

$$x(x-4) - (x-2) = -4$$

2. Solve the new equation.

$$x^2 - 4x - x + 2 = -4$$
$$x^2 - 5x + 6 = 0$$
$$(x-2)(x-3) = 0$$
$$x = 2 \quad \text{or} \quad x = 3$$

3. Substitute 2 and 3 for x in the original equation to determine whether or not they are solutions of it.

$x = 2$: The number 2 is not a solution of the original equation. Two terms in the original equation are not defined when $x = 2$ because their denominators are zero for $x = 2$.

$x = 3$: $\dfrac{3-4}{3-2} - \dfrac{1}{3} = \dfrac{-4}{3(3-2)}$

$-1 - \dfrac{1}{3} = \dfrac{-4}{3}$ True

The number 3 is the only solution of the original equation.

EXAMPLE 4

Solve the equation

$$\frac{x-1}{x+1} + \frac{6}{x-1} = \frac{x^2+11}{x^2-1}.$$

Solution

The LCD of all the fractions is $x^2 - 1 = (x-1)(x+1)$.

1. Multiply through by $x^2 - 1 = (x-1)(x+1)$.

$$\frac{(x-1)\cancel{(x+1)}(x-1)}{\cancel{x+1}} + \frac{6\cancel{(x-1)}(x+1)}{\cancel{x-1}} = \frac{(x^2+11)\cancel{(x^2-1)}}{\cancel{x^2-1}}$$

$$(x-1)(x-1) + 6(x+1) = x^2 + 11$$

$$x^2 - 2x + 1 + 6x + 6 = x^2 + 11$$

$$4x + 7 = 11$$

2. Solve the new equation.

$$4x = 4$$

$$x = 1$$

3. The number 1 is the only candidate for a solution of the original equation, but it is *not* a solution. Two terms of the original equation are not defined for $x = 1$. The original equation has no solutions; its solution set is empty.

Exercise 1.8, 26–35

Certain equations and inequalities involving absolute value appear frequently in calculus. Inequalities will be treated in the next section. An example of an equation involving an absolute value is

$$|x - 5| = 2.$$

We use the term *absolute value equation* for equations involving absolute values of variable expressions.

To work with equations such as $|x - 5| = 2$, we need to be familiar with the definition of absolute value. In Section 1.1 we defined the absolute value of real number x as follows:

$$|x| = x, \text{ if } x \text{ is positive or zero,}$$
$$|x| = -x, \text{ if } x \text{ is negative.}$$

We noted that $|x|$ represents the number of units between the point representing x on a number line and the point representing zero. It is likewise true that for real number a, $|x - a|$ represents the number of units between the point representing x on a number line and the point representing a. Furthermore,

$$|x - a| = x - a, \qquad \text{if } x - a \text{ is positive or zero (if } x - a \geq 0),$$
$$|x - a| = -(x - a), \qquad \text{if } x - a \text{ is negative (if } x - a < 0).$$

Specifically,

$$|7 - 3| = 7 - 3 = 4,$$

Here 4 is the number of units between the point representing 7 and the point representing 3.

$$|3 - 7| = -(3 - 7) = 4.$$

We can solve an equation such as $|x - 5| = 2$ by "inspection" or by employing the definition of absolute value and then solving the resulting two linear equations. In particular, since the equation $|x - 5| = 2$ states that the point for x must be located 2 units from the point for 5, by inspection we see that x must be $5 + 2 = 7$ or $5 - 2 = 3$. (See Figure 1.7) We get the same solutions if we employ the definition of absolute value. Indeed, to say that

$$|x - 5| = 2$$

means that

$$x - 5 = 2 \quad \text{or} \quad -(x - 5) = 2.$$

Solving these two equations, we have

$$x = 7 \quad \text{or} \quad -x + 5 = 2$$
$$-x = -3$$
$$x = 3.$$

The numbers 7 and 3 are the solutions of $|x - 5| = 2$. Each of the numbers 7 and 3 represents a point on the number line located 2 units from the point representing 5.

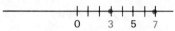

FIGURE 1.7 The solutions of $|x - 5| = 2$

EXAMPLE 5

The following example utilizes the fact that the absolute value of a product of real numbers is the product of their absolute values. In other words, if a and b are real numbers,

$$|ab| = |a|\,|b|.$$

Solution

Method 1

Solve $|3x - 5| = 6$, and graph the set of solutions on a number line. In order to solve the given equation by inspection, we must write it in the form $|x - a| = c$ where a and c are real numbers. To do this, we write $3x - 5 = 3(x - \frac{5}{3})$ and proceed as follows:

$$|3x - 5| = 6$$
$$|3(x - \tfrac{5}{3})| = 6$$
$$|3|\,|x - \tfrac{5}{3}| = 6 \qquad |ab| = |a|\,|b|$$
$$3|x - \tfrac{5}{3}| = 6$$
$$|x - \tfrac{5}{3}| = 2$$

The equation $|x - \frac{5}{3}| = 2$ states that the point for x must be located 2 units from the point for $\frac{5}{3}$. It follows that x must be $\frac{5}{3} + 2 = \frac{11}{3}$ or $\frac{5}{3} - 2 = \frac{-1}{3}$.

Method 2 To say that

$$|3x - 5| = 6$$

means that $\qquad 3x - 5 = 6 \quad \text{or} \quad -(3x - 5) = 6.$

Solving these, we have

$$3x = 11 \qquad -3x + 5 = 6$$
$$x = \tfrac{11}{3} \qquad\qquad -3x = 1$$
$$x = \tfrac{-1}{3}.$$

The numbers $\frac{11}{3}$ and $\frac{-1}{3}$ are the solutions of the given equation. They are represented by points on a number line as shown in Figure 1.8.

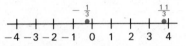

FIGURE 1.8 The solutions of $|3x - 5| = 6$

In the second method that was demonstrated in Example 5, we had to solve *two* linear equations; there were *two* cases to consider. In general, in using the second method to solve $|ax + b| = c$ where a, b, and positive c are specific real numbers and $a \neq 0$, we have to solve two linear equations because $|ax + b|$ is equal to either $ax + b$ or $-(ax + b)$.

Exercise 1.8, 36–45   If $|ax + b| = c$, then $ax + b = c$ or $ax + b = -c$.

Likewise, if $P(x)$ is any polynomial in x with real-number coefficients and if c is a specific positive real number, to solve $|P(x)| = c$ we must consider two equations, because $|P(x)|$ is equal to either $P(x)$ or $-P(x)$.

If $|P(x)| = c$, then $P(x) = c$ or $P(x) = -c$.

EXAMPLE 6 Solve $|x^2 - 5| = 4$, and graph the set of solutions on a number line.

Solution To say that $|x^2 - 5| = 4$ means that

$$(1)\ \ x^2 - 5 = 4 \quad \text{or} \quad (2)\ \ x^2 - 5 = -4.$$

Exercise 1.8, 46–50
Additional exercises,
Exercise 1.8, 51–55
In Case (1), $x^2 = 9$ so $x = 3$ or $x = -3$. In Case (2), $x^2 = 1$ so $x = 1$ or $x = -1$. The solutions of $|x^2 - 5| = 4$ are 3, -3, 1, and -1. They are graphed in Figure 1.9.

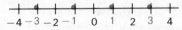

FIGURE 1.9 The solutions of $|x^2 - 5| = 4$

EXERCISE 1.8 RADICAL, FRACTIONAL, AND
ABSOLUTE VALUE EQUATIONS

▶ Solve each of the following.

EXAMPLE

$x = 5 + \sqrt{x - 3}$

Solution

$x - 5 = \sqrt{x - 3}$ Substitute in original equation.
$x^2 - 10x + 25 = x - 3$ $x = 4$: $4 = 5 + \sqrt{4 - 3}$ False
$x^2 - 11x + 28 = 0$ $x = 7$: $7 = 5 + \sqrt{7 - 3}$ True
$(x - 4)(x - 7) = 0$ The only solution is 7.
$x = 4, \qquad x = 7$

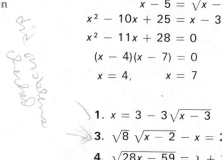

1. $x = 3 - 3\sqrt{x - 3}$ 2. $\sqrt{5x - 19} - 2\sqrt{x - 4} = 1$

3. $\sqrt{8}\,\sqrt{x - 2} - x = 2$

4. $\sqrt{28x - 59} = 1 + 2\sqrt{6x - 9}$

5. $x = 1 + \sqrt{x - 19}$

6. $\sqrt{65 - 16x} - 2\sqrt{14 - 5x} = 1$

7. $x = 8 + \sqrt{8x - 16}$ 8. $\sqrt{12x - 15} = 1 + 2\sqrt{2x}$

9. $x - 2 = 3\sqrt{x - \frac{8}{9}}$ 10. $x = 2\sqrt{6 - x} - 2$

▶ Write each of the following in the form

$$\frac{x^2}{a^2} + \frac{y^2}{b^2} = 1$$

where a and b are positive real numbers. Give the values of a and b.

11. $\sqrt{(x - 4)^2 + y^2} + \sqrt{(x + 4)^2 + y^2} = 10$

12. $\sqrt{(x - 5)^2 + y^2} + \sqrt{(x + 5)^2 + y^2} = 26$

13. $\sqrt{(x - 12)^2 + y^2} + \sqrt{(x + 12)^2 + y^2} = 26$

14. $\sqrt{(x - 8)^2 + y^2} + \sqrt{(x + 8)^2 + y^2} = 20$

15. $\sqrt{(x - 6)^2 + y^2} + \sqrt{(x + 6)^2 + y^2} = 20$

▶ Each of the following represents a curve called a hyperbola. Write each in the form

$$\frac{x^2}{a^2} - \frac{y^2}{b^2} = 1$$

where a and b are positive real numbers. Give the values of a and b.

16. $\sqrt{(x + 5)^2 + y^2} - \sqrt{(x - 5)^2 + y^2} = 6$

17. $\sqrt{(x + 13)^2 + y^2} - \sqrt{(x - 13)^2 + y^2} = 24$

18. $\sqrt{(x + 5)^2 + y^2} - \sqrt{(x - 5)^2 + y^2} = 8$

19. $\sqrt{(x + 13)^2 + y^2} - \sqrt{(x - \sqrt{13})^2 + y^2} = 4$

20. $\sqrt{(x + \sqrt{5})^2 + y^2} - \sqrt{(x - \sqrt{5})^2 + y^2} = 4$

▶ Each of the following represents a curve called a parabola. Where p is a real number, write each in the form $y^2 = 2px$ or the form $x^2 = 2py$, as indicated. Give the value of $p/2$.

EXAMPLE

$\sqrt{(x - 3)^2 + y^2} = x + 3; \ y^2 = 2px$

Solution

$$(x - 3)^2 + y^2 = (x + 3)^2$$
$$\cancel{x^2} - 6x + \cancel{9} + y^2 = \cancel{x^2} + 6x + \cancel{9}$$
$$y^2 = 12x$$
$$2p = 12$$
$$p = 6, \quad \frac{p}{2} = 3$$

21. $\sqrt{(x - 5)^2 + y^2} = x + 5; \ y^2 = 2px$

22. $\sqrt{x^2 + (y - 3)^2} = y + 3; \ x^2 = 2py$

23. $\sqrt{x^2 + (y + 5)^2} = y - 5; \ x^2 = 2py$

24. $\sqrt{(x - 7)^2 + y^2} = x + 7; \ y^2 = 2px$

25. $\sqrt{(x + 2)^2 + y^2} = x - 2; \ y^2 = 2px$

▶ Solve each of the following.

EXAMPLE

$$\frac{x - 2}{x + 2} + \frac{5}{x - 2} = \frac{20}{x^2 - 4}$$

Solution

The LCD is $x^2 - 4 = (x - 2)(x + 2)$.

$$\frac{(x - 2)(x + 2)(x - 2)}{x + 2} + \frac{5(x - 2)(x + 2)}{x - 2} = \frac{20(x^2 - 4)}{x^2 - 4}$$

$$x^2 - 4x + 4 + 5x + 10 = 20 \qquad x = 2: \text{ The number 2 is not a solution of}$$
$$x^2 + x - 6 = 0 \qquad\qquad\qquad \text{ the original equation.}$$
$$(x - 2)(x + 3) = 0$$
$$x = 2, \qquad x = -3 \qquad x = -3: \ \frac{-5}{-1} + \frac{5}{-5} = \frac{20}{9 - 4}$$
$$5 - 1 = 4 \qquad \text{True}$$

The only solution is -3.

26. $\dfrac{x - 1}{x + 1} + \dfrac{3}{x - 1} = \dfrac{6}{x^2 - 1}$

27. $\dfrac{x - 1}{x - 3} = \dfrac{x + 3}{x - 3} - \dfrac{4}{x - 1}$

28. $1 + \dfrac{1}{x^2 - 4} + \dfrac{1}{x + 2} + \dfrac{1}{x - 2} = 0$

29. $\dfrac{x - 2}{x(x + 2)} - \dfrac{2}{x(x - 2)} + \dfrac{8}{x^3 - 4x} = 0$

30. $\dfrac{x + 3}{x - 3} - \dfrac{x - 3}{x + 3} = 1 - \dfrac{x^2}{x^2 - 9}$

31. $1 + \dfrac{1}{x} - \dfrac{8}{x(x + 3)} = 0$

32. $\dfrac{x - 3}{x - 1} + \dfrac{1}{x - 3} = \dfrac{2}{x^2 - 4x + 3}$

33. $2 + \dfrac{x - 2}{x - 1} = \dfrac{2}{x - 1}$

34. $\dfrac{x - 2}{x + 2} - \dfrac{1}{x + 2} + \dfrac{8}{x^2 + 5x + 6} = 0$

35. $\dfrac{x - 2}{x + 1} - \dfrac{3(x - 2)}{x^2 + 3x + 2} = \dfrac{6}{(x + 1)(x + 2)}$

▶ Solve each of the following, and graph the set of solutions on a number line.

EXAMPLE

$|x + 1| = 3$

Solution

$x + 1 = 3$ or $-(x + 1) = 3$
$\qquad x = 2 \qquad\qquad -x - 1 = 3$
$\qquad\qquad\qquad\qquad\qquad -x = 4$
$\qquad\qquad\qquad\qquad\qquad\ x = -4$

(number line graph with points at -4 and 2; marks at $-4\ -3\ -2\ -1\ \ 0\ \ 1\ \ 2$)

36. $|x + 1| = 5$ **37.** $|2x + 2| = 5$ **38.** $|x - 5| = 8$

39. $|3x - 4| = 6$ **40.** $|x + 9| = 11$ **41.** $|2x + 9| = 11$

42. $|3x - 7| = 4$ **43.** $|2x + 8| = 6$ **44.** $|3x - 10| = 0$

45. $|6x| = 12$

▶ Find the set of real-number solutions of each of the following, and graph it on a number line.

EXAMPLE

$|x^2 - 3x - 1| = 3$

Solution

$x^2 - 3x - 1 = 3$ or $x^2 - 3x - 1 = -3$
$x^2 - 3x - 4 = 0 \qquad\qquad x^2 - 3x + 2 = 0$
$(x - 4)(x + 1) = 0 \qquad\ (x - 1)(x - 2) = 0$
$x = 4$ or $x = -1 \qquad\quad x = 1$ or $x = 2$

(number line graph with points at $-1,\ 1,\ 2,\ 4$)

46. $|x^2 - 13| = 3$ **47.** $|x^2 + 3x - 10| = 12$

48. $|x^2 - 7| = 2$ **49.** $|x^2 - 6| = 3$

50. $|x^2 + 2x - 1| = 2$

▶ In each of the following, h, k, a, and b are real numbers, and a and b are positive. Put the given equation in the form indicated and give the values of h, k, a, and b.

51. $\sqrt{(x + 1)^2 + (y - 1)^2} + \sqrt{(x - 5)^2 + (y - 1)^2} = 10$;

$$\frac{(x - h)^2}{a^2} + \frac{(y - k)^2}{b^2} = 1$$

52. $\sqrt{(x + 6)^2 + (y - 1)^2} - \sqrt{(x - 4)^2 + (y - 1)^2} = 6$;

$$\frac{(x - h)^2}{a^2} - \frac{(y - k)^2}{b^2} = 1$$

53. $\sqrt{(x + 6)^2 + (y + 2)^2} - \sqrt{(x - 4)^2 + (y + 2)^2} = 8$;

$$\frac{(x - h)^2}{a^2} - \frac{(y - k)^2}{b^2} = 1$$

54. $\sqrt{(x - 3)^2 + (y - 1)^2} + \sqrt{(x + 5)^2 + (y - 1)^2} = 12$;

$$\frac{(x - h)^2}{a^2} + \frac{(y - k)^2}{b^2} = 1$$

55. Write the equation $\sqrt{(x - 4)^2 + (y - 2)^2} = x + 4$ in the form $(y - k)^2 = 2p(x - h)$ where h, k, and p are real numbers. Give the value of $p/2$.

CHECK LIST FOR THE STUDENT

If you have learned the material in Section 1.8, you should be able to do the following:

1. Tell what is meant by a radical equation.

2. Solve one-variable radical equations in which the radicals are square roots.

3. Write certain radical equations in prescribed standard forms without radicals.

4. Tell what is meant by a fractional equation.

5. Solve fractional equations in one variable.

6. Tell what is meant by an absolute value equation.

7. Solve any equation of the form $|ax + b| = c$ where nonzero a, b, and positive c are specific real numbers; graph the set of solutions on a number line.

8. Solve equations of the form $|P(x)| = c$ where $P(x)$ is a polynomial in x with real-number coefficients and c is a specific positive real number; graph the set of solutions on a number line.

SECTION QUIZ 1.8

1. Solve the following equation.

$$x + \sqrt{x + 28} - 2 = 0$$

2. Write the following equation in the form

$$\frac{x^2}{a^2} + \frac{y^2}{b^2} = 1$$

where a and b are positive real numbers. Give the values of a and b.

$$\sqrt{(x - \sqrt{3})^2 + y^2} + \sqrt{(x + \sqrt{3})^2 + y^2} = 4$$

3. Solve the following equation.

$$\frac{x - 1}{x + 2} - \frac{2(x - 1)}{(x + 2)^2} - \frac{6}{(x + 2)^2} = 0$$

Solve each of the following equations, and graph its set of solutions on a number line.

4. $|3x - 7| = 8$

5. (a) $|x^2 - 2| = 3$
   (b) $|x^2 + x - 2| = 10$

1.9 Linear and Absolute Value Inequalities

The purpose of this section is to review properties of inequalities and to consider the solution of linear inequalities in one variable and the solution of certain inequalities involving absolute values. Examples at the end of the section illustrate absolute value inequalities encountered in calculus.

A statement that one expression is less than ($<$), greater than ($>$), less than or equal to ($\leq$), or greater than or equal to ($\geq$) another expression is called an **inequality**. Just as in the case for equations, the expression on the left of the inequality symbol is called its *left side;* the expression on the right is called its *right side.* A solution of an inequality in one variable is a real number such that the inequality becomes a true statement when the variable is replaced by that number. A solution of an inequality is said to *satisfy* the inequality. The set of all solutions of an inequality is sometimes called its *solution set.* To *solve* an inequality is to find its solution set. Two inequalities are said to be **equivalent** and each is said to be an *equivalent form* of the other if they have exactly the same solution set.

Inequalities have properties that are very similar to some of the properties of equations. These properties are listed in the following paragraphs. The major difference between the properties of inequalities and the similar properties of equations is that in an inequality, if both sides are multiplied or divided by the same *negative* number, then the sense of the inequality is reversed (changed from $<$ to $>$, etc.).

Some properties of inequalities are as follows:

Property 1 If the same number or polynomial in the variable(s) of an inequality is added to or subtracted from both sides of an inequality, the result is an inequality with exactly the same solution set as the original inequality.

Example: $3x + 4 < 7 - 5x$

has the same solution set as

$$3x + 5x < 7 - 4 \qquad \text{Adding } 5x - 4 \text{ to both sides}$$

Property 2 If both sides of an inequality are multiplied or divided by the same *positive* number, the result is an inequality with exactly the same solution set as the original inequality.

Example: $3x < 6$

has the same solution set as

$$x < 2 \qquad \text{Dividing both sides by the same positive number}$$

Property 3 If both sides of an inequality are multiplied or divided by the same *negative* number *and the sense of the inequality is reversed,* the result is an inequality with exactly the same solution set as the original inequality.

Example: $-3x < 6$

has the same solution set as

$x > -2$ Dividing both sides by the same negative
number and reversing the sense of the
inequality

A **linear inequality in one variable** is an inequality in which the one variable occurs to the first power and no other power. It can be written in one of the forms

$$ax + b < 0 \qquad ax + b > 0$$
$$ax + b \leq 0 \qquad ax + b \geq 0$$

where $a \neq 0$ and a and b are real numbers. Such inequalities may be solved by using the properties of inequalities, as illustrated in the following example.

EXAMPLE 1

Solution

Exercise 1.9, 1–10

Solve $-2x - 2 \leq 4 + x$, and graph the set of solutions on a number line.

$-2x - 2 \leq 4 + x$

$-2x - x \leq 4 + 2$ Adding $2 - x$ to both sides

$-3x \leq 6$

$x \geq -2$ Dividing both sides by the same negative number and reversing the sense of the inequality

The graph is shown in Figure 1.10. As in Section 1.1, a heavy dot has been used to show that the endpoint -2 is included on the graph.

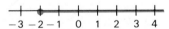

$-3\ -2\ -1\quad 0\quad 1\quad 2\quad 3\quad 4$

FIGURE 1.10   The solutions of $-2x - 2 \leq 4 + x$; $\{x \mid x \geq -2\}$

In calculus we frequently encounter certain inequalities and equations involving absolute values. Some absolute value equations were considered in Section 1.8. Here we shall consider some *absolute value inequalities*—that is, inequalities involving absolute values of variable expressions.

In Section 1.1 we noted that for any real number x, $|x|$ represents the number of units between the point representing x on a number line and the point representing zero. Thus to say that $|x| \leq c$ where c is some specific positive real number is to say that x is located no more than c units from the zero point. In other words, x must be between c and $-c$ or

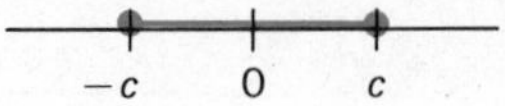

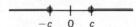

FIGURE 1.11 $|x| \le c$; $-c \le x \le c$

at most *at c or −c*. Otherwise stated, if $|x| \le c$, then $-c \le x \le c$.

$$|x| \le c \quad \text{provided that} \quad -c \le x \le c$$

This is shown in Figure 1.11.

To say that $|x| \ge c$ where c is positive is to say that x is located more than c units from the zero point or possibly exactly c units from the zero point. In other words, x must be to the right of c, or to the left of $-c$, or possibly at c or $-c$. Otherwise stated, if $|x| \ge c$, then $x \ge c$ or $x \le -c$.

$$|x| \ge c \quad \text{provided that} \quad x \ge c \quad \text{or} \quad x \le -c$$

This is shown in Figure 1.12.

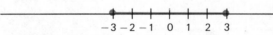

FIGURE 1.12 $|x| \ge c$; $x \ge c$ or $x \le -c$

EXAMPLE 2

Solve $|x| \le 3$ and $|x| \ge 3$, and graph the set of solutions of each of them on a separate number line.

Solution

If $|x| \le 3$, then $-3 \le x \le 3$. The graph is shown in Figure 1.13.

If $|x| \ge 3$, then $x \ge 3$ or $x \le -3$. The graph is shown in Figure 1.14.

FIGURE 1.13 $|x| \le 3$

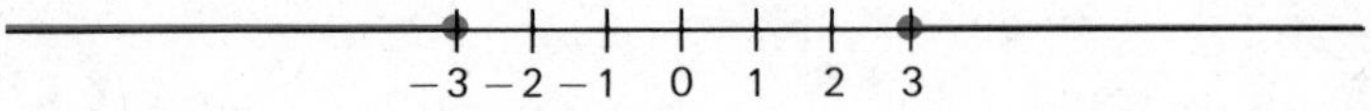

FIGURE 1.14 $|x| \ge 3$

If a, b, and c are specific real numbers, $a \ne 0$, and c is positive, then $|ax + b| \le c$ means that $(ax + b)$ is between c and $-c$ on a number line or at most *at c or −c*.

$$|ax + b| \le c \quad \text{provided that} \quad -c \le ax + b \le c$$

In other words, $|ax + b| \le c$ means that

$$ax + b \le c \quad \text{and also} \quad ax + b \ge -c.$$

Since $|ax + b|$ equals either $ax + b$ or $-(ax + b)$, to say that $|ax + b|$ $\geq c$ means that $ax + b \geq c$ or else $-(ax + b) \geq c$ so that $(ax + b) \leq -c$.

$$|ax + b| \geq c \quad \text{provided that} \quad ax + b \geq c \text{ or } ax + b \leq -c$$

In other words, $|ax + b| \geq c$ means that *either*

$$ax + b \geq c \quad \text{or else} \quad ax + b \leq -c.$$

EXAMPLE 3

Solve $|3x - 2| \leq 4$, and graph the set of solutions on a number line.

Solution

If $|3x - 2| \leq 4$, then

$$-4 \leq 3x - 2 \leq 4.$$

In other words,

$$3x - 2 \leq 4 \quad \text{and also} \quad 3x - 2 \geq -4$$
$$3x \leq 6 \qquad\qquad\qquad 3x \geq -2$$
$$x \leq 2 \qquad\qquad\qquad x \geq \tfrac{-2}{3}$$

The solutions are those numbers x such that $x \leq 2$ and also $x \geq \tfrac{-2}{3}$. Otherwise written,

$$\tfrac{-2}{3} \leq x \leq 2.$$

The set of solutions is graphed in Figure 1.15.

FIGURE 1.15 The solutions of $|3x - 2| \leq 4$; $-\tfrac{2}{3} \leq x \leq 2$

In this example, the solution set could also have been found by treating the two inequalities simultaneously, as follows:

$$-4 \leq 3x - 2 \leq 4$$
$$-4 + 2 \leq 3x \leq 4 + 2 \qquad \text{Adding 2 to all sides.}$$
$$-2 \leq 3x \leq 6$$
$$\tfrac{-2}{3} \leq x \leq 2 \qquad\qquad \text{Dividing all sides by 3.}$$

EXAMPLE 4

Solve $|3x - 2| \geq 4$, and graph the set of solutions on a number line.

Solution

If $|3x - 2| \geq 4$, then *either*

$$3x - 2 \geq 4 \quad \text{or else} \quad 3x - 2 \leq -4.$$
$$3x \geq 6 \qquad\qquad\qquad 3x \leq -2$$
$$x \geq 2 \qquad\qquad\qquad x \leq \tfrac{-2}{3}$$

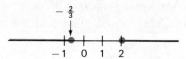

FIGURE 1.16 The solutions of $|3x - 2| \geq 4$; $x \geq 2$ or $x \leq -\frac{2}{3}$

Exercise 1.9, 11–20

The solutions are those numbers x such that $x \geq 2$ or else $x \leq \frac{-2}{3}$. This set of solutions is graphed in Figure 1.16.

In calculus we are sometimes called upon to work with absolute value inequalities using the Greek letters ϵ (epsilon) and δ (delta) as illustrated in the next examples.

EXAMPLE 5

Let δ and ϵ represent positive real numbers. Show that if $\delta \leq \epsilon/5$, then $|(5x - 3) - 7| < \epsilon$ whenever $|x - 2| < \delta$.

Solution

For the prescribed δ (a positive real number less than or equal to $\epsilon/5$), we must show that $|(5x - 3) - 7| < \epsilon$ whenever $|x - 2| < \delta$. To show this, we may begin by working with $|(5x - 3) - 7|$.

$$|(5x - 3) - 7| = |5x - 10|$$
$$= 5|x - 2|$$
$$< 5\delta \quad \text{whenever} \quad |x - 2| < \delta$$

That is,

$$|(5x - 3) - 7| < 5\delta \quad \text{whenever} \quad |x - 2| < \delta.$$

But $\delta \leq \epsilon/5$, so $5\delta \leq 5(\epsilon/5)$ or

$$5\delta \leq \epsilon$$

It follows that

$$|(5x - 3) - 7| < \epsilon \quad \text{whenever} \quad |x - 2| < \delta.$$

EXAMPLE 6

Let ϵ represent any positive real number. Find positive number δ in terms of ϵ so that if $|x - 4| < \delta$, then $|(3x - 2) - 10| < \epsilon$.

Solution

We want to find δ so that whenever $|x - 4| < \delta$, we will have $|(3x - 2) - 10| < \epsilon$. That is, since $|(3x - 2) - 10| = |3x - 12|$, we want to find δ so that whenever $|x - 4| < \delta$, we will have

$$|3x - 12| < \epsilon,$$

or

$$3|x - 4| < \epsilon.$$

In other words, since $3|x - 4| < \epsilon$ is equivalent to $|x - 4| < \epsilon/3$, we want to find δ so that whenever $|x - 4| < \delta$, we will have

$$|x - 4| < \epsilon/3.$$

We see that as long as δ is $\epsilon/3$ or less, then whenever $|x - 4| < \delta$, we will have $|x - 4| < \epsilon/3$, and hence we will have $|(3x - 2) - 10| < \epsilon$. Checking this out, note that if we have $|x - 4| < \epsilon/3$, then indeed, we do have

$$3|x - 4| < \epsilon$$
$$|3x - 12| < \epsilon$$
$$|(3x - 2) - 10| < \epsilon.$$

The required δ must be any positive number less than or equal to $\epsilon/3$.

$$\delta \leq \epsilon/3$$

As a double check, we can show that if $\delta \leq \epsilon/3$, then $|(3x - 2) - 10| < \epsilon$ whenever $|x - 4| < \delta$.

$$|(3x - 2) - 10| = |3x - 12|$$
$$= 3|x - 4|$$
$$< 3\delta \quad \text{whenever} \quad |x - 4| < \delta$$

But if $\delta \leq \epsilon/3$, then $3\delta \leq \epsilon$ so that

$$|(3x - 2) - 10| < \epsilon \quad \text{whenever} \quad |x - 4| < \delta,$$

Exercise 1.9, 21–30 as we wanted to show.

EXERCISE 1.9 LINEAR AND ABSOLUTE VALUE INEQUALITIES

▶ Solve each of the following, and graph the set of solutions on a number line.

EXAMPLE

$-5x - 10 \leq 14 + 3x$

Solution

$-8x \leq 24$

$x \geq -3$

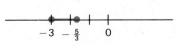

$-3 \ -2 \ -1 \quad 0$

1. $-3x - 3 \leq 5 + 2x$ 2. $2x - 6 \geq 7x + 8$

3. $5x + 5 \leq 4x + 5$ 4. $5x + 8 > 4x + 7$

5. $3x + 1 \geq 4x - 4$ 6. $6 + x < 8 + 2x$

7. $2x - 5 < 3x - 4$ 8. $3x + 1 \geq 2x - 2$

9. $8x - 1 \geq 7x + 2$ 10. $4x - 3 \leq 3x - 5$

EXAMPLE

$|3x + 7| \leq 2$

Solution

$-2 \leq 3x + 7 \leq 2$

$-9 \leq 3x \quad\quad \leq -5$

$-3 \leq x \quad\quad \leq -\frac{5}{3}$

$-3 \quad -\frac{5}{3} \quad 0$

11. $|3x + 7| \geq 2$ **12.** $|2x - 1| \leq 3$ **13.** $|2x - 1| \geq 3$

14. $|4x| > 8$ **15.** $|3x| < 6$ **16.** $|4x + 3| \geq 1$

17. $|4x + 3| \leq 1$ **18.** $|2x - 5| \leq 3$ **19.** $|2x - 5| \geq 3$

20. $|5x - 1| \leq 11$

▶ In each of the following, let δ and ϵ represent positive real numbers.

EXAMPLE

Show that if $\delta \leq \epsilon/2$, then

$$|(2x - 4) - 6| < \epsilon \quad \text{whenever} \quad |x - 5| < \delta$$

Solution

$$|(2x - 4) - 6| = |2x - 10|$$
$$= 2|x - 5|$$
$$< 2\delta \quad \text{whenever} \quad |x - 5| < \delta.$$

But if $\delta \leq \epsilon/2$, then $2\delta \leq \epsilon$ so that

$$|(2x - 4) - 6| < \epsilon \quad \text{whenever} \quad |x - 5| < \delta.$$

21. Show that if $\delta \leq \epsilon/2$, then $|(2x - 5) - 1| < \epsilon$ whenever $|x - 3| < \delta$.

22. Show that if $\delta \leq \epsilon/4$, then $|(4x - 1) - 11| < \epsilon$ whenever $|x - 3| < \delta$.

23. Show that if $\delta \leq \epsilon/3$, then $|(3x + 7) - 10| < \epsilon$ whenever $|x - 1| < \delta$.

24. Show that if $\delta \leq \epsilon/3$, then $|(3x + 7) - 4| < \epsilon$ whenever $|x + 1| < \delta$.

25. Show that if $\delta \leq \epsilon/4$, then $|(4x + 1) + 7| < \epsilon$ whenever $|x + 2| < \delta$.

▶ In each of the following let ϵ represent any positive real number, and find positive number δ in terms of ϵ so that the given statement is true.

EXAMPLE

If $|x - 7| < \delta$, then $|(3x - 1) - 20| < \epsilon$.

Solution

Since $|(3x - 1) - 20| = |3x - 21|$, we want δ so that whenever $|x - 7| < \delta$, $|3x - 21| = 3|x - 7|$ will be less than ϵ. We want δ so that $3|x - 7| < \epsilon$ whenever $|x - 7| < \delta$—that is, so that $|x - 7| < \epsilon/3$ whenever $|x - 7| < \delta$. We will have this for any $\delta \leq \epsilon/3$.

26. If $|x - 5| < \delta$, then $|(3x - 1) - 14| < \epsilon$.

27. If $|x - 2| < \delta$, then $|(3x - 1) - 5| < \epsilon$.

28. If $|x - 1| < \delta$, then $|(2x + 3) - 5| < \epsilon$.

29. If $|x + 1| < \delta$, then $|(2x + 3) - 1| < \epsilon$.

30. If $|x + 2| < \delta$, then $|(4x - 1) + 9| < \epsilon$.

CHECK LIST FOR THE STUDENT

If you have learned the material in Section 1.9, you should be able to do the following:

1. Define inequality, a solution of an inequality, equivalent inequalities, and an equivalent form of an inequality.

2. Tell how the solutions of a given inequality are related to the solutions of a new inequality obtained from the given inequality by doing each of the following:

(a) Adding or subtracting the same polynomial in its variable(s) to or from both sides

(b) Multiplying or dividing both sides by the same positive number

(c) Multiplying or dividing both sides by the same negative number and reversing the sense of the inequality.

3. Give an example of a linear inequality in one variable; solve any linear inequality in one variable, and graph the set of solutions on a number line.

4. Tell what is meant by an absolute value inequality.

5. Solve inequalities of the form $|ax + b| \leq c$ or of the form $|ax + b| \geq c$ where nonzero a, b, and positive c are specific real numbers; graph the set of solutions on a number line.

SECTION QUIZ 1.9

Solve each of the following, and graph its set of solutions on a number line.

1. $x - 2 < 7$ **2.** $-4x \geq 8$ **3.** $-9x - 2 \leq 6 - 5x$

4. $|2x - 5| < 7$ **5** $|3x - 7| \geq 2$

1.10 Simultaneous Linear Equations

The purpose of this section is to show how to solve two simultaneous linear equations in two variables, to demonstrate the solution of n simultaneous linear equations in n variables for $n > 2$, and to indicate where the solution of such simultaneous equations is needed in calculus.

A **linear equation in two variables** is a first-degree polynomial equation containing exactly two variables. The equation

$$3x + 2y + 5 = 0$$

is a linear equation in x and y. The equation

$$5x - 6y = 0$$

is also a linear equation in x and y.

A **linear equation in n variables** is a first-degree polynomial equation containing exactly n variables. The equation

$$7x - 8y + 4z - 12 = 0$$

is a linear equation in x, y, and z. The equation

$$7A + 5B - C + D - 3E = -11$$

is a linear equation in A, B, C, D, and E.

A **solution** of a linear equation in n variables is a set of values for the n variables that makes the equation a true statement. A solution of a linear equation is said to *satisfy* the equation. For example, $x = 1, y = -4$ is a solution of

$$3x + 2y + 5 = 0$$

because $\qquad\qquad 3(1) + 2(-4) + 5 = 0$

is a true statement. Also, $A = 1, B = -2, C = 3, D = 1$, and $E = 2$ is a solution of

$$7A + 5B - C + D - 3E = -11$$

because $\qquad 7(1) + 5(-2) - 3 + 1 - 3(2) = -11$

is a true statement.

Any single linear equation in n variables with $n > 1$ has an unlimited number of solutions. A solution may be found simply by assigning values to $(n - 1)$ of the variables and solving the resulting linear equation in one variable for that variable. For example, we may find a solution of

$$7x - 8y + 4z - 12 = 0$$

by assigning values to x and y and then solving for z. In particular, if we let $x = 0$ and $y = 0$, we have

$$0 - 0 + 4z - 12 = 0$$
$$4z = 12$$
$$z = 3.$$

Thus $x = 0$, $y = 0$, $z = 3$ is one solution of $7x - 8y + 4z - 12 = 0$. Again, if we let $x = 0$ and $y = 1$, we have

$$0 - 8 + 4z - 12 = 0$$
$$4z = 20$$
$$z = 5.$$

The values $x = 0$, $y = 1$, and $z = 5$ constitute another solution of $7x - 8y + 4z - 12 = 0$.

By a *system of n simultaneous linear equations in n variables* with $n > 1$, we mean a set of n linear equations collectively containing terms of first degree in each of the n variables and containing no other variable terms. The pair of equations

$$\begin{cases} 3x + 2y + 5 = 0 \\ 5x - 4y + 12 = 0 \end{cases}$$

is an example of a system of two simultaneous equations in the two variables x and y. The pair of equations

$$\begin{cases} 3x + 2y + 5 = 0 \\ 8x \quad\quad - 6 = 0 \end{cases}$$

is also an example of a system of two simultaneous equations in the two variables x and y. The four equations

$$\begin{cases} 3A - 2B + C - D = 7 \\ B + 3C - 4D = 5 \\ A - 3B = 4 \\ 4A + 2B - 5D = 2 \end{cases}$$

make up a system of four simultaneous linear equations in the four variables A, B, C, and D.

In working with simultaneous linear equations, it is customary to call the variables "unknowns." We may say that the system of four equations in the previous paragraph is a system of four simultaneous linear equations in the four *unknowns* A, B, C, and D. Similarly, we may say that the system

$$\begin{cases} 3x + 2y + 5 = 0 \\ 5x - 4y + 12 = 0 \end{cases}$$

is a system of two simultaneous linear equations in the two *unknowns* x and y.

A solution of a system of *n* simultaneous linear equations in *n* unknowns with $n > 1$ is a set of values of the unknowns that satisfies each and every one of the *n* equations. For example, $x = -2$, $y = \frac{1}{2}$ is a solution of the system

$$\begin{cases} 3x + 2y + 5 = 0 \\ 5x - 4y + 12 = 0 \end{cases}$$

because $x = -2$, $y = \frac{1}{2}$ satisfies both equations.

$$\begin{cases} 3(-2) + 2(\frac{1}{2}) + 5 = 0 & \text{True} \\ 5(-2) - 4(\frac{1}{2}) + 12 = 0 & \text{True} \end{cases}$$

Also, $A = 1$, $B = -1$, $C = 2$, $D = 0$ is a solution of the system

$$\begin{cases} 3A - 2B + C - D = 7 \\ B + 3C - 4D = 5 \\ A - 3B = 4 \\ 4A + 2B - 5D = 2 \end{cases}$$

because these values satisfy each and every equation of the system.

$$\begin{cases} 3(1) - 2(-1) + 2 - 0 = 7 & \text{True} \\ (-1) + 3(2) - 4(0) = 5 & \text{True} \\ 1 - 3(-1) = 4 & \text{True} \\ 4(1) + 2(-1) - 5(0) = 2 & \text{True} \end{cases}$$

It can be shown that a system of *n* simultaneous linear equations in *n* unknowns has either

1. Exactly one solution,

2. No solution, or

3. An unlimited number of solutions.

Criteria for determining which of these three applies to any given system of *n* simultaneous linear equations can be developed. In this section we shall consider systems having exactly one solution, and we shall show how to solve them.

First let us review two methods of solving a system of two simultaneous linear equations in two unknowns. The idea in each of the two methods is to eliminate one of the variables and thereby reduce the problem to that of solving a linear equation in one variable.

Method 1 *Substitution*

Step 1 Solve either one of the equations for one unknown in terms of the other.

Step 2 Substitute for that one unknown in the other equation.

Step 3 Solve the resulting linear equation in one variable.

Step 4 Substitute the value obtained in Step 3 in either one of the given equations, and solve for the remaining unknown.

To illustrate, let us solve the following system by substitution.

$$\begin{cases} 4x + 3y + 7 = 0 \\ 5x - 6y + 12 = 0 \end{cases}$$

Step 1 Suppose that we solve the first equation for y in terms of x.

$$3y = -4x - 7$$
$$y = \tfrac{-4}{3}x - \tfrac{7}{3}$$

Step 2 Substitute $\tfrac{-4}{3}x - \tfrac{7}{3}$ for y in the second equation.

$$5x - 6(\tfrac{-4}{3}x - \tfrac{7}{3}) + 12 = 0$$

Step 3 Solve for x.

$$5x - 6(\tfrac{-4}{3})x - 6(\tfrac{-7}{3}) + 12 = 0$$
$$5x + 8x + 14 + 12 = 0$$
$$13x = -26$$
$$x = -2$$

Step 4 Substitute -2 for x in the first equation of the given system, and solve for y.

$$4(-2) + 3y + 7 = 0$$
$$-8 + 3y + 7 = 0$$
$$3y = 1$$
$$y = \tfrac{1}{3}$$

The solution is $x = -2$, $y = \tfrac{1}{3}$. The solution should be checked by substituting it in both the given equations.

Check $\begin{cases} 4x + 3y + 7 = 0 & 4(-2) + 3(\tfrac{1}{3}) + 7 = 0 & \text{True} \\ 5x - 6y + 12 = 0 & 5(-2) - 6(\tfrac{1}{3}) + 12 = 0 & \text{True} \end{cases}$

Method 2 *Choice of Multiplier and Addition*

Step 1 Decide which unknown is to be eliminated.

Step 2 Choose a multiplier for each equation so that after each equation is multiplied through by its multiplier, the coefficients of the unknown to be eliminated will be the same except for sign.

Step 3 Multiply the equations through by their multipliers, and add. (This will eliminate the unknown to be eliminated.)

Step 4 Solve for the remaining variable.

Step 5 Substitute the value obtained in Step 4 in either of the given equations and solve for the other variable.

For example, let us solve the system

$$\begin{cases} 4x + 3y + 7 = 0 \\ 5x - 6y + 12 = 0 \end{cases}$$

by choice of multiplier and addition.

Step 1 Suppose that we decide to eliminate y.

Step 2 Multipliers 2 (for the first equation) and 1 (for the second equation) will make the coefficients of y become 6 and -6, the same except for sign. (Other multipliers will also make the coefficients of y become the same except for sign. For example, multipliers 4 for the first equation and 2 for the second equation will make the coefficients of y become 12 and -12. We may use any convenient multipliers that make the coefficients of the unknown to be eliminated become the same except for sign.)

Step 3

$$\begin{array}{ll} 2\,|\,4x + 3y + 7 = 0 & \quad 8x + \cancel{6y} + 14 = 0 \\ 1\,|\,5x - 6y + 12 = 0 & \quad \underline{5x - \cancel{6y} + 12 = 0} \\ & \quad 13x \qquad\;\; + 26 = 0 \end{array}$$

Step 4

$$13x + 26 = 0$$
$$13x = -26$$
$$x = -2$$

Step 5 Substitute -2 for x in the first of the given equations, and solve for y.

$$4(-2) + 3y + 7 = 0$$
$$-8 + 3y + 7 = 0$$
$$3y = 1$$
$$y = \tfrac{1}{3}$$

The solution is $x = -2$, $y = \tfrac{1}{3}$.

Exercise 1.10, 1–12 This agrees with the solution found by using Method 1. It was checked by substituting it in both the given equations.

If a system of three simultaneous linear equations in three unknowns has exactly one solution, it can be solved by adapting the elimination methods just used and systematically eliminating one of the three unknowns in order to reduce the problem to that of solving a pair of simultaneous linear equations in two unknowns. This is illustrated in the next example.

EXAMPLE 1

Find the solution of the following system.

$$\begin{cases} 3x - 2y + 6z = -6 \\ -3x + 10y + 11z = 13 \\ x - 2y - z = -3 \end{cases}$$

Solution

Suppose that we decide to eliminate z from the given system and reduce the problem to that of solving a pair of linear equations in x and y.

Step 1   Decide to eliminate z. We may now use either the method of substitution or the method of choice of multiplier and addition to eliminate z from the first and second, or from the first and third, or from the second and third of the given equations. We only need to eliminate z from two of these three pairs, and we will have a new pair of linear equations in the two unknowns x and y.

Step 2   Choose two pairs of the given equations, and eliminate z from each pair. Suppose we choose the first and third and also the second and third, and eliminate z from each of these two pairs.

$$\begin{array}{r|l} 1 & 3x - 2y + 6z = -6 \\ 6 & x - 2y - z = -3 \end{array} \qquad \begin{array}{r|l} 1 & -3x + 10y + 11z = 13 \\ 11 & x - 2y - z = -3 \end{array}$$

$$\begin{array}{r} 3x - 2y + 6z = -6 \\ 6x - 12y - 6z = -18 \\ \hline 9x - 14y = -24 \end{array} \qquad \begin{array}{r} -3x + 10y + 11z = 13 \\ 11x - 22y - 11z = -33 \\ \hline 8x - 12y = -20 \end{array}$$

Step 3   Solve the resulting pair of linear equations in x and y.

$$\begin{cases} 9x - 14y = -24 \\ 8x - 12y = -20 \end{cases}$$

1. Eliminate x; solve for y.

$$\begin{array}{r|l} 8 & 9x - 14y = -24 \\ -9 & 8x - 12y = -20 \end{array} \qquad \begin{array}{r} 72x - 112y = -192 \\ -72x + 108y = 180 \\ \hline -4y = -12 \\ y = 3 \end{array}$$

2. Substitute 3 for y in $9x - 14y = -24$, and solve for x.

$$\begin{array}{r} 9x - 14(3) = -24 \\ 9x - 42 = -24 \\ 9x = 18 \\ x = 2 \end{array}$$

The solution of the resulting pair of equations in x and y is $x = 2, y = 3$.

Check $\begin{cases} 9x - 14y = -24 & \quad 9(2) - 14(3) = -24 & \quad \text{True} \\ 8x - 12y = -20 & \quad 8(2) - 12(3) = -20 & \quad \text{True} \end{cases}$

Step 4   Substitute the values found for x and y in any one of the original three equations, and solve for z.

$$3x - 2y + 6z = -6$$
$$3(2) - 2(3) + 6z = -6$$
$$6 - 6 + 6z = -6$$
$$z = -1$$

The solution of the given system of three equations in three unknowns is $x = 2$, $y = 3$, $z = -1$.

Check

$3x - 2y + 6z = -6$	$3(2) - 2(3) + 6(-1) = -6$ True
$-3x + 10y + 11z = 13$	$-3(2) + 10(3) + 11(-1) = 13$ True
$x - 2y - z = -3$	$(2) - 2(3) - (-1) = -3$ True

Exercise 1.10, 13–18

If a system of n simultaneous linear equations in n unknowns has exactly one solution, even when n is greater than three the system can be solved by adapting the elimination methods just presented. This can be done by systematically eliminating one of the unknowns in order to reduce the problem to that of solving $(n - 1)$ simultaneous linear equations in $(n - 1)$ unknowns. This process can be repeated until the problem is reduced to that of solving a pair of linear equations in two unknowns. In Part 4 of this text, more sophisticated and usually more direct methods for solving n equations in n unknowns for $n > 2$ will be presented. The elimination methods presented in this section are apt to be quite tedious for any large n. If some of the unknowns do not appear in some of the given equations, however, these elimination methods might produce the solution quite readily. This is illustrated for $n = 5$ in the next example.

EXAMPLE 2

Find the solution of the following system.

$$\begin{cases} A + C = 4 \\ 3A + B + 2C + D = 6 \\ 3A + 3B + C + D + E = 3 \\ A + 3B = -5 \\ B = -2 \end{cases}$$

Solution

We can solve this system by repeatedly using the substitution method. From the last equation we see that $B = -2$. Substituting -2 for B in the fourth equation and solving for A, we have

$$A + 3(-2) = -5$$
$$A - 6 = -5$$
$$A = 1.$$

Substituting 1 for A in the first equation and solving for C, we have

$$1 + C = 4$$
$$C = 3.$$

Now substituting $A = 1$, $B = -2$, $C = 3$ in the second equation and solving for D, we have

$$3(1) + (-2) + 2(3) + D = 6$$
$$3 - 2 + 6 + D = 6$$
$$D = -1.$$

Finally, substituting $A = 1$, $B = -2$, $C = 3$, and $D = -1$ in the third equation, we can solve for the remaining unknown E.

$$3(1) + 3(-2) + 3 + (-1) + E = 3$$
$$3 - 6 + 3 - 1 + E = 3$$
$$E = 4$$

The solution of the given system is $A = 1$, $B = -2$, $C = 3$, $D = -1$, $E = 4$. This can be checked by substituting in each of the given equations.

Check

$A + C = 4$	$1 + 3 = 4$	True
$3A + B + 2C + D = 6$	$3(1) + (-2) + 2(3) + (-1) = 6$	True
$3A + 3B + C + D + E = 3$	$3(1) + 3(-2) + 3 + (-1) + 4 = 3$	True
$A + 3B = -5$	$1 + 3(-2) = -5$	True
$B = -2$	$-2 = -2$	True

Exercise 1.10, 19–25 ——— 4 – 7

Systems of n simultaneous linear equations in n unknowns such as the one presented in the previous example arise in integral calculus in performing certain manipulations called "integration by partial fractions." In order to perform such an integration, it is necessary to be able to decompose a given rational expression into a sum of partial fractions—that is, fractions with lower-degree denominators than the given rational expression. For example, the so-called *partial fraction theorem*, to be presented in Section 4.4, assures us that the expression

$$\frac{6x^2 + 7x - 4}{x(x - 1)(x + 2)}$$

can be expressed as a sum of partial fractions of the form

$$\frac{A}{x} + \frac{B}{x - 1} + \frac{C}{x + 2}$$

where A, B, and C represent particular real numbers. The actual values of

EXAMPLE 3

A, B, and C may be found by solving a system of three simultaneous linear equations in three unknowns. This is illustrated in the next example.

Assume that

$$\frac{6x^2 + 7x - 4}{x(x - 1)(x + 2)} = \frac{A}{x} + \frac{B}{x - 1} + \frac{C}{x + 2},$$

and find the values of A, B, and C.

Solution

Combining the fractions on the right side of the given equation, we have

$$\frac{6x^2 + 7x - 4}{x(x - 1)(x + 2)} = \frac{A(x - 1)(x + 2) + Bx(x + 2) + Cx(x - 1)}{x(x - 1)(x + 2)}.$$

This equation states that two fractions with the same denominator are equal. In order for this equality to be true, we must have the two numerators equal. Equating the numerators, we have

$$6x^2 + 7x - 4 = A(x - 1)(x + 2) + Bx(x + 2) + Cx(x - 1).$$

We may now determine the values of A, B, and C by either one of two methods. The method to be presented first happens to be the longer method for this particular problem, but it is the method that applies to more general situations than the situation dictated by the denominator in this particular problem.

Method 1   *Equating Coefficients of Like Powers*
First multiply out the right side of

$$6x^2 + 7x - 4 = A(x - 1)(x + 2) + Bx(x + 2) + Cx(x - 1)$$

and collect like terms.

$$6x^2 + 7x - 4 = A(x^2 + x - 2) + Bx^2 + 2Bx + Cx^2 - Cx$$
$$= (A + B + C)x^2 + (A + 2B - C)x - 2A$$

In order for this to be true, the numbers A, B, and C must be such that the coefficients of like powers are equal. Equating the coefficients of x^2, equating the coefficients of x, and equating the constant terms, we get a system of three simultaneous linear equations in the three unknowns A, B, and C.

$$\begin{cases} A + B + C = 6 \\ A + 2B - C = 7 \\ \qquad -2A = -4 \end{cases}$$

Now we must solve this system for A, B, and C. From the third equation we get

$$A = 2.$$

Substituting 2 for A in the first two equations, we get a pair of linear equations in the two unknowns B and C.

$$\begin{cases} B + C = 4 \\ 2B - C = 5 \end{cases}$$

Solving these by choice of multiplier and addition, we have

$$
\begin{array}{r|l}
1 & B + C = 4 \\
1 & 2B - C = 5 \\
\hline
 & 3B = 9 \\
 & B = 3 \\
 & 3 + C = 4 \\
 & C = 1.
\end{array}
$$

It follows that $A = 2$, $B = 3$, and $C = 1$ so that

$$\frac{6x^2 + 7x - 4}{x(x - 1)(x + 2)} = \frac{2}{x} + \frac{3}{x - 1} + \frac{1}{x + 2}.$$

Method 2   *Substituting Convenient Values*
In the equation

$$6x^2 + 7x - 4 = A(x - 1)(x + 2) + Bx(x + 2) + Cx(x - 1)$$

that we obtained when we first equated the numerators, if we successively let x take on values that make each of the three distinct linear factors x, $(x - 1)$, and $(x + 2)$ become zero, we will be able to compute A, B, and C rather easily. This process will work easily as long as the denominator in the original fraction that is to be decomposed into partial fractions has only unrepeated linear factors, as it does in this example. To illustrate, first let $x = 0$, then let $x = 1$, then let $x = -2$.

$x = 0$:

$$6(0)^2 + 7(0) - 4 = A(-1)(2) + B(0)(2) + C(0)(-1)$$
$$-4 = -2A$$
$$2 = A$$

$x = 1$:

$$6(1)^2 + 7(1) - 4 = A(0)(3) + B(1)(3) + C(1)(0)$$
$$9 = 3B$$
$$3 = B$$

$x = -2$:

$$6(-2)^2 + 7(-2) - 4 = A(-3)(0) + B(-2)(0) + C(-2)(-3)$$
$$24 \quad - \quad 14 \quad - 4 = 6C$$
$$6 = 6C$$
$$1 = C$$

Thus we find that $A = 2$, $B = 3$, and $C = 1$ in agreement with the results of Method 1. As was stated earlier, although Method 1 is longer for this particular example, it can be used in more general situations than Method 2.

Exercise 1.10, 26–35

EXAMPLE 4

A man is on a diet that requires him to consume exactly 1500 calories per day in exactly three meals. The number of calories he consumes at dinner must equal the total number of calories he consumes at breakfast and lunch. The number of calories he consumes at lunch must be 150 more than twice the number he consumes at breakfast. How many calories must he consume at each meal?

Solution

Let x, y, and z be the number of calories he consumes daily at breakfast, lunch, and dinner, respectively. According to the requirements of the diet,

$$x + y + z = 1500$$
$$z = x + y$$
$$y = 2x + 150.$$

Rewriting these equations, we have the following system of three linear equations in the three unknowns x, y, and z.

$$\begin{cases} x + y + z = 1500 \\ x + y - z = 0 \\ -2x + y \quad\;\; = 150 \end{cases}$$

We must solve this system for x, y, and z. Suppose that we decide to eliminate z. Adding the first two equations, we have

$$2x + 2y = 1500.$$

This equation and the third equation of the system, $-2x + y = 150$, make up a system of two linear equations in the two unknowns x and y.

$$\begin{cases} 2x + 2y = 1500 \\ -2x + y = 150 \end{cases}$$

Eliminating x by adding, we have

$$3y = 1650$$
$$y = 550.$$

Substituting this value of y in the equation $-2x + y = 150$ and solving for x, we have

$$-2x + 550 = 150$$
$$-2x = -400$$
$$x = 200.$$

Substituting $x = 200$ and $y = 550$ in the second equation of the original system and solving for z, we have

$$200 + 550 - z = 0$$
$$z = 750.$$

The man must consume 200 calories at breakfast, 550 calories at lunch, and 750 calories at dinner.

Exercise 1.10, 36–45

EXERCISE 1.10 SIMULTANEOUS LINEAR EQUATIONS

▶ Find the solution of each of the following systems.

EXAMPLE

$$\begin{cases} A + B + C = 3 \\ 2A - 3B + 4C = 31 \\ 6A = 6 \end{cases}$$

Solution

$$\begin{aligned} 6A &= 6 \\ A &= 1 \end{aligned} \quad \begin{cases} 1 + B + C = 3 \\ 2 - 3B + 4C = 31 \end{cases} \quad \begin{cases} B + C = 2 \\ -3B + 4C = 29 \end{cases} \begin{array}{c} 3 \\ 1 \end{array} \Bigg| \begin{array}{l} B + C = 2 \\ -3B + 4C = 29 \end{array}$$

$$\begin{array}{r} 3B + 3C = 6 \\ -3B + 4C = 29 \\ \hline 7C = 35 \\ C = 5 \end{array}$$

$$A + B + C = 3$$
$$1 + B + 5 = 3$$
$$B = -3 \qquad \text{The solution is } A = 1, B = -3, C = 5.$$

1. $\begin{cases} 2x - 7y = 16 \\ 3x + 4y = -5 \end{cases}$ **2.** $\begin{cases} 2A + 3B = -2 \\ -A - 2B = 2 \end{cases}$

3. $\begin{cases} 4A - 6B = -36 \\ 3A + 8B = 24 \end{cases}$ **4.** $\begin{cases} x - 6y - 20 = 0 \\ 2x + 5y = -11 \end{cases}$

5. $\begin{cases} 2A + 4B - 2 = 0 \\ 3A + 9B - 6 = 0 \end{cases}$ **6.** $\begin{cases} 5x + 4y = -6 \\ -2x + 3y - 7 = 0 \end{cases}$

7. $\begin{cases} x + y = -1 \\ x - y = 17 \end{cases}$ **8.** $\begin{cases} 7A - 8B = -5 \\ -5A + 4B = 1 \end{cases}$

9. $\begin{cases} 8x - 5y - 17 = 0 \\ 4x + 3y - 3 = 0 \end{cases}$

10. $\begin{cases} 2A + B - 8 = 0 \\ 3A = 23 + 4B \end{cases}$

11. $\begin{cases} 11A + 8B + 52 = 0 \\ 9A = B - 35 \end{cases}$

12. $\begin{cases} 15A - 27B + 3 = 0 \\ 3A + 9B = 9 \end{cases}$

13. $\begin{cases} 3x - 2y + z = -7 \\ 4x + 5y + 2z = 8 \\ x + 3y - 4z = 19 \end{cases}$

14. $\begin{cases} A - 2B + 3C = -4 \\ 4A + 3B - 2C = 20 \\ A + 3B - 5C = 14 \end{cases}$

15. $\begin{cases} A + B + C = 1 \\ 2A - 3B - 1 = 0 \\ A + 4B - C = -8 \end{cases}$

16. $\begin{cases} 2x - 5y + 3z = 18 \\ x + 7y - 2z = -5 \\ -x + 4y - 3z = -15 \end{cases}$

17. $\begin{cases} 3x + 2y - 5z + 20 = 0 \\ -x - y + 4z = 15 \\ 2x + 3y = 5 \end{cases}$

18. $\begin{cases} x + y + a = 5 \\ x + y = 0 \\ y + z = 1 \end{cases}$

19. $\begin{cases} x + y + 2z - w = 4 \\ -x + 2y + z + w = 2 \\ 2x - y + 3z + 2w = 16 \\ 2x - 3y + 2z + 2w = 15 \end{cases}$

20. $\begin{cases} A + B = 4 \\ B + 3C + D = 15 \\ 2A + 3B - C - D = 3 \\ C + 4D = 12 \end{cases}$

21. $\begin{cases} A + B + C + D = 18 \\ B + C + D = 11 \\ A + B + C = 16 \\ A - 2B = -1 \end{cases}$

22. $\begin{cases} 3x - 2y + z + 2w = 23 \\ 3y - z - w = -6 \\ z + w = 9 \\ 2x + 4y - 3z - 2w = -10 \end{cases}$

23. $\begin{cases} 3A + B = 2 \\ A - 2B + 3C - D + 2E = 17 \\ B - 3C + 2D + 4E = 1 \\ A + 4C = 9 \\ 3C = 6 \end{cases}$

24. $\begin{cases} A - B + C - D + E = 0 \\ 2E = 2 \\ -C + D - E = 4 \\ B - 2C = 6 \\ A + 3C = -5 \end{cases}$

25. $\begin{cases} x + y + z + w + q = 0 \\ y + z + w + q = -1 \\ z + w + q = -1 \\ w + q = 0 \\ q = -2 \end{cases}$

▶ In each of the following, find the values of the capital letters by two different methods.

EXAMPLE

$$\frac{x + 21}{(3x - 1)(x + 5)} = \frac{A}{3x - 1} + \frac{B}{x + 5}$$

Solution

Method 1

$$x + 21 = A(x + 5) + B(3x - 1)$$
$$= (A + 3B)x + (5A - B)$$

1	$A + 3B = 1$	$A + \cancel{3B} = 1$	$4 + 3B = 1$
3	$5A - B = 21$	$15A - \cancel{3B} = 63$	$B = -1$

$$16A = 64$$
$$A = 4$$

Method 2

$$x + 21 = A(x + 5) + B(3x - 1)$$

$x = -5$: $16 = 0 + B(-16)$

$\qquad\qquad -1 = B$

$x = \frac{1}{3}$: $21\frac{1}{3} = A(5\frac{1}{3}) + 0$

$\qquad\qquad \frac{64}{3} = \frac{16}{3}A$

$\qquad\qquad 4 = A$

26. $\dfrac{8x + 29}{(2x - 1)(3x + 4)} = \dfrac{A}{2x - 1} + \dfrac{B}{3x + 4}$

27. $\dfrac{-30x + 8}{(3x - 7)(5x + 9)} = \dfrac{A}{3x - 7} + \dfrac{B}{5x + 9}$

28. $\dfrac{2x^2 + 16x - 12}{x(x + 3)(x - 4)} = \dfrac{A}{x} + \dfrac{B}{x + 3} + \dfrac{C}{x - 4}$

29. $\dfrac{5x^2 - 30x - 30}{x(x - 2)(x + 5)} = \dfrac{A}{x} + \dfrac{B}{x - 2} + \dfrac{C}{x + 5}$

30. $\dfrac{-6x^2 - 17x + 70}{(x - 2)(x + 2)(x - 3)} = \dfrac{A}{x - 2} + \dfrac{B}{x + 2} + \dfrac{C}{x - 3}$

31. $\dfrac{4x^2 + 6x + 26}{(x - 1)(x + 1)(x + 5)} = \dfrac{A}{x - 1} + \dfrac{B}{x + 1} + \dfrac{C}{x + 5}$

32. $\dfrac{4x^2 - 16x + 24}{(x - 1)(x - 2)(x - 3)} = \dfrac{A}{x - 1} + \dfrac{B}{x - 2} + \dfrac{C}{x - 3}$

33. $\dfrac{6x^3 + 17x^2 + x - 4}{x(x - 1)(x + 1)(x + 4)} = \dfrac{A}{x} + \dfrac{B}{x - 1} + \dfrac{C}{x + 1} + \dfrac{D}{x + 4}$

34. $\dfrac{2x^3 - 34x^2 + 28x + 40}{x(x - 2)(x + 2)(x - 5)} = \dfrac{A}{x} + \dfrac{B}{x - 2} + \dfrac{C}{x + 2} + \dfrac{D}{x - 5}$

35. $\dfrac{7x^4 + x^3 - 82x^2 + 6x + 36}{x(x - 2)(x + 2)(x - 3)(x + 3)} = \dfrac{A}{x} + \dfrac{B}{x - 2} + \dfrac{C}{x + 2} + \dfrac{D}{x - 3}$

$$+ \dfrac{E}{x + 3}$$

36. One serving of a cereal contains one less than 12 times as many grams of carbohydrate as of protein. If one serving contains 25 grams of protein and carbohydrate, how many grams of each kind does it contain?

37. A rice cereal is fortified with three more essential vitamins than a corn cereal. The number of vitamins in the rice cereal is two less than twice the number in the corn cereal. How many vitamins are there in each cereal?

38. A dance record has 16 different numbers consisting of Latins and non-Latins. There are two more non-Latins than Latins. How many dances of each kind are there on the record?

39. One year Professor Smiley gave 50 more than 10 times the number of lectures given by Professor Carberry. If Carberry had given 10 more lectures than he did give, Smiley would have given 5 times as many lectures as Carberry. How many lectures were given that year by each of these professors?

40. One serving of a rice cereal contains the same number of calories as one serving of a corn cereal and contains 20 less calories than one serving of a wheat cereal. If there are 310 calories in a bowl of cereal consisting of one serving of each kind, how many calories are there in one serving of each kind of cereal?

41. Professor Douglas spends 2 hours per class-hour preparing his lectures. Each week he spends 2 hours more in advising students than he spends in class, and each week he works 10 additional hours studying, etc. If he works 60 hours per week in class, preparing lectures, advising, studying, etc., how many hours per week does he spend in class?

42. During the first 56 years of their marriage, Mary and Roland purchased 3 more major appliances than cars. If they had purchased 3 more cars or major appliances during that time, they would have purchased either a car or a major appliance for each 2 years of their marriage. How many cars and how many major appliances did they purchase in that 56 years?

43. Fred plays the tuba or piccolo at hockey and football games, and he does not switch instruments at a game. One season there were 39 games. If he had played the tuba at 3 fewer games, he would have played the tuba at 5 times as many games as he played the piccolo. At how many games did he play each instrument that season?

44. Tom plays the French horn and piano and takes a course in French. Each week he spends 22 hours practicing horn and piano and studying French. He spends 3 times as much time studying French as practicing the piano, and each week he practices the horn 7 hours longer than the piano. How many hours does he spend each week practicing each instrument and studying French?

45. Marianne had wheat-ear pennies, new pennies, and silver bicentennial coins. She had twice as many pennies as silver coins, and she had 1 more new penny than silver coins. If she gave away half of her wheat-ears, she would have 18 coins left. How many coins of each kind did she have originally?

CHECK LIST FOR THE STUDENT

If you have learned the material in Section 1.10, you should be able to do the following:

1. Give an example of a linear equation in the two variables x and y and an example of a linear equation in the four variables A, B, C, and D.

2. Define the solution of a linear equation in n variables where n is an integer greater than 1.

3. Find several solutions of a single linear equation in two or more variables.

4. Give an example of a system of two simultaneous linear equations in the two variables x and y; give an example of a system of four simultaneous linear equations in the four variables A, B, C, and D.

5. Define the solution of a system of n simultaneous linear equations in n unknowns where n is an integer greater than 1.

6. State the three possibilities as far as the number of solutions of a system of n simultaneous linear equations in n unknowns is concerned.

7. Solve a system of two simultaneous linear equations in two unknowns by substitution and also by choice of multiplier and addition.

8. Solve a system of n simultaneous linear equations in n unknowns where $n \geq 3$ by systematically eliminating unknowns.

9. Find the numerators in a given partial-fraction decomposition of a given rational expression by equating coefficients of like powers and solving a system of simultaneous equations or by substituting convenient values.

10. Solve some verbal problems leading to systems of simultaneous linear equations.

SECTION QUIZ 1.10

1. Find the four solutions of the equation $3x - 2y + 4 = 0$ for which

$$x = 0, y = 0, x = 1, \text{ and } y = 1.$$

2. Solve the following system of equations by substitution.

$$3x - 4y = -2$$
$$x + 2y = -4$$

3. Solve the following system of equations by choice of multiplier and addition.

$$5x - 3y = 3$$
$$4x + 2y = 20$$

4. Solve the following system of equations.

$$3x - 4y + z + 8 = 0$$
$$2x + y - z + 3 = 0$$
$$x - 2y + 3z - 4 = 0$$

5. Assume that

$$\frac{7x^2 - 11x - 6}{x(x + 1)(x - 2)} = \frac{A}{x} + \frac{B}{x + 1} + \frac{C}{x - 2},$$

and find the values of A, B, and C by two different methods.

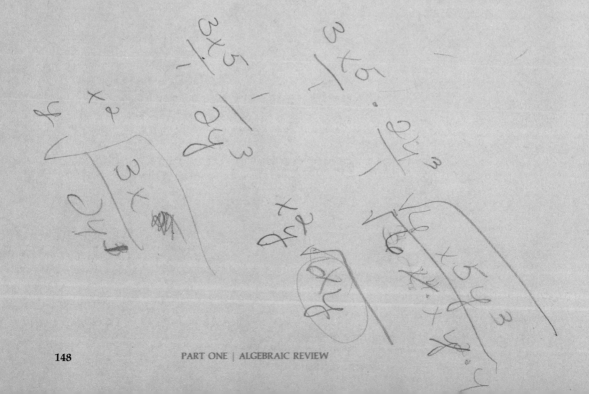

Part 1 Final Test

1. (a) Use braces and write an expression for the set of nonnegative integers.
   (b) Use set-builder notation and write an expression for the set of all positive real numbers. Make a graph of this set on a number line.

2. Let A be the set
$$\{-5, \; 0, \; |-5|, \; 0.5, \; \tfrac{1}{5}, \; \sqrt{5}, \; 2.\overline{51}, \; \pi, \; \tfrac{22}{7}, \; 3.14, \; \sqrt{2}, \; \tfrac{-4}{5}\}.$$
   (a) List all the rational numbers of A.
   (b) List all the irrational numbers of A.

3. Find the value of each of the following.
   (a) 3^5 (b) 3^1 (c) 3^0 (d) 3^{-5} (e) 0^3

4. In each of the following, if the expression is defined and is a real number, give its value. If the expression is not defined or is not real, write "not defined" or "not real."
   (a) $(\sqrt{3})^2$ (b) $\tfrac{3}{0}$ (c) $\sqrt{-16}$ (d) $\sqrt{(-5)^2}$ (e) $(\tfrac{2}{3})^{-1}$

5. Let x and y represent real numbers, and express each of the following in a simplified form containing no negative exponents. State any restrictions that must be placed on x and y so that the given expression will be defined.
   (a) $(x^4 y^{-2})(x^2 y^{-5})$ (b) $\left(\dfrac{3x^{-1} y^3}{x^{-3} y^5}\right)^2$

6. Find the value of each of the following.
   (a) $(-8)^{4/3}$ (b) $25^{1/2}$ (c) $8^{-1/3}$ (d) $\sqrt[3]{(27)^2}$ (e) $0^{1/3}$

7. Using the exponent laws, express each of the following in a form containing no radicals or negative exponents. Assume all variables are positive.
   (a) $(x^{6/5} y^{-3/10})^{1/3}(x^{-2/5} y^{11/10})$ (b) $\left(\dfrac{x^{-2} y^{-2}}{x^2 y^2}\right)^{-1/4}$

8. Assume that all variables are positive, and tell what belongs in the blank parentheses in each of the following.
   (a) $\sqrt{12x^7 y^3} = (\quad)\sqrt{3xy}$ (b) $\sqrt{\dfrac{3x^5}{2y^3}} = (\quad)\sqrt{6xy}$

9. (a) Let $P(x) = 3x^3 - 2x^2 + x - 7$ and $Q(x) = 5x^3 + 2x^2 - 2x - 3$.   Find $P(x) - Q(x)$.
   (b) Let $P(x) = 3x^2 - 2x + 1$ and $Q(x) = x - 7$. Find the product, $P(x) \times Q(x)$.

10. Find the quotient and remainder.
$$\frac{4x^3 - 2x^2 + x - 5}{x + 3}$$

11. Let $P(x) = 5x^2 - 3x - 7$. Express $P(x + h)$ as a polynomial in x and h.

12. If possible, express each of the following as a product of prime polynomials with integral coefficients. If such a factorization is not possible, write "not factorable."

(a) $25x^2 - 36$ (b) $25x^2 + 20x + 4$
(c) $27x^3 + 8$ (d) $25x^2 - 10x - 8$

13. Tell what number should be put in the parentheses to make the following expression become a perfect-square trinomial, and then factor that trinomial.

$$x^2 - 26x + (\quad)$$

14. (a) Find what belongs in the blank parentheses.

$$\frac{x - 2}{x + 3} = \frac{(\qquad)}{x^2 - 9}$$

(b) Reduce the following to lowest terms.

$$\frac{3x^3 - 12x}{x^4 - x^3 - 6x^2}$$

15. Express each of the following as a single fraction; if possible, reduce to lowest terms.

(a) $\dfrac{1}{x} - \dfrac{3}{x - 1} + \dfrac{5}{(x - 1)(x + 3)}$ (b) $\left(\dfrac{x^2 - 16}{x^2 + 3x - 4}\right) \div \left(\dfrac{x + 4}{x^2 - 2x + 1}\right)$

16. Express the following as a fraction having no fractions in the numerator or denominator.

$$\frac{\dfrac{3}{\sqrt{2x}} - \dfrac{\sqrt{2}}{x^2}}{\dfrac{\sqrt{2}}{x^3} + \dfrac{1}{\sqrt{2}}}$$

17. Solve each of the following equations.
(a) $2x - 5 = -7x + 2$ (b) $x(x + 2)(x - 5) = 0$

18. (a) Solve the equation $3x^2 - 5x = 0$ by factoring.
(b) Solve the equation $9x^2 - 30x + 23 = 0$.

Solve each of the following equations.
19. $x + \sqrt{3x + 15} - 1 = 0$

20. $\dfrac{x - 1}{x + 4} - \dfrac{4(x - 1)}{(x + 4)^2} - \dfrac{20}{(x + 4)^2} = 0$

Solve each of the following, and graph its set of solutions on a number line.
21. (a) $|5x - 6| = 9$ (b) $|x^2 - 3x + 1| = 1$

22. $-11x - 3 \geq 5 - 2x$

23. $2x + 5 < 3x - 1$

24. $|8x - 3| \leq 4$

25. Solve the following system of equations.

$$5x - 7y = -15$$
$$10x + 3y = \quad 4$$

Part 2 Elementary Functions and Analytic Geometry

Calculus consists mainly of the study of certain functions and of geometric or physical applications involving these functions. Calculus students must be familiar with the function concept and must be well enough grounded in analytic geometry so that they can readily sketch the graphs of commonly encountered functions and equations. Part 2 of this text provides an introduction to the function concept and to analytic geometry. It demonstrates the graphs of functions and equations that occur frequently in calculus.

2.1 Functions, Relations, Cartesian Coordinates, and Graphs

The purpose of this section is to introduce functions and relations, to introduce Cartesian coordinates, and to define and illustrate the graphs of functions, relations, or equations.

Any particular function is defined provided that the following two things are given:

1. A set called the **domain**

2. A *rule* assigning a single thing (usually a number) to each element in the domain

The set of all things assigned to the domain elements by the rule of assignment is called the **range** of the function. Functions are commonly denoted by lower-case letters such as f, g, h, and k.

EXAMPLE 1

Let f be the function such that

1. The domain is $\{-2, -1, 0, 1, 2, 3\}$.

2. The rule assigns to each domain element the square of that element.

(a) What number is assigned to each of the domain elements?

(b) What is the range of f?

Solution

(a) To -2 the rule assigns $(-2)^2 = 4$.
To -1 the rule assigns $(-1)^2 = 1$.
To 0 the rule assigns $0^2 = 0$.
To 1 the rule assigns $1^2 = 1$.
To 2 the rule assigns $2^2 = 4$.
To 3 the rule assigns $3^2 = 9$.

(b) The range of f is the set $\{4, 1, 0, 9\}$. Note that, as required of a function, the rule of this f assigns a single thing to each element in the domain, but it is permissible and possible for the rule of a function to assign the same single thing to two different domain elements. In this example the

Exercise 2.1, 1–5 .

rule assigns the single number 4 to both -2 and 2, and it also assigns the single number 1 to both -1 and 1. In stating the range, however, it is unnecessary for us to list 4 or 1 or any assignment more than once.

There are several common ways of giving the rule of assignment of any function. One way is to state the rule in words, as was done in Example 1. Another way is to use arrows to indicate the assignments. For example, the rule of f in Example 1 could be written

Rule of f:

$$-2 \rightarrow 4$$
$$-1 \rightarrow 1$$
$$0 \rightarrow 0$$
$$1 \rightarrow 1$$
$$2 \rightarrow 4$$
$$3 \rightarrow 9$$

If there are more than just a few elements in the domain, this way of giving the rule becomes cumbersome and impractical. In many cases it can be abbreviated by letting the letter x represent any element in the domain. To illustrate, the rule of f in Example 1 can be written

Rule of f:

$$x \rightarrow x^2$$

This can be condensed still more by writing

$$f: \qquad x \rightarrow x^2.$$

The most common and usually the most useful way of giving the rule of assignment of a function is to let the letter x represent any element in the domain and to let the symbol $f(x)$, read "f of x," represent the single thing assigned to x by the rule of f. Thus the rule of f in Example 1 could be written

$$f(x) = x^2.$$

Note that the symbol $f(x)$ does *not* mean that f is in some way multiplied by x. The symbol $f(x)$ stands for the single number assigned to x by the rule of f. The statement $f(x) = x^2$ can be read "f of x equals x squared" or "the number that f assigns to x is x squared."

EXAMPLE 2

Let g be the function such that

1. The domain is the set of all nonnegative real numbers.
2. $g(x) = x^2$.

Find $g(0)$, $g(5)$, $g(\frac{1}{2})$, $g(\sqrt{3})$, $g(a)$, and $g(a + h)$ where a is any nonnegative real number and h is positive.

Since $g(x) = x^2$, we can find $g(0)$ by replacing x by 0, we can find $g(5)$ by replacing x by 5, etc.

$$g(0) = 0^2 = 0$$
$$g(5) = 5^2 = 25$$
$$g(\tfrac{1}{2}) = (\tfrac{1}{2})^2 = \tfrac{1}{4}$$
$$g(\sqrt{3}) = (\sqrt{3})^2 = 3$$
$$g(a) = a^2$$
$$g(a + h) = (a + h)^2 = a^2 + 2ah + h^2$$

Exercise 2.1, 6–20

Let us look again at the function f of Example 1. The domain of that f is the set $\{-2, -1, 0, 1, 2, 3\}$, and the rule of f may be written

$$-2 \to 4$$
$$-1 \to 1$$
$$0 \to 0$$
$$1 \to 1$$
$$2 \to 4$$
$$3 \to 9.$$

Note that associated with this f there is a set of *ordered pairs,* namely the set $\{(-2, 4), (-1, 1), (0, 0), (1, 1), (2, 4), (3, 9)\}$. An **ordered pair** is a pair in which order of listing makes a difference so that the pair $(2, 4)$, for example, is not the same as the pair $(4, 2)$. The first member of each ordered pair is a domain element, and the second member is that single thing that is assigned to the first member by the rule of assignment of the function. The set of all first members is the domain, and the set of all second members is the range of the function. Note that since the rule of assignment of a function must assign exactly one single thing to each element in the domain, no two different ordered pairs have the same first element.

Just as there is a certain set of ordered pairs associated with the function f of Example 1, there is such a set of ordered pairs associated with any function. For this reason, many authors like to define a function as

A set of ordered pairs such that no two different ordered pairs have the same first member

or

A set of ordered pairs such that to each first member there corresponds a unique second member.

Certainly, if we are given the set of ordered pairs associated with some function, we are actually given the two essential things: domain and rule of assignment. The domain is simply the set of all first members, and the rule of assignment assigns to each domain element the second member of its ordered pair.

EXAMPLE 3

Does the set $\{(-1, -1), (1, -1), (4, -2), (9, -3)\}$ represent a function? If so, what is the domain, how can the rule of assignment be written with arrows, and what is the range?

Solution

The given set does represent a function; no two of its ordered pairs have the same first member.

The domain is the set $\{-1, 1, 4, 9\}$.

Using arrows, the rule of assignment can be written

$$-1 \rightarrow -1$$
$$1 \rightarrow -1$$
$$4 \rightarrow -2$$
$$9 \rightarrow -3.$$

Exercise 2.1, 21–25

The range is $\{-1, -2, -3\}$. (It is unnecessary to list the -1 twice.)

We have seen that every function must have a rule of assignment. For this reason, some authors prefer to define a function as

A correspondence or rule that associates exactly one thing with each element of a set called the domain.

But whether a function is defined as a certain set of ordered pairs or as a certain correspondence, the two essential ingredients, domain and rule of assignment, must be given.

Sometimes it is convenient not to state the domain explicitly but rather to give the domain according to the following convention.

Domain Convention

If the domain of a function is not explicitly stated, it is understood to be the set of all real numbers except those to which the rule of assignment would not assign a real number.

EXAMPLE 4

According to the domain convention, what is the domain of h such that $h(x) = \sqrt{x}$?

Solution

The domain of h is the set of all real numbers x except those for which $\sqrt{x}$ is not a real number. Since $\sqrt{x}$ is real for all real numbers x except negative numbers x, the domain of h is the set of all nonnegative numbers x. Otherwise written,

Domain of $h = \{x | x$ is a nonnegative real$\}$

or   Domain of $h = \{x | x \geq 0\}$.

EXAMPLE 5

According to the domain convention, what is the domain of k such that $k(x) = 1/(x - 5)$?

Solution

When $x = 5$, $1/(x - 5)$ is not defined so it is not a real number. Since $1/(x - 5)$ is a real number for all other real numbers x, the domain of k is the set of all real numbers except 5.

Exercise 2.1, 26–30

$$\text{Domain of } k = \{x \mid x \neq 5\}$$

The functions encountered in calculus usually have ranges that are sets of real numbers. Such functions are called *real-valued* functions. Functions whose domains *and* ranges are both sets of real numbers are called **real** functions. All the functions in our Examples 1—5 are real functions. We shall limit our discussion almost exclusively to real or real-valued functions.

Real functions can be represented geometrically on a *Cartesian coordinate* system. The configuration in Figure 2.1 is a Cartesian coordinate system. It consists of two mutually perpendicular number lines intersecting at their zero points, with the positive numbers increasing up and to the right. The common zero point is called the **origin**. The horizontal number line is usually called the **x-axis**, and the vertical number line is usually called the **y-axis**. The arrows indicate the directions in which the positive numbers increase. The two axes split the plane into four parts called **quadrants** labeled counterclockwise I, II, III, and IV, as shown in Figure 2.1. The axes themselves do not belong to any quadrant.

By means of a Cartesian coordinate system, any point in the plane is represented by exactly one ordered pair of real numbers, and conversely, any ordered pair of real numbers represents exactly one point in the plane. For example, the point located 2 units to the *right* of the y-axis and 3 units *below* the x-axis is represented by the pair $(2, -3)$. The pair $(-2, 3)$ represents the point located by beginning at the origin and moving 2 units *left* (the first member is negative) and 3 units *up* (the second member is positive). The points $(2, -3)$, $(-2, 3)$, and others are shown in Figure 2.1. To *plot* a point is to locate it by using its associated ordered pair of real numbers. To plot a point, begin at the origin and proceed right (if the first member is positive) or left (if the first member is negative) and then

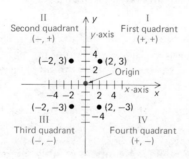

FIGURE 2.1 A Cartesian coordinate system

up (if the second member is positive) or down (if the second member is negative).

The members of the ordered pair of real numbers representing a point are called the **coordinates** of the point. The first member is called the **first coordinate**, the **x-coordinate**, or the **abscissa**. The second member is called the **second coordinate**, the **y-coordinate**, or the **ordinate**. The signs of the coordinates of all points in each of the four quadrants are shown in Figure 2.1.

Exercise 2.1, 31–40

A real function is represented on a Cartesian coordinate system by means of a *graph*. The **graph** of a function f is the set of all points (x, y) such that x is in the domain of f and $y = f(x)$. In other words, it is the set of all points associated with the ordered pairs of the function.

EXAMPLE 6

Let f be the function such that

1. The domain is $\{-2, -1, 0, 1, 2, 3\}$.
2. $f(x) = x^2$.

Sketch the graph of f. (Note that this is precisely the function f of Example 1.)

Solution

The graph of f will consist of the points $(-2, 4)$, $(-1, 1)$, $(0, 0)$, $(1, 1)$, $(2, 4)$, and $(3, 9)$. They are shown in Figure 2.2. Note that we do *not* connect these points because the graph of f consists *only* of these six points. Since there are only six elements in the domain, there are only six points on the graph.

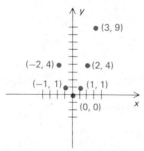

FIGURE 2.2 Graph of f: Domain of $f = \{-2, -1, 0, 1, 2, 3\}$
$$f(x) = x^2$$
(Graph of $y = x^2$ where $x = -2, -1, 0, 1, 2,$ or 3)

EXAMPLE 7

Let g be the function such that

1. The domain is $\{x \mid x \geq 0\}$.
2. $g(x) = x^2$.

Sketch the graph of g. (Note that this is precisely the function g of Example 2.)

Solution

Here the domain is the set of all nonnegative real numbers, so the graph consists of an unlimited number of points—one point for each nonnegative real number. Since we cannot possibly plot an unlimited number of points, we will just plot enough points so that our intuition will enable us to sketch the graph. Let us begin by making a "table of values" that lists the coordinates of points associated with several domain elements. *Any* nonnegative real number x will give us a point (x, y) on the graph, where $y = g(x)$. It will probably be easier for us to guess what the graph looks like, however, if we start with 0 and then take a few convenient successively larger nonnegative numbers.

x	0	$\frac{1}{2}$	1	$\frac{3}{2}$	2	$\frac{5}{2}$	3
$y = g(x) = x^2$	0	$\frac{1}{4}$	1	$\frac{9}{4}$	4	$\frac{25}{4}$	9

These points are plotted in Figure 2.3. If we were to plot more and more points on this graph, we would see that the points all appear to be strung together like beads on the curve shown in Figure 2.3. The curve terminates at $(0, 0)$, but it continues on indefinitely beyond $(3, 9)$. We shall see in Section 2.5 that our graph is the right half of a curve called a parabola.

Note that function g of this example has the same rule of assignment as function f of Example 6, but the two functions are different functions because their domains are different.

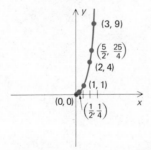

FIGURE 2.3 Graph of g: Domain of $g = \{x | x \geq 0\}$
$$g(x) = x^2$$
(Graph of $y = x^2$ where $x \geq 0$)

EXAMPLE 8

Let h be the function such that

1. The domain is $\{x | x \geq 0\}$.
2. $h(x) = -\sqrt{x}$.

Sketch the graph of h.

Solution

Since the domain consists of an unlimited number of elements, there will be an unlimited number of points on the graph. We will plot a few of them and then sketch the graph. First let us make a table of values for h.

x	0	$\frac{1}{4}$	1	$\frac{9}{4}$	4	9
$y = h(x) = -\sqrt{x}$	0	$-\frac{1}{2}$	-1	$-\frac{3}{2}$	-2	-3

The points whose coordinates are listed in our table of values are plotted in Figure 2.4. They all appear to be strung together like beads on the curve shown in that figure. The curve terminates at (0, 0) and continues on indefinitely beyond (9, -3). We shall see in Section 2.5 that the graph of h is the lower half of a parabola that opens to the right.

Note that the function h of Example 8 has the same domain as the function g of Example 7, but the functions h and g are different functions because their rules of assignment are different. Two functions are the same or equal if and only if both their domains and their rules of assignment are the same.

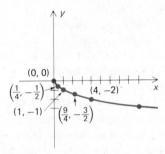

FIGURE 2.4   Graph of h: Domain of $h = \{x \mid x \geq 0\}$
$$h(x) = -\sqrt{x}$$
(Graph of $y = -\sqrt{x}$ where $x \geq 0$)

EXAMPLE 9

Let k be the function such that

1. The domain is the set of all real numbers.
2. $k(x) = x^2$.

Sketch the graph of k.

Let us make a table of values for k.

Solution

x	-3	-2	-1	0	2	3
$y = k(x) = x^2$	9	4	1	0	4	9

The points whose coordinates are listed in this table of values are plotted in Figure 2.5. They all appear to be strung together like beads on the

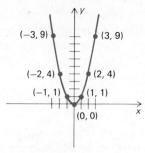

FIGURE 2.5 Graph of k: Domain of $k = \{x \mid x \text{ is a real number}\}$
$$k(x) = x^2$$
(Graph of $y = x^2$)

curve shown in Figure 2.5. In Section 2.5 we shall see that the graph of k is a parabola.

Compare the graphs of function f in Example 6, function g in Example 7, and function k in Example 9. The graph of k is the whole parabola; the graph of g is the right half of the parabola; the graph of f consists of exactly six points on the parabola.

The **graph of an equation in two variables,** say x and y, is the set of all points (x, y) whose coordinates satisfy the equation. Thus the graph of any function f is that part of the graph of the equation $y = f(x)$ for which x is in the domain of f. The graph of function f in Example 6 consists of that part of the graph of $y = x^2$ for which $x = -2, -1, 0, 1, 2,$ or 3; the graph of function g in Example 7 consists of that part of the graph of $y = x^2$ for which $x \geq 0$; the graph of function k in Example 9 is all of the graph of $y = x^2$. If the graph of an equation in x and y is the graph of a function, we say that the equation *defines* that function. The equation $y = x^2$ defines the function k in Example 9. In such a case the domain of the function is determined by the domain convention.

Exercise 2.1, 41–60

Any function is a special kind of a more general concept called a *relation.* A relation is defined provided that a set called the domain is given and provided that there is given a rule of assignment that assigns one *or more* things to each element in the domain. Thus a function is a special relation in which the rule of assignment assigns exactly *one* thing to each element in the domain. A **relation,** then, may be defined as any set of ordered pairs—without any restriction on the first members. Unless otherwise specified, we shall restrict our attention to relations in which the ordered pairs are pairs of real numbers.

The **graph of a relation** is the set of all points associated with its ordered pairs. The graph of any equation in two variables is the graph of a relation, so any equation in two variables may be thought of as representing a relation. For example, the graph of $y^2 = x$ is the graph of a relation. It is shown in Figure 2.6.

PART TWO | ELEMENTARY FUNCTIONS AND ANALYTIC GEOMETRY

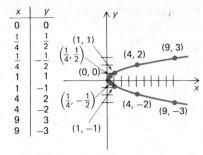

x	y
0	0
$\frac{1}{4}$	$\frac{1}{2}$
$\frac{1}{4}$	$-\frac{1}{2}$
1	1
1	-1
4	2
4	-2
9	3
9	-3

FIGURE 2.6 Graph of $y^2 = x$

Note that although the graph of $y^2 = x$ is the graph of a relation, it is not the graph of a function. Indeed, there exist distinct points on the graph with the same first coordinates. For example, the points (4, 2) and (4, -2) are both on the graph. Both of these points could not possibly be on the graph of a function, because the rule of assignment of a function assigns only a single number to each element in the domain; the rule of a function could not assign both 2 and -2 to the domain element 4.

There is an easy way to tell from looking at the graph whether or not the graph represents a function. Think of sweeping a vertical line across the graph. If the sweeping vertical line never intersects the graph in more than one point at a time, then the graph is the graph of a function. In other words, if no vertical line ever intersects the graph in more than one point, then the graph is the graph of a function. Applying the ''sweeping vertical line'' test in Figure 2.7, we see that the graph of $y^2 = x$ is not the graph of a function, while the graph of $y = x^2$ *is* the graph of a function.

Exercise 2.1, 61–80

The notation $f(x)$ that we have been using is used extensively in calculus. The most basic concept of differential calculus is usually defined

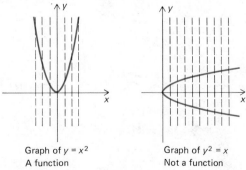

Graph of $y = x^2$
A function

Graph of $y^2 = x$
Not a function

FIGURE 2.7 Sweeping vertical line test

in terms of it. Beginning calculus students must use this notation as they work with expressions such as

$$\frac{f(x + h) - f(x)}{h}$$

for nonzero h, or expressions such as

$$\frac{f(x_1 + \Delta x) - f(x_1)}{\Delta x}$$

where x_1 represents a real number and where the symbol Δx, read "delta x," represents a nonzero real number. Specific examples of such expressions appear in the next example.

EXAMPLE 10

(a) If $f(x) = x^2$ and h is nonzero, simplify the expression

$$\frac{f(x + h) - f(x)}{h}.$$

(b) If $f(x) = x^2 - 3x + 4$, if x_1 represents a real number in the domain of f, and if the symbol Δx represents a nonzero real number, simplify the expression

$$\frac{f(x_1 + \Delta x) - f(x_1)}{\Delta x}.$$

Solution

(a) $f(x) = x^2$

$$f(x + h) = (x + h)^2$$

$$\frac{f(x + h) - f(x)}{h} = \frac{(x + h)^2 - x^2}{h}$$

$$= \frac{x^2 + 2xh + h^2 - x^2}{h}$$

$$= \frac{2xh + h^2}{h}$$

$$= 2x + h$$

(b) $f(x) = x^2 - 3x + 4$
$$f(x_1 + \Delta x) = (x_1 + \Delta x)^2 - 3(x_1 + \Delta x) + 4$$
(The symbol Δx is printed in color to emphasize that it represents one single number.)

$$f(x_1) = x_1^2 - 3x_1 + 4$$

$$\frac{f(x_1 + \Delta x) - f(x_1)}{\Delta x} = \frac{(x_1 + \Delta x)^2 - 3(x_1 + \Delta x) + 4 - (x_1^2 - 3x_1 + 4)}{\Delta x}$$

$$= \frac{x_1^2 + 2x_1\,\Delta x + (\Delta x)^2 - 3x_1 - 3\,\Delta x + 4 - x_1^2 + 3x_1 - 4}{\Delta x}$$

$$= \frac{2x_1\,\Delta x + (\Delta x)^2 - 3\,\Delta x}{\Delta x}$$

$$= 2x_1 + \Delta x - 3$$

Exercise 2.1, 81–110

EXERCISE 2.1 FUNCTIONS, RELATIONS, CARTESIAN COORDINATES, AND GRAPHS

▶ In each of the following, tell what number is assigned to each of the domain elements, and give the range of the function.

EXAMPLE

Domain of $f = \{-2, 1, \frac{1}{3}\}$

$$f(x) = \frac{1}{x}$$

Solution

To -2 the rule assigns $\frac{-1}{2}$.

To 1 the rule assigns $\frac{1}{1} = 1$.

To $\frac{1}{3}$ the rule assigns $\dfrac{1}{\frac{1}{3}} = 3$.

The range is $\{\frac{-1}{2}, 1, 3\}$.

1. Domain of $f = \{-3, \frac{-1}{2}, 0, \frac{1}{2}, 3\}$
   $f(x) = x^2 + 1$

2. Domain of $g = \{-3, \frac{-1}{2}, 0, \frac{1}{2}, 3\}$
   $g(x) = x^2 + 2x - 1$

3. Domain of $h = \{0, 1, \frac{9}{4}, 4, 9\}$
   $h(x) = \sqrt{x}$

4. Domain of $k = \{-3, -2, -1, 0, 1, 2\}$
   $k(x) = x^3$

5. Domain of $f = \{\frac{-1}{8}, -1, 0, 1, 8\}$
   $f(x) = \sqrt[3]{x}$

▶ In each of the following, let a and $a + h$ be in the domain of f.

EXAMPLE

If $f(x) = 3x^2 - 2x + 1$, find $f(0)$, $f(-1)$, $f(4)$, $f(a)$, and $f(a + h)$.

Solution

$$f(0) = 3(0)^2 - 2(0) + 1 = 1$$
$$f(-1) = 3(-1)^2 - 2(-1) + 1 = 3 + 2 + 1 = 6$$
$$f(4) = 3(4)^2 - 2(4) + 1 = 48 - 8 + 1 = 41$$
$$f(a) = 3a^2 - 2a + 1$$
$$f(a + h) = 3(a + h)^2 - 2(a + h) + 1$$
$$= 3(a^2 + 2ah + h^2) - 2a - 2h + 1$$
$$= 3a^2 + 6ah + 3h^2 - 2a - 2h + 1$$

6. If $f(x) = 2x^2 - 4x + 5$, find $f(0)$, $f(-1)$, $f(3)$, $f(a)$, and $f(a + h)$.

7. If $f(x) = x^3 - x^2 + 1$, find $f(0)$, $f(-1)$, $f(1)$, $f(a)$, and $f(a + h)$.

8. If $f(x) = \sqrt{2x} + 3$, find $f(2)$, $f(8)$, $f(0)$, $f(a)$, and $f(a + h)$.

9. If $f(x) = \dfrac{1}{2x} + 1$, find $f(-1)$, $f(3)$, $f(\frac{1}{2})$, $f(a)$, and $f(a + h)$.

10. If $f(x) = x^{1/3}$, find $f(0)$, $f(\frac{1}{27})$, $f(-8)$, $f(a)$, and $f(a + h)$.

▶ In each of the following let Δx represent a single number, let x_1 represent a number, and let $x_1 + \Delta x$ be a number in the domain of f. Express $f(x_1 + \Delta x)$ as a polynomial in x_1 and Δx.

EXAMPLE

$$f(x) = 3x^2 - 2x + 1$$

Solution

$$f(x_1 + \Delta x) = 3(x_1 + \Delta x)^2 - 2(x_1 + \Delta x) + 1$$
$$= 3(x_1^2 + 2x_1 \Delta x + (\Delta x)^2) - 2x_1 - 2 \Delta x + 1$$
$$= 3x_1^2 + 6x_1 \Delta x + 3(\Delta x)^2 - 2x_1 - 2 \Delta x + 1$$

11. $f(x) = x^2 - 5x + 3$ **12.** $f(x) = 2x^2 + 3x - 4$

13. $f(x) = 4x^2 - 2x + 1$ **14.** $f(x) = 5x^2 - x + 7$

15. $f(x) = 2x^3$

▶ In each of the following let $x + h$ be in the domain of f, and express $f(x + h)$ as a polynomial in x and h.

EXAMPLE

$$f(x) = 3x^2 - 2x + 1$$

Solution

$$f(x + h) = 3(x + h)^2 - 2(x + h) + 1$$
$$= 3(x^2 + 2xh + h^2) - 2x - 2h + 1$$
$$= 3x^2 + 6xh + 3h^2 - 2x - 2h + 1$$

16. $f(x) = x^2 - 5x + 3$ **17.** $f(x) = 2x^2 + 3x - 4$

18. $f(x) = 4x^2 - 2x + 1$ **19.** $f(x) = 5x^2 - x + 7$

20. $f(x) = 2x^3$

▶ In each of the following, state whether or not the given set represents a function. If it does represent a function, give the domain, show how the rule of assignment can be written with arrows, and give the range.

EXAMPLE $\{(\frac{-1}{2}, 0), (\frac{1}{2}, 1), (\frac{1}{3}, 1), (\frac{3}{2}, 2)\}$

Solution   The set represents a function.

Domain $= \{\frac{-1}{2}, \frac{1}{2}, \frac{1}{3}, \frac{3}{2}\}$

Rule $\frac{-1}{2} \to 0$

$\qquad \frac{1}{2} \to 1$

$\qquad \frac{1}{3} \to 1$

$\qquad \frac{3}{2} \to 2$

Range $= \{0, 1, 2\}$

21. $\{(\frac{-1}{2}, -1), (\frac{1}{2}, 0), (\frac{1}{3}, 0), (\frac{3}{2}, 1)\}$

22. $\{(-1, \frac{-1}{2}), (0, \frac{1}{2}), (0, \frac{1}{3}), (1, \frac{3}{2})\}$

23. $\{(-2, -8), (0, 0), (-1, -1), (2, 8)\}$

24. $\{(-2, 2), (2, 2), (-3, 3), (3, 3)\}$

25. $\{(16, 4), (16, -4), (9, 3), (25, 5)\}$

▶ According to the domain convention, what is the domain of the function in each of the following?

EXAMPLES   (a) $f(x) = 1/x$ (b) $g(x) = \sqrt{x - 3}$

Solutions   (a) Domain of $f = \{x | x \neq 0\}$ (b) Domain of $g = \{x | x \geq 3\}$

26. $f(x) = \dfrac{1}{x - 1}$ **27.** $g(x) = \sqrt{x - 1}$

28. $f(x) = \dfrac{1}{x^2}$ **29.** $g(x) = \sqrt{x + 2}$

30. $f(x) = \dfrac{1}{x - 2} + \sqrt{x + 1}$

▶ In each of the following, plot and label the points on a Cartesian coordinate system.

EXAMPLE

Solution

$(-2, 1), (2, -1), (2, 1), (-2, -1)$

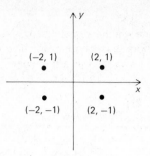

31. $(3, 5), (3, -5), (-3, 5), (-3, -5), (0, 0)$

32. $(0, 1), (0, -1), (1, 0), (-1, 0)$

33. $(0, 0), (1, 1), (2, 2), (3, 3), (-1, -1), (-2, -2), (-3, -3)$

34. $(0, 3), (1, 3), (2, 3), (3, 3), (-1, 3), (-2, 3)$

35. $(0, 0), (1, 1), (2, 2), (3, 3), (-1, 1), (-2, 2), (-3, 3)$

▶ In each of the following give the coordinates of the point which is located as described.

EXAMPLE

Solution

3 units left of *y*-axis, 4 units above *x*-axis

$(-3, 4)$

36. 3 units right of *y*-axis, 4 units below *x*-axis

37. 5 units left of *y*-axis, 2 units below *x*-axis

38. 4 units right of *y*-axis, 2 units above *x*-axis

39. 2 units left of *y*-axis, on *x*-axis

40. on *y*-axis, 3 units below *x*-axis

▶ In each of the following, sketch the graph of the function.

EXAMPLE

(1) Domain of $f = \{x \mid x > -2\}$
(2) $f(x) = x^2$

Solution

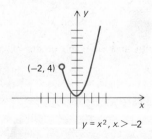

$y = x^2, x > -2$

The graph of f is the graph of $y = x^2$ where $x > -2$. An open circle is placed at the endpoint $(-2, 4)$ to show that it is not included on the graph.

41. (1) Domain of $f = \{-3, -2, -1, 0, 1\}$
    (2) $f(x = x^2$

42. (1) Domain of $f = \{x \mid x \le 0\}$
    (2) $f(x) = x^2$

43. (1) Domain of $f = \{x \mid x < 0\}$
    (2) $f(x) = x^2$

44. (1) Domain of $f = \{x \mid x > 0\}$
    (2) $f(x) = x^2$

45. (1) Domain of $f = \{x \mid x \ge -1\}$
    (2) $f(x) = x^2$

46. (1) Domain of $f = \{x \mid x < -1\}$
    (2) $f(x) = x^2$

47. (1) Domain of $f = \{x \mid -1 \le x \le 2\}$
    (2) $f(x) = x^2$

48. (1) Domain of $f = \{x \mid -2 \le x < 3\}$
    (2) $f(x) = x^2$

49. (1) Domain of $f = \{0, \frac{9}{4}, 4\}$
    (2) $f(x) = \sqrt{x}$

50. (1) Domain of $f = \{x \mid x \ge 0\}$
    (2) $f(x) = \sqrt{x}$

51. (1) Domain of $f = \{x \mid x \ge 1\}$
    (2) $f(x) = -\sqrt{x}$

52. (1) Domain of $f = \{x \mid 0 \le x \le 9\}$
    (2) $f(x) = -\sqrt{x}$

53. (1) Domain of $f = \{x \mid 0 \le x \le 9\}$
    (2) $f(x) = \sqrt{x}$

54. (1) Domain of $f = \{x \mid 1 \le x \le 9\}$
    (2) $f(x) = \sqrt{x}$

55. (1) Domain of $f = \{x \mid 1 \le x \le 4\}$
    (2) $f(x) = -\sqrt{x}$

In each of the following, complete the table of values and plot the points on a Cartesian coordinate system.

EXAMPLE

x	$y = f(x) = \lvert x \rvert$
-2	
-1	
0	
1	
2	

Solution

x	$y = f(x) = \lvert x \rvert$
-2	2
-1	1
0	0
1	1
2	2

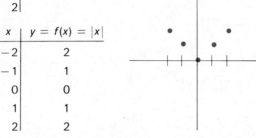

56.

x	$y = f(x) = \lvert x - 1 \rvert$
-1	
0	
1	
2	
3	

57.

x	$y = f(x) = \lvert x + 1 \rvert$
-3	
-2	
-1	
0	
1	

58.

x	$y = f(x) = x$
-3	
-2	
-1	
0	
1	
2	
3	

59.

x	$y = f(x) = x + 2$
-3	
-2	
-1	
0	
1	
2	
3	

60.

x	$y = f(x) = x^3$
-3	
-2	
-1	
0	
1	
2	
3	

▶ Sketch the graph of each of the following relations and state whether or not the relation is a function.

EXAMPLE $\{(x, y) \mid y^2 = x \text{ and } 0 \le x \le 4\}$

Solution Not a function.

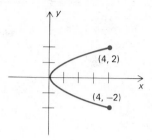

61. $\{(x, y) \mid y^2 = x \text{ and } 0 \le x \le 9\}$

62. $\{(x, y) \mid y^2 = x \text{ and } 0 \le x < 1\}$

63. $\{(x, y) \mid y = x^2 \text{ and } 0 \le x \le 1\}$

64. $\{(x, y) \mid y = x^2 \text{ and } -1 \le x \le 1\}$

65. $\{(9, 3), (9, -3), (4, 2), (1, 1)\}$

▶ In each of the following, tell whether or not the graph represents a function.

EXAMPLE

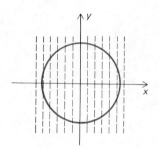

Solution Not a function. (There exists a vertical line that intersects the graph in more than one point. In fact, there exist an unlimited number of such vertical lines.)

66.

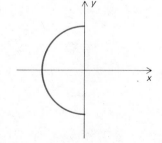

67.

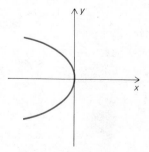

68.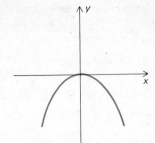

69.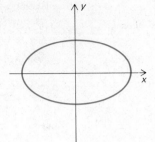

70.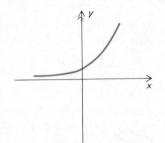

71.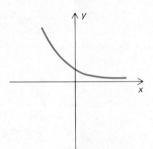

72.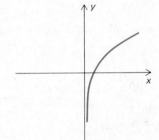

73.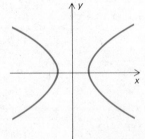

74.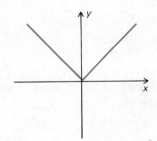

75.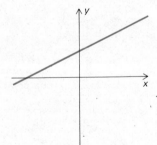

76.

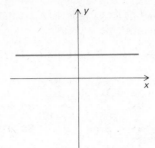

77.

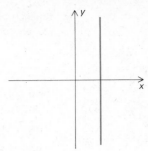

78.

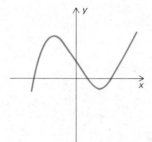

79.

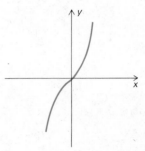

80.

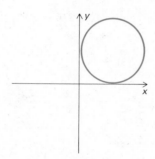

The following exercises provide practice with notation found in some of the most widely used calculus texts.

▶ In each of the following, let h be nonzero and write the expression

$$\frac{f(x + h) - f(x)}{h}$$

as a polynomial in x and h.

81. $f(x) = 3x^2 - 2x + 7$ **82.** $f(x) = x^3$

83. $f(x) = 4x^2 + 3x - 8$ **84.** $f(x) = x^3 - 2x$

85. $f(x) = 7x^2 - 4x + 1$

► In each of the following, let x_1 represent a real number in the domain of f, let Δx be nonzero, and write the expression

$$\frac{f(x_1 + \Delta x) - f(x_1)}{\Delta x}$$

as a polynomial in x_1 and Δx.

86. $f(x) = 3x^2 - 2x + 7$ **87.** $f(x) = x^3$

88. $f(x) = 4x^2 + 3x - 8$ **89.** $f(x) = x^3 - 2x$

90. $f(x) = 7x^2 - 4x + 1$

► If $x \neq a$, in each of the following write

$$\frac{f(x) - f(a)}{x - a}$$

as a polynomial in x and a.

EXAMPLE $f(x) = 3x^2 - 2x$

Solution

$$\frac{f(x) - f(a)}{x - a} = \frac{3x^2 - 2x - 3a^2 + 2a}{x - a} = \frac{3(x^2 - a^2) - 2(x - a)}{x - a}$$

$$= \frac{3(x - a)(x + a) - 2(x - a)}{x - a} = 3(x + a) - 2$$

$$= 3x + 3a - 2$$

91. $f(x) = 3x^2 - 2x + 7$ **92.** $f(x) = x^3$

93. $f(x) = 4x^2 + 3x - 8$ **94.** $f(x) = x^3 - 2x$

95. $f(x) = 7x^2 - 4x + 1$

► If $x \neq x_1$, in each of the following write

$$\frac{f(x_1) - f(x)}{x_1 - x}$$

as a polynomial in x_1 and x.

EXAMPLE $f(x) = 3x^2 - 2x$

Solution

$$\frac{f(x_1) - f(x)}{x_1 - x} = \frac{3x_1^2 - 2x_1 - 3x^2 + 2x}{x_1 - x} = \frac{3(x_1^2 - x^2) - 2(x_1 - x)}{x_1 - x}$$

$$= \frac{3(x_1 - x)(x_1 + x) - 2(x_1 - x)}{x_1 - x} = 3(x_1 + x) - 2$$

$$= 3x_1 + 3x - 2$$

96. $f(x) = 3x^2 - 2x + 7$ **97.** $f(x) = x^3$

98. $f(x) = 4x^2 + 3x - 8$ **99.** $f(x) = x^3 - 2x$

100. $f(x) = 7x^2 - 4x + 1$

▶ If $h \neq 0$, in each of the following write

$$\frac{f(x_1 + h) - f(x_1)}{h}$$

as a polynomial in x_1 and h.

EXAMPLE $f(x) = 3x^2 - 2x$

Solution

$$\frac{f(x_1 + h) - f(x_1)}{h} = \frac{3(x_1 + h)^2 - 2(x_1 + h) - 3x_1^2 + 2x_1}{h}$$

$$= \frac{3x_1^2 + 6x_1 h + 3h^2 - 2x_1 - 2h - 3x_1^2 + 2x_1}{h}$$

$$= 6x_1 + 3h - 2$$

101. $f(x) = 3x^2 - 2x + 7$ **102.** $f(x) = x^3$

103. $f(x) = 4x^2 + 3x - 8$ **104.** $f(x) = x^3 - 2x$

105. $f(x) = 7x^2 - 4x + 1$

▶ If $r \neq x$, in each of the following write

$$\frac{f(r) - f(x)}{r - x}$$

as a polynomial in r and x.

EXAMPLE $f(x) = 3x^2 - 2x$

Solution

$$\frac{f(r) - f(x)}{r - x} = \frac{3r^2 - 2r - 3x^2 + 2x}{r - x} = \frac{3(r^2 - x^2) - 2(r - x)}{r - x}$$

$$= \frac{3(r - x)(r + x) - 2(r - x)}{r - x} = 3(r + x) - 2$$

$$= 3r + 3x - 2$$

106. $f(x) = 3x^2 - 2x + 7$ **107.** $f(x) = x^3$

108. $f(x) = 4x^2 + 3x - 8$ **109.** $f(x) = x^3 - 2x$

110. $f(x) = 7x^2 - 4x + 1$

CHECK LIST FOR THE STUDENT

If you have learned the material in Section 2.1, you should be able to do the following:

1. State the two things that must be given in the definition of any particular function.

2. State the definition of the range of a function.

3. Illustrate several common ways of giving the rule of assignment of a function.

4. Find $f(a)$, if $f(x)$ is given and a is in the domain of f.

5. Simplify expressions such as

$$\frac{f(x + h) - f(x)}{h} \quad \text{or} \quad \frac{f(x_1 + \Delta x) - f(x_1)}{\Delta x}$$

where $f(x)$ is a specific second- or third-degree polynomial.

6. Recognize that there is a set of ordered pairs associated with any function.

7. Identify the domain and range and give the rule of assignment of a function that is given as a set of ordered pairs.

8. Recognize that some authors define a function as a certain set of ordered pairs.

9. Recognize that some authors define a function as a certain correspondence or rule.

10. State and use the domain convention for determining the domain of a function whose domain is not explicitly stated.

11. State the definition of a real function.

12. Plot points on a Cartesian coordinate system.

13. State the names of the coordinates of a point on a Cartesian coordinate system.

14. State the definition of the graph of a real function f.

15. Sketch the graph of any real function f with a given domain and with $f(x) = x^2$.

16. Make a table of values for a given real function.

17. Sketch the graph of any real function f with a given domain and with $f(x) = \sqrt{x}$ or with $f(x) = -\sqrt{x}$.

18. State the provisions under which two functions are equal.

19. State the definition of the graph of an equation in two variables.

20. State the definition of a relation.

21. Tell from looking at the graph of a relation whether or not it represents a function.

SECTION QUIZ 2.1

1. Let f be the function whose domain is $\{-3, 0, 1, 2\}$ and which is such that $f(x) = x^2 + 3$. Find the range of f.

2. If 0, 2, and real number a are in the domain of f, and if $f(x) = x^2 - 3x + 5$, find $f(0)$, $f(2)$, and $f(a)$.

3. State whether or not the following set represents a function. If it does represent a function, give the domain, show how the rule of assignment can be written with arrows, and give the range.

$$\{(\tfrac{1}{3}, -4), (1, -2), (0, -5), (2, -2)\}$$

4. According to the domain convention, what is the domain of the function f in each of the following?
   (a) $f(x) = 1/(x + 5)$ (b) $f(x) = \sqrt{x + 5}$

5. (a) What is the sign of the first coordinate of a point in the third quadrant, and what is the sign of the second coordinate of a point in the fourth quadrant? No
   (b) Let $f(x) = |x|$, let $y = f(x)$, and plot the points (x, y) on the graph of f for $x = -3, -1, 0, 1,$ and 3.

6. Let f be the function whose domain is $\{-1, 0, 1, 2, 3\}$ and which is such that $f(x) = x^2$. Sketch the graph of f.

7. Let f be the function whose domain is $\{x \mid x \geq -1\}$ and which is such that $f(x) = x^2$. Sketch the graph of f.

8. Sketch the graph of each of the following relations, and state whether or not the relation is a function.
   (a) $\{(x, y) \mid y = x^2 \quad \text{and} \quad 0 \leq x \leq 4\}$
   (b) $\{(x, y) \mid y^2 = x \quad \text{and} \quad 0 \leq x \leq 9\}$

9. Tell whether or not each of the following graphs represents a function.

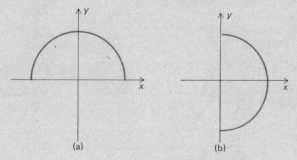

(a)   (b)

10. Let $f(x) = x^2 - 5x + 2$, let h be nonzero, and let $x + h$ be in the domain of f. Express

$$\frac{f(x + h) - f(x)}{h}$$

as a polynomial in x and h.

2.2 Lines; Linear Functions

The purpose of this section is to show that a given linear equation in one or two variables algebraically represents a line, to show how an equation can be found to represent a given line, to define the slope of a line, and to define and graph linear functions.

We have said (Section 2.1) that the graph of an equation in the two variables x and y is the set of all points (x, y) whose coordinates satisfy the equation. The graph of any linear equation in x and y is a line that is not parallel to either axis. Such a line is called an **oblique** line. For example, the graph of

$$3x + 2y - 12 = 0$$

is an oblique line. In general, if A, B, and C are real numbers with A and B not zero, it will be proved later in this section that the graph of

$$Ax + By + C = 0$$

is an oblique line. We say that $Ax + By + C = 0$ is an equation *of* the line that is its graph, and sometimes we refer to the equation itself as the line.

The graph of $3x + 2y - 12 = 0$ is shown in Figure 2.8. Since two points determine a line, to sketch the graph of a linear equation in two variables we need only to plot two points whose coordinates satisfy the equation and then draw the line through them. Any number of points on the line may be found simply by assigning convenient values to one of the variables and then solving for the other. In the table of values in Figure 2.8, first the value 0 was assigned to x, and y was found to be 6.

$$3 \cdot 0 + 2y - 12 = 0$$
$$2y = 12$$
$$y = 6$$

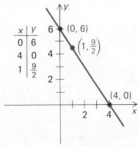

FIGURE 2.8   Graph of $3x + 2y - 12 = 0$

Next the value 0 was assigned to y, and x was found to be 4.

$$3x + 2 \cdot 0 - 12 = 0$$
$$3x = 12$$
$$x = 4$$

Finally, as a "check," the value 1 was assigned to x, and y was found to be $\frac{9}{2}$.

$$3 \cdot 1 + 2y - 12 = 0$$
$$2y = 9$$
$$y = \frac{9}{2}$$

If the check point $(1, \frac{9}{2})$ had not seemed to lie on the line drawn through the first two points plotted, we would have gone back and checked our computations.

EXAMPLE 1

Find the point of intersection of the lines $2x - 3y + 10 = 0$ and $3x - y + 1 = 0$. Sketch the graphs of the lines, and label the point of intersection.

Solution

To find the point of intersection, we must find the point whose coordinates satisfy both equations. We can do this by solving the system of simultaneous equations consisting of the two given equations. Using the method of choice of multiplier and addition that was presented in Section 1.10, let us first choose multipliers and eliminate y.

$$
\begin{array}{r|l}
1 & 2x - 3y + 10 = 0 \\
-3 & 3x - y + 1 = 0
\end{array}
$$

$$
\begin{array}{r}
2x - 3y + 10 = 0 \\
-9x + 3y - 3 = 0 \\
\hline
-7x + 7 = 0
\end{array}
$$

$$- 7x = -7$$
$$x = 1$$

Substituting 1 for x in the first equation and solving for y, we find that $y = 4$.

$$2 - 3y + 10 = 0$$
$$- 3y = -12$$
$$y = 4$$

Check

$$2x - 3y + 10 = 0 \qquad\qquad 3x - y + 1 = 0$$
$$2(1) - 3(4) + 10 = 0 \quad \text{True} \qquad 3(1) - 4 + 1 = 0 \quad \text{True}$$

The solution is $x = 1$, $y = 4$. In other words, the coordinates of the point (1, 4) satisfy both equations, so the point (1, 4) lies on both lines. The point of intersection is (1, 4). It is shown with the lines in Figure 2.9.

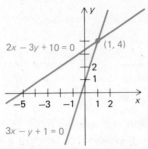

FIGURE 2.9 Graphs of $2x - 3y + 10 = 0$ and $3x - y + 1 = 0$

The points in which a line intersects the coordinate axes are called its **intercepts.** The point in which a line intersects the x-axis is called its **x-intercept.** The point in which a line intersects the y-axis is called its **y-intercept.** To find the x-intercept of a line whose equation is given, let $y = 0$ and solve for x. To find the y-intercept, let $x = 0$ and solve for y. This was done in sketching the graph of the equation $3x + 2y - 12 = 0$ in Figure 2.8. The x-intercept of that line is (4, 0), and the y-intercept is (0, 6). Sometimes just the first coordinate of the x-intercept is called the x-intercept, and sometimes just the second coordinate of the y-intercept is called the y-intercept. We might say that the x-intercept of $3x + 2y - 12 = 0$ is 4 and the y-intercept of $3x + 2y - 12 = 0$ is 6. In any case, it should be clear from context whether the word intercept refers to the point itself, the coordinates of the point, or just the pertinent coordinate of the point.

Exercise 2.2, 1–10

The graph of any equation in the one variable x is the set of all points (x, y) whose coordinates satisfy the equation. Since no y appears in the equation, any point whose x-coordinate satisfies the equation will be on the graph, and the graph consists of all such points. The graph of a *linear* equation in x is a line parallel to the y-axis. For example, the graph of the equation $3x - 6 = 0$ is shown in Figure 2.10. It is the line every point of which has x-coordinate 2.

$$3x - 6 = 0$$
$$3x = 6$$
$$x = 2$$

In other words, the graph of $3x - 6 = 0$ is the graph of the equation $x = 2$, and the graph of $x = 2$ is a line parallel to the y-axis and two units to the right of it.

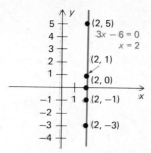

FIGURE 2.10 Graph of $3x - 6 = 0$
(Graph of $x = 2$)

In general, if a is a nonzero real number,

$x = a$ represents a line parallel to the y-axis.

$x = 0$ represents the y-axis.

Note that the graph of $x = a$ is not just a single point. It is the *line* every point of which has x-coordinate a.

EXAMPLE 2

Sketch the graph of the equation $x = -2$. What are its intercepts?

Solution

The equation $x = -2$ represents a line parallel to the y-axis and two units to the left of it. It is shown in Figure 2.11. Its x-intercept is $(-2, 0)$. It has no y-intercept.

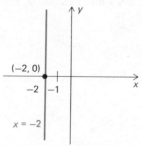

FIGURE 2.11 Graph of $x = -2$

The graph of any equation in the one variable y is the set of all points (x, y) whose coordinates satisfy the equation. Since no x appears in the equation, any point whose y-coordinate satisfies the equation will be on the graph. The graph of a *linear* equation in y is a line parallel to the x-axis. For example, the graph of $2y + 8 = 0$ is the line every point of which has y-coordinate -4.

$$2y + 8 = 0$$
$$2y = -8$$
$$y = -4$$

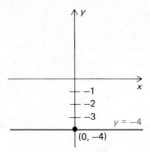

FIGURE 2.12   Graph of $2y + 8 = 0$
(Graph of $y = -4$)

That is, the graph of $2y + 8 = 0$ is the graph of $y = -4$. It is shown in Figure 2.12.

In general, if b is a nonzero real number,

$y = b$ represents a line parallel to the x-axis.

$y = 0$ represents the x-axis.

Note that the graph of $y = b$ is *not* just a point. It is the line every point of which has y-coordinate b.

EXAMPLE 3

Solution

Exercise 2.2, 11–15

Sketch the graph of the equation $y = 4$. What are its intercepts?

The graph of $y = 4$ is the line parallel to the x-axis and four units above it. It is shown in Figure 2.13. Its y-intercept is (0, 4). It has no x-intercept.

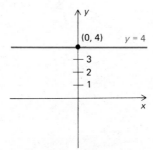

FIGURE 2.13   Graph of $y = 4$

We have noted that any given linear equation in one or two variables represents a line, and we have seen how to sketch the graph of that line. Conversely, if we are given a line, we can find a linear equation in one or two variables to represent that line. Sometimes a line is given by giving two points on it, and we shall see later in this section how to find an equation for a line if we are given two of its points. But first let us define the *slope* of a line because sometimes a line is given by giving just one of its points together with the number called its slope.

The slope of a segment of a nonvertical line is defined in terms of the coordinates of its endpoints. The **slope** of the nonvertical line segment between point P_1 with coordinates (x_1, y_1) and point P_2 with coordinates (x_2, y_2) is

$$\frac{y_2 - y_1}{x_2 - x_1}.$$

To get a feeling for the geometric meaning of slope, consider Figures 2.14 and 2.15. In those figures lines have been drawn through P_1 and P_2 parallel to the x- and y-axes, respectively. In each figure the point of intersection of these drawn parallels has been labeled M and has coordinates (x_2, y_1). In each figure, if we move along the line from P_1 to P_2, the numerator of the slope formula indicates how many units we move in the y-direction (positive, if up; negative, if down), and the denominator indicates how many units we move in the x-direction (positive, if right; negative, if left).

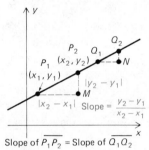

Slope of $\overline{P_1 P_2}$ = Slope of $\overline{Q_1 Q_2}$

FIGURE 2.14   Slope of line segment

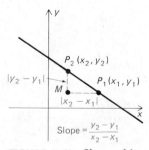

FIGURE 2.15   Slope of line segment

It can be shown by constructing similar right triangles that the slope of any segment of an oblique line is the same as the slope of any other segment of it. Figure 2.14 suggests how this can be shown. In that figure let

Q_1 and Q_2 be the endpoints of some other segment of the line. Note that the triangles P_1MP_2 and Q_1NQ_2 are similar triangles, so the ratios of the lengths of corresponding sides are equal and

$$\frac{MP_2}{P_1M} = \frac{NQ_2}{Q_1N}.$$

The slope of the segment with endpoints Q_1 and Q_2 must be the same as the slope of the segment with endpoints P_1 and P_2. Similar arguments show that

The slope of any segment of a nonvertical line is the same as the slope of any other segment of it.

Consequently, we define the **slope of a nonvertical line** as the slope of any segment of the line. The slope of a line is usually denoted by the lower-case letter m. Thus if (x_1, y_1) and (x_2, y_2) are any two points on a nonvertical line, we may say that

$$m = \frac{y_2 - y_1}{x_2 - x_1}. \qquad \text{Slope of nonvertical line containing } (x_1, y_1) \text{ and } (x_2, y_2)$$

The slope of a vertical line is not defined. Indeed, if we tried to apply the slope formula to two points on a vertical line, we would obtain a zero in the denominator, because all points on any vertical line have the same x-coordinate and hence the $x_2 - x_1$ would be zero.

The slope of any horizontal line is zero. For example, the slope of the horizontal line $y = 3$ is zero. This is consistent with the slope formula because the y-coordinates of any two distinct points on $y = 3$ are both 3 so that the numerator $y_2 - y_1$ is $3 - 3 = 0$. In particular, the points $(1, 3)$ and $(5, 3)$ are both on the line $y = 3$. Using the slope formula, the slope of $y = 3$ is

$$\frac{3 - 3}{5 - 1} = \frac{0}{4} = 0.$$

EXAMPLE 4

Show that $(3, 1)$ and $(8, 4)$ are both points on the line $3x - 5y - 4 = 0$. Find the slope, and sketch the line.

Solution

To show that the given points are on the line represented by the given equation, we must show that their coordinates satisfy the equation.

$3x - 5y - 4 = 0$	$3x - 5y - 4 = 0$
$3(3) - 5(1) - 4 = 0$	$3(8) - 5(4) - 4 = 0$
$9 - 5 - 4 = 0$ True	$24 - 20 - 4 = 0$ True
$(3,1)$ is on the line.	$(8, 4)$ is on the line.

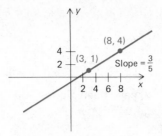

FIGURE 2.16   Graph of $3x - 5y - 4 = 0$

The slope of the line containing (3, 1) and (8, 4) is

$$\frac{4 - 1}{8 - 3} = \frac{3}{5}.$$

The points (3, 1) and (8, 4) have been plotted and the line they determine has been drawn in Figure 2.16.

Note that in using the slope formula to find the slope of a line through two given points, it does not matter which of the points we call P_1 and which we call P_2. In Example 4, if we call (3, 1) the point P_1 and (8, 4) the point P_2, then

$$m = \frac{4 - 1}{8 - 3} = \frac{3}{5}.$$

If we call (8, 4) the point P_1 and (3, 1) the point P_2, we still get the same number for the slope.

$$m = \frac{1 - 4}{3 - 8} = \frac{-3}{-5} = \frac{3}{5}$$

EXAMPLE 5

Show that the points (3, 5) and (8, 2) are both on the line $3x + 5y - 34 = 0$. Find the slope, and sketch the line.

Solution

Let us first show that (3, 5) and (8, 2) satisfy the given equation.

$$3x + 5y - 34 = 0 \qquad\qquad 3x + 5y - 34 = 0$$
$$3(3) + 5(5) - 34 = 0 \qquad\qquad 3(8) + 5(2) - 34 = 0$$
$$9 + 25 - 34 = 0 \quad \text{True} \qquad 24 + 10 - 34 = 0 \quad \text{True}$$

(3, 5) is on the line.   (8, 2) is on the line.

Using the slope formula, we can call (3, 5) the point P_1 and (8, 2) the point P_2.

$$m = \frac{2 - 5}{8 - 3} = \frac{-3}{5} = -\frac{3}{5}$$

If we call (8, 2) the point P_1 and (3, 5) the point P_2, we still get $-\frac{3}{5}$ for the slope.

$$m = \frac{5 - 2}{3 - 8} = \frac{3}{-5} = -\frac{3}{5}$$

The points (3, 5) and (8, 2) have been plotted and the line they determine has been sketched in Figure 2.17.

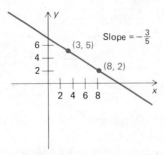

FIGURE 2.17   Graph of $3x + 5y - 34 = 0$

The line in Figure 2.16 slopes up from left to right, and its slope is positive. The line in Figure 2.17 slopes down from left to right, and its slope is negative. Any line that slopes up from left to right has positive slope. Any line that slopes down from left to right has negative slope. The next example utilizes notation that is frequently used in calculus.

EXAMPLE 6

Find an expression for the slope of the line segment with one end at the point with coordinates $(x, f(x))$ and with the other end at the point with coordinates $(x + h, f(x + h))$ where $h \neq 0$.

Solution

Using the slope formula, the slope is

$$\frac{f(x + h) - f(x)}{x + h - x} = \frac{f(x + h) - f(x)}{h}.$$

Exercise 2.2, 16–25

If two nonvertical lines are parallel, they have exactly the same slope. More formally, if nonvertical lines ℓ_1 and ℓ_2 are parallel and have slopes m_1 and m_2 respectively,

$$m_1 = m_2. \qquad \text{Lines } \ell_1 \text{ and } \ell_2 \text{ are parallel.}$$

The slopes of perpendicular oblique lines are also related in a simple way. If two oblique lines are perpendicular, it can be shown that the slope of either one of them is the negative reciprocal of the slope of the other. For example, if two lines are perpendicular and the slope of one of them is $\frac{2}{3}$, then the slope of the other one must be $-\frac{3}{2}$. Another way of describing this relationship is to say that the product of the slopes of perpendicular oblique lines is -1. In more general terms, if oblique lines ℓ_1 and ℓ_2 are perpendicular with slopes m_1 and m_2 respectively, then

$$m_1 m_2 = -1. \qquad \text{Lines } \ell_1 \text{ and } \ell_2 \text{ are perpendicular.}$$

In other words,
$$m_1 = \frac{-1}{m_2}$$

and
$$m_2 = \frac{-1}{m_1}.$$

If we know one point on a line and we also know the slope of the line, then we can draw the line. For example, suppose that the point is $(1, 4)$ and the slope is $-\frac{2}{3}$. To draw the line, all we need to do is to plot the given point and locate another point on the line by starting at the given point and moving three units to the right (the positive x-direction) and then moving two units down (the negative y-direction) in accordance with the given slope of $-\frac{2}{3} = \frac{-2}{3}$. Then we just draw the line through the given point and the newly located point. This was done in Figure 2.18. We could also have located a second point on the line by considering the slope $-\frac{2}{3}$ to be $\frac{2}{-3}$ and moving from the given point three units left followed by two units up.

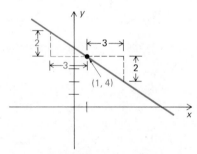

FIGURE 2.18 The line through $(1, 4)$ with slope $-\frac{2}{3}$

We may now show how to find an equation of a line if we are given one of its points together with its slope. Suppose that the given point is $(1, 4)$ and the slope is $\frac{-2}{3}$, as in Figure 2.18. To find an equation satisfied by any point (x, y) on the line, first employ the slope formula and acknowledge that the slope of the segment from $(1, 4)$ to any other point (x, y) must be equal to $\frac{-2}{3}$. In other words, if $x \neq 1$,

$$\frac{y - 4}{x - 1} = \frac{-2}{3}.$$

Multiplying both sides of this equation by $(x - 1)$, we have

$$y - 4 = \frac{-2}{3}(x - 1).$$

This is an equation of the line through $(1, 4)$ with slope $\frac{-2}{3}$. It can be written in the standard form $Ax + By + C = 0$ by first multiplying both sides through by 3.

$$3(y - 4) = -2(x - 1)$$
$$3y - 12 = -2x + 2$$
$$2x + 3y - 14 = 0$$

This is the standard form of an equation of the line through $(1, 4)$ with slope $\frac{-2}{3}$.

To obtain a general formula for an equation of the line through a given point with a given slope, suppose that the given point is the point (x_1, y_1) and that the given slope is m. Then, just as we did in the particular example in the preceding paragraph, to find an equation satisfied by any point (x, y) on the line, first employ the slope formula and acknowledge that the slope of the segment from (x_1, y_1) to any other point (x, y) must be equal to m. In other words, if $x \neq x_1$,

$$\frac{y - y_1}{x - x_1} = m.$$

Multiplying both sides of this equation by $(x - x_1)$, we have

$$y - y_1 = m(x - x_1). \qquad \text{Point-slope form}$$

This is called the **point-slope form** of an equation of a line. It is an equation of the line through (x_1, y_1) with slope m.

Exercise 2.2, 26–40

If the given point is the y-intercept, the point-slope form reduces to a very useful formula. For example, suppose that the given point is $(0, 2)$ and that the slope is $\frac{3}{5}$. Using the point-slope form, an equation of the line through $(0, 2)$ with slope $\frac{3}{5}$ is

$$y - 2 = \tfrac{3}{5}(x - 0),$$
or
$$y = \tfrac{3}{5}x + 2.$$

In this form, note that the coefficient of x is the slope and the nonvariable term is the y-intercept (that is, it is the second coordinate of the y-intercept).

To obtain a general formula for an equation of the line with a given slope and a given y-intercept, suppose that the given slope is m and that the given y-intercept is b—in other words, the y-intercept is $(0, b)$. Using the point-slope form just as we did in the particular example in the preceding paragraph,

$$y - b = m(x - 0),$$

or $\qquad\qquad y = mx + b \qquad$ Slope—y-intercept form

Slope $\qquad$ y-intercept

$\qquad m \qquad\qquad (0, b)$

The slope—y-intercept form is particularly useful if we are given a linear equation in x and y and are required to find the slope of the line it represents. All we need do is write the given linear equation in the form $y = mx + b$ and read off the coefficient of x. In other words, given a linear equation in the two variables x and y, to find the slope of its line, solve for y in terms of x and read off the coefficient of x.

EXAMPLE 7

Find the slope of $3x - 5y + 7 = 0$.

Solution

Solving for y in terms of x, we have

$$-5y = -3x - 7$$
$$y = \tfrac{3}{5}x + \tfrac{7}{5}.$$

Comparing this with

$$y = mx + b,$$

we see that the slope m is $\tfrac{3}{5}$. (We also see that the y-intercept is $(0, \tfrac{7}{5})$.)

We can also find the slope of $3x - 5y + 7 = 0$ by locating any two points on it and using the definition of slope, but this involves more computation than simply writing the equation in the slope—y-intercept form.

It is now easy to prove that if A, B, and C are real numbers with A and B not zero, then, as was asserted earlier in this section, the graph of

$$Ax + By + C = 0$$

is an oblique line. All we need do is solve for y in terms of x and note that we have the slope—y-intercept form of an equation of a line.

$$By = -Ax - C$$
$$y = \frac{-A}{B}x - \frac{C}{B}$$

Thus $Ax + By + C = 0$ represents a line with slope $-A/B$ and with y-intercept $(0, -C/B)$.

EXAMPLE 8

Find the slope of a line parallel to $2x - 7y + 5 = 0$, and find the slope of a line perpendicular to $2x - 7y + 5 = 0$.

First let us find the slope of the given line.

$$-7y = -2x - 5$$
$$y = \tfrac{2}{7}x + \tfrac{5}{7}$$

The slope of $2x - 7y + 5 = 0$ is $\tfrac{2}{7}$. The slope of any line parallel to $2x - 7y + 5 = 0$ is also $\tfrac{2}{7}$. The slope of any line perpendicular to $2x - 7y + 5 = 0$ is the negative reciprocal of $\tfrac{2}{7}$ or $-\tfrac{7}{2}$.

If we are given two points on a nonvertical line, we can find an equation of the line by first computing its slope and then using the point-slope form with either one of the given points together with the computed slope. For example, suppose that the two given points are $(-1, 2)$ and $(4, 8)$. The slope of the line through $(-1, 2)$ and $(4, 8)$ is

$$\frac{8 - 2}{4 - (-1)} = \frac{6}{5}.$$

Using the given point $(-1, 2)$ and the slope $\tfrac{6}{5}$, the point-slope form yields an equation of the line.

$$y - 2 = \tfrac{6}{5}(x + 1).$$

Writing this in standard form, we have

$$5(y - 2) = 6(x + 1)$$
$$5y - 10 = 6x + 6$$
$$6x - 5y + 16 = 0.$$

If we had used the other given point, $(4, 8)$, we would have obtained the same standard form. (Students should verify this for themselves.)

EXAMPLE 9

Find an equation of the line with intercepts $(0, -2)$ and $(3, 0)$.

Solution

Here we have a choice of ways to proceed. We can first compute the slope and then use the slope–y-intercept form, or we can first compute the slope and then use the point-slope form. The slope is

$$\frac{0 - (-2)}{3 - 0} = \frac{2}{3}.$$

Since the y-intercept is $(0, -2)$, if we now use the slope–y-intercept form, we have

$$y = \tfrac{2}{3}x - 2.$$

Writing this in standard form, we have

$$3y = 2x - 6$$
$$2x - 3y - 6 = 0.$$

If we use the point-slope form and choose the point (3, 0), we have

$$y - 0 = \tfrac{2}{3}(x - 3).$$

Writing this in standard form should produce the same result as the one we obtained by using the slope–y-intercept form.

$$y = \tfrac{2}{3}(x - 3)$$
$$3y = 2x - 6$$
$$2x - 3y - 6 = 0$$

The two results do agree.

EXAMPLE 10

If a and b are nonzero real numbers, find an equation of the line with intercepts $(0, b)$ and $(a, 0)$.

Solution

We can proceed just as we did in Example 9 by first finding the slope and then using either the slope–y-intercept form or the point-slope form. The slope is

$$\frac{0 - b}{a - 0} = \frac{-b}{a}.$$

Since the y-intercept is $(0, b)$, the slope–y-intercept form gives us

$$y = \frac{-b}{a}x + b.$$

If we divide this through by b and isolate the nonvariable term on the right side of the equation, we will have it written in a form that is often seen in classical analytic geometry texts.

$$\frac{y}{b} = \frac{-bx}{ba} + \frac{b}{b}$$

$$\frac{y}{b} = \frac{-x}{a} + 1$$

$$\frac{x}{a} + \frac{y}{b} = 1$$

Exercise 2.2, 51–65

This last form is sometimes called the *intercept form* of the equation of a line.

A **linear function** is a function f such that

1. the domain is the set of all real numbers
2. $f(x) = mx + b$

where m and b are specific real numbers. Since the graph of linear function f is the graph of the equation

$$y = mx + b,$$

we see that the graph of a linear function f is a line with slope m and y-intercept $(0, b)$.

EXAMPLE 11

Sketch the graph of linear function f such that $f(x) = 3x - 2$.

Solution

The graph of f is the graph of the equation $y = 3x - 2$ (see Figure 2.19). The graph is the line with slope 3 and y-intercept $(0, -2)$.

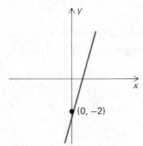

$(0, -2)$

FIGURE 2.19   Graph of f: $f(x) = 3x - 2$
(Graph of $y = 3x - 2$)

We saw in Section 2.1 that any function is associated with a set of ordered pairs and that some authors define a function to be its associated set of ordered pairs. If a function is defined as a certain set of ordered pairs, set-builder notation can be used to define a linear function as

$$\{(x, y)\,|\,y = mx + b\}$$

where m and b are specific real numbers. Using this notation, the linear function f of Example 11 could be written as follows:

$$f = \{(x, y)\,|\,y = 3x - 2\}.$$

We may say that any linear equation in two variables, say x and y, *defines* a linear function because the graph of the equation is the graph of a linear function. We can solve such an equation for y in terms of x to obtain the rule of assignment, and we can use the domain convention to determine that the domain is the set of all real numbers. For example, if f is the function defined by the equation

$$3x - 7y + 2 = 0,$$

then $f(x) = \tfrac{3}{7}x + \tfrac{2}{7}$

because if we solve the equation for y in terms of x, we get

$$-7y = -3x - 2$$
$$y = \tfrac{3}{7}x + \tfrac{2}{7}.$$

Defining a function as a certain set of ordered pairs, we could say that

Exercise 2.2, 66–80

$$f = \{(x, y) \mid y = \tfrac{3}{7}x + \tfrac{2}{7}\}.$$

In this section we introduced a formula for the slope of a line segment. The next example utilizes this formula together with notation and algebraic manipulations that are frequently encountered in calculus. Similar exercises, using notation found in the most widely used calculus texts, are listed in Exercise 2.2.

EXAMPLE 12

If $f(x) = 5x^2 + 3x - 7$ and $h \neq 0$, find an expression for the slope of the line segment with endpoints $(x, f(x))$ and $(x + h, f(x + h))$ and write this expression for the slope as a polynomial in x and h.

Solution

The slope is

$$\frac{f(x + h) - f(x)}{x + h - x} = \frac{f(x + h) - f(x)}{h}.$$

Since

$$f(x) = 5x^2 + 3x - 7,$$
$$f(x + h) = 5(x + h)^2 + 3(x + h) - 7.$$

It follows that the slope of the line segment with endpoints $(x, f(x))$ and $(x + h, f(x + h))$ can be written as follows:

$$\frac{f(x + h) - f(x)}{h} = \frac{5(x + h)^2 + 3(x + h) - 7 - (5x^2 + 3x - 7)}{h}$$

$$= \frac{5(x^2 + 2xh + h^2) + 3x + 3h - 7 - 5x^2 - 3x + 7}{h}$$

$$= \frac{5x^2 + 10xh + 5h^2 + 3x + 3h - 7 - 5x^2 - 3x + 7}{h}$$

$$= \frac{10xh + 5h^2 + 3h}{h}$$

$$= 10x + 5h + 3$$

That is, where $f(x) = 5x^2 + 3x - 7$, the slope of the line segment from $(x, f(x))$ to $(x + h, f(x + h))$ is

$$\frac{f(x + h) - f(x)}{h} = 10x + 5h + 3.$$

Exercise 2.2, 81–95

▶ Sketch the graph of each of the following and label its intercepts.

EXAMPLE $-2x - y + 6 = 0$

Solution

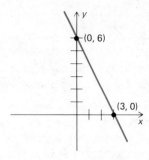

1. $x - 3y + 6 = 0$ **2.** $5x - 2y - 10 = 0$

3. $3x - 4y = 7$ **4.** $2x = 5y + 3$

5. $7x - 8y + 1 = 0$

▶ In each of the following, find the point of intersection of the given lines, sketch the graphs of the lines, and label the point of intersection.

EXAMPLE

$x - 4y + 7 = 0$

$3x + y - 5 = 0$

Solution

$$
\begin{aligned}
x - 4y + 7 &= 0 \\
12x + 4y - 20 &= 0 \\
\hline
13x \qquad\quad - 13 &= 0 \\
x &= 1 \\
1 \ - 4y + \ 7 &= 0 \\
-4y &= -8 \\
y &= 2
\end{aligned}
$$

6. $x - 3y - 10 = 0, \ 2x + y + 1 = 0$

7. $x + y + 3 = 0, \ x - y + 1 = 0$

8. $5x - 2y = 4, \ 3x + 4y - 18 = 0$

9. $2x = 7y - 32, \ 4x = 16 - 5y$

10. $3x + 7y = 5, \ 9x - 2y = -31$

▶ Sketch the graph of each of the following and label its intercept.

EXAMPLE $2y + 2 = 0$

Solution
$2y + 2 = 0$

$y = -1$

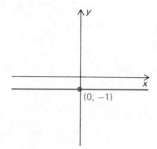

(0, −1)

11. $3y - 6 = 0$ **12.** $3x + 6 = 0$

13. $x = -1$ **14.** $y = 3$

15. $x = 0$

▶ In each of the following, find the slope of the line containing the given points.

EXAMPLE $(2, -4), (-5, 3)$

Solution
$$m = \frac{3 - (-4)}{-5 - 2} = \frac{7}{-7} = -1$$

16. $(3, -5), (-4, 2)$ **17.** $(3, 0), (-2, 1)$

18. $(1, -2), (7, -2)$ **19.** $(0, 4), (3, 0)$

20. $(-2, 5), (-1, -4)$

▶ In each of the following, show that the given points lie on the given line, find the slope, and sketch the line.

EXAMPLE $(-1, 2), (3, -4), 3x + 2y - 1 = 0$

Solution
$3(-1) + 2(2) - 1 = 0$ $3(3) + 2(-4) - 1 = 0$

$-3 + 4 \quad - 1 = 0$ True $9 - 8 \quad - 1 = 0$ True

$(-1, 2)$ is on the line. $(3, -4)$ is on the line.

$$m = \frac{-4 - 2}{3 - (-1)} = \frac{-6}{4}$$

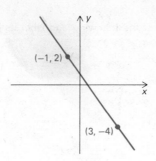

21. $(1, 2)$, $(3, 4)$, $x - y + 1 = 0$

22. $(1, -2)$, $(4, 5)$, $7x - 3y - 13 = 0$

23. $(2, 1)$, $(-3, 4)$, $3x + 5y - 11 = 0$

24. $(5, 0)$, $(0, -3)$, $3x - 5y - 15 = 0$

25. $(-3, 2)$, $(1, 1)$, $x + 4y - 5 = 0$

▶ In each of the following, sketch the graph of the line through the given point with the given slope.

EXAMPLE $(4, 5)$, $\frac{-3}{2}$

Solution

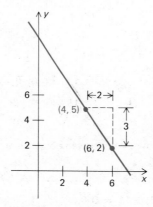

26. $(3, 5)$, $\frac{-4}{3}$ **27.** $(2, 7)$, $\frac{-3}{5}$ **28.** $(-1, 3)$, $\frac{3}{4}$

29. $(3, 3)$, 1 **30.** $(-1, 3)$, -1

▶ In each of the following, find the standard form of an equation of the line through the given point and with the given slope.

EXAMPLES   (a) $(2, 3)$, $\frac{-1}{2}$ (b) $(0, 2)$, $\frac{3}{4}$

Solutions   (a) $y - 3 = \frac{-1}{2}(x - 2)$ (b) $y = \frac{3}{4}x + 2$

$\qquad\qquad\qquad\qquad 2y - 6 = -x + 2 \qquad\qquad 4y = 3x + 8$

$\qquad\qquad\qquad\qquad x + 2y - 8 = 0 \qquad\qquad 3x - 4y + 8 = 0$

31. $(3, 4)$, $\frac{-2}{5}$ **32.** $(-2, 4)$, $\frac{3}{7}$ **33.** $(-3, 2)$, -1

34. $(-2, -3)$, $\frac{-4}{5}$ **35.** $(-1, 5)$, $\frac{3}{8}$ **36.** $(0, 4)$, $\frac{2}{3}$

37. $(0, -5)$, $\frac{-3}{4}$ **38.** $(0, 6)$, $\frac{3}{7}$ **39.** $(0, 3)$, 1

40. $(0, -1)$, $\frac{-4}{7}$

▶ Write each of the following lines in the slope–y-intercept form and then find the slope and the y-intercept.

EXAMPLE $2x - 3y + 12 = 0$

Solution $-3y = -2x - 12$

$\qquad\qquad y = \frac{2}{3}x + 4$

Slope $= \frac{2}{3}$; y-intercept is $(0, 4)$.

41. $2x - 7y + 28 = 0$ **42.** $x - 6y + 12 = 0$

43. $5x + y - 11 = 0$ **44.** $x + y - 3 = 0$

45. $5x - 8y + 12 = 0$

▶ Find the slope of a line parallel to each of the given lines, and then find the slope of a line perpendicular to it.

EXAMPLE $3x - 7y + 2 = 0$

Solution $-7y = -3x - 2$ Slope of parallel line $= \frac{3}{7}$.

$\qquad\qquad y = \frac{3}{7}x + \frac{2}{7}$ Slope of perpendicular line $= \frac{-7}{3}$.

46. $5x - 3y + 1 = 0$ **47.** $2x + 4y - 5 = 0$

48. $3x + 7y + 8 = 0$ **49.** $2x + y - 6 = 0$

50. $3x + 4y + 9 = 0$

▶ In each of the following, find the standard form of an equation of the line through the two given points.

EXAMPLE $(2, 3)$, $(4, -5)$

Solution   Slope $= \dfrac{-5 - 3}{4 - 2} = \dfrac{-8}{2} = -4$

$$y - 3 = -4(x - 2)$$
$$y - 3 = -4x + 8$$
$$4x + y - 11 = 0$$

51. $(3, -4), (2, 1)$ **52.** $(2, 5), (3, 7)$

53. $(-1, -3), (4, -2)$ **54.** $(2, -3), (4, -3)$

55. $(-2, 6), (7, 11)$ **56.** $(0, 4), (-5, 0)$

57. $(0, 1), (1, 0)$ **58.** $(0, -2), (-3, 0)$

59. $(0, 3), (2, 0)$ **60.** $(0, \frac{1}{2}), (\frac{-1}{3}, 0)$

61. Find the standard form of an equation of the line through $(0, 0)$ and parallel to $3x - 4y + 1 = 0$.

62. Find the standard form of an equation of the line through $(0, 0)$ and perpendicular to $4x - 3y + 7 = 0$.

63. Find the standard form of an equation of the line through $(0, 0)$ and the point of intersection of $x = 2$ and $3x - 8y + 2 = 0$.

64. Find the standard form of an equation of the line with y-intercept 7 and perpendicular to $y = -3x + 4$.

65. Find the standard form of an equation of the line with x-intercept -2 and with y-intercept 2.

▶ In each of the following, sketch the graph of the linear function f.

EXAMPLES

Solutions

(a) $f(x) = 2x - 3$ (b) $f = \{(x, y)|y = x + 3\}$

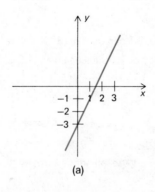

(a)

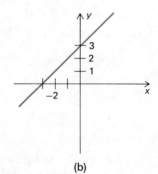

(b)

66. $f(x) = 5x + 2$ **67.** $f(x) = -5x + 1$

68. $f(x) = x + 5$ **69.** $f(x) = -x + 5$

70. $f(x) = 4x + 3$ **71.** $f = \{(x, y)|y = 3x + 2\}$

72. $f = \{(x, y) | y = \frac{2}{3}x - 1\}$ **73.** $f = \{(x, y) | y = -x + 2\}$

74. $f = \{(x, y) | y = x + 2\}$ **75.** $f = \{(x, y) | y = \frac{-2}{3}x + 1\}$

▶ In each of the following, let f be the function defined by the given equation and find $f(x)$, $f(1)$, and $f(0)$.

EXAMPLE $2x - 5y + 6 = 0$

Solution $-5y = -2x - 6$

$y = \frac{2}{5}x + \frac{6}{5}$

$f(x) = \frac{2}{5}x + \frac{6}{5}$

$f(1) = \frac{8}{5}$

$f(0) = \frac{6}{5}$

76. $3x - 7y + 10 = 0$ **77.** $2x - y + 18 = 0$

78. $3x + 7y - 21 = 0$ **79.** $x + y - 6 = 0$

80. $-x - 3y + 8 = 0$

The following exercises utilize notation found in some of the most widely used calculus texts. In these exercises the letters, h, a, r, c, and x represent real numbers, the symbol x_1 represents a real number, and the symbol Δx represents a real number.

▶ In each of the following, find the slope of the line segment having the given endpoints.

EXAMPLE $(x_1, f(x_1)), (x_1 + \Delta x, f(x_1 + \Delta x))$

Solution $\text{Slope} = \dfrac{f(x_1 + \Delta x) - f(x_1)}{x_1 + \Delta x - x_1} = \dfrac{f(x_1 + \Delta x) - f(x_1)}{\Delta x}$

81. $(x, f(x)), (a, f(a))$ **82.** $(c, f(c)), (c + h, f(c + h))$

83. $(x, f(x)), (x_1, f(x_1))$ **84.** $(x, f(x)), (x + \Delta x, f(x + \Delta x))$

85. $(x_1, f(x_1)), (x_1 + h, f(x_1 + h))$

▶ In each of the following, express the slope of the line segment with the given endpoints as a polynomial in the letters or symbols indicated. (The numbers h and Δx are nonzero.)

EXAMPLE $f(x) = 3x^2$;   endpoints $(x_1, f(x_1))$ and $(x_1 + \Delta x, f(x_1 + \Delta x))$; polynomial in x_1 and Δx.

Solution

$$\text{Slope} = \frac{f(x_1 + \Delta x) - f(x_1)}{\Delta x}$$

$$= \frac{3(x_1 + \Delta x)^2 - 3x_1^2}{\Delta x}$$

$$= \frac{3(x_1^2 + 2x_1(\Delta x) + (\Delta x)^2) - 3x_1^2}{\Delta x}$$

$$= \frac{3x_1^2 + 6x_1(\Delta x) + 3(\Delta x)^2 - 3x_1^2}{\Delta x}$$

$$= 6x_1 + 3\,\Delta x$$

86. $f(x) = 5x^2 + 3x$; endpoints $(x_1, f(x_1))$ and $(x_1 + \Delta x, f(x_1 + \Delta x))$; polynomial in x_1 and Δx.

87. $f(x) = 3x^2 - 5x + 7$; endpoints $(x_1, f(x_1))$ and $(x_1 + \Delta x, f(x_1 + \Delta x))$; polynomial in x_1 and Δx.

88. $f(x) = 3x^2 - 2x + 6$; endpoints $(x, f(x))$ and $(x + h, f(x + h))$; polynomial in x and h.

89. $f(x) = 2x^2 + 4x - 7$; endpoints $(x, f(x))$ and $(a, f(a))$; polynomial in x and a $(x \neq a)$.

90. $f(x) = 5x^2 - x + 3$; endpoints $(x, f(x))$ and $(x_1, f(x_1))$; polynomial in x and x_1 $(x \neq x_1)$.

91. $f(x) = 7x^2 + 3x - 5$; endpoints $(x, f(x))$ and $(r, f(r))$; polynomial in x and r $(x \neq r)$.

92. $f(x) = x^2 - 4x + 3$; endpoints $(x_1, f(x_1))$ and $(x_1 + h, f(x_1 + h))$; polynomial in x_1 and h.

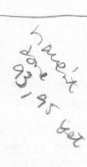

93. $f(x) = 4x^2 - 3x + 2$; endpoints $(c, f(c))$ and $(c + h, f(c + h))$; polynomial in c and h.

94. $f(x) = x^3$; endpoints $(x, f(x))$ and $(x + h, f(x + h))$; polynomial in x and h.

95. $f(x) = x^3$; endpoints $(x_1, f(x_1))$ and $(x_1 + \Delta x, f(x_1 + \Delta x))$; polynomial in x_1 and Δx.

CHECK LIST FOR THE STUDENT

If you have learned the material in Section 2.2, you should be able to do the following:

1. Sketch the graph of a linear equation in two variables.

2. Sketch the graph of a linear equation in one variable.

3. Find the point of intersection of two lines if their equations are given.

4. Find the intercepts of a line if its equation is given.

5. Find the slope of a nonvertical line through two given points.

6. Show whether or not a given point lies on a line whose equation is given.

7. Tell from looking at the graph of a line whether its slope is positive, negative, zero, or not defined.

8. Find the slope of a line parallel to a line with known slope.

9. Find the slope of a line perpendicular to a line with known slope.

10. Draw the line with a given slope through a given point.

11. Find an equation of the line with a given slope through a given point.

12. Write an equation of a line in standard form.

13. Find an equation of the line with a given slope and a given y-intercept.

14. Write an equation of a line in slope—y-intercept form and find its slope and y-intercept.

15. Find an equation of the line through two given points.

16. Find an equation of a line whose intercepts are given.

17. Define a linear function.

18. Sketch the graph of a linear function.

19. Find the rule of assignment of the linear function defined by a given linear equation in two variables.

20. Use functional notation and the slope formula and write simplified expressions for the slope of a line segment whose endpoints have coordinates written as they appear in widely used calculus texts.

SECTION QUIZ 2.2

1. Sketch the graph of $2x + 5y - 10 = 0$, and label its intercepts.

2. Find the point of intersection of the following lines, sketch the graphs of the lines, and label the point of intersection.

$$2x - 3y + 10 = 0$$
$$x - 4y + 14 = 0$$

3. Sketch the graph of each of the following, and label its intercept.
   (a) $3x + 3 = 0$ (b) $y = 4$

4. (a) Find the slope of the line containing the points $(3, -2)$ and $(-1, 5)$.
   (b) Find an expression for the slope of the line containing the points $(a, f(a))$ and $(a + h, f(a + h))$ where $h \neq 0$.

5. Sketch the graph of the line through $(6, 2)$ with slope $\frac{-1}{3}$.

6. Find the standard form of an equation of the line through the point $(3, -1)$ and with slope $\frac{2}{5}$.

7. Write the line $3x - 2y + 10 = 0$ in the slope–y-intercept form and then find the slope and the y-intercept.

8. (a) Find the slope of a line parallel to $2x - 9y + 4 = 0$.
   (b) Find the slope of a line perpendicular to $2x - 9y + 4 = 0$.

9. Find the standard form of an equation of the line through the points $(3, 5)$ and $(1, -3)$.

10. (a) Sketch the graph of the linear function f such that $f(x) = -x + 2$.
   (b) Let f be the function defined by the equation $3x - 6y + 5 = 0$. Find $f(x)$, $f(0)$, and $f(1)$.

2.3 Some Properties of Graphs

The purpose of this section is to demonstrate some properties of graphs that may provide information that can be used in sketching a graph. In particular, intercepts, symmetry, asymptotes, and excluded areas will be discussed and illustrated.

INTERCEPTS

In Section 2.2 we saw that the points in which a line intersects the coordinate axes are called its intercepts. Likewise, the point(s) in which any graph intersects the x-axis is (are) called its x-intercept(s), and the point(s) in which any graph intersects the y-axis is (are) called its y-intercept(s). To find the x-intercept(s) of the graph of an equation in x and y, let $y = 0$ and solve for x. To find the y-intercept(s), let $x = 0$ and solve for y.

EXAMPLE 1   Find the intercepts of the graph of $4x^2 + 9y^2 = 36$.

Solution   To find the x-intercepts, let $y = 0$ and solve for x.

$$4x^2 = 36$$
$$x^2 = 9$$
$$x = \pm 3$$

The x-intercepts are $(3, 0)$ and $(-3, 0)$. To find the y-intercepts, let $x = 0$ and solve for y.

$$9y^2 = 36$$
$$y^2 = 4$$
$$y = \pm 2$$

The y-intercepts are $(0, 2)$ and $(0, -2)$.

SYMMETRY

Some graphs are symmetric with respect to one or both of the coordinate axes. Roughly speaking, a graph is symmetric with respect to a line provided that if the paper were folded on the line, the two halves would exactly match. The graph in Figure 2.20 is symmetric with respect to the x-axis. Technically speaking, **a graph is symmetric with respect to the x-axis** provided that whenever the point (x, y) is on the graph, the point $(x, -y)$ is also on the graph. In other words, for any point (x, y) on the graph, there is another point $(x, -y)$ on the graph such that the x-axis is the perpendicular bisector of the segment joining the two points. There is an easy way to test an equation to see whether or not its graph is symmetric with respect to the x-axis.

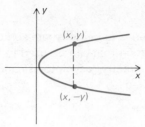

FIGURE 2.20 Graph of $y^2 = x$
(Symmetric with respect to x-axis)

TEST FOR SYMMETRY WITH RESPECT TO x-AXIS

Construct an equation by replacing y by $-y$ in the given equation. If the constructed equation is equivalent to the given equation, then the graph of the given equation is symmetric with respect to the x-axis. Otherwise there is no symmetry with respect to the x-axis.

EXAMPLE 2

Test $y^2 = x$ for symmetry with respect to the x-axis.

Solution

Replacing y by $-y$ in $y^2 = x$, we obtain an equivalent equation.

$$(-y)^2 = x$$

or

$$y^2 = x$$

It follows that for any point (x, y) on the graph, the point $(x, -y)$ is also on the graph, so the graph is symmetric with respect to the x-axis. The graph of $y^2 = x$ is shown in Figure 2.20.

A graph is symmetric with respect to the y-axis provided that whenever the point (x, y) is on the graph, the point $(-x, y)$ is also on the graph. The graph in Figure 2.21 is symmetric with respect to the y-axis.

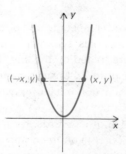

FIGURE 2.21 Graph of $y = x^2$
(Symmetric with respect to y-axis)

The test for symmetry with respect to the y-axis is similar to the x-axis symmetry test.

TEST FOR SYMMETRY WITH RESPECT TO y-AXIS

Construct an equation by replacing x by $-x$ in the given equation. If the constructed equation is equivalent to the given equation, then the graph of the given equation is symmetric with respect to the y-axis. Otherwise there is no symmetry with respect to the y-axis.

EXAMPLE 3

Test $y = x^2$ for symmetry with respect to the y-axis and with respect to the x-axis.

Solution

To y-axis Replacing x by $-x$ in $y = x^2$, we obtain an equivalent equation.

$$y = (-x)^2$$
or
$$y = x^2$$

It follows that for any point (x, y) on the graph, the point $(-x, y)$ is also on the graph, so the graph is symmetric with respect to the y-axis.

To x-axis Replacing y by $-y$ in $y = x^2$ to test for symmetry with respect to the x-axis, we obtain

$$(-y) = x^2$$
or
$$y = -x^2.$$

This equation is not equivalent to the given equation, so the graph is not symmetric with respect to the x-axis. The graph of $y = x^2$ is shown in Figure 2.21.

EXAMPLE 4

Test $y = x^3$ for symmetry with respect to the axes.

Solution

To x-axis Replacing y by $-y$, we obtain

$$(-y) = x^3$$
or
$$y = -x^3.$$

This equation is not equivalent to the given equation. The graph is not symmetric with respect to the x-axis.

To y-axis Replacing x by $-x$, we again obtain an equation that is not equivalent to the given equation.

$$y = (-x)^3$$
or
$$y = -x^3$$

The graph is not symmetric with respect to the *y*-axis. The graph is not symmetric with respect to either axis, and it is shown in Figure 2.22.

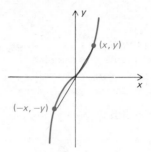

FIGURE 2.22   Graph of $y = x^3$
(Symmetric with respect to the origin)

Some graphs are symmetric with respect to the origin. The graph in Figure 2.22 is symmetric with respect to the origin. **A graph is symmetric with respect to the origin** provided that whenever (x, y) is on the graph, the point $(-x, -y)$ is also on the graph. In other words, for any point (x, y) on the graph, there is another point $(-x, -y)$ on the graph such that the origin is the midpoint of the segment joining them.

TEST FOR SYMMETRY WITH RESPECT TO ORIGIN

Construct an equation by replacing x by $-x$ and y by $-y$ in the given equation. If the constructed equation is equivalent to the given equation, then the graph of the given equation is symmetric with respect to the origin. Otherwise there is no symmetry with respect to the origin.

EXAMPLE 5

Test $y = x^3$ for symmetry with respect to the origin.

Solution

Replacing x by $-x$ and y by $-y$ in $y = x^3$, we obtain an equivalent equation.

$$(-y) = (-x)^3$$

or

$$-y = -x^3$$

or

$$y = x^3$$

It follows that for any point (x, y) on the graph, the point $(-x, -y)$ is also on the graph, so the graph is symmetric with respect to the origin.

EXAMPLE 6

Test $4x^2 + 9y^2 = 36$ for symmetry with respect to the origin and the axes.

Solution

To x-axis   Replacing y by $-y$, we obtain an equivalent equation.

$$4x^2 + 9(-y)^2 = 36$$

or $\qquad\qquad 4x^2 + 9y^2 = 36$

The graph of $4x^2 + 9y^2 = 36$ is symmetric with respect to the x-axis.

To y-axis   Replacing x by $-x$, we obtain an equivalent equation.

$$4(-x)^2 + 9y^2 = 36$$

or $\qquad\qquad 4x^2 + 9y^2 = 36$

The graph of $4x^2 + 9y^2 = 36$ is symmetric with respect to the y-axis.

To origin   Replacing x by $-x$ and y by $-y$, we obtain an equivalent equation.

$$4(-x)^2 + 9(-y)^2 = 36$$

or $\qquad\qquad 4x^2 + 9y^2 = 36$

The graph of $4x^2 + 9y^2 = 36$ is symmetric with respect to the origin.

The graph is shown in Figure 2.23. Its intercepts were found in Example 1. It will be discussed more fully in Example 9, and we shall learn in Section 2.6 that it is called an *ellipse*.

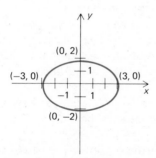

FIGURE 2.23   Graph of $4x^2 + 9y^2 = 36$
(Symmetric with respect to both axes and the origin)

Note that if a graph is symmetric with respect to *both* axes, then it must be symmetric with respect to the origin, but Examples 4 and 5 show that a graph can be symmetric with respect to the origin without being symmetric with respect to *either* axis.

Intercepts and Symmetry
Exercise 2.3, 1–18, 41–45

ASYMPTOTES

Some graphs have one or more *asymptotes*. An **asymptote** is a line that a graph appears to approach in the sense that as we go farther and farther out on the graph, the graph looks more and more like the line. The graph in Figure 2.24 has the line $x = 1$ as a vertical asymptote, and it has the x-axis as a horizontal asymptote.

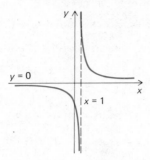

FIGURE 2.24 Graph of $y = \dfrac{1}{x - 1}$

Asymptotes: $x = 1$ and $y = 0$

To get the feeling that the graph in Figure 2.24 really does approach the line $x = 1$, consider a table of values such as Table 2.1, in which x successively takes on values nearer and nearer 1 through values larger than 1. Also consider a table of values such as Table 2.2, in which x successively takes on values nearer and nearer 1 through values smaller than 1. Then plot the points from each of these tables of values, and note that the points appear to be strung together like beads on a graph that more nearly assumes the position of the line $x = 1$ as the values of x get nearer and nearer 1. Plot points, taking x nearer and nearer 1, first to the right of 1 and then to the left of 1, until you see that you can make the points on the graph be as close to the line $x = 1$ as you please simply by taking values of x sufficiently close to 1.

To see intuitively that the graph of $y = 1/(x - 1)$ approaches the x-axis, consider a table of values such as Table 2.3, in which x succes-

TABLE 2.1

x	$y = \dfrac{1}{x - 1}$
2	1
$1\frac{1}{2}$	2
$1\frac{1}{4}$	4
$1\frac{1}{8}$	8
$1\frac{1}{16}$	16
$1\frac{1}{32}$	32

TABLE 2.2

x	$y = \dfrac{1}{x - 1}$
0	-1
$1 - \frac{1}{2}$	-2
$1 - \frac{1}{4}$	-4
$1 - \frac{1}{8}$	-8
$1 - \frac{1}{16}$	-16
$1 - \frac{1}{32}$	-32

TABLE 2.3	
x	$y = \dfrac{1}{x-1}$
3	$\frac{1}{2}$
4	$\frac{1}{3}$
5	$\frac{1}{4}$
10	$\frac{1}{9}$
100	$\frac{1}{99}$

TABLE 2.4	
x	$y = \dfrac{1}{x-1}$
-1	$\frac{-1}{2}$
-2	$\frac{-1}{3}$
-3	$\frac{-1}{4}$
-4	$\frac{-1}{5}$
-10	$\frac{-1}{11}$
-100	$\frac{-1}{101}$

sively takes on larger positive values. Also consider a table such as Table 2.4, in which x successively takes on smaller negative values—that is, x successively becomes larger in absolute value but is negative. Then note in Figure 2.24 that if we plot the points from each of these tables, the points appear to be strung together like beads on a graph that more nearly assumes the position of the x-axis as the values of x get larger and larger positively or as the values of x are negative and get larger and larger in absolute value. The students should plot points, extending Tables 2.3 and 2.4, if necessary, until they see that they can make the points on the graph of $y = 1/(x - 1)$ be as close to the x-axis as they please simply by taking sufficiently large positive values of x or by taking negative values of x with sufficiently large absolute values.

Precise definitions of vertical and horizontal asymptotes and ways of finding them can be given early in the study of calculus. Some unsophisticated ways of finding vertical or horizontal asymptotes for the graphs of some equations of the form $y = P(x)/D(x)$ are given and illustrated below. The veracity of these methods can be proved in beginning calculus.

TO FIND VERTICAL ASYMPTOTES FOR $y = P(x)/D(x)$

Find any real numbers x for which $D(x) = 0$. If a is a real number such that $D(a) = 0$ and $P(a) \neq 0$, then $x = a$ is a vertical asymptote. If $D(x) = 0$ has no real solutions, there are no vertical asymptotes.

EXAMPLE (i)

$$y = \frac{3x + 2}{4x - 1}$$

$$4x - 1 = 0$$

$$x = \tfrac{1}{4}$$

The line $x = \tfrac{1}{4}$ is the vertical asymptote.

EXAMPLE (ii)

$$y = \frac{8}{\sqrt{x^2 - 4}}$$

Here the denominator $D(x)$ is zero when $x = 2$ or -2. The lines $x = 2$ and $x = -2$ are the vertical asymptotes.

TO FIND HORIZONTAL ASYMPTOTES FOR $y = P(x)/D(x)$
WHERE $P(x)$ AND $D(x)$ ARE POLYNOMIALS IN x

Case 1   *The degree of P(x) is lower than the degree of D(x).* The horizontal asymptote is the x-axis, $y = 0$.

Example: $y = \dfrac{1}{x - 1}$

The degree of $P(x)$ is zero; the degree of $D(x)$ is 1. The horizontal asymptote is the x-axis, $y = 0$.

Case 2   *The degree of P(x) is the same as the degree of D(x).* Suppose that $P(x)$ and $D(x)$ are both of degree n. Then the horizontal asymptote is $y = A/B$ where A is the coefficient of x^n in $P(x)$ and B is the coefficient of x^n in $D(x)$.

Example: $y = \dfrac{3x + 2}{4x - 1}$

$P(x)$ and $D(x)$ are both of degree 1. The coefficient of x in $P(x)$ is 3; the coefficient of x in $D(x)$ is 4. The horizontal asymptote is $y = \frac{3}{4}$.

Case 3   *The degree of P(x) is higher than the degree of D(x).* There are no horizontal asymptotes.

Example: $y = \dfrac{x^2}{x - 1}$

The degree of $P(x)$ is 2; the degree of $D(x)$ is 1. In any table of values in which x successively takes on larger absolute values, y also takes on successively larger absolute values and fails to approach any definite number. There are no horizontal asymptotes.

If the equation $y = P(x)/D(x)$ can be solved for x in terms of y so that $x = Q(y)/E(y)$, then any horizontal asymptotes that the graph of $y = P(x)/D(x)$ may have can be found by considering $E(y)$. In particular, if b is a real number for which $E(y) = 0$ and $Q(y) \neq 0$, then $y = b$ is a horizontal asymptote. If no real values of y make $E(y)$ equal to zero, then there are no horizontal asymptotes.

EXAMPLE 7   Discuss the graph of

$$y = \frac{3x + 2}{4x - 1},$$

taking into consideration intercepts, symmetry with respect to axes and origin, and vertical and horizontal asymptotes. Sketch the graph.

Solution

Intercepts   If $x = 0$, $y = -2$. The y-intercept is $(0, -2)$.
If $y = 0$, $3x + 2 = 0$, so $x = \frac{-2}{3}$. The x-intercept is $(\frac{-2}{3}, 0)$.

Symmetry

To x-axis   Replacing y by $-y$, we obtain an equation that is not equivalent to the given equation.

$$-y = \frac{3x + 2}{4x - 1} \quad \text{or} \quad y = \frac{-3x - 2}{4x - 1}$$

There is no symmetry with respect to the x-axis.

To y-axis   Replacing x by $-x$, we obtain an equation that is not equivalent to the given equation.

$$y = \frac{3(-x) + 2}{4(-x) - 1} \quad \text{or} \quad y = \frac{-3x + 2}{-4x - 1}$$

There is no symmetry with respect to the y-axis.

To origin   Replacing x by $-x$ and y by $-y$, we obtain an equation that is not equivalent to the given equation.

$$-y = \frac{3(-x) + 2}{4(-x) - 1} \quad \text{or} \quad -y = \frac{-3x + 2}{-4x - 1} \quad \text{or} \quad y = \frac{3x - 2}{-4x - 1}$$

There is no symmetry with respect to the origin.

Asymptotes

Vertical $4x - 1 = 0$

$x = \frac{1}{4}$ is the vertical asymptote.

Horizontal   The numerator and denominator are both polynomials of degree 1 with the coefficients of their first-degree terms being 3 and 4, respectively. As we saw in the illustration of Case 2, the horizontal asymptote is $y = \frac{3}{4}$. In this example the horizontal asymptote can also readily be determined by solving the given equation for x in terms of y and then examining the new denominator.

$$y = \frac{3x + 2}{4x - 1}$$

$$4xy - y = 3x + 2$$

$$4xy - 3x = y + 2$$

$$x(4y - 3) = y + 2$$

$$x = \frac{y + 2}{4y - 3}$$

The new denominator $4y - 3$ equals zero when $y = \frac{3}{4}$. The horizontal asymptote is $y = \frac{3}{4}$, as we have already discovered.

EXCLUDED AREAS

The intercepts and asymptotes for

$$y = \frac{3x + 2}{4x - 1}$$

have been marked in Figure 2.25. It is apparent that more information will be needed in order to sketch the graph. In this example, as is often the case, areas can be excluded by considering what signs the variables must have on each side of each asymptote or intercept. We shall now demonstrate this technique.

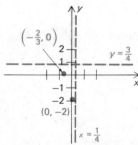

FIGURE 2.25 Intercepts and asymptotes of $y = \dfrac{3x + 2}{4x - 1}$

In this case, the x-intercept is at $(\frac{-2}{3}, 0)$ and the vertical asymptote is at $x = \frac{1}{4}$. We see that if x is greater than $\frac{1}{4}$, the denominator $4x - 1$ is positive and the numerator $3x + 2$ is also positive, so y must be positive.

$$\text{If } x > \tfrac{1}{4}, \text{ then } y > 0.$$

We can therefore cross out the portion of the plane for $x > \frac{1}{4}$ and y negative, because there is no part of the graph there. This has been done in Figure 2.26.

If x is between $\frac{1}{4}$ and $\frac{-2}{3}$ (that is, $x < \frac{1}{4}$ and $x > \frac{-2}{3}$), the denominator $4x - 1$ is negative and the numerator $3x + 2$ is positive, so y must be negative.

$$\text{If } \tfrac{-2}{3} < x < \tfrac{1}{4}, \text{ then } y < 0.$$

This means that y is not positive for any x between $\frac{-2}{3}$ and $\frac{1}{4}$. We can therefore cross out the portion of the plane where y would be positive between $x = \frac{-2}{3}$ and $x = \frac{1}{4}$, because there is no part of the graph there.

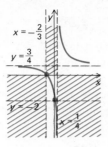

FIGURE 2.26   Graph of $y = \dfrac{3x + 2}{4x - 1}$

Finally, if x is less than $\frac{-2}{3}$, the numerator $3x + 2$ is negative and the denominator $4x - 1$ is also negative; since the numerator and denominator are of the same sign, y must be positive.

$$\text{If } x < \tfrac{-2}{3}, \text{ then } y > 0.$$

In other words, y cannot be negative for $x < \frac{-2}{3}$. The portion of the plane to the left of $\frac{-2}{3}$ where y is negative can be crossed out, because there is no part of the graph there.

We could now make a table of values and plot a few points and sketch the graph. In this example, however, as we saw earlier, it happens to be particularly easy to solve for x in terms of y and rule out portions of the plane by looking at the signs that x must have on each side of the y-intercept $(0, -2)$ and each side of the horizontal asymptote $y = \frac{3}{4}$. Let us do just that.

Solving the given equation for x in terms of y, we found that

$$x = \frac{y + 2}{4y - 3}.$$

From this we see that the signs of the numerator $y + 2$ and the denominator $4y - 3$ change at $y = -2$ and $y = \frac{3}{4}$. If $y > \frac{3}{4}$ (for points above the horizontal asymptote), the denominator $4y - 3$ is positive, and the numerator $y + 2$ is also positive so x must be positive.

$$\text{If } y > \tfrac{3}{4}, \text{ then } x > 0.$$

We can rule out the portion of the plane above the horizontal asymptote where x would be negative. Some of this was already ruled out before and was crossed out in Figure 2.26.

For points between the horizontal asymptote and the y-intercept (that is, for $-2 < y < \frac{3}{4}$), the numerator $y + 2$ is positive and the denominator is negative, so x must be negative.

2.3 SOME PROPERTIES OF GRAPHS

If $-2 < y < \frac{3}{4}$, then $x < 0$.

That means that x cannot be positive for $-2 < y < \frac{3}{4}$. We can therefore rule out all the portion of the plane to the right of the y-axis between $y = -2$ and $y = \frac{3}{4}$ because there is no part of the graph there.

For points below the y-intercept (that is, for $y < -2$), the numerator $y + 2$ is negative and the denominator is also negative, so x must be positive.

If $y < -2$, then $x > 0$.

In other words, x cannot be negative for $y < -2$, so we can rule out all that part of the plane below $y = -2$ where x would be negative.

It is now fairly evident that the graph must look like the one sketched. Constructing a table of values and plotting points would verify this.

EXAMPLE 8   Discuss the graph of

$$y = \frac{3x^2}{x - 1},$$

taking into consideration intercepts, symmetry with respect to axes and origin, vertical and horizontal asymptotes, and excluded areas. Sketch the graph.

Solution

Intercepts   If $x = 0$, $y = 0$. The y-intercept is $(0, 0)$.
If $y = 0$, $3x^2 = 0$, so $x = 0$. The x-intercept is $(0, 0)$.

Symmetry
To x-axis   Replacing y by $-y$, we obtain an equation not equivalent to the given equation. There is no symmetry with respect to the x-axis.

To y-axis   Replacing x by $-x$, we obtain

$$y = \frac{3x^2}{-x - 1},$$

which is not equivalent to the given equation. There is no symmetry with respect to the y-axis.

To origin   Replacing x by $-x$ and y by $-y$, we obtain

$$-y = \frac{3x^2}{-x - 1} \quad \text{or} \quad y = \frac{3x^2}{x + 1},$$

which is not equivalent to the given equation. There is no symmetry with respect to the origin.

Vertical $x - 1 = 0$

$x = 1$ is the vertical asymptote.

Horizontal   The numerator and denominator are polynomials, and the degree of the numerator is higher than the degree of the denominator. There are no horizontal asymptotes.

Excluded areas   In this example the numerator $3x^2$ is zero when $x = 0$, but otherwise the numerator is always positive. The denominator $x - 1$ is positive if $x > 1$; the denominator is negative if $x < 1$. It follows that

if $x > 1$, y is positive;

if $x < 1$ and $x \neq 0$, y is negative;

if $x = 0$, $y = 0$.

We can therefore *exclude* those areas for which

$x > 1$ and y is negative,

and *exclude* those areas for which

$x < 1$, $x \neq 0$, and y is positive

because there is no part of the graph there. These areas have been excluded in Figure 2.27.

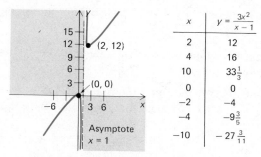

x	$y = \dfrac{3x^2}{x-1}$
2	12
4	16
10	$33\frac{1}{3}$
0	0
−2	−4
−4	$-9\frac{3}{5}$
−10	$-27\frac{3}{11}$

FIGURE 2.27   Graph of $y = \dfrac{3x^2}{x-1}$

We know from our considerations that $x = 1$ is an asymptote and that the graph of

$$y = \frac{3x^2}{x - 1}$$

goes through $(0, 0)$, although it does not cross the x-axis there. We could now tediously make an extensive table of values and sketch the graph as it is shown in Figure 2.27. We could see from an extensive table that the point (2, 12) appears to be the lowest point of the branch of the graph in the first quadrant. We could also see from an extensive table that if we take larger and larger values of x beyond 2, the corresponding values of y increase without bound and the graph rises to the right as shown.

An extensive table of values would also indicate that (0, 0) is the highest point on the branch of the graph to the left of $x = 1$ and that if we successively take negative values of x with larger and larger absolute values, the corresponding values of y are negative and also successively take on larger and larger absolute values. The graph would fall to the left as shown.

In Example 8, the precise location and identification of the points (2, 12) and (0, 0) as a relative low point and a relative high point can be readily accomplished by using elementary calculus. Here we hope to prepare for that study by providing some experience with common properties of graphs that can be examined with algebraic methods.

We have seen that in sketching graphs, we can sometimes exclude areas by considering the signs of the variables involved. At other times, we can exclude intervals and hence areas because values of one of the variables in those intervals would produce imaginary values for the other variable. This is illustrated in the next example.

EXAMPLE 9

Discuss the graph of $4x^2 + 9y^2 = 36$, taking into consideration intercepts, symmetry with respect to axes and origin, vertical and horizontal asymptotes, and excluded areas. Sketch the graph.

Solution

Intercepts We saw in Example 1 that the x-intercepts are (3, 0) and $(-3, 0)$ and that the y-intercepts are (0, 2) and (0, -2).

Symmetry We saw in Example 6 that the graph of $4x^2 + 9y^2 = 36$ is symmetric with respect to the x-axis, the y-axis, and the origin.

Asymptotes
Vertical If we solve $4x^2 + 9y^2 = 36$ for y in terms of x, we have

$$9y^2 = 36 - 4x^2$$

$$y^2 = \frac{36 - 4x^2}{9}$$

$$y = \frac{\pm \sqrt{36 - 4x^2}}{3}$$

$$= \frac{\pm 2\sqrt{9 - x^2}}{3}.$$

There are no real values of x for which the denominator is zero, so there are no vertical asymptotes.

Horizontal   If we solve $4x^2 + 9y^2 = 36$ for x in terms of y, we have

$$4x^2 = 36 - 9y^2$$

$$x^2 = \frac{36 - 9y^2}{4}$$

$$x = \frac{\pm\sqrt{36 - 9y^2}}{2}$$

$$= \frac{\pm 3\sqrt{4 - y^2}}{2}.$$

There are no values of y for which the denominator is zero, so there are no horizontal asymptotes.

Excluded areas   When we solved the given equation for y in terms of x, we found that

$$y = \frac{\pm 2\sqrt{9 - x^2}}{3}.$$

We have agreed to restrict ourselves to relations in which the ordered pairs (x, y) are pairs of real numbers. Consequently, we must *exclude* all values of x for which $9 - x^2$ is negative, because for such values of x the corresponding values of y would be imaginary.

Exclude x such that $9 - x^2 < 0$ or $9 < x^2$.

In other words, we must exclude the portion of the plane for which $x^2 > 9$, that is, the areas where $x > 3$ or $x < -3$.

When we solved the given equation for x in terms of y, we obtained

$$x = \frac{\pm 3\sqrt{4 - y^2}}{2}.$$

Thus to avoid imaginary values for x, we must *exclude* the portion of the plane for which $4 - y^2$ is negative.

Exclude y such that $4 - y^2 < 0$ or $4 < y^2$.

In other words, exclude the portion of the plane for which $y^2 > 4$, that is, the areas where $y > 2$ or $y < -2$.

The intercepts and excluded areas are shown, together with the graph, in Figure 2.28. By virtue of the symmetry with respect to both axes, in order to obtain a sketch of the entire graph it is only necessary to sketch the part in the first quadrant corresponding to values of x between 0 and 3. The rest of the graph can be obtained by reflection in the x-axis

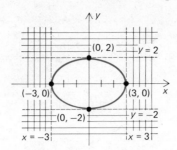

FIGURE 2.28 Graph of $4x^2 + 9y^2 = 36$
Excluded areas: $|x| > 3$ and $|y| > 2$

Excluded Areas
Exercise 2.3, 41–45

Miscellaneous
Exercise 2.3, 1–50

and then in the y-axis. As we noted in Example 6, this graph is called an ellipse.

EXERCISE 2.3 SOME PROPERTIES OF GRAPHS

▶ Discuss the graph of each of the following, taking into consideration intercepts, symmetry with respect to axes and origin, vertical and horizontal asymptotes, and excluded areas. Sketch the graph. Label any intercepts or vertical or horizontal asymptotes that the graph has.

1. $y^2 = x - 1$ 2. $y^2 = x + 1$

3. $y^2 = 4x$ 4. $y^2 = 6x$

5. $y^2 = 4(x - 1)$ 6. $y = x^2 + 1$

7. $y = x^2 - 1$ 8. $y = 3x^2$

9. $4y = x^2$ 10. $4(y - 1) = x^2$

11. $y = 2x^3$ 12. $y = -x^3$

13. $y = -2x^3$ 14. $y = x^3 + 1$

15. $y = x^5$ 16. $y = x^7$

17. $y = -x^5$ 18. $y = -x^7$

19. $y = \dfrac{1}{x + 1}$ 20. $y = \dfrac{1}{x - 3}$

21. $y = \dfrac{1}{x + 3}$ 22. $y = \dfrac{2}{x - 4}$

23. $y = \dfrac{3}{x - 1}$ 24. $xy = 1$

25. $xy = -2$ 26. $y = \dfrac{2x + 3}{4x - 1}$

27. $y = \dfrac{3x - 6}{2x + 1}$

28. $y = \dfrac{x - 5}{x + 1}$

29. $y = \dfrac{5x + 10}{x - 3}$

30. $y = \dfrac{x}{x + 5}$

31. $y = \dfrac{2x - 8}{3x + 6}$

32. $y = \dfrac{x - 1}{2x}$

33. $y = \dfrac{x^2}{3x^2 - 12}$

34. $y = \dfrac{x^2}{x^2 - 9}$

35. $y = \dfrac{3x^2}{x^2 - 4}$

36. $y = \dfrac{4x^2}{x - 3}$

37. $y = \dfrac{x^2}{x - 1}$

38. $y = \dfrac{2x^2}{x - 3}$

39. $y = \dfrac{3x^2}{x + 1}$

40. $y = \dfrac{4x^2}{x + 3}$

41. $9x^2 + 4y^2 = 36$

42. $4x^2 + 4y^2 = 36$

43. $9x^2 + y^2 = 9$

44. $9x^2 + 9y^2 = 9$

45. $x^2 + 9y^2 = 9$

46. $9x^2 - 4y^2 = 36$

47. $4x^2 - 9y^2 = 36$

48. $25x^2 + 16y^2 = 400$

49. $16x^2 + 25y^2 = 400$

50. $25x^2 - 16y^2 = 400$

CHECK LIST FOR THE STUDENT

If you have learned the material in Section 2.3, you should be able to do the following:

1. Find the intercepts of the graphs of some equations.

2. Test for symmetry with respect to the x-axis.

3. Test for symmetry with respect to the y-axis.

4. Test for symmetry with respect to the origin.

5. Find vertical asymptotes for $y = P(x)/D(x)$ provided that any exist and provided that you can solve $D(x) = 0$.

6. Find horizontal asymptotes for $y = P(x)/D(x)$ where $P(x)$ and $D(x)$ are polynomials in x provided that any horizontal asymptotes exist.

7. Determine excluded areas for some equations by considering signs of the variables.

8. Determine excluded areas for some equations by ruling out imaginary values of the variables.

9. Sketch the graphs of equations similar to those illustrated in the examples of Section 2.3.

SECTION QUIZ 2.3

1. Find the intercepts of the graph of $4x^2 + 25y^2 = 100$.

2. Determine whether or not the graph of each of the following is symmetric with respect to the x-axis.
   (a) $y^2 = 3x$ (b) $y = 3x^2$ (c) $4x^2 + 25y^2 = 100$ (d) $y = 2x^3$

3. Determine whether or not the graph of each of the following is symmetric with respect to the y-axis.
   (a) $y^2 = 3x$ (b) $y = 3x^2$ (c) $4x^2 + 25y^2 = 100$ (d) $y = 2x^3$

4. Determine whether or not the graph of each of the following is symmetric with respect to the origin.
   (a) $y^2 = 3x$ (b) $y = 3x^2$ (c) $4x^2 + 25y^2 = 100$ (d) $y = 2x^3$

5. In each of the following, find any vertical asymptote(s) that the graph has.

   (a) $y = \dfrac{x - 1}{x^2 - 9}$ (b) $y = \dfrac{x - 1}{x^2 + 9}$ (c) $y = \dfrac{x - 1}{x - 9}$

6. In each of the following, find any horizontal asymptote(s) that the graph has.

   (a) $y = \dfrac{3x}{x^2 + 4}$ (b) $y = \dfrac{2x - 3}{5x + 1}$ (c) $y = \dfrac{x^2 + 4}{3x + 1}$

7. Determine areas from which the graph of

$$y = \frac{5x + 5}{x - 4}$$

is excluded by considering signs of the variables.

8. Determine areas from which the graph of $4x^2 + 25y^2 = 100$ is excluded by ruling out imaginary values of the variables.

Discuss the graph of each of the following, taking into consideration intercepts, symmetry with respect to axes and origin, vertical and horizontal asymptotes, and excluded areas. Sketch the graph. Label any intercepts or vertical or horizontal asymptotes that the graph has.

9. $25x^2 + 4y^2 = 100$

10. $y = \dfrac{3x + 6}{2x - 8}$

2.4 Distance and Midpoint Formulas; Circles; Translation

The purpose of this section is to present formulas for the distance between two points and for the midpoint of a line segment, to consider equations of circles, and to show how to transform an equation of a graph by translating the coordinate axes.

DISTANCE AND MIDPOINT FORMULAS

The distance between two points on a line parallel to the x-axis is the absolute value of the difference in their x-coordinates. For example, the distance between (2, 3) and (7, 3) on the line $y = 3$ is $|7 - 2| = 5$. In Figure 2.29 we see that $|7 - 2|$ is the number of units between (2, 3) and (7, 3). Note that $|7 - 2| = |2 - 7|$.

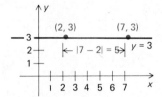

FIGURE 2.29 $|7 - 2| = |2 - 7| = 5$ The distance between (2, 3) and (7, 3)

The distance between two points on a line parallel to the y-axis is the absolute value of the difference in their y-coordinates. In Figure 2.30 the distance between (2, 1) and (2, 5) on the line $x = 2$ is $|5 - 1| = 4$. Note that $|5 - 1| = |1 - 5|$.

FIGURE 2.30 $|5 - 1| = |1 - 5| = 4$ The distance between (2, 1) and (2, 5)

It is easy to find the distance between two points on an oblique line by constructing a right triangle and making a direct application of the theorem from geometry called the *Pythagorean theorem*. This theorem states that in any right triangle, the square of the length of the hypotenuse is equal to the sum of the squares of the lengths of the other two sides. To demonstrate, let us use the Pythagorean theorem to find the distance between (1, 2) and (4, 6). Let P be the point (1, 2), and let Q be the point (4, 6), as shown in Figure 2.31. Draw a line through P parallel to the x-axis and a line through Q parallel to the y-axis. Let the point of intersection of these two parallels be M. Then M has coordinates (4, 2), and the triangle PMQ is a right triangle with hypotenuse $\overline{PQ}$. The length of the hypotenuse is the distance d between P and Q. The length of $\overline{PM}$ is $|4 - 1| = 3$; the length of $\overline{MQ}$ is $|6 - 2| = 4$. Using the Pythagorean theorem,

$$d^2 = 3^2 + 4^2$$
$$= 9 + 16$$
$$= 25.$$

Since the distance between two points is nonnegative, $d = 5$.

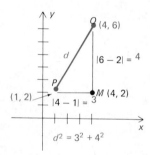

FIGURE 2.31   The distance between (1, 2) and (4, 6)

Precisely the same technique can be used to derive a general formula for the distance between point P_1 with coordinates (x_1, y_1) and point P_2 with coordinates (x_2, y_2) where P_1 and P_2 are on an oblique line. In Figure 2.32, triangle P_1MP_2 is a right triangle with hypotenuse $\overline{P_1P_2}$. The coordinates of M are (x_2, y_1); the length of $\overline{P_1M}$ is $|x_2 - x_1|$; the length of $\overline{MP_2}$ is $|y_2 - y_1|$. By the Pythagorean theorem,

$$d^2 = |x_2 - x_1|^2 + |y_2 - y_1|^2.$$

But $|x_2 - x_1|^2$ is the same as $(x_2 - x_1)^2$, and $|y_2 - y_1|^2$ is the same as $(y_2 - y_1)^2$, so we can write

$$d^2 = (x_2 - x_1)^2 + (y_2 - y_1)^2.$$

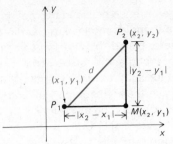

FIGURE 2.32 $d^2 = (x_2 - x_1)^2 + (y_2 - y_1)^2$

And since the distance between two points is nonnegative,

$$d = \sqrt{(x_2 - x_1)^2 + (y_2 - y_1)^2}.$$ Distance between (x_1, y_1) and (x_2, y_2)

EXAMPLE 1   Find the distance between $(-2, 3)$ and $(10, 8)$.

Solution   If we call $(-2, 3)$ the point P_1 and $(10, 8)$ the point P_2, we have

$$d = \sqrt{(10 - (-2))^2 + (8 - 3)^2}$$
$$= \sqrt{12^2 + 5^2}$$
$$= \sqrt{144 + 25}$$
$$= \sqrt{169}$$
$$= 13.$$

Note that we get the same answer if we call $(10, 8)$ the point P_1 and $(-2, 3)$ the point P_2. That is because $(x_2 - x_1)^2$ is the same as $(x_1 - x_2)^2$, and $(y_2 - y_1)^2$ is the same as $(y_1 - y_2)^2$.

$$d = \sqrt{(-2 - 10)^2 + (3 - 8)^2}$$
$$= \sqrt{(-12)^2 + (-5)^2}$$
$$= \sqrt{144 + 25}$$
$$= 13$$

Exercise 2.4, 1–10

We can construct appropriate similar triangles and use a geometric argument to prove that the x-coordinate of the midpoint of a line segment is the *average* of the x-coordinates of the endpoints, and the y-coordinate of the midpoint is the *average* of the y-coordinates of the endpoints. More precisely, as displayed in Figure 2.33, the midpoint of the line segment with endpoints at (x_1, y_1) and (x_2, y_2) is

$$\left(\frac{x_1 + x_2}{2}, \frac{y_1 + y_2}{2} \right).$$ Midpoint of segment with endpoints (x_1, y_1) and (x_2, y_2)

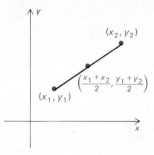

FIGURE 2.33 Midpoint: $\left(\dfrac{x_1 + x_2}{2}, \dfrac{y_1 + y_2}{2}\right)$

EXAMPLE 2

Use the midpoint formula to find the coordinates of the midpoint of the line segment with endpoints $(-2, 3)$ and $(4, -5)$. Use the distance formula to verify that the midpoint is equidistant from each end.

Solution

The midpoint of the line segment with endpoints $(-2, 3)$ and $(4, -5)$ is

$$\left(\frac{-2 + 4}{2}, \frac{3 + (-5)}{2}\right).$$

Simplifying these coordinates, we obtain $(\frac{2}{2}, -\frac{2}{2})$ or

$$(1, -1).$$

Using the distance formula, we can show that the distance between $(1, -1)$ and $(-2, 3)$ is the same as the distance between $(1, -1)$ and $(4, -5)$. The distance between $(1, -1)$ and $(-2, 3)$ is

$$\sqrt{(-2 - 1)^2 + (3 - (-1))^2} = \sqrt{3^2 + 4^2} = \sqrt{25} = 5.$$

The distance between $(1, -1)$ and $(4, -5)$ is

$$\sqrt{(4 - 1)^2 + (-5 - (-1))^2} = \sqrt{3^2 + (-4)^2} = \sqrt{25} = 5.$$

Exercise 2.4, 11–15

The point $(1, -1)$ *is* equidistant from each end.

CIRCLES

A **circle** is the set of all points in a plane equally distant from a fixed point. The fixed point is called the **center**. The distance from the center to any point on the circle is called the **radius**. Twice the radius is called the **diameter**.

We can use the distance formula to find an equation of the circle with a given center and radius. That is, we can find an equation that is satisfied by every point on the circle and by no other point. To demonstrate, let us find an equation of the circle with center at $(-2, 3)$ and with radius 5. In

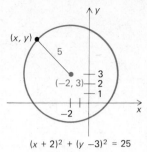

$$(x + 2)^2 + (y - 3)^2 = 25$$

FIGURE 2.34 Circle with center $(-2, 3)$ and radius 5

Figure 2.34, let (x, y) be any point on the circle. Then the distance from (x, y) to the center $(-2, 3)$ must be 5. Using the distance formula,

$$\sqrt{(x + 2)^2 + (y - 3)^2} = 5.$$

This is an equation of the circle with center at $(-2, 3)$ and radius 5. We can write the equation in an equivalent form by squaring both sides.

$$(x + 2)^2 + (y - 3)^2 = 25$$

Using precisely the same technique we just used for the specific circle in the previous paragraph, we can derive a general form for an equation of a circle with center at (h, k) and radius r where h, k, and nonnegative r represent specific real numbers. To do this, we again let (x, y) be any point on the circle, and we use the distance formula to express the fact that the distance from (x, y) to the center (h, k) must be equal to the radius.

$$\sqrt{(x - h)^2 + (y - k)^2} = r$$

Squaring both sides, we obtain the "standard" form

$$(x - h)^2 + (y - k)^2 = r^2 \qquad \text{Circle: Center } (h, k), \text{ radius } r$$

If the center is at the origin, $h = 0$ and $k = 0$ so that the standard form becomes much simpler.

Exercise 2.4, 16–20

$$x^2 + y^2 = r^2 \qquad \text{Circle: Center } (0, 0), \text{ radius } r$$

EXAMPLE 3

What are the center and radius of the circle $(x - 1)^2 + (y + 4)^2 = 36$?

Solution

Comparing the given equation with the standard form, we see that $h = 1$, $k = -4$, and $r^2 = 36$. The center is therefore at $(1, -4)$ and the radius is 6. (The radius must be nonnegative.)

If we multiply out the standard form $(x - h)^2 + (y - k)^2 = r^2$ and collect terms, we obtain the equation

$$x^2 + y^2 - 2hx - 2ky + (h^2 + k^2 - r^2) = 0.$$

We see that this is a special case of the second-degree equation

$$Ax^2 + By^2 + Cx + Dy + E = 0$$

where A, B, C, D, and E represent specific real numbers. In our special case, $A = B = 1$, $C = 2h$, $D = 2k$, and $E = h^2 + k^2 - r^2$. If we are given such a second-degree equation with $A = B$ and A and B nonzero, we can complete squares and write the equation in standard form as illustrated in Exercise 1.7 and in the next example. This next example shows that such a second-degree equation represents either a circle, a point, or no graph at all, depending on whether the constant term on the right side of the standard form is positive, zero, or negative.

Exercise 2.4, 21–25

EXAMPLE 4

Write each of the following equations in the form $(x - h)^2 + (y - k)^2 = r^2$. If the equation has a graph, sketch or plot the graph; if the equation has no graph, simply state "No graph."

(a) $2x^2 + 2y^2 - 16x - 20y + 64 = 0$

(b) $2x^2 + 2y^2 - 16x - 20y + 82 = 0$

(c) $2x^2 + 2y^2 - 16x - 20y + 100 = 0$

Solution

(a) $2x^2 + 2y^2 - 16x - 20y + 64 = 0$

We will use the completing-the-square process discussed in Sections 1.5 and 1.7 to complete the square in x and to complete the square in y.

1. Divide through by the common coefficients of x and y.

$$x^2 + y^2 - 8x - 10y + 32 = 0$$

2. Isolate the nonvariable term on the right; write the terms in x and the terms in y together.

$$x^2 - 8x + y^2 - 10y = -32$$

3. To both sides of the equation add the square of half the coefficient of x and also the square of half the coefficient of y.

$$x^2 - 8x + (-4)^2 + y^2 - 10y + (-5)^2 = -32 + (-4)^2 + (-5)^2$$
$$x^2 - 8x + 16 + y^2 - 10y + 25 = -32 + 16 + 25$$
$$x^2 - 8x + 16 + y^2 - 10y + 25 = 9$$

4. Factor the first three terms; factor the second three terms.

$$(x - 4)^2 + (y - 5)^2 = 9$$

This is the standard form of an equation of a circle with center at (4, 5) and radius 3. It is sketched in Figure 2.35.

(b) $2x^2 + 2y^2 - 16x - 20y + 82 = 0$

Completing the square in x and also in y we have the following.

$$x^2 - 8x \qquad + y^2 - 10y \qquad = -41$$
$$x^2 - 8x + 16 + y^2 - 10y + 25 = -41 + 16 + 25$$
$$(x - 4)^2 \qquad + (y - 5)^2 \qquad = 0$$

This equation represents the single point (4, 5). It is shown in Figure 2.35.

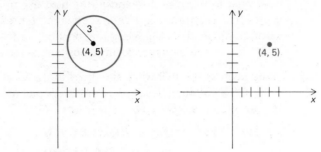

FIGURE 2.35   Circle: $(x - 4)^2 + (y - 5)^2 = 9$
Point: $(x - 4)^2 + (y - 5)^2 = 0$

(c) $2x^2 + 2y^2 - 16x - 20y + 100 = 0$

Again completing squares, we have:

$$x^2 - 8x \qquad + y^2 - 10y \qquad = -50$$
$$x^2 - 8x + 16 + y^2 - 10y + 25 = -50 + 16 + 25$$
$$(x - 4)^2 \qquad + (y - 5)^2 \qquad = -9$$

Since the sum of two squares is nonnegative, there are no points (x, y) whose coordinates satisfy this equation. This equation has no graph.

Exercise 2.4, 26–35

TRANSLATION

We have seen that the equation of a circle is much simpler when the origin of the coordinate system is at the center of the circle. An equation of a circle or of any graph depends on the orientation or placement of the graph relative to the coordinate axes. In calculus, as well as in most applications, calculations can frequently be simplified by proper location of coordinate axes. The circle with radius 5 in Figure 2.34 has the equation $(x + 2)^2 + (y - 3)^2 = 25$ because the coordinate axes are labeled the x-axis and the y-axis and are located so that the center of the circle is at $(-2, 3)$. If the coordinate axes were labeled x'-y' and were located so that the center of the circle was at the origin, that same circle would have the simpler equation $x'^2 + y'^2 = 25$. Figure 2.36 demonstrates this. In that figure we see that the center Q has coordinates $(-2, 3)$ relative to the x-y coordinate system, but Q has coordinates $(0, 0)$ relative to the x'-y' coordinate system.

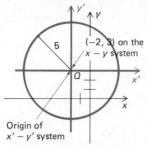

FIGURE 2.36   The circle $(x + 2)^2 + (y - 3)^2 = 25$
(The circle $x'^2 + y'^2 = 25$)

The process of transforming an equation of a graph by referring the graph to a new set of axes parallel to and having the same directions as the original axes is called **translation of axes**. We can transform the equation $(x + 2)^2 + (y - 3)^2 = 25$ of the circle in Figure 2.34 by replacing $(x + 2)$ by x' and replacing $(y - 3)$ by y'. These algebraic replacements effectively refer the circle to a new set of axes parallel to the original axes, having the same directions as the original axes, and having the new origin at the old point $(-2, 3)$. In general, to translate axes so that the new axes are labeled x'-y' and have their origin at the old point (h, k), replace $(x - h)$ by x' and replace $(y - k)$ by y'.

$$x - h = x' \qquad y - k = y'$$

In other words, to translate axes so that the new origin is at (h, k), let

$$x = x' + h \quad \text{and} \quad y = y' + k.$$

These are called the **translation equations**. The coordinates of P in Figure 2.37 are shown relative to both the x-y axes and the x'-y' axes.

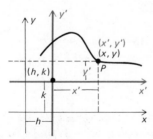

FIGURE 2.37 $x = x' + h$
$y = y' + k$
Translation of axes

EXAMPLE 5

Write the equation

$$3x^2 + 3y^2 - 18x + 30y + 90 = 0$$

in standard form; find the center and radius; transform the equation by translating axes so that the new origin is at the center of the circle; draw and label both sets of axes; sketch the graph.

Solution

$$3x^2 + 3y^2 - 18x + 30y + 90 = 0$$
$$x^2 + y^2 - 6x + 10y + 30 = 0$$
$$x^2 - 6x + y^2 + 10y = -30$$
$$x^2 - 6x + (\tfrac{-6}{2})^2 + y^2 + 10y + (\tfrac{10}{2})^2 = -30 + (\tfrac{-6}{2})^2 + (\tfrac{10}{2})^2$$
$$x^2 - 6x + 9 + y^2 + 10y + 25 = -30 + 9 + 25$$
$$(x - 3)^2 + (y + 5)^2 = 4$$

This is the standard form.

Center: $(3, -5)$
Radius: 2

To translate axes so that the new origin is at $(3, -5)$, let $x = x' + 3$ and $y = y' - 5$, so that $x - 3 = x'$ and $y + 5 = y'$. The standard form becomes

$$x'^2 + y'^2 = 4.$$

The axes and graph are shown in Figure 2.38.

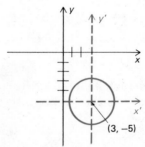

FIGURE 2.38 $3x^2 + 3y^2 - 18x + 30y + 90 = 0$
$$(x - 3)^2 + (y + 5)^2 = 4$$
$$x'^2 + y'^2 = 4$$

EXAMPLE 6

Transform and simplify the equation $y - x^2 - 6x - 9 = 0$ by translating the axes so that the new origin is at $(-3, 0)$. Draw and label both sets of axes, and sketch the graph.

Solution

The translation equations are $x = x' - 3$ and $y = y' + 0$. The given equation becomes

$$y' - (x' - 3)^2 - 6(x' - 3) - 9 = 0.$$

This may be simplified as follows:

$$y' - (x'^2 - 6x' + 9) - 6x' + 18 - 9 = 0$$
$$y' - x'^2 + 6x' - 9 - 6x' + 18 - 9 = 0$$
$$y' = x'^2$$

The graph of this equation was sketched in Figure 2.5 (Section 2.1) and again in Figure 2.21 (Section 2.3). It is shown with both sets of axes in Figure 2.39. Here the x and x' axes coincide, but the new origin is three units to the left of the original origin. The graph will be identified as a parabola in Section 2.5.

Exercises 2.4, 36–65

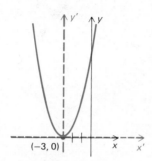

FIGURE 2.39 $y - x^2 - 6x - 9 = 0$
$$y' = x'^2$$

EXERCISE 2.4 DISTANCE AND MIDPOINT FORMULAS; CIRCLES; TRANSLATION

▶ In each of the following, find the distance between the two given points.

EXAMPLE

$(-1, 2), (11, 7)$

Solution

$$\sqrt{(11 - (-1))^2 + (7 - 2)^2} = \sqrt{144 + 25}$$
$$= 13$$

1. $(-1, 2), (7, 2)$ **2.** $(4, -3), (4, 5)$

3. $(-2, 5), (4, 12)$ **4.** $(1, -2), (7, 5)$

5. $(2, -4), (6, 0)$ **6.** $(-3, 1), (6, 13)$

7. $(-2, \frac{3}{5}), (10, \frac{7}{5})$ **8.** $(5, -7), (10, 5)$

9. $(-3, \frac{-2}{3}), (\frac{2}{3}, 10)$ **10.** $(-4, 3), (0, 6)$

▶ In each of the following, find the midpoint of the segment joining the two given points.

EXAMPLE $(2, -3), (7, 1)$

Solution $\left(\dfrac{2 + 7}{2}, \dfrac{-3 + 1}{2}\right)$ or $(\frac{9}{2}, -1)$

11. $(3, 7), (9, 1)$ **12.** $(1, -2), (6, 0)$

13. $(-3, 1), (4, 6)$ **14.** $(\frac{-2}{7}, \frac{3}{4}), (\frac{10}{7}, \frac{5}{4})$

15. $(-3, -2), (-4, 3)$

▶ In each of the following, find the standard form of an equation of the circle with the given center and radius.

EXAMPLE   Center $(3, -7)$, radius 6

Solution $(x - 3)^2 + (y + 7)^2 = 36$

16. Center $(-2, 3)$, radius 4   **17.** Center $(-4, -5)$, radius $\frac{3}{4}$

18. Center $(-3, \frac{1}{2})$, radius 8   **19.** Center $(5, 0)$, radius $\frac{7}{4}$

20. Center $(2, -5)$, radius 9

▶ Find the center and radius of each of the following circles.

EXAMPLE $(x + 2)^2 + (y - 7)^2 = 19$

Solution   Center $(-2, 7)$, radius $\sqrt{19}$

21. $(x + 3)^2 + (y - 5)^2 = 36$ **22.** $(x - 7)^2 + (y - 8)^2 = 13$

23. $(x - 6)^2 + (y + 2)^2 = 80$ **24.** $(x + \frac{1}{2})^2 + (y - \frac{5}{2})^2 = 10$

25. $(x - 7)^2 + (y + 8)^2 = 2$

▶ Write each of the following in the form $(x - h)^2 + (y - k)^2 = r^2$. If the equation has a graph, sketch or plot the graph; otherwise state "No graph."

EXAMPLE $x^2 + y^2 - 10x + 6y - 15 = 0$

Solution $x^2 - 10x + y^2 + 6y = 15$

$x^2 - 10x + 25 + y^2 + 6y + 9 = 15 + 34$

$(x - 5)^2 + (y + 3)^2 = 49$

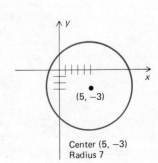

Center $(5, -3)$
Radius 7

26. $x^2 + y^2 - 8x - 2y - 8 = 0$

27. $x^2 + y^2 + 2x - 6y + 10 = 0$

28. $x^2 + y^2 - 4x + 10y + 65 = 0$

29. $x^2 + y^2 - 12x - 2y = -33$

30. $x^2 + y^2 + 14x + 12y + 76 = 0$

31. $4x^2 + 4y^2 - 4x + 24y + 56 = 0$

32. $16x^2 + 16y^2 + 8x - 160y + 145 = 0$

33. $2x^2 + 2y^2 - 16x - 12y + 50 = 0$

34. $64x^2 + 64y^2 - 16x + 128y = 191$

35. $2x^2 + 2y^2 = 8$

▶ In each of the following, transform the equation by translating axes so that the new origin is at the center of the circle. Draw and label both sets of axes, and sketch the graph.

EXAMPLE

$(x - 7)^2 + (y + 6)^2 = 16$

Solution

$x' = x - 7 \qquad y' = y + 6$

$x'^2 + y'^2 = 16$

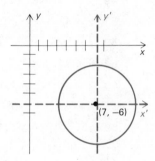

$(7, -6)$

36. $(x - 8)^2 + (y - 7)^2 = 25$

37. $(x + 1)^2 + (y + 4)^2 = 9$

38. $(x - 1)^2 + y^2 = 4$

39. $(x + 2)^2 + (y - 5)^2 = 1$

40. $x^2 + (y + 4)^2 = 9$

41. $(x - 5)^2 + (y + 2)^2 = 49$

42. $x^2 + (y + 4)^2 = 25$

43. $(x + 3)^2 + y^2 = 36$

44. $(x + 1)^2 + (y + 2)^2 = 81$

45. $(x + 6)^2 + (y - 8)^2 = 64$

▶ In each of the following, transform and simplify the given equation by translating the axes so that the new origin is at the given point. Draw and label both sets of axes, and sketch the graph.

EXAMPLE

$y - 1 = -\sqrt{x - 2}, \ (2, 1)$

Solution

$x = x' + 2$

$y = y' + 1$

$y' + 1 - 1 = -\sqrt{x' + 2 - 2}$

$y' = -\sqrt{x'}$ (See Example 8, Section 2.1)

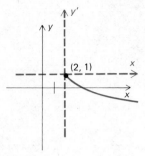

46. $y - x^2 - 2x - 3 = 0$, $(-1, 2)$

47. $y = x^2 + 3$, $(0, 3)$

48. $y - 3 = -\sqrt{x - 4}$, $(4, 3)$

49. $y - x^2 + 6x = 9$, $(3, 0)$

50. $y - 1 = \sqrt{x - 2}$, $(2, 1)$

51. $y = x^2 + 6x + 8$, $(-3, -1)$

52. $y - 2 = |x + 3|$, $(-3, 2)$

53. $y = |x + 3|$, $(-3, 0)$

54. $y = x^3 + 6x^2 + 12x + 8$, $(-2, 0)$

55. $y + 1 = (x - 2)^3$, $(2, -1)$

56. $(y - 1)^2 = x - 2$, $(2, 1)$

57. $y = 3x + 4$, $(0, 4)$

58. $y = 3x - 4$, $(0, -4)$

59. $y = -|x - 2|$, $(2, 0)$

60. $y + 3 = -|x - 4|$, $(4, -3)$

61. $y = \dfrac{1}{x - 1}$, $(1, 0)$

62. $y = \dfrac{1}{x + 1}$, $(-1, 0)$

63. $y + 2 = \dfrac{1}{x - 3}$, $(3, -2)$

64. $y - 2 = \dfrac{1}{x + 3}$, $(-3, 2)$

65. $y + 1 = \dfrac{3(x - 2)^2}{x - 3}$, $(2, -1)$

CHECK LIST FOR THE STUDENT

If you have learned the material in Section 2.4, you should be able to do the following:

1. Find the distance between two given points.

2. Find the midpoint of a line segment whose endpoints are given.

3. Find an equation of a circle whose center and radius are given.

4. Find the center and radius of a circle whose equation is given in standard form.

5. Complete squares and write an equation of the form $Ax^2 + By^2 + Cx + Dy + E = 0$ in the form $(x - h)^2 + (y - k)^2 = r^2$. If the equation has a graph, sketch or plot the graph; otherwise state "No graph."

6. Transform an equation in x and y by translating axes so that the new origin is at a given point.

7. Translate axes, and draw and label both sets of axes.

SECTION QUIZ 2.4

1. (a) Find the distance between the points $(-2, 5)$ and $(2, 8)$.
   (b) Find the midpoint of the segment joining the points $(5, -3)$ and $(3, 9)$.

2. Find the standard form of an equation of the circle with center at $(2, -5)$ and radius 7.

3. Find the center and radius of the circle $(x + 4)^2 + (y - 1)^2 = 36$.

4. Write each of the following in the form $(x - h)^2 + (y - k)^2 = r^2$. If the equation has a graph, sketch or plot the graph; otherwise state "No graph."
   (a) $x^2 + y^2 - 14x + 4y - 53 = 0$
   (b) $x^2 + y^2 - 2x + 4y + 14 = 0$

5. Transform the equation $(x + 6)^2 + (y - 4)^2 = 9$ by translating axes so that the new origin is at $(-6, 4)$. Draw and label both sets of axes. Sketch the graph.

2.5 Parabolas; Quadratic Functions

The purpose of this section is to define a parabola and derive equations for parabolas, to consider the graphs of certain parabolas, and to define and graph quadratic functions. An example at the end of the section indicates one way in which this material is put to use in calculus.

A **parabola** is the set of all points equally distant from a fixed point and a fixed line. The fixed point is called the **focus**. The fixed line is called the **directrix**. The line through the focus and perpendicular to the directrix is called the **axis** of the parabola. The point on the axis midway between the focus and the directrix is called the **vertex**. These points and lines are labeled for the parabola shown in Figure 2.40.

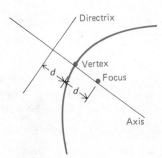

FIGURE 2.40   A parabola

If the coordinate axes are located so that the focus is on one of them and so that the directrix is parallel to the other with the origin halfway between the focus and the directrix, we can use the distance formula and find an equation of the parabola in a relatively simple form. For example, suppose that the focus is the point $(3, 0)$ on the x-axis and that the directrix is the line $x = -3$ parallel to the y-axis, as shown in Figure 2.41. Let (x, y) be

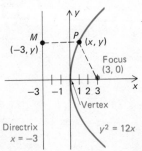

FIGURE 2.41   Parabola with focus $(3, 0)$ and directrix $x = -3$

any point on the parabola, and call (x, y) the point P. Then the distance from P to the directrix must equal the distance from P to the focus. By the distance from a point to a line, we mean the distance along the perpendicular to the line, so the distance from P to the directrix must be the distance between P and M shown in Figure 2.41. This distance is $|x - (-3)| = |x + 3|$.

$$|x + 3| \qquad \text{Distance from } (x, y) \text{ to}$$
$$\text{directrix in Figure 2.41}$$

By the distance formula, the distance from P to the focus is

$$\sqrt{(x - 3)^2 + y^2} \qquad \text{Distance from } (x, y) \text{ to}$$
$$\text{focus in Figure 2.41}$$

Since the distance from P to the directrix must equal the distance from P to the focus, we have

$$|x + 3| = \sqrt{(x - 3)^2 + y^2}.$$

This equation can be simplified by squaring both sides and collecting terms. (Note that $|x + 3|^2$ is the same as $(x + 3)^2$.)

$$|x + 3|^2 = (x - 3)^2 + y^2$$
$$x^2 + 6x + 9 = x^2 - 6x + 9 + y^2$$
$$12x - y^2 = 0$$
$$y^2 = 12x$$

The equation $y^2 = 12x$ is called the *standard form* of an equation of the parabola with focus at $(3, 0)$ and with directrix $x = -3$. The graph is shown in Figure 2.41. We see that the vertex is at the origin and that the parabola opens to the right.

Precisely the same technique can be used to derive an equation for any parabola whose focus is on the x-axis, whose vertex is at the origin, and whose directrix is perpendicular to the x-axis. For example, suppose that the focus is on the x-axis to the *right* of the origin with coordinates $(p, 0)$, as shown in Figure 2.42. Since the focus is to the *right* of the origin, p must be *positive*. Since the origin is midway between the focus and the directrix, the directrix must be the line $x = -p$. Now, again let (x, y) be any point on the parabola and call (x, y) the point P. Then in Figure 2.42, point M has coordinates $(-p, y)$, and the distance from P to the directrix is the distance from P to M, or $|x - (-p)| = |x + p|$.

$$|x + p| \qquad \text{Distance from } (x, y) \text{ to}$$
$$\text{directrix in Figure 2.42}$$

By the distance formula, the distance from (x, y) to the focus is

$$\sqrt{(x - p)^2 + (y - 0)^2} \qquad \text{Distance from } (x, y) \text{ to}$$
$$\text{focus in Figure 2.42}$$

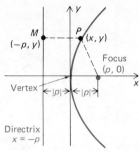

FIGURE 2.42 $y^2 = 4px$; $p > 0$, opens right

Since the distance from (x, y) to the directrix must equal the distance from (x, y) to the focus, we have

$$|x + p| = \sqrt{(x - p)^2 + (y - 0)^2}.$$

Squaring both sides, using the fact that $|x + p|^2 = (x + p)^2$, and collecting terms, we have the following.

$$(x + p)^2 = (x - p)^2 + (y - 0)^2$$
$$x^2 + 2px + p^2 = x^2 - 2px + p^2 + y^2$$
$$4px - y^2 = 0$$
$$y^2 = 4px$$

Any point whose coordinates satisfy this equation is equally distant from the focus $(p, 0)$ and the directrix $x = -p$. Notice that with the focus to the *right* of the origin, p is positive and the parabola opens to the *right*.

If the directrix is perpendicular to the x-axis and the focus is on the x-axis to the *left* of the origin with coordinates $(p, 0)$, p must be negative and the parabola opens to the *left*, as shown in Figure 2.43. In either case, $|p|$ is the distance between the vertex and the directrix.

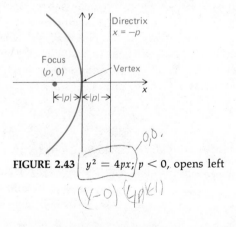

FIGURE 2.43 $y^2 = 4px$; $p < 0$, opens left

$(y - 0)(4p|x|)$

The equation

$$y^2 = 4px \qquad \text{Parabola: Vertex: } (0, 0)$$
$$\text{Axis: } x\text{-axis}$$
$$p > 0, \text{ opens right}$$
$$p < 0, \text{ opens left}$$

is called the standard form of an equation of a parabola with vertex at the origin, with focus on the x-axis at $(p, 0)$, with axis along the x-axis, and with directrix $x = -p$ perpendicular to the x-axis.

EXAMPLE 1

Find the vertex, axis, focus, and directrix of $y^2 = 12x$. Sketch the graph, plot the focus, and draw the directrix.

Comparing $y^2 = 12x$ with the standard form $y^2 = 4px$, we see that $y^2 = 12x$ represents a parabola with vertex at the origin and with axis along the x-axis. We also see that

$$4p = 12$$
$$p = 3.$$

Thus the focus must be $(3, 0)$ and the directrix must be $x = -3$. The graph was sketched in Figure 2.41.

If the x-y coordinate axes are located so that a parabola with axis parallel to or along the x-axis has vertex at (h, k) instead of at the origin, we can translate the axes and readily write the standard form of an equation of the parabola by referring to a new set of axes labeled x'-y' with origin at the old point (h, k). Relative to the x'-y' axes, the parabola would have the equation

$$y'^2 = 4px'^2.$$

Using the translation equations

$$x - h = x'$$
$$y - k = y'$$

developed in the last section, on the x-y coordinate system the parabola must have the equation

$$(y - k)^2 = 4p(x - h). \qquad \text{Parabola: Vertex: } (h, k)$$
$$\text{Axis parallel to or along } x\text{-axis.}$$
$$p > 0, \text{ opens right}$$
$$p < 0, \text{ opens left}$$

Figure 2.44 shows such a parabola.

EXAMPLE 2

Find the vertex, axis, focus, and directrix of $(y - 5)^2 = 12(x + 4)$. Transform this equation by translating axes so that the new origin is at

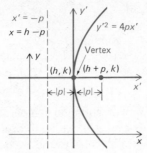

FIGURE 2.44 $(y - k)^2 = 4p(x - h);$ $p > 0$
$$y'^2 = 4px'$$

$(-4, 5)$. Draw and label both sets of axes, plot the focus, draw the directrix, and sketch the graph.

Solution

Comparing

$$(y - 5)^2 = 12(x + 4)$$

with

$$(y - k)^2 = 4p(x - h),$$

we see that the vertex is at $(-4, 5)$ and that $4p = 12$ so $p = 3$. Since the axis is parallel to the x-axis and goes through $(-4, 5)$, the axis must be the line $y = 5$. The focus is located on the axis a distance of $|p| = 3$ units to the right of the vertex and so must be at the point $(-4 + 3, 5)$, or $(-1, 5)$. The directrix is perpendicular to the x-axis and is located $|p| = 3$ units to the *left* of the vertex, so the directrix must be the line $x = -4 - 3$, or $x = -7$. The parabola is shown in Figure 2.45.

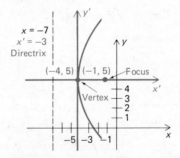

FIGURE 2.45 $(y - 5)^2 = 12(x + 4)$
$$y'^2 = 12x'$$

If we transform the equation $(y - 5)^2 = 12(x + 4)$ by translating axes so that the new origin is at $(-4, 5)$, the translation equations are

$$x + 4 = x'$$
$$y - 5 = y',$$

and the equation becomes

$$y'^2 = 12x'.$$

This is an equation of the parabola with vertex at the origin of the x'-y' coordinate system—that is, at the old point $(-4, 5)$. We could have sketched the graph by first translating the axes and then sketching the graph of $y'^2 = 12x'$ on the x'-y' coordinate system. The equation $y^2 = 12x$ was discussed in Example 1 and sketched in Figure 2.41.

We have seen that if the coordinate axes are located so that the vertex of a parabola is at the origin and the axis is along the x-axis, the parabola has the equation $y^2 = 4px$ where $|p|$ is the distance between the vertex and the directrix. In like manner, if the coordinate axes are located so that the vertex is at the origin and the axis is along the y-axis, the roles of x and y are interchanged, and the parabola has the equation

$$x^2 = 4py \qquad \text{Parabola: Vertex: } (0, 0)$$
$$\text{Axis: } y\text{-axis}$$
$$p > 0, \text{ opens up}$$
$$p < 0, \text{ opens down}$$

where $|p|$ is again the distance between the vertex and the directrix, where the focus is on the y-axis at $(0, p)$, and where the directrix is the line $y = -p$. If p is positive, the parabola $x^2 = 4py$ opens up; if p is negative, it opens down. Figures 2.46 and 2.47 show both cases.

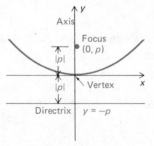

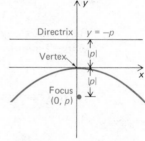

FIGURE 2.46 $x^2 = 4py$; $p > 0$, opens up

FIGURE 2.47 $x^2 = 4py$; $p < 0$, opens down

EXAMPLE 3

Find the vertex, axis, focus, and directrix of $x^2 = 12y$. Plot the focus, draw the directrix, and sketch the graph.

Solution

Comparing

$$x^2 = 12y$$

with

$$x^2 = 4py,$$

we see that $4p = 12$ so $p = 3$. The vertex is at the origin, the axis is along the y-axis, and since p is positive the parabola opens up. Since the focus is located on the axis $|p| = 3$ units above the origin, the focus must be at $(0, 3)$. Since the directrix is perpendicular to the y-axis and $|p| = 3$ units below the origin, the directrix must be the line $y = -3$. The graph is shown in Figure 2.48.

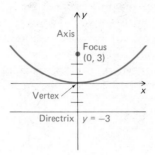

FIGURE 2.48 $x^2 = 12y$

If the x-y coordinate axes are located so that a parabola with axis parallel to or along the y-axis has vertex at (h, k) instead of at the origin, on the x-y coordinate system the parabola has the equation

$(x - h)^2 = 4p(y - k).$ Parabola: Vertex: (h, k)
  Axis parallel to or along y-axis
  $p > 0$, opens up
  $p < 0$, opens down

Relative to x'-y' axes with origin at the old point (h, k), the parabola would have the equation

$$x'^2 = 4py'.$$

Such a parabola is shown with both sets of axes in Figure 2.49.

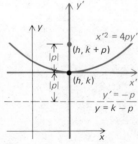

FIGURE 2.49 $(x - h)^2 = 4p(y - k);\quad p > 0$
$\qquad\qquad\quad x'^2 = 4py'$

PART TWO | ELEMENTARY FUNCTIONS AND ANALYTIC GEOMETRY

EXAMPLE 4

Find the vertex, axis, focus, and directrix of $(x + 4)^2 = -12(y - 5)$. Transform this equation by translating axes so that the new origin is at $(-4, 5)$. Draw and label both sets of axes, plot the focus, draw the directrix, and sketch the graph.

Solution

Comparing

$$(x + 4)^2 = -12(y - 5)$$

with

$$(x - h)^2 = 4p(y - k),$$

we see that the vertex is at $(-4, 5)$ and that $4p = -12$ so $p = -3$. Since the axis is parallel to the y-axis and goes through $(-4, 5)$, the axis must be the line $x = -4$. In Figure 2.50 the focus is located on the axis a distance of $|p| = 3$ units below the vertex and so must be at the point $(-4, 5 - 3)$, or $(-4, 2)$. The directrix is perpendicular to the y-axis and is located $|p| = 3$ units above the vertex, so the directrix must be the line $y = 5 + 3$, or $y = 8$.

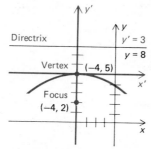

FIGURE 2.50 $(x + 4)^2 = -12(y - 5)$
$$x'^2 = -12y'$$

If we transform the equation $(x + 4)^2 = -12(y - 5)$ by translating axes so that the new origin is at $(-4, 5)$, the translation equations are

$$x + 4 = x'$$
$$y - 5 = y',$$

and the equation becomes

$$x'^2 = -12y'.$$

This is an equation of the parabola with vertex at the origin of the x'-y' coordinate system—that is, at the old point $(-4, 5)$. We could have sketched the graph by first translating the axes and then sketching the graph of $x'^2 = -12y'$ on the x'-y' coordinate system.

Exercise 2.5, 1–20

We have seen that parabolas with axes parallel to or along the co-ordinate axes have equations of the form

$$(y - k)^2 = 4p(x - h)$$

or of the form

$$(x - h)^2 = 4p(y - k).$$

These are called *standard forms*. If we multiply out the standard forms, we obtain the following second-degree equations.

$$y^2 - 4px - 2ky + (k^2 - 4ph) = 0$$
$$x^2 - 2hx - 4py + (h^2 + 4pk) = 0$$

The first of these second-degree equations is of the form

$$By^2 + Cx + Dy + E = 0$$

where B, C, D, and E represent specific real numbers. In this special case,

$$B = 1, \quad C = -4p, \quad D = -2k, \quad \text{and} \quad E = k^2 + 4ph.$$

The second of these second-degree equations is of the form

$$Ax^2 + Cx + Dy + E = 0$$

where in this case

$$A = 1, \quad C = -2h, \quad D = -4p, \quad \text{and} \quad E = h^2 + 4pk.$$

Any second-degree equation of the form

$$By^2 + Cx + Dy + E = 0$$

or of the form

$$Ax^2 + Cx + Dy + E = 0$$

where the letters A, B, C, D, and E represent specific real numbers is an equation of a parabola with its axis parallel to or along a coordinate axis. Such a second-degree equation can be written in a standard form by completing the square. This is illustrated in the next example.

EXAMPLE 5

Write the equation $2x^2 - 4x - 3y - 10 = 0$ in standard form. Find the vertex and axis and sketch the graph.

Solution

We begin to complete the square in x by factoring the coefficient of x^2 out of the terms involving x and by isolating the remaining terms on the other side of the equation.

$$2(x^2 - 2x \quad) = 3y + 10$$

To complete the square in x, we must insert the number 1 in the paren-

PART TWO | ELEMENTARY FUNCTIONS AND ANALYTIC GEOMETRY

theses. This is equivalent to adding $2 \times 1 = 2$, so we insert the 1 in the parentheses and add 2 to the other side of the equation.

$$2(x^2 - 2x + 1) = 3y + 10 + 2$$
$$2(x - 1)^2 = 3(y + 4)$$
$$(x - 1)^2 = \tfrac{3}{2}(y + 4)$$

This is in standard form. Comparing it with

$$(x - h)^2 = 4p(y - k),$$

we see that the vertex is at $(1, -4)$. The axis is parallel to the y-axis and goes through the vertex, so the axis is the line $x = 1$. The graph is shown in Figure 2.51.

Exercise 2.5, 21–30

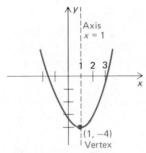

FIGURE 2.51 $2x^2 - 4x - 3y - 10 = 0$
$(x - 1)^2 = \tfrac{3}{2}(y + 4)$

Parabolas with axes along or parallel to the y-axis are graphs of *quadratic functions*. A **quadratic function** f is a function such that

1. The domain is the set of all real numbers,

and

2. $f(x) = ax^2 + bx + c$,

where a, b, and c represent specific real numbers and $a \neq 0$.

The graph of quadratic function f is the graph of the equation

$$y = ax^2 + bx + c.$$

This can be written $ax^2 + bx - y + c = 0$, so it is a second-degree equation of one of the two forms we have just been considering. Since we can complete the square in x and write this equation in the standard form

$$(x - h)^2 = 4p(y - k),$$

we know that the graph of quadratic function f is a parabola with axis parallel to or along the y-axis and opening either up or down. Furthermore, when we complete the square to put the equation in standard form, we see that the $4p$ of the standard form is $1/a$ where a is the coefficient of x^2 in the equation $y = ax^2 + bx + c$. Thus the sign of coefficient a determines whether the graph of f opens up or down. In short, we can say that the graph of the quadratic function f such that

$$f(x) = ax^2 + bx + c$$

is a parabola with axis parallel to or along the y-axis; the parabola opens up if a is positive, and it opens down if a is negative. This is illustrated in the next example.

EXAMPLE 6

Find the vertex and axis and sketch the graph of the function f such that $f(x) = -2x^2 + 12x - 17$.

Solution

The given function f is a quadratic function with the coefficient of x^2 negative, so we know in advance that the graph is a parabola with axis parallel to the y-axis and opening down. To find the vertex and axis, we complete the square in x and put the equation $y = -2x^2 + 12x - 17$ in standard form.

$$-2(x^2 - 6x \qquad) = y + 17$$
$$-2(x^2 - 6x + 9) = y + 17 - 18$$
$$-2(x - 3)^2 = y - 1$$
$$(x - 3)^2 = -\tfrac{1}{2}(y - 1)$$

The vertex is at $(3, 1)$, and the axis is the line $x = 3$. The graph is shown in Figure 2.52.

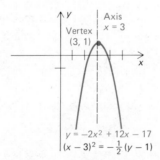

FIGURE 2.52 Graph of f: $f(x) = -2x^2 + 12x - 17$

In the previous example we found the vertex of the graph of the quadratic function f by completing the square in x and writing the equation $y = f(x)$ in standard form. Early in beginning calculus a more efficient method for locating the vertex of the graph of a quadratic function is encountered. That method is used more generally to locate relative high and low points on the graphs of functions other than quadratic functions.

One of the most common applications of calculus has to do with finding the speed with which a certain body moves provided that the distance s that it moves in time t is known. For example, if a projectile such as a bullet is fired from the ground, its distance s (in feet) above the ground t seconds later is given approximately by the equation

$$s = -16t^2 + kt$$

where k is a constant depending on the initial velocity and the angle from the horizontal to the direction of fire. Using a bit of calculus, the speed of the projectile at any time t can be found from this equation. Analyzing such a problem sometimes requires the ability to sketch the graph of the equation relating s and t. This is demonstrated for a particular case of the projectile problem in the next example.

EXAMPLE 7

A projectile is fired from the ground and its distance s above the ground at t seconds later is given by the equation $s = -16t^2 + 64t$. Sketch the graph of this equation on a Cartesian coordinate system having the t-axis for its horizontal axis and the s-axis for its vertical axis. At what time t does the projectile reach its maximum height? At what time t does the projectile return to the ground?

Solution

The function f such that $f(t) = -16t^2 + 64t$ is a quadratic function. The graph of $s = -16t^2 + 64t$ on a coordinate system having t as the horizontal axis is a parabola opening down. To find the vertex of the parabola, we complete the square in t and put the equation in standard form.

$$-16(t^2 - 4t \qquad) = s$$
$$-16(t^2 - 4t + 4) = s - 64$$
$$-16(t - 2)^2 = s - 64$$
$$(t - 2)^2 = -\tfrac{1}{16}(s - 64)$$

The vertex is at $t = 2$, $s = 64$; the axis is $t = 2$. Note that one t-intercept is at $t = 0$, $s = 0$. To find the other, let $s = 0$ and solve for t.

$$-16t^2 + 64t = 0$$
$$-16t(t - 4) = 0$$
$$t = 0 \quad \text{or} \quad t = 4$$

As shown in Figure 2.53, the graph crosses the t-axis at $(0, 0)$ and also at $(4, 0)$.

We see in Figure 2.53 that the maximum height (the largest s) is reached at the vertex. At the vertex, $t = 2$. Thus it takes the projectile 2 seconds to reach its maximum height.

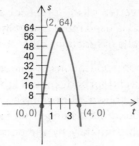

FIGURE 2.53 Graph of $s = -16t^2 + 64t$
(Graph of $(t - 2)^2 = -\frac{1}{16}(s - 64)$)

Exercise 2.5, 36–40

The projectile is on the ground when $s = 0$; that is, it is on the ground at $t = 0$ and also at $t = 4$. Therefore it takes the projectile 4 seconds to return to the ground.

For convenience of reference, the various forms of equations of parabolas considered in this section are summarized as follows:

$y^2 = 4px$ **Parabola:** Vertex: $(0, 0)$
                                      Axis: x-axis
                                      $p > 0$, opens right
                                      $p < 0$, opens left

$x^2 = 4py$ **Parabola:** Vertex: $(0, 0)$
                                      Axis: y-axis
                                      $p > 0$, opens up
                                      $p < 0$, opens down

$(y - k)^2 = 4p(x - h)$ **Parabola:** Vertex: (h, k)
                                      Axis parallel to or along
                                         x-axis
                                      $p > 0$, opens right
                                      $p < 0$, opens left

$(x - h)^2 = 4p(y - k)$ **Parabola:** Vertex: (h, k)
                                      Axis parallel to or along y-axis
                                      $p > 0$, opens up
                                      $p < 0$, opens down

$By^2 + Cx + Dy + E = 0$ **Parabola:** Axis parallel to or along x-axis
$Ax^2 + Cx + Dy + E = 0$ **Parabola:** Axis parallel to or along y-axis
$y = ax^2 + bx + c$ **Parabola:** Axis parallel to or along y-axis
                                      $a > 0$, opens up
                                      $a < 0$, opens down

EXERCISE 2.5 PARABOLAS; QUADRATIC FUNCTIONS

▶ In each of the following, find the vertex, axis, focus, and directrix. Sketch the graph, plot the focus, and draw the directrix.

EXAMPLE

$x^2 = -8y$

Solution

$x^2 = 4py$

$p = -2$, opens down
Vertex: (0, 0)
Axis: y-axis
Focus: (0, −2)
Directrix: $y = 2$

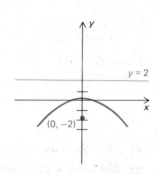

at origin

1. $y^2 = -8x$ 2. $x^2 = 8y$ 3. $y^2 = 8x$ 4. $y^2 = 4x$

5. $y^2 = -4x$ 6. $x^2 = 4y$ 7. $x^2 = -4y$ 8. $x^2 = -y$

9. $y^2 = -x$ 10. $x^2 = -\frac{1}{4}y$

▶ In each of the following, find the vertex, axis, focus, and directrix. Transform the equation by translating axes so that the new origin is at the vertex. Draw and label both sets of axes, plot the focus, draw the directrix, and sketch the graph.

EXAMPLE

$y^2 = -16(x - 5)$

Solution

Vertex: (5, 0)
Axis: x-axis
$p = -4$, opens left
Focus: (1, 0)
Directrix: $x = 9$
$y'^2 = -16x'$

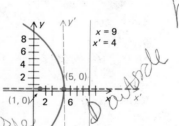

not at origin

11. $y^2 = -12(x - 5)$ 12. $y^2 = 16(x - 5)$
13. $x^2 = -16(y - 5)$ 14. $(x - 1)^2 = 4(y + 3)$
15. $(y - 3)^2 = 8(x + 1)$ 16. $(x - 4)^2 = -20(y + 7)$
17. $(y - 2)^2 = 2(x + 3)$ 18. $(x - 3)^2 = -(y + 4)$
19. $(y + 1)^2 = (x + 4)$ 20. $(x - 3)^2 = 3(y + 2)$

▶ Write each of the following equations in standard form. Find the vertex and axis and sketch the graph.

EXAMPLE

$y^2 - 2x + 2y + 7 = 0$

Solution

$y^2 + 2y = 2x - 7$

$y^2 + 2y + 1 = 2x - 6$

$(y + 1)^2 = 2(x - 3)$

Vertex: $(3, -1)$
Axis: $y = -1$

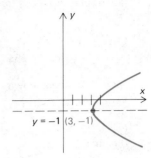

21. $x^2 - 2x - 4y - 7 = 0$

22. $4x^2 - 8x - 3y - 2 = 0$

23. $4y^2 - 3x + 8y + 10 = 0$

24. $2x^2 - 8x - 3y - 4 = 0$

25. $4y^2 - 7x - 40y + 142 = 0$

26. $x^2 - 6x - 7y + 2 = 0$

27. $2y^2 - 5x - 20y + 30 = 0$

28. $y^2 - 2x - 12y + 28 = 0$

29. $3x^2 - 12x - 5y + 7 = 0$

30. $2x^2 - 28x - 9y + 53 = 0$

▶ In each of the following, find the vertex and axis and sketch the graph of the function f.

EXAMPLE

$f(x) = 4x^2 - 40x + 101$

Solution

$y = 4x^2 - 40x + 101$

$4x^2 - 40x = y - 101$

$4(x^2 - 10x \quad\quad) = y - 101$

$4(x^2 - 10x + 25) = y - 1$

$(x - 5)^2 = \frac{1}{4}(y - 1)$

Vertex: $(5, 1)$
Axis: $x = 5$

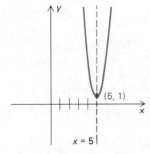

31. $f(x) = \frac{1}{4}x^2 - \frac{5}{2}x + \frac{29}{4}$

32. $f(x) = 4x^2 + 40x + 102$

33. $f(x) = -2x^2 - 16x - 29$

34. $f(x) = -x^2 + 2x - 4$

35. $f(x) = \frac{2}{7}x^2 + \frac{8}{7}x + \frac{15}{7}$

▶ In each of the following, assume that a projectile is fired from the ground and that its distance s above the ground at t seconds later is given by the equation $s = -16t^2 + kt$. For each given value of k, sketch the graph of $s = -16t^2 + kt$, determine the time it takes the projectile to reach its maximum height, and determine the time it takes the projectile to return to the ground.

EXAMPLE

$k = 96$

Solution

$s = -16t^2 + 96t$
$-16(t^2 - 6t) = s$
$\quad (t - 3)^2 = \frac{-1}{16}(s - 144)$

Vertex at $t = 3$, $s = 144$
If $s = 0$, $t = 0$ or 6.
Reaches maximum height at $t = 3$.
Returns to ground at $t = 6$.

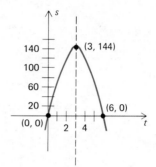

36. $k = 128$ **37.** $k = 160$ **38.** $k = 192$ **39.** $k = 48$ **40.** $k = 80$

CHECK LIST FOR THE STUDENT

If you have learned the material in Section 2.5, you should be able to do the following:

1. Define a parabola.

2. Recognize the standard form of an equation of a parabola.

3. Find the vertex, axis, focus, and directrix of a parabola from its equation in standard form.

4. Tell from looking at the standard form of an equation of a parabola whether it opens up, down, right, or left.

5. Sketch the graph of a parabola if its equation is given in standard form.

6. Transform the equations $(y - k)^2 = 4p(x - h)$ and $(x - h)^2 = 4p(y - k)$ by translating axes so that the new origin is at (h, k). Draw and label both sets of axes, and sketch the graphs of the equations.

7. Recognize equations of the form

$$By^2 + Cx + Dy + E = 0 \text{ or } Ax^2 + Cx + Dy + E = 0$$

as equations of parabolas and write such equations in standard form. (A, B, C, D, and E represent specific real numbers.)

8. Define a quadratic function.

9. Sketch the graph of a given quadratic function.

10. Tell from looking at the coefficient of x^2 in $f(x) = ax^2 + bx + c$ whether the graph of f opens up or down.

11. Sketch the graph of the equation $s = -16t^2 + kt$ for a given real number k.

12. Find the time it takes a projectile fired from the ground to reach its maximum height if its distance s above the ground at t seconds after firing is given by $s = -16t^2 + kt$ for some given constant k.

SECTION QUIZ 2.5

In each of the following, find the vertex, axis, focus, and directrix. Sketch the graph, plot the focus, and draw the directrix.

1. $x^2 = 20y$

2. $y^2 = -20x$

In each of the following, label the vertex and sketch the graph.

3. $(x + 3)^2 = -20(y - 2)$

4. $(y + 3)^2 = -20(x - 2)$

Write each of the following equations in standard form. Label the vertex and sketch the graph.

5. $y^2 - 6y + 9 - 20x = 0$

6. $x^2 - 2x - 8y - 7 = 0$

7. In each of the following, name the graph of the function f and tell whether it opens up or down.
   (a) $f(x) = 2x^2 - 12x + 17$
   (b) $f(x) = -2x^2 + 12x - 19$

8. If $f(x) = -x^2 + 1$, sketch the graph of f.

9. Sketch the graph of $y = x^2 + 2$.

10. A projectile is fired from the ground, and its distance s above the ground at t seconds after firing is given by $s = -16t^2 + 224t$. Find the time it takes the projectile to reach its maximum height.

2.6 Ellipses and Hyperbolas

The purpose of this section is to define ellipses and hyperbolas, to derive equations for ellipses and hyperbolas, and to demonstrate their graphs. Examples at the end of the section show some of what is done with these graphs in calculus.

ELLIPSES

An **ellipse** is the set of all points such that the *sum* of the distances from each of the points to two given fixed points is a constant. Each of the two fixed points is called a **focus** of the ellipse. The midpoint of the line segment joining the two foci (plural of focus) is called the **center** of the ellipse. An ellipse with foci F and F' is shown in Figure 2.54.

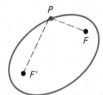

FIGURE 2.54   Ellipse: $PF + PF' = $ constant

If the coordinate axes are located so that they intersect at the center and so that one of the axes goes through the foci, we can use the distance formula and obtain an equation of the ellipse in a relatively simple form. For example, suppose that the axes are located as shown in Figure 2.55 so that the foci are on the x-axis at the points F and F' with coordinates

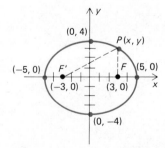

FIGURE 2.55   Ellipse with foci (3, 0) and (−3, 0)

$$PF + PF' = 10$$

$$\frac{x^2}{25} + \frac{y^2}{16} = 1$$

(3, 0) and (−3, 0). Further suppose that the sum of the distances from any point P on the ellipse to F and to F' is equal to the constant 10.

$$PF + PF' = 10$$

According to the distance formula,

$$PF = \sqrt{(x - 3)^2 + y^2} \quad \text{and} \quad PF' = \sqrt{(x + 3)^2 + y^2},$$

so $\quad \sqrt{(x - 3)^2 + y^2} + \sqrt{(x + 3)^2 + y^2} = 10.$

This equation represents the ellipse in Figure 2.55; it is satisfied by every point P on the ellipse and by no other point. It is customary to simplify such an equation by writing it in a form containing no radicals, as was done in Example 2 of Section 1.8. (Recall that to do this, we isolate a radical on one side, square both sides, collect like terms, and repeat the procedure.) Without radicals, this equation can be simplified to

$$16x^2 + 25y^2 = 400.$$

It can now be written in *standard form*

$$\frac{x^2}{a^2} + \frac{y^2}{b^2} = 1$$

by dividing both sides through by 400.

$$\frac{x^2}{25} + \frac{y^2}{16} = 1$$

Some properties of graphs that were discussed in Section 2.3 and that apply to the ellipse shown in Figure 2.55 can be readily deduced from the equation $x^2/25 + y^2/16 = 1$. (See Example 9 of Section 2.3.) For example, the ellipse is symmetric with respect to both axes and the origin. (Replacing x by $-x$ produces an equivalent equation, and replacing y by $-y$ produces an equivalent equation.) The intercepts are (5, 0), (−5, 0), (0, 4), and (0, −4). All points (x, y) for which $x > 5$ or for which $x < -5$ are excluded from the graph, because for such values of x the corresponding values of y would be imaginary. Likewise, all points (x, y) for which $y > 4$ or for which $y < -4$ are excluded from the graph, because for such values of y the corresponding values of x would be imaginary.

In general, suppose that the axes are located as shown in Figure 2.56 so that the foci are at the points F and F' with coordinates $(c, 0)$ and $(-c, 0)$ rather than at the specific points (3, 0) and (−3, 0). Also suppose that the sum of the distances from any point on the ellipse to F and F' equals the positive constant $2a$ where $a > c$, rather than the specific constant 10. (Specifying that $a > c$ makes the constant $2a$ greater than the distance between F and F'.) Then precisely the same type of algebraic manipulations as were performed in Example 2, Section 1.8, for our

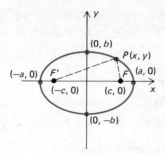

FIGURE 2.56 Ellipse with foci $(c, 0)$ and $(-c, 0)$

$$PF + PF' = 2a$$

$$\frac{x^2}{a^2} + \frac{y^2}{b^2} = 1,\ b^2 = a^2 - c^2$$

specific case will produce an equation of the ellipse in the standard form

$$\frac{x^2}{a^2} + \frac{y^2}{b^2} = 1 \qquad \text{Ellipse}$$

Ellipse
Foci at $(c, 0)$ and $(-c, 0)$

where $b^2 = a^2 - c^2$. (Note that since $a^2 = b^2 + c^2$, a^2 is the larger of the two denominators in the standard form.)

If the axes are located as shown in Figure 2.57 so that the foci are at the points F and F' with coordinates $(0, 3)$ and $(0, -3)$, and if $PF + PF' = 10$, an equation of the ellipse can be shown to be

$$\frac{x^2}{16} + \frac{y^2}{25} = 1.$$

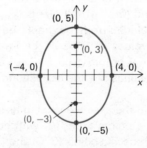

FIGURE 2.57 Ellipse with foci $(0, 3)$ and $(0, -3)$

$$PF + PF' = 10$$

$$\frac{x^2}{16} + \frac{y^2}{25} = 1$$

In general, as in Figure 2.58, if the foci are at $(0, c)$ and $(0, -c)$, and if $PF + PF' = 2a$ for positive $a > c$, an equation of the ellipse is

$$\frac{x^2}{b^2} + \frac{y^2}{a^2} = 1 \qquad \text{Ellipse} \\ \text{Foci at } (0, c) \text{ and } (0, -c)$$

where again $b^2 = a^2 - c^2$ and a^2 is the larger of the two denominators in the standard form.

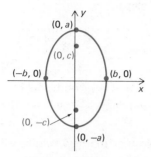

FIGURE 2.58 Ellipse with foci $(0, c)$ and $(0, -c)$

$$PF + PF' = 2a$$

$$\frac{x^2}{b^2} + \frac{y^2}{a^2} = 1, \ b^2 = a^2 - c^2$$

The line segment through the foci of an ellipse and with endpoints on the ellipse is called the **major axis** of the ellipse. The line segment through the center, perpendicular to the major axis, and with endpoints on the ellipse is called the **minor axis**. The major axis is longer than the minor axis. We see in Figure 2.56 that if the center is at the origin and the major axis is on the x-axis, the larger denominator a^2 is under x^2 in the standard form. Likewise, we see in Figure 2.58 that if the center is at the origin and the major axis is on the y-axis, the larger denominator a^2 is under y^2. If the center is at the origin, the ends of the axes are the intercepts. Using the equation in standard form, it is easy to sketch the graph simply by locating the intercepts and hence the ends of the axes.

Exercise 2.6, 1–10

If the center of an ellipse is at some point (h, k) rather than necessarily at the origin, and if the axes of the ellipse are parallel to the coordinate axes, the standard-form equation becomes

$$\frac{(x - h)^2}{a^2} + \frac{(y - k)^2}{b^2} = 1, \ a > b \qquad \text{Ellipse} \\ \text{Center: } (h, k) \\ \text{Major axis parallel to or} \\ \text{along } x\text{-axis}$$

or $\dfrac{(x-h)^2}{b^2} + \dfrac{(y-k)^2}{a^2} = 1,\ a > b.$ Ellipse
Center: (h, k)
Major axis parallel to or
along y-axis

Note that the larger denominator a^2 is under $(x-h)^2$ if the major axis is parallel to or along the x-axis, and the larger denominator a^2 is under $(y-k)^2$ if the major axis is parallel to or along the y-axis.

EXAMPLE 1

Locate the center of the ellipse

$$\frac{(x-2)^2}{25} + \frac{(y-3)^2}{16} = 1.$$

Transform the equation by translating coordinate axes so that the new origin is at the center. Draw and label both sets of coordinate axes. Label the ends of the axes of the ellipse. Sketch the graph.

Solution

The center is at $(2, 3)$.
  To translate the axes so that the new origin will be at $(2, 3)$, let

$$x' = x - 2 \quad \text{and} \quad y' = y - 3$$

or $\qquad\qquad x' + 2 = x \quad \text{and} \quad y' + 3 = y$

so that the equation becomes

$$\frac{x'^2}{25} + \frac{y'^2}{16} = 1.$$

This is an equation of the ellipse with center at the new origin and with major axis of length $2\sqrt{25} = 10$ and minor axis of length $2\sqrt{16} = 8$. In other words, the intercepts on the x'-y' axes are at $x' = 5$ and $y' = 0$, at $x' = -5$ and $y' = 0$, at $x' = 0$ and $y' = 4$, and at $x' = 0$ and $y' = -4$. On the x-y axes those are the points $(7, 3)$, $(-3, 3)$, $(2, 7)$, and $(2, -1)$, respectively. The graph and both sets of axes are sketched in Figure 2.59. We can check the location of the ends of the major axis by starting at the center $(2, 3)$ and counting off $\sqrt{25} = 5$ units to the right and then to the left on the x'-axis (the line $y = 3$).

  If the standard forms

$$\frac{(x-h)^2}{a^2} + \frac{(y-k)^2}{b^2} = 1 \quad \text{and} \quad \frac{(x-h)^2}{b^2} + \frac{(y-k)^2}{a^2} = 1$$

are multiplied out, the equations respectively become

$$b^2x^2 + a^2y^2 - 2b^2hx - 2a^2ky + b^2h^2 + a^2k^2 - a^2b^2 = 0$$
$$\ \ \ \|\qquad\quad \|$$
$$\ \ \ A\qquad\quad B$$

and

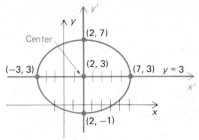

FIGURE 2.59 $\dfrac{(x - 2)^2}{25} + \dfrac{(y - 3)^2}{16} = 1$

$$\frac{x'^2}{25} + \frac{y'^2}{16} = 1$$

$$a^2 x^2 + b^2 y^2 - 2a^2hx - 2b^2ky + a^2h^2 + b^2k^2 - a^2b^2 = 0.$$
$$\underset{A}{\parallel} \qquad \underset{B}{\parallel}$$

These are both of the form

$$Ax^2 + By^2 + Cx + Dy + E = 0$$

where A and B are different since $a^2 > b^2$ and either $A = b^2$ and $B = a^2$, as they are in the first equation, or $A = a^2$ and $B = b^2$, as they are in the second equation. Note also that A and B are both of the same sign.

Where A, B, C, D, and E are all specific real numbers, any second degree equation of the form

$$Ax^2 + By^2 + Cx + Dy + E = 0, \qquad A \neq B \text{ and } A \text{ and } B \text{ same sign:}$$
$$\text{Ellipse, point, or no graph}$$

where $A \neq B$ and A and B are of the same sign, represents an ellipse, a single point, or no graph. The equation can be written in standard form by completing the squares in x and y, as was done in an illustrative example in Exercise 1.7 and as is illustrated again in the next example. After the squares are completed, if the constant term is isolated on one side of the equation and is positive, the equation represents an ellipse; if the isolated constant term is zero, the equation represents a single point; if the isolated constant term is negative, the equation represents no graph at all. Each of these cases is illustrated in the next example.

EXAMPLE 2

Write each of the following equations in standard form. If the equation has a graph, sketch it. If the equation has no graph, simply state "No graph."

(a) $16x^2 + 9y^2 + 160x - 36y + 292 = 0$

(b) $16x^2 + 9y^2 + 160x - 36y + 436 = 0$

(c) $16x^2 + 9y^2 + 160x - 36y + 580 = 0$

Solution

(a) $16x^2 + 160x \quad + 9y^2 - 36y \quad = -292$

$16(x^2 + 10x \quad) + 9(y^2 - 4y \quad) = -292$

Completing the squares in both parentheses by adding the appropriate numbers to both sides of the equation produces the following:

$$16(x^2 + 10x + 25) + 9(y^2 - 4y + 4) = -292 + 400 + 36$$

$$16(x + 5)^2 \quad + 9(y - 2)^2 \quad = 144$$

$$\frac{(x + 5)^2}{9} + \frac{(y - 2)^2}{16} = 1$$

This is an ellipse with center at $(-5, 2)$ and major axis parallel to the y-axis. The graph is sketched in Figure 2.60.

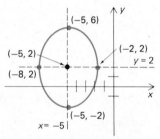

FIGURE 2.60   Ellipse: $16x^2 + 9y^2 + 160x - 36y + 292 = 0$

$$\frac{(x + 5)^2}{9} + \frac{(y - 2)^2}{16} = 1$$

(b) $16x^2 + 9y^2 + 160x - 36y + 436 = 0$

$$16(x^2 + 10x + 25) + 9(y^2 - 4y + 4) = -436 + 400 + 36$$

$$16(x + 5)^2 \quad + 9(y - 2)^2 \quad = 0$$

$$\frac{(x + 5)^2}{9} + \frac{(y - 2)^2}{16} = 0$$

The only point that satisfies this equation is the point $(-5, 2)$. To sketch the graph we need only to plot this single point. The graph is shown in Figure 2.61.

(c) $16x^2 + 9y^2 + 160x - 36y + 580 = 0$

$$16(x^2 + 10x + 25) + 9(y^2 - 4y + 4) = -580 + 400 + 36$$

$$16(x + 5)^2 \quad + 9(y - 2)^2 \quad = -144$$

$$\frac{(x + 5)^2}{9} + \frac{(y - 2)^2}{16} = -1$$

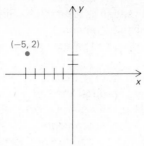

FIGURE 2.61 $16x^2 + 9y^2 + 160x - 36y + 436 = 0$

$$\frac{(x + 5)^2}{9} + \frac{(y - 2)^2}{16} = 0$$

The point $(-5, 2)$

This equation states that the sum of two squares is negative. There are no real numbers x and y for which this is true. There is no graph.

We have seen that if A and B are of the same sign but $A \neq B$, the equation $Ax^2 + By^2 + Cx + Dy + E = 0$ represents an ellipse, a single point, or no graph. Note that if A and B are equal, the equation represents a circle, a single point, or no graph, because then the equation can be divided through by the common value of A and B and the resulting equation can be put in the form $(x - h)^2 + (y - k)^2 = r^2$, as was illustrated in Example 4 of Section 2.4.

Exercise 2.6, 11–30

HYPERBOLAS

A **hyperbola** is the set of all points such that the absolute value of the *difference* of the distances from each of the points to two given fixed points is a constant. As in an ellipse, each of the two fixed points is called a **focus**. The midpoint of the line segment joining the two foci is called the **center**. A hyperbola with foci F and F' is shown in Figure 2.62. Notice

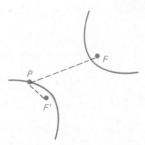

FIGURE 2.62 Hyperbola: $|PF - PF'| = $ constant

that a hyperbola has two separate branches. Since $|PF - PF'|$ is a constant (call it C), one of the branches consists of the points P such that $PF - PF' = C$, and the other branch consists of the points P such that $PF - PF' = -C$.

If the coordinate axes are located so that they intersect at the center and so that one of the axes goes through the foci, we can use the distance formula and obtain an equation of the hyperbola as we did for an ellipse. For example, suppose that the axes are located as shown in Figure 2.63 so that the foci are on the *x*-axis at the points F and F' with coordinates (5, 0) and (−5, 0). Also suppose that the absolute value of the difference of the distances from any point P on the hyperbola to F and to F' is equal to the constant 8.

$$|PF - PF'| = 8$$

Then, since $PF = \sqrt{(x - 5)^2 + y^2}$ and $PF' = \sqrt{(x + 5)^2 + y^2}$, an equation of the hyperbola is

$$|\sqrt{(x - 5)^2 + y^2} - \sqrt{(x + 5)^2 + y^2}| = 8.$$

In other words,

$$\sqrt{(x - 5)^2 + y^2} - \sqrt{(x + 5)^2 + y^2} = \pm 8$$

where the $+8$ is associated with points on one branch and the -8 is associated with points on the other branch. By systematically isolating a radical and squaring both sides as we did for an ellipse, we may write this as a single equation in standard form containing no radicals and no double signs. Let us outline these manipulations and show that we do obtain a single equation satisfied by points on both branches of the hyperbola.

Isolating a radical:

$$\sqrt{(x - 5)^2 + y^2} = \pm 8 + \sqrt{(x + 5)^2 + y^2}$$

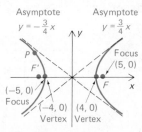

FIGURE 2.63 Hyperbola: $|PF - PF'| = 8$

$$\frac{x^2}{16} - \frac{y^2}{9} = 1$$

Squaring both sides: (Note that $(\pm 8)^2 = 64$.)

$$(x - 5)^2 + y^2 = 64 \pm 16\sqrt{(x + 5)^2 + y^2} + (x + 5)^2 + y^2$$

Collecting like terms and isolating a radical:

$$-64 - 20x = \pm 16\sqrt{(x + 5)^2 + y^2}$$

or

$$16 + 5x = \mp 4\sqrt{(x + 5)^2 + y^2}$$

Squaring both sides: (Note that $(\mp 4)^2 = 16$.)

$$(16 + 5x)^2 = 16[(x + 5)^2 + y^2]$$

Collecting like terms:

$$9x^2 - 16y^2 = 144$$

This equation may be written in the standard form

$$\frac{x^2}{a^2} - \frac{y^2}{b^2} = 1$$

by dividing through by 144.

$$\frac{x^2}{16} - \frac{y^2}{9} = 1$$

Some properties of the graph of the hyperbola considered in the previous paragraph are fairly evident. We see from the equation $x^2/16 - y^2/9 = 1$ that the graph is symmetric with respect to both coordinate axes and with respect to the origin. The x-intercepts are $(4, 0)$ and $(-4, 0)$. There are no y-intercepts, because if we let $x = 0$ we have $y^2 = -9$, and there are no such real numbers y. We can locate an area excluded by points on the graph by solving for y in terms of x and noting that some values of x would produce imaginary values of y. In particular, if we do solve for y in terms of x, we obtain

$$\frac{y^2}{9} = \frac{x^2}{16} - 1$$

$$y^2 = \frac{9}{16}(x^2 - 16)$$

$$y = \pm \frac{3}{4}\sqrt{x^2 - 16}.$$

From this we see that there are no points on the graph for which $-4 < x < 4$, because for such values of x the radicand $x^2 - 16$ is negative and y is imaginary.

The two points on a hyperbola that are nearest its center are called its **vertices** (plural of *vertex*). The vertices are always on the line through the foci. In Figure 2.63 the vertices are also the intercepts (4, 0) and $(-4, 0)$. Note that in the equation $x^2/16 - y^2/9 = 1$, the coefficient of x^2 is positive and the vertices and foci are on the x-axis. It is true in general that if the vertices and foci are on the x-axis, then the coefficient of x^2 in the standard form is positive.

In Figure 2.63 the lines $y = \frac{3}{4}x$ and $y = -\frac{3}{4}x$ appear to be asymptotes of the hyperbola. It can be shown that these lines *are* its asymptotes. It can also be shown that an easy mechanical way to find equations of the asymptotes is to factor the left side of the standard form and set each of the factors equal to zero. That is, factor $x^2/16 - y^2/9$, and set each of its factors equal to zero.

$$\frac{x^2}{16} - \frac{y^2}{9} = \left(\frac{x}{4} - \frac{y}{3}\right)\left(\frac{x}{4} + \frac{y}{3}\right)$$

$$\frac{x}{4} - \frac{y}{3} = 0 \qquad \frac{x}{4} + \frac{y}{3} = 0$$

$$\frac{y}{3} = \frac{x}{4} \qquad \frac{y}{3} = -\frac{x}{4}$$

$$y = \frac{3}{4}x \qquad y = -\frac{3}{4}x$$

Knowing these asymptotes makes it much easier to sketch the graph of the hyperbola.

We have just considered a particular hyperbola with foci at F and F', with coordinates (5, 0) and $(-5, 0)$, and such that $|PF - PF'| = 8$. Its center was at the origin, and in standard form its equation became

$$\frac{x^2}{16} - \frac{y^2}{9} = 1;$$

its vertices were located four units from the center on the x-axis at its only intercepts. In general, if the foci are at $(c, 0)$ and $(-c, 0)$, and if $|PF - PF'| = 2a$ with $a < c$, the center is at the origin and its equation in standard form is

$$\frac{x^2}{a^2} - \frac{y^2}{b^2} = 1 \qquad \begin{array}{l}\text{Hyperbola} \\ \text{Vertices on } x\text{-axis}\end{array}$$

where the number $c^2 - a^2$ is denoted by b^2. Specifying that $a < c$ makes the constant $2a$ less than the distance between F and F'.

$$c^2 - a^2 = b^2$$

(Note that a^2 is not prohibited from being equal to b^2. This is contrary to the case in the standard form of an equation of an ellipse, where $a^2 > b^2$.) As in the specific example, the graph is symmetric with respect to both axes, the only intercepts are on the x-axis, and the vertices are located a units from the center at the intercepts. Again as in the specific example, the asymptotes may be found mechanically by factoring $x^2/a^2 - y^2/b^2$ and setting each factor equal to zero. The asymptotes are

$$y = \frac{b}{a}x \quad \text{and} \quad y = -\frac{b}{a}x.$$

The hyperbola is sketched in Figure 2.64. Note that, as in the specific example, the coefficient of x^2 in the standard form is positive and the vertices and foci are on the x-axis.

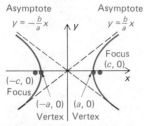

FIGURE 2.64 $\dfrac{x^2}{a^2} - \dfrac{y^2}{b^2} = 1$

If the foci of a hyperbola are on the y-axis at $(0, c)$ and $(0, -c)$, the standard form becomes

$$\frac{y^2}{a^2} - \frac{x^2}{b^2} = 1 \qquad \text{Hyperbola}$$
Vertices on y-axis

Again the graph is symmetric with respect to both axes, but this time the only intercepts are on the y-axis. The intercepts are at $(0, a)$ and $(0, -a)$. The vertices are located on the line through the foci and are a units from the center, so the vertices are also the points $(0, a)$ and $(0, -a)$. Here again we may use our mechanical technique to determine the asymptotes. The asymptotes are

$$\frac{y}{a} - \frac{x}{b} = 0 \quad \text{and} \quad \frac{y}{a} + \frac{x}{b} = 0.$$

That is, the asymptotes are

$$y = \frac{a}{b}x \quad \text{and} \quad y = -\frac{a}{b}x.$$

The graph is sketched in Figure 2.65. Note that in the equation $y^2/a^2 - x^2/b^2 = 1$, the coefficient of y^2 is positive and the vertices and foci are on the y-axis.

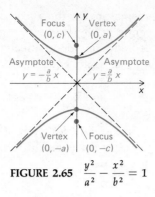

FIGURE 2.65 $\dfrac{y^2}{a^2} - \dfrac{x^2}{b^2} = 1$

If the center of a hyperbola is at the origin, it is easy to sketch the graph from the standard form by first locating the vertices (the intercepts) and drawing the asymptotes.

Exercise 2.6, 31–40

If the center of a hyperbola is at some point (h, k) rather than necessarily at the origin, and if the foci are on a line parallel to the x-axis or to the y-axis, then the standard-form equation becomes

$$\frac{(x - h)^2}{a^2} - \frac{(y - k)^2}{b^2} = 1$$

Hyperbola
Center: (h, k)
Vertices on line parallel to or along x-axis

or $\quad \dfrac{(y - k)^2}{a^2} - \dfrac{(x - h)^2}{b^2} = 1.$

Hyperbola
Center: (h, k)
Vertices on line parallel to or along y-axis

(Note that when the coefficient of $(x - h)^2$ is positive, the vertices are on a line parallel to the x-axis; when the coefficient of $(y - k)^2$ is positive, the vertices are on a line parallel to the y-axis.) The same mechanical technique that we used to determine the asymptotes of hyperbolas with centers at the origin and vertices on a coordinate axis can be used here. That is, to find equations for the asymptotes, factor the left side of the standard form and set each factor equal to zero.

EXAMPLE 3

Locate the center and vertices of the hyperbola

$$\frac{(y - 5)^2}{4} - \frac{(x + 2)^2}{9} = 1.$$

Transform the equation by translating coordinate axes so that the new origin is at the center. Draw and label both sets of coordinate axes. Label

PART TWO | ELEMENTARY FUNCTIONS AND ANALYTIC GEOMETRY

the vertices. Find equations of the asymptotes. Draw the asymptotes and sketch the graph.

The center is at $x = -2$, $y = 5$. The vertices are on a line through the center parallel to the y-axis and $\sqrt{4} = 2$ units from the center. The vertices must be at $x = -2$, $y = 7$ and at $x = -2$, $y = 3$.

To transform the equation by translating axes so that the new origin is at the center, let

$$x' = x + 2 \quad \text{and} \quad y' = y - 5.$$

The equation becomes

$$\frac{y'^2}{4} - \frac{x'^2}{9} = 1.$$

One way to find equations for the asymptotes is to factor $y'^2/4 - x'^2/9$ and set each factor equal to zero. Doing this, we have

$$\frac{y'}{2} - \frac{x'}{3} = 0 \quad \text{and} \quad \frac{y'}{2} + \frac{x'}{3} = 0,$$

or

$$y' = \frac{2}{3}x' \quad \text{and} \quad y' = -\frac{2}{3}x'.$$

Since $x' = x + 2$ and $y' = y - 5$, equations of the asymptotes referred to the x-y-axes are

$$y - 5 = \frac{2}{3}(x + 2) \quad \text{and} \quad y - 5 = -\frac{2}{3}(x + 2).$$

Another way to find equations of the asymptotes referred to the x-y axes is to factor $(y - 5)^2/4 - (x + 2)^2/9$ and set each factor equal to zero. If we do this, we have

$$\frac{y - 5}{2} - \frac{x + 2}{3} = 0 \quad \text{and} \quad \frac{y - 5}{2} + \frac{x + 2}{3} = 0,$$

or $y - 5 = \frac{2}{3}(x + 2)$ and $y - 5 = -\frac{2}{3}(x + 2)$, as we found before. The graph is sketched in Figure 2.66.

If the standard forms

$$\frac{(x - h)^2}{a^2} - \frac{(y - k)^2}{b^2} = 1 \quad \text{and} \quad \frac{(y - k)^2}{a^2} - \frac{(x - h)^2}{b^2} = 1$$

are multiplied out, the equations respectively become

$$\underset{\substack{\| \\ A}}{b^2x^2} - \underset{\substack{\| \\ B}}{a^2y^2} - 2b^2hx + 2a^2ky + b^2h^2 - a^2k^2 - a^2b^2 = 0$$

and

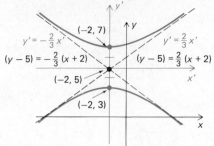

FIGURE 2.66 $\dfrac{(y-5)^2}{4} - \dfrac{(x+2)^2}{9} = 1$

$$\dfrac{y'^2}{4} - \dfrac{x'^2}{9} = 1$$

$$-a^2x^2 + b^2y^2 + 2a^2hx - 2b^2ky + b^2k^2 - a^2h^2 - a^2b^2 = 0.$$
$$\underset{A}{\parallel} \qquad \underset{B}{\parallel}$$

These are both of the form

$$Ax^2 + By^2 + Cx + Dy + E = 0$$

where A and B are of opposite signs because either $A = b^2$ and $B = -a^2$ or $A = -a^2$ and $B = b^2$. (Recall that in an ellipse, A and B are different but of the same sign. Here A and B might possibly have the same absolute values, but A and B are of opposite sign.) Where A, B, C, D, and E are specific real numbers and A and B have opposite signs, any equation of the form

$$Ax^2 + By^2 + Cx + Dy + E = 0 \qquad$$ A and B opposite sign
Hyperbola or two intersecting lines

represents a hyperbola or a pair of intersecting lines. The equation can be put in standard form by completing squares in x and y. When the squares are completed, if the constant term is isolated on one side of the equation and is not zero, the equation represents a hyperbola; if the isolated constant term is zero, the equation represents two intersecting lines. Both of these cases are illustrated in the next example.

EXAMPLE 4

Write each of the following equations in standard form and sketch the graph.

(a) $9x^2 - 4y^2 - 90x - 16y + 173 = 0$

(b) $9x^2 - 4y^2 - 90x - 16y + 209 = 0$

Solution

(a)
$$9x^2 - 90x \qquad\quad - (4y^2 + 16y \quad) = -173$$
$$9(x^2 - 10x \qquad) - 4(y^2 + 4y \quad) = -173$$
$$9(x^2 - 10x + 25) - 4(y^2 + 4y + 4) = -173 + 225 - 16$$
$$9(x - 5)^2 - 4(y + 2)^2 \qquad\quad = 36$$
$$\frac{(x - 5)^2}{4} - \frac{(y + 2)^2}{9} \qquad = 1$$

This is a hyperbola with center at $(5, -2)$ and with vertices located two units from the center on a line parallel to the x-axis. The graph is shown in Figure 2.67. Contrast this with the graph in Figure 2.66.

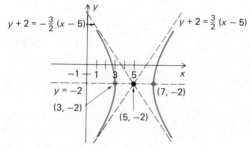

FIGURE 2.67 $\quad 9x^2 - 4y^2 - 90x - 16y + 173 = 0$
$$\frac{(x - 5)^2}{4} - \frac{(y + 2)^2}{9} = 1$$

(b)
$$9x^2 - 90x \qquad\quad - (4y^2 + 16y \quad) = -209$$
$$9(x^2 - 10x \qquad) - 4(y^2 + 4y \quad) = -209$$
$$9(x^2 - 10x + 25) - 4(y^2 + 4y + 4) = -209 + 225 - 16$$
$$9(x - 5)^2 - 4(y + 2)^2 \qquad\quad = 0$$
$$\frac{(x - 5)^2}{4} - \frac{(y + 2)^2}{9} \qquad = 0$$

The left side may be factored as the difference of two squares.

$$\left(\frac{x - 5}{2} - \frac{y + 2}{3}\right)\left(\frac{x - 5}{2} + \frac{y + 2}{3}\right) = 0$$

This equation is satisfied by all points (x, y) for which

$$\frac{x - 5}{2} - \frac{y + 2}{3} = 0 \quad \text{or} \quad \frac{x - 5}{2} + \frac{y + 2}{3} = 0.$$

Exercise 2.6, 41–60
Miscellaneous exercises,
Exercise 2.6, 61–70
These equations represent two lines intersecting at $(5, -2)$. These two lines constitute the graph of the given equation. The graph is shown in Figure 2.68.

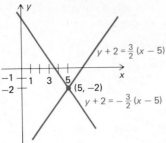

$$y + 2 = \frac{3}{2}(x - 5)$$

$$y + 2 = -\frac{3}{2}(x - 5)$$

(5, −2)

FIGURE 2.68 $9x^2 - 4y^2 - 90x - 16y + 209 = 0$

$$\frac{(x - 5)^2}{4} - \frac{(y + 2)^2}{9} = 0$$

In calculus one considers such problems as finding the volume of the solid generated by revolving an ellipse about one of its axes. In setting up such a problem, one might be led to consider the area of a circular cross section, as demonstrated in the next example.

EXAMPLE 5

If the ellipse $x^2/9 + y^2/4 = 1$ is revolved about the x-axis, what is the area of the circular cross section generated by the line segment from $(x, 0)$ on the x-axis to the point (x, y) on the ellipse? Express this cross-sectional area in terms of x. If (x, y) must be on the ellipse and x must be nonnegative, what values may x take on?

Solution

Let us first sketch the ellipse so that we can visualize this problem. The center is at the origin. The intercepts are at (3, 0), (−3, 0), (0, 2), and (0, −2). The sketch is shown in Figure 2.69.

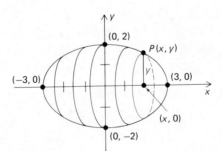

(0, 2)

$P(x, y)$

(−3, 0)

(3, 0)

(x, 0)

(0, −2)

FIGURE 2.69 $\dfrac{x^2}{9} + \dfrac{y^2}{4} = 1$, revolved about x-axis

As the ellipse is revolved about the x-axis, the line segment from $(x, 0)$ to (x, y) on the ellipse does generate a circular cross section. The radius of the circular cross section is y. The area of the circular cross section is therefore πy^2.

$$\text{Area of cross section} = \pi y^2$$

To express this area in terms of x, we note that the point (x, y) is on the ellipse so that x and y are related by the equation

$$\frac{x^2}{9} + \frac{y^2}{4} = 1.$$

Solving this for y^2 in terms of x, we have

$$\frac{y^2}{4} = 1 - \frac{x^2}{9}$$

$$y^2 = 4\left(1 - \frac{x^2}{9}\right).$$

It follows that in terms of x,

$$\text{Area of cross section} = 4\pi\left(1 - \frac{x^2}{9}\right).$$

If (x, y) is on the ellipse and x is nonnegative, x must be between 0 and 3 inclusive. That is, $0 \leq x \leq 3$.

Another type of problem considered in calculus is that of finding the area of a surface generated by revolving a graph or part of a graph about a line. For example, one might be required to find the area of the surface generated by revolving part of a hyperbola about the line through its vertices. In setting up a problem like this, one might be led to consider the circumference of a circle, as illustrated in the next example.

EXAMPLE 6

If the hyperbola $x^2/9 - y^2/4 = 1$ is revolved about the x-axis, what is the circumference of the circle generated by the point (x, y) on the hyperbola in the first quadrant? Express this circumference in terms of x. If (x, y) must be on the hyperbola and x is nonnegative, and if x must be less than or equal to 7, what values may x take on?

Solution

Let us first sketch the hyperbola. The center is at the origin. The vertices are on the x-axis at $(3, 0)$ and $(-3, 0)$. The sketch is shown in Figure 2.70.

As the hyperbola is revolved about the x-axis, the point (x, y) generates a circle of radius y. The circumference of the generated circle is therefore $2\pi y$.

$$\text{Circumference of generated circle} = 2\pi y$$

To express this circumference in terms of x, we use the fact that (x, y) is on the hyperbola so that

$$\frac{x^2}{9} - \frac{y^2}{4} = 1.$$

2.6 ELLIPSES AND HYPERBOLAS

269

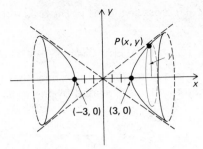

FIGURE 2.70 $\dfrac{x^2}{9} - \dfrac{y^2}{4} = 1$, revolved about x-axis

We now solve this for y in terms of x, and recall that (x, y) is in the first quadrant or on the x-axis so that y must be nonnegative.

$$\frac{-y^2}{4} = 1 - \frac{x^2}{9}$$

$$\frac{y^2}{4} = \frac{x^2}{9} - 1$$

$$y^2 = \frac{4}{9}(x^2 - 9)$$

$$y = \frac{2}{3}\sqrt{x^2 - 9}$$

Circumference of generated circle $= \dfrac{4\pi}{3}\sqrt{x^2 - 9}$

If (x, y) is on the hyperbola and in the first quadrant or on the x-axis with x nonnegative, then $x \geq 3$. If it is also true that $x \leq 7$, then $3 \leq x \leq 7$.

Exercise 2.6, 71–80

The various forms of equations of ellipses and hyperbolas considered in this section and similar equations encountered in Sections 2.4 and 2.5 are summarized in the following list. The graphs of all of these equations are called **conic sections,** because each is the cross section of a right circular cone cut by a plane. (See Figure 2.71.)

$$\frac{x^2}{a^2} + \frac{y^2}{b^2} = 1, \; a > b$$

Ellipse
Center: (0, 0)
Major axis on x-axis

$$\frac{x^2}{b^2} + \frac{y^2}{a^2} = 1, \; a > b$$

Ellipse
Center: (0, 0)
Major axis on y-axis

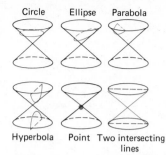

Circle   Ellipse   Parabola

Hyperbola Point Two intersecting
lines

FIGURE 2.71   The conic sections

$$\frac{(x-h)^2}{a^2} + \frac{(y-k)^2}{b^2} = 1, \ a > b$$

Ellipse
Center: (h, k)
Major axis parallel
to or along x-axis

$$\frac{(x-h)^2}{b^2} + \frac{(y-k)^2}{a^2} = 1, \ a > b$$

Ellipse
Center: (h, k)
Major axis parallel
to or along y-axis

$$\frac{x^2}{a^2} - \frac{y^2}{b^2} = 1$$

Hyperbola
Center: $(0, 0)$
Vertices on x-axis

$$\frac{y^2}{a^2} - \frac{x^2}{b^2} = 1$$

Hyperbola
Center: $(0, 0)$
Vertices on y-axis

$$\frac{(x-h)^2}{a^2} - \frac{(y-k)^2}{b^2} = 1$$

Hyperbola
Center: (h, k)
Vertices on line
parallel to or along
x-axis

$$\frac{(y-k)^2}{a^2} - \frac{(x-h)^2}{b^2} = 1$$

Hyperbola
Center: (h, k)
Vertices on line
parallel to or along
y-axis

$$y^2 = 4px$$

Parabola
Vertex: $(0, 0)$
Axis: x-axis

$$x^2 = 4py$$

Parabola
Vertex: $(0, 0)$
Axis: y-axis

$$(y - k)^2 = 4p(x - h)$$

Parabola
Vertex: (h, k)
Axis parallel to or along x-axis

$$(x - h)^2 = 4p(y - k)$$

Parabola
Vertex: (h, k)
Axis parallel to or along y-axis

$$y = ax^2 + bx + c$$

Parabola
Axis parallel to or along y-axis

$Ax^2 + By^2 + Cx + Dy + E = 0$, $A \neq B$ and A and B of same sign — Ellipse or Point or No graph

$Ax^2 + By^2 + Cx + Dy + E = 0$, A and B of opposite sign — Hyperbola or Two intersecting lines

$Ax^2 + By^2 + Cx + Dy + E = 0$, $A = B \neq 0$ — Circle or Point or No graph

$By^2 + Cx + Dy + E = 0$, $C \neq 0$, $B \neq 0$ — Parabola

$Ax^2 + Cx + Dy + E = 0$, $D \neq 0$, $A \neq 0$ — Parabola

EXERCISE 2.6 ELLIPSES AND HYPERBOLAS

▶ Find the center, label the ends of the axes, and sketch the graph of each of the following.

EXAMPLE

$$\frac{4}{9}x^2 + \frac{16}{81}y^2 = 1$$

Solution

$$\frac{x^2}{\frac{9}{4}} + \frac{y^2}{\frac{81}{16}} = 1$$

Center: $(0, 0)$

1. $4x^2 + 9y^2 = 36$　　　　　　　　　**2.** $9x^2 + 4y^2 = 36$

3. $9x^2 + 4y^2 = 1$　　　　　　　　　**4.** $4x^2 + 9y^2 = 1$

5. $\frac{16}{81}x^2 + \frac{4}{9}y^2 = 1$　　　　　　　**6.** $9x^2 + 16y^2 = 144$

7. $16x^2 + 9y^2 = 144$　　　　　　　**8.** $9x^2 + y^2 = 9$

9. $x^2 + 9y^2 = 9$　　　　　　　　　**10.** $4x^2 + 100y^2 = 400$

▶ In each of the following, locate the center and then transform the equation by translating coordinate axes so that the new origin is at the center. Draw and label both sets of coordinate axes. Label the ends of the axes of the ellipse. Sketch the graph.

EXAMPLE

$$\frac{(x-2)^2}{16} + \frac{(y-3)^2}{25} = 1$$

Solution

Center $(2, 3)$

$$\frac{x'^2}{16} + \frac{y'^2}{25} = 1$$

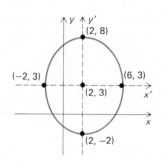

11. $\dfrac{(x+3)^2}{9} + \dfrac{(y-4)^2}{16} = 1$　　　**12.** $\dfrac{(x-3)^2}{1} + \dfrac{(y-2)^2}{9} = 1$

13. $\dfrac{(x+5)^2}{4} + \dfrac{(y+2)^2}{1} = 4$　　　**14.** $9(x-6)^2 + (y+3)^2 = 9$

15. $(x-1)^2 + 9(y+2)^2 = 9$　　　**16.** $\frac{1}{9}(x-1)^2 + \frac{1}{4}(y+4)^2 = 1$

17. $\frac{16}{25}(x+3)^2 + \frac{4}{9}(y-1)^2 = 1$

18. $100(x-5)^2 + 4(y-4)^2 = 400$

19. $16(x+3)^2 + 4(y-5)^2 = 64$　　**20.** $4(x+3)^2 + 16(y-5)^2 = 64$

▶ Write each of the following equations in standard form. If the equation has a graph, sketch it. If the equations has no graph, simply state "No graph."

EXAMPLE

$$4x^2 + 9y^2 - 16x + 90y + 242 = 0$$

Solution

$$4(x^2 - 4x \quad) + 9(y^2 + 10y \quad) = -242$$

$$4(x-2)^2 + 9(y+5)^2 = -242 + 16 + 225$$

$$4(x-2)^2 + 9(y+5)^2 = -1 \qquad \text{No graph}$$

21. $4x^2 + 9y^2 - 16x + 90y + 240 = 0$

22. $4x^2 + 9y^2 - 16x + 90y + 241 = 0$

23. $4x^2 + 9y^2 - 24x + 18y + 9 = 0$

24. $4x^2 + 9y^2 - 24x + 18y + 81 = 0$

25. $4x^2 + 9y^2 - 24x + 18y + 45 = 0$

26. $4x^2 + 25y^2 + 24x - 100y + 36 = 0$

27. $x^2 + 9y^2 - 10x - 56 = 0$

28. $4x^2 + 25y^2 + 24x - 100y + 236 = 0$

29. $x^2 + 9y^2 - 72y + 63 = 0$

30. $4x^2 + 25y^2 + 24x - 100y + 136 = 0$

▶ Find the center, vertices, and equations of the asymptotes of each of the following. Draw the asymptotes and sketch the graph.

EXAMPLE

$$\frac{y^2}{16} - \frac{x^2}{9} = 1$$

Solution

Center: $(0, 0)$
Vertices: $(0, 4)$ and $(0, -4)$

$$\frac{y^2}{16} - \frac{x^2}{9} = 0$$

$$\left(\frac{y}{4} - \frac{x}{3}\right)\left(\frac{y}{4} + \frac{x}{3}\right) = 0$$

Asymptotes: $y = \frac{4}{3}x$ and $y = -\frac{4}{3}x$

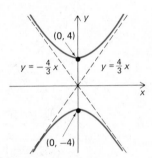

31. $4x^2 - 9y^2 = 36$ **32.** $9x^2 - 4y^2 = 36$

33. $9y^2 - 4x^2 = 1$ **34.** $4y^2 - 9x^2 = 1$

35. $\frac{16}{81}x^2 - \frac{4}{9}y^2 = 1$ **36.** $\frac{16}{81}y^2 - \frac{4}{9}x^2 = 1$

37. $4y^2 - 100x^2 = 400$ **38.** $100x^2 - 4y^2 = 400$

39. $x^2 - 9y^2 = 9$ **40.** $y^2 - 9x^2 = 9$

▶ Find the center and vertices of each of the following. Transform each equation by translating coordinate axes so that the new origin is at the center. Draw and label both sets of coordinate axes. Find equations of the asymptotes referred to the new coordinate axes, and also find equations of the asymptotes referred to the original coordinate axes. Draw the asymptotes and sketch the graph.

EXAMPLE

$$\frac{(x + 2)^2}{4} - \frac{(y - 5)^2}{9} = 1$$

PART TWO | ELEMENTARY FUNCTIONS AND ANALYTIC GEOMETRY

Solution

Center: $(-2, 5)$

$$\frac{x'^2}{4} - \frac{y'^2}{9} = 1$$

Vertices: $x' = 2$, $y' = 0$ and $x' = -2$, $y' = 0$

$x = 0$, $y = 5$ and $x = -4$, $y = 5$

Asymptotes: $y' = \pm \frac{3}{2}x'$

$$y - 5 = \pm \frac{3}{2}(x + 2)$$

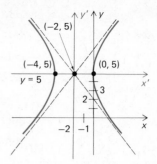

41. $\dfrac{(x + 3)^2}{9} - \dfrac{(y - 4)^2}{16} = 1$ **42.** $\dfrac{(y - 4)^2}{16} - \dfrac{(x + 3)^2}{9} = 1$

43. $\dfrac{(x + 5)^2}{4} - \dfrac{(y + 2)^2}{1} = 4$ **44.** $\dfrac{(y + 2)^2}{1} - \dfrac{(x + 5)^2}{4} = 4$

45. $9(x - 6)^2 - (y + 3)^2 = 9$ **46.** $(y + 3)^2 - 9(x - 6)^2 = 9$

47. $\frac{1}{9}(x - 1)^2 - \frac{1}{4}(y + 4)^2 = 1$

48. $100(x - 5)^2 - 4(y - 4)^2 = 400$

49. $4(y - 4)^2 - 100(x - 5)^2 = 400$

50. $4(x + 3)^2 - 16(y - 5)^2 = 64$

▶ Write each of the following equations in standard form and sketch the graph.

EXAMPLE

$$4x^2 - 9y^2 - 16x - 90y - 209 = 0$$

Solution

$$4(x^2 - 4x \quad) - 9(y^2 + 10y \quad) = 209$$
$$4(x - 2)^2 - 9(y + 5)^2 = 209 + 16 - 225$$
$$4(x - 2)^2 - 9(y + 5)^2 = 0$$
$$(y + 5) = \pm \frac{2}{3}(x - 2)$$

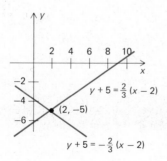

51. $4x^2 - 9y^2 - 16x - 90y - 210 = 0$

52. $4x^2 - 9y^2 - 24x - 18y + 27 = 0$

53. $-4x^2 + 24x + 9y^2 + 18y - 28 = 0$

54. $4x^2 - 9y^2 - 24x - 18y + 26 = 0$

55. $x^2 - y^2 + 2x + 1 = 0$

56. $y^2 - x^2 - 10x - 25 = 0$

57. $x^2 - y^2 + 2x = 0$

58. $4x^2 - 9y^2 - 24x - 18y - 9 = 0$

59. $-4x^2 + 9y^2 + 24x + 18y - 63 = 0$

60. $x^2 - y^2 - 2x + 2y = 0$

▶ In each of the following, if the equation has a graph, identify it. If the equation has no graph, state "No graph."

EXAMPLES	Solutions
(a) $x^2 + 9y^2 = 36$	(a) Ellipse
(b) $x^2 - 9y^2 = 36$	(b) Hyperbola
(c) $x^2 + 9y^2 = -36$	(c) No graph
(d) $x^2 - 9y^2 = -36$	(d) Hyperbola
(e) $9y^2 - x^2 = 36$	(e) Hyperbola
(f) $9x^2 + 9y^2 = 36$	(f) Circle
(g) $9x^2 - 9y^2 = 36$	(g) Hyperbola
(h) $x^2 + 9y^2 = 0$	(h) Point
(i) $x^2 - 9y^2 = 0$	(i) Two intersecting lines
(j) $x^2 - 9y = 0$	(j) Parabola
(k) $9y^2 - x = 0$	(k) Parabola

61. $4x^2 + 25y^2 = 100$　　　　**62.** $4x^2 - 25y^2 = 100$

63. $4x^2 + 4y^2 = 100$　　　　**64.** $4x^2 + 25y^2 = -100$

65. $4x^2 - 25y^2 = -100$　　　**66.** $4x^2 + 25y^2 = 0$

67. $4x^2 - 25y^2 = 0$　　　　**68.** $25y^2 - 100x = 0$

69. $4x^2 + 4y^2 - 40x + 100 = 0$　　**70.** $25x^2 + 100x + 4y = 0$

▶ In each of the following, the graph of the given equation is to be revolved about the x-axis. Express in terms of x the area of the circular cross section generated by the line segment from $(x, 0)$ to the point (x, y) on the graph. If (x, y) must be on the graph and also in the first quadrant, what values may x take on?

EXAMPLE　　　　$x^2 + y^2 = 25$

Solution

Cross-sectional area $= \pi y^2$

$$= \pi(25 - x^2),\ 0 < x < 5$$

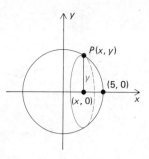

71. $\dfrac{x^2}{4} + \dfrac{y^2}{9} = 1$ 72. $\dfrac{x^2}{25} + y^2 = 1$ 73. $\dfrac{x^2}{4} + \dfrac{y^2}{4} = 1$

74. $\dfrac{x^2}{36} + \dfrac{y^2}{9} = 1$ 75. $9x^2 + 9y^2 = 1$

▶ In each of the following, the part of the graph indicated is to be revolved about the x-axis so that the point (x, y) on the graph generates the circumference of a circle. Express this circumference in terms of x, and state what values x may take on.

EXAMPLE

$y^2 = 4x$, (x, y) in first quadrant or on x-axis and $x \le 3$

Solution

Circumference $= 2\pi y$

$$= 2\pi \sqrt{4x},\ 0 \le x \le 3$$

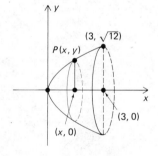

76. $x^2 + y^2 = 25$, (x, y) in first quadrant or on an axis with x nonnegative

77. $\dfrac{x^2}{25} + \dfrac{y^2}{9} = 1$, (x, y) in first quadrant or on x-axis and $x \ge 1$

78. $\dfrac{x^2}{25} - \dfrac{y^2}{9} = 1$, (x, y) in first quadrant or on x-axis and $x \le 8$

79. $x^2 = 4y$, x nonnegative and $x \le 3$

80. $3x - y - 6 = 0$, y nonnegative and $x \le 5$

CHECK LIST FOR THE STUDENT

If you have learned the material in Section 2.6, you should be able to do the following:

1. Define an ellipse.

2. Recognize the standard form of an equation of an ellipse.

3. Find the center and the ends of the axes of an ellipse from its equation in standard form.

4. Sketch the graph of an ellipse if its equation is given in standard form.

5. Transform the standard form of an equation of an ellipse with center at (h, k) by translating axes so that the new origin is at (h, k). Draw and label both sets of axes and sketch the graph.

6. Write in standard form an equation of the form $Ax^2 + By^2 + Cx + Dy + E = 0$ with A and B of the same sign and $A \neq B$.

7. Define a hyperbola.

8. Recognize the standard form of an equation of a hyperbola.

9. Find the center, vertices, and asymptotes of a hyperbola from its equation in standard form.

10. Sketch the graph of a hyperbola if its equation is given in standard form.

11. Transform the standard form of an equation of a hyperbola with center at (h, k) by translating axes so that the new origin is at (h, k). Draw and label both sets of axes and sketch the graph.

12. Write in standard form an equation of the form $Ax^2 + By^2 + Cx + Dy + E = 0$ with A and B of opposite sign.

13. Identify an equation of the form $Ax^2 + By^2 + Cx + Dy + E = 0$ with A and B not both zero as a circle, parabola, ellipse, hyperbola, point, two lines, or no graph. Sketch the graph if it exists.

14. Find an expression in terms of x for the cross-sectional area generated by a line segment from $(x, 0)$ to (x, y) on an ellipse with center at the origin, or from $(x, 0)$ to (x, y) on a hyperbola with center at the origin and axis along the x-axis, as the graph or part of the graph is revolved about the x-axis.

15. Find an expression in terms of x for the circumference of the circle generated by the point (x, y) on an ellipse with center at the origin, or by the point (x, y) on a hyperbola with center at the origin and axis along the x-axis, as the graph or part of the graph is revolved about the x-axis.

1. Find the center, label the ends of the axes, and sketch the graph of

$$\frac{x^2}{16} + \frac{y^2}{9} = 1.$$

2. Transform the equation

$$\frac{(x - 3)^2}{9} + \frac{(y - 6)^2}{25} = 1$$

by translating coordinate axes so that the new origin is at the center. Draw and label both sets of coordinate axes. Label the ends of the axes of the ellipse. Sketch the graph.

Write each of the following equations in standard form and identify it.
3. $4x^2 + 9y^2 - 24x + 72y + 144 = 0$

4. $4x^2 - 9y^2 - 24x - 72y - 144 = 0$

5. Find the center, vertices, and equations of the asymptotes of

$$\frac{x^2}{25} - \frac{y^2}{9} = 1.$$

Draw the asymptotes and sketch the graph.

6. Transform the equation

$$\frac{(y - 6)^2}{25} - \frac{(x - 3)^2}{9} = 1$$

by translating coordinate axes so that the new origin is at the center. Draw and label both sets of coordinate axes. Find equations of the asymptotes referred to the new coordinate axes, and also find equations of the asymptotes referred to the original coordinate axes. Draw the asymptotes and sketch the graph.

In each of the following, if the equation has a graph, identify it. If the equation has no graph, state "No graph."
7. (a) $4x^2 + 9y^2 = 36$ (b) $4x^2 + 4y^2 = 36$

8. (a) $4x^2 - 9y^2 = 36$ (b) $4x^2 - 9y^2 = 0$

9. (a) $4x^2 + 9y^2 = -36$ (b) $4x^2 = 9y$

10. (a) $4x^2 - 9y^2 = -36$ (b) $9y^2 = 4x$

2.7 Inverse Relations and One-to-One Functions

The purpose of this section is to reinforce the definitions of relation and function, to define inverse relations and one-to-one functions, and to demonstrate that the inverse of any one-to-one function is itself a function.

We noted in Section 2.1 that a *relation* is defined provided that a set called the domain is given and provided that there is given a rule of assignment that assigns one or more things to each element in the domain. We saw that a relation may be defined as any set of ordered pairs. We saw that if the ordered pairs are pairs of real numbers, the graph of the relation is the set of all points associated with its ordered pairs.

The lines, circles, parabolas, ellipses, and hyperbolas that we considered in the last few sections are all graphs of relations. For example, the ellipse in Figure 2.55 is the graph of the relation

$$\left\{ (x,\ y) \left| \frac{x^2}{25} + \frac{y^2}{16} = 1 \right. \right\}.$$

We say that the equation $x^2/25 + y^2/16 = 1$ *defines* this relation.

We also saw in Section 2.1 that a *function* is defined provided that a set called the domain is given and provided that there is given a rule of assignment that assigns a *single* thing to each element in the domain. In other words, a function is a special relation in which the rule of assignment assigns exactly one thing to each element in the domain. Thus a function may be defined as a set of ordered pairs with the restriction that no two different ordered pairs have the same first member.

The circles, ellipses, hyperbolas, and many of the parabolas that we considered in the last few sections are graphs of relations but are *not* graphs of functions. The "sweeping vertical line" test demonstrated in Figure 2.7 verifies this. Parabolas that open up or down, however, *are* graphs of functions. For example, the parabola opening up in Figure 2.7 is the graph of the function

$$\{(x,\ y)|y = x^2\}.$$

We saw in Section 2.5 that such a function is called a quadratic function.

Parts of circles, ellipses, hyperbolas, or parabolas opening left or right might represent functions. For example, the lower half of the parabola $y^2 = x$ in Figure 2.7 is the graph of the function

$$\{(x,\ y)|y = -\sqrt{x} \text{ and } x \geq 0\}.$$

Exercise 2.7, 1–10   (This is the function h sketched in Figure 2.4.)

The **inverse** of any given relation is the relation obtained by interchanging the members in each of its ordered pairs. If a relation is defined by a given equation in x and y, an equation defining its inverse can be found simply by interchanging x and y in the given equation. The inverse of the relation

$$\left\{ (x,\, y)\,\middle|\, \frac{x^2}{25} + \frac{y^2}{16} = 1 \right\}$$

is the relation

$$\left\{ (x,\, y)\,\middle|\, \frac{y^2}{25} + \frac{x^2}{16} = 1 \right\}.$$

The inverse of a relation r is called r-inverse and is denoted symbolically by r^{-1}. Here the symbol -1 stands for "inverse"; it does *not* mean "negative one."

$$r^{-1} \text{ represents } r\text{-inverse}$$

EXAMPLE 1 Sketch the graph of the relation $r = \{(x,\, y)\,|\,2x + 3y - 18 = 0\}$. State whether or not this relation is a function. Find the inverse of r. State whether or not the inverse of r is a function. Sketch the graph of r^{-1} on the same coordinate system as the graph of r.

Solution The graph of r is the oblique line shown in Figure 2.72 with intercepts $(0,\, 6)$ and $(9,\, 0)$. We saw in Section 2.2 that any linear equation in x and y defines a linear function, so the relation r is a linear function.

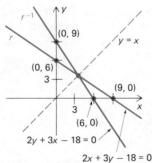

FIGURE 2.72 $r = \{(x,\, y)\,|\,2x + 3y - 18 = 0\}$
$r^{-1} = \{(x,\, y)\,|\,2y + 3x - 18 = 0\}$

The inverse of r is the relation

$$r^{-1} = \{(x,\, y)\,|\,2y + 3x - 18 = 0\}.$$

The relation r^{-1} is also a linear function. Its graph is the line in Figure 2.72 with intercepts $(6,\, 0)$ and $(0,\, 9)$. The sweeping vertical line test verifies that both r and r^{-1} are functions.

EXAMPLE 2

We have seen that the relation $f = \{(x, y)|y = x^2\}$ is a function. Find the inverse of f. State whether or not the inverse of f is a function.

Solution

$$f = \{(x, y)|y = x^2\}$$
$$f^{-1} = \{(x, y)|x = y^2\}$$

Exercise 2.7, 11–30

Here f^{-1} is not a function. Its graph is shown in Figure 2.7, where the sweeping vertical line test verifies that it is not a function.

Examples 1 and 2 demonstrate that the inverse of a relation that is a function may or may not be a function. The reason for this is that although the rule of assignment of a function must assign only one single thing to each domain element, that rule may assign the same single thing to more than one domain element. For example, the function f of Example 2 assigns the single number 9 to both 3 and -3. That is, the points $(3, 9)$ and $(-3, 9)$ are both on the graph of f. This forces the points $(9, 3)$ and $(9, -3)$ both to be on the graph of f^{-1}, so the graph of f^{-1} does not represent a function. The relation f^{-1} has different ordered pairs with the same first member, so f^{-1} is not a function.

If the rule of assignment of a function does not assign the same single number to more than one domain element, the function is said to be a *one-to-one function,* and its inverse has to be a one-to-one function. The function r of Example 1 is a one-to-one function. A **one-to-one function** is a relation whose rule of assignment assigns exactly one thing to each domain element *and* each of whose range elements is the assignment made to exactly one domain element. There is what is known as a one-to-one correspondence between the domain and the range of a one-to-one function. No two different ordered pairs have the same first member, *and* no two different ordered pairs have the same second member.

It is easy to tell from the graph of a function whether or not it is one-to-one by applying a "sweeping horizontal line" test similar to the sweeping vertical line test that we use to determine whether or not a relation is a function. If no horizontal line intersects the graph of a function in more than one point, then the function is one-to-one.

EXAMPLE 3

Use the sweeping horizontal line test to determine whether or not the function $g = \{(x, y)|y = \sqrt{x}$ and $x \geq 0\}$ is one-to-one.

Solution

The graph of g is shown in Figure 2.73. It is the upper half of the parabola $y^2 = x$. The sweeping vertical line test verifies that g is a function. Furthermore, no horizontal line intersects the graph of g in more than one point, so g is a one-to-one function.

Exercise 2.7, 31–50

We have said that the inverse of a one-to-one function is itself a one-to-one function. In Example 1 we found an equation defining the inverse of a given one-to-one function simply by interchanging x and y in the

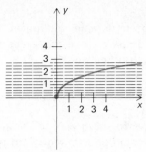

FIGURE 2.73   A one-to-one function, $g = \{(x, y) \mid y = \sqrt{x} \text{ and } x \geq 0\}$

equation defining the given function. Such interchanging is equivalent to interchanging the domain and range and "reversing" the rule of assignment. For example, let f be the function such that

$$f(x) = 2x - 4,$$
$$\text{Domain of } f = \{x \mid x \geq 0\}.$$

The graph of f is that part of the line $y = 2x - 4$ for which $x \geq 0$. In Figure 2.74 it is clear that the range of f is $\{y \mid y \geq -4\}$.

$$\text{Range of } f = \{y \mid y \geq -4\}$$

To find the domain and range of f^{-1}, we simply interchange the domain and range of f. As usual, we use the letter x to represent any domain element, and we use the letter y to represent the corresponding range element.

$$\text{Domain of } f^{-1} = \{x \mid x \geq -4\}$$
$$\text{Range of } f^{-1} = \{y \mid y \geq 0\}$$

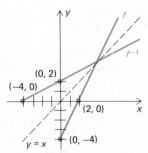

FIGURE 2.74 $f(x) = 2x - 4, \; x \geq 0$

$$f^{-1}(x) = \frac{x + 4}{2}, \; x \geq -4$$

To find the rule of assignment of f^{-1}, we "reverse" the rule of f. Since f takes any element x from its domain, multiplies it by 2, and then subtracts 4, the rule of f^{-1} must reverse this procedure by first adding 4 and then dividing by 2. In other words,

$$f^{-1}(x) = \frac{x + 4}{2}.$$

The graph of f^{-1} is shown, together with the graph of f, in Figure 2.74.

If f is a one-to-one function and the equation $y = f(x)$ can easily be solved for x in terms of y, the rule of assignment of function f^{-1} can easily be found mechanically by interchanging x and y and then solving for y in terms of x. For example, if $f(x) = 2x - 4$, we can find $f^{-1}(x)$ mechanically as follows:

$$f(x) = 2x - 4$$

1. Let $y = f(x)$. $\qquad\qquad$ $y = 2x - 4$ $\quad$ Defines f where $y = f(x)$

2. Interchange x and y. $\qquad$ $x = 2y - 4$ $\quad$ Defines f^{-1} where $y = f^{-1}(x)$

3. Solve for y. $\qquad\qquad\quad$ $2y = x + 4$

$$y = \frac{x + 4}{2}$$

$$f^{-1}(x) = \frac{x + 4}{2}$$

The next example demonstrates that when this mechanical procedure is utilized to find $f^{-1}(x)$, the algebraic manipulations involved must take the range of f^{-1} into consideration.

EXAMPLE 4

Let f be the function such that

$$\text{Domain of } f = \{x \mid x \leq -2\},$$
$$f(x) = x^2.$$

Sketch the graph of f. Find the domain and range of f^{-1}. Find $f^{-1}(x)$. Sketch the graph of f^{-1} on the same coordinate system as the graph of f.

Solution

The graph of f is shown in Figure 2.75. It is that part of the parabola $y = x^2$ to the left of and including $(-2, 4)$.

The range of f is $\{y \mid y \geq 4\}$, so the domain of f^{-1} is $\{x \mid x \geq 4\}$. The range of f^{-1} is the domain of f, so the range of f^{-1} is $\{y \mid y \leq -2\}$.

To find $f^{-1}(x)$, interchange x and y in $y = x^2$ and then solve for y, taking into account the range of f^{-1}.

$$x = y^2$$
$$y = \pm\sqrt{x}$$

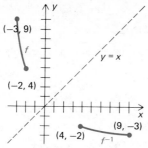

FIGURE 2.75　$f(x) = x^2,\ x \le -2$
$f^{-1}(x) = -\sqrt{x},\ x \ge 4$

Either $y = +\sqrt{x}$ or $y = -\sqrt{x}$ will satisfy the equation $x = y^2$. But the range of f^{-1} is a set of negative numbers, in fact it is the set of all numbers $y \le -2$, so for the rule of f^{-1} we need the minus sign.

$$f^{-1}(x) = -\sqrt{x}$$

The graph of f^{-1} is shown in Figure 2.75. It is that part of the lower half of the parabola $y^2 = x$ to the right of and including $(4, -2)$.

In each of Figures 2.73, 2.74, and 2.75 we have sketched the graphs of a one-to-one function and its inverse on the same coordinate system. In each of these figures the line $y = x$ has been drawn as a dotted line to show that in each case the graph of f^{-1} is the reflection of the graph of f in the line $y = x$. In other words, if the paper were folded on the line $y = x$, the graphs of f and f^{-1} would exactly match.

EXAMPLE 5

Let $f(x) = x^2$, and let the domain of f be $\{x \mid x \ge 0\}$. Sketch the graphs of f and f^{-1} on the same coordinate system.

Solution

The graph of f is the right half of the parabola $y = x^2$. It is shown in Figure 2.76. The line $y = x$ has been drawn as a dotted line in Figure 2.76. The reflection of f in $y = x$ is the graph of f^{-1}. The graph of f^{-1} is the upper half of the parabola $x = y^2$.

Exercise 2.7, 51–60

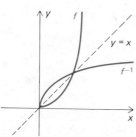

FIGURE 2.76　$f(x) = x^2,\ x \ge 0$
$f^{-1}(x) = \sqrt{x},\ x \ge 0$

EXERCISE 2.7 INVERSE RELATIONS AND ONE-TO-ONE FUNCTIONS

▶ Sketch the graph of each of the following, and state whether or not it is the graph of a function.

EXAMPLE

$\{(x, y) \mid x^2 + y^2 = 9, x \geq 0, \text{ and } y \geq 0\}$

Solution

Graph of a function.

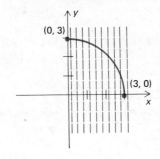

1. $\{(x, y) \mid x^2 + y^2 = 9 \text{ and } x \geq 0\}$

2. $\{(x, y) \mid x^2 + y^2 = 9, x \leq 0, \text{ and } y \geq 0\}$

3. $\{(x, y) \mid x^2 + y^2 = 9\}$

4. $\{(x, y) \mid x^2 + y^2 = 9 \text{ and } y \geq 0\}$

5. $\{(x, y) \mid x^2 + y^2 = 9, x \geq 0, \text{ and } y \leq 0\}$

6. $\{(x, y) \mid x^2 + y^2 = 9, x \leq 0, \text{ and } y \leq 0\}$

7. $\left\{(x, y) \mid \dfrac{y^2}{9} - \dfrac{x^2}{16} = 1 \text{ and } y \geq 0\right\}$

8. $\left\{(x, y) \mid \dfrac{y^2}{9} + \dfrac{x^2}{16} = 1 \text{ and } y \geq 0\right\}$

9. $\left\{(x, y) \mid \dfrac{x^2}{9} - \dfrac{y^2}{16} = 1 \text{ and } x \geq 0\right\}$

10. $\left\{(x, y) \mid \dfrac{x^2}{9} - \dfrac{y^2}{16} = 1 \text{ and } y \geq 0\right\}$

▶ Find the inverse of each of the following relations:

EXAMPLE

$r = \{(x, y) \mid 9x^2 - 4y^2 = 36\}$

Solution

$r^{-1} = \{(x, y) \mid 9y^2 - 4x^2 = 36\}$

11. $r = \{(x, y) \mid 7x - 8y + 9 = 0\}$

12. $r = \{(x, y) \mid 9x^2 + 4y^2 = 36\}$

13. $r = \{(x, y) \mid x^2 + y^2 = 9\}$

14. $r = \{(x, y) \mid y = 2x\}$

15. $r = \{(x, y) \mid y = x^2 - 6x + 9\}$

16. $r = \{(x, y) \mid y^2 - 8y + 16 = x\}$

17. $r = \{(x, y) \mid y = 10^x\}$

18. $r = \{(x, y) \mid y = 2^x\}$

19. $r = \{(x, y) \mid y = e^x\}$

20. $r = \{(x, y) \mid y = \sqrt{1 - x^2}\}$

▶ Find the inverse of each of the following functions, and state whether or not the inverse is a function.

EXAMPLES

Solutions

(a) $f = \{(x, y) \mid y = 3x\}$ (b) $f = \{(x, y) \mid y = 3x^2\}$

(a) $f^{-1} = \{(x, y) \mid x = 3y\}$ A function
(b) $f^{-1} = \{(x, y) \mid x = 3y^2\}$ Not a function

21. $f = \{(x, y) \mid y = 3x - 7\}$ **22.** $f = \{(x, y) \mid y = 3x^2 - 6x + 3\}$

23. $f = \{(x, y) \mid 3x - 7y + 10 = 0\}$

24. $f = \{(x, y) \mid y = x^3\}$ **25.** $f = \{(x, y) \mid y = x^4\}$

26. $f = \{(x, y) \mid y = x^5\}$ **27.** $f = \{(x, y) \mid y = \sqrt{x}\}$

28. $f = \{(x, y) \mid y = -\sqrt{x}\}$ **29.** $\{(x, y) \mid y = -x\}$

30. $\{(x, y) \mid y = 5x^2 + 10x + 5\}$

▶ State whether or not the function in each of the following is one-to-one.

EXAMPLES

Solutions

(a) $\{(x, y) \mid y = 3x^2\}$ (b) $\{(x, y) \mid y = 3x^2 \text{ and } x \geq 0\}$

(a) Not one-to-one   (b) One-to-one

31. $\{(x, y) \mid y = 3x\}$ **32.** $\{(x, y) \mid y = x^2 - 6x + 9\}$

33. $\{(x, y) \mid y = x^2 - 6x + 9 \text{ and } x \geq 0\}$

34. $\{(x, y) \mid y = x^2 - 6x + 9 \text{ and } x \geq 3\}$

35. $\{(x, y) \mid x^2 + y^2 = 9, x \geq 0 \text{ and } y \geq 0\}$

36. Domain of $f = \{x \mid x \geq 0\}$

   $f(x) = 3x$ for $x \leq 1$

   $f(x) = 3$ for $x > 1$

37. Domain of $f = \{x \mid x \geq -1\}$

   $f(x) = x^2$

38. Domain of $f = \{x \mid -1 \leq x \leq 1\}$
$f(x) = x^2$

39. Domain of $f = \{x \mid x \geq 1\}$
$f(x) = x^2$

40. Domain of $f = \{x \mid -4 \leq x \leq 4\}$
$f(x) = x^2$ for $x \leq 1$
$f(x) = 3$ for $x > 1$

▶ Use the sweeping horizontal line test to determine whether or not each of the following graphs represents a one-to-one function.

41.

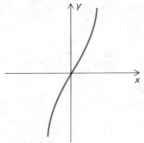

42.

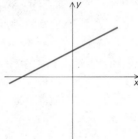

43.

44.

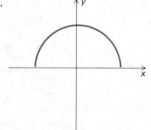

45.

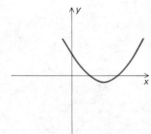

46.

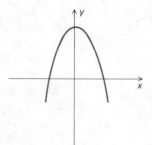

47.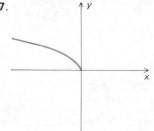

48.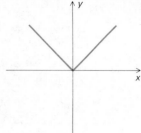

49.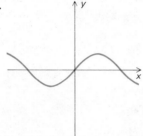

50.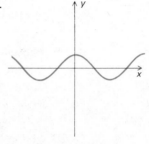

▶ In each of the following, find the domain and range of *f*-inverse, and find $f^{-1}(x)$. Sketch the graphs of *f* and f^{-1} on the same coordinate system. (If the domain is not explicitly stated, use the domain convention introduced in Section 2.1.)

EXAMPLE

$f(x) = 3x^2$

Domain of $f = \{x \mid x \geq 1\}$

Solution

Range of $f = \{y \mid y \geq 3\}$
Domain of $f^{-1} = \{x \mid x \geq 3\}$
Range of $f^{-1} = \{y \mid y \geq 1\}$

$$y = 3x^2$$
$$x = 3y^2$$
$$y = +\sqrt{\frac{x}{3}}, \; f^{-1}(x) = \sqrt{\frac{x}{3}}$$

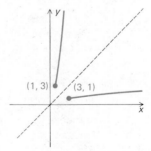

51. $f(x) = 3x + 5$
Domain of $f = \{x \mid x \geq -1\}$

52. $f(x) = 3x + 5$

53. $f(x) = x^2$
Domain of $f = \{x \mid x \geq 2\}$

54. $f(x) = \sqrt{x}$
Domain of $f = \{x \mid x \geq 9\}$

55. $f(x) = 2x - 5$
Domain of $f = \{x \mid x \geq 0\}$

56. $f(x) = 2x - 5$

57. $f(x) = x^2$
Domain of $f = \{x \mid x \leq 0\}$

58. $f(x) = -x$
Domain of $f = \{x \mid x \leq 0\}$

59. $f(x) = 3x - 4$

60. $f(x) = -\sqrt{x}$

CHECK LIST FOR THE STUDENT

If you have learned the material in Section 2.7, you should be able to do the following:

1. Tell from looking at the graph of a relation whether or not the relation is a function.

2. Define the inverse of a relation.

3. Find an equation defining the inverse of a relation that is defined by a given equation.

4. Use the notation r^{-1} for the inverse of a relation r.

5. Recognize that the inverse of a function may or may not be a function.

6. Define a one-to-one function.

7. Recognize that the inverse of a one-to-one function is itself a function.

8. Tell from looking at the graph of a function whether or not the function is one-to-one.

9. Find the domain, range, and rule of assignment of the inverse of a given one-to-one function.

10. Sketch the graph of the inverse of a one-to-one function whose graph is given by reflecting the given graph in the line $y = x$.

SECTION QUIZ 2.7

1. Sketch the graph of each of the following, and state whether or not it is the graph of a function.
   (a) $\{(x, y) \mid x^2 + y^2 = 25 \text{ and } x \geq 0\}$
   (b) $\{(x, y) \mid x^2 + y^2 = 25, x \geq 0, \text{ and } y \geq 0\}$

2. Find the inverse of each of the following, and state whether or not the inverse is a function.
   (a) $\{(x, y) \mid y = 2x + 7\}$ (b) $\{(x, y) \mid y = 8x^2\}$

3. State whether or not each of the following functions is one-to-one.
   (a) $\{(x, y) \mid y = 5x\}$ (b) $\{(x, y) \mid y = 5x^2\}$

4. In each of the following, tell whether or not the graph represents a one-to-one function.

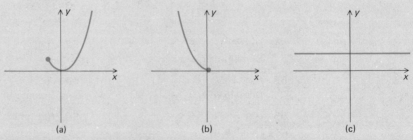

(a)   (b)   (c)

5. Let the domain of f be $\{x \mid x \geq 0\}$, and let $f(x) = 2x + 3$. Find the domain and range of f-inverse, and find $f^{-1}(x)$. Sketch the graphs of f and f^{-1} on the same coordinate system.

2.8 Exponential and Logarithmic Functions

The purpose of this section is to define and graph exponential functions and to introduce logarithmic functions as the inverses of exponential functions.

If b is any positive real number other than the number 1, then the function f such that

$$f(x) = b^x$$

is called the **exponential function** with base b. In Sections 1.2 and 1.3 we defined b^x for integral x and more generally for rational x. We saw in those sections that integral and rational exponents satisfy five basic exponent laws. Using a concept usually taken up in calculus, the expression b^x can also be defined for irrational numbers x so that b^x represents a real number and so that the five exponent laws hold for real numbers x whether they are rational or irrational. The expression b^x is called a **power** of b. Let us agree that the exponent laws hold for real-number exponents and that b^x represents a real number for any real number x. Then according to our domain convention, the domain of an exponential function is the set of all real numbers.

$$f(x) = b^x$$

Domain of f is the set of all real numbers

For example, the function f such that

$$f(x) = 2^x$$

is the exponential function with base 2. Its domain is the set of all real numbers.

The graph of function f such that $f(x) = 2^x$ is the graph of the equation $y = 2^x$. If we make a table of values and plot points on the graph of $y = 2^x$, we see that the points all appear to be strung together like beads as shown in Figure 2.77. Note that the graph lies entirely above the

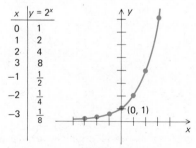

x	$y = 2^x$
0	1
1	2
2	4
3	8
−1	$\frac{1}{2}$
−2	$\frac{1}{4}$
−3	$\frac{1}{8}$

FIGURE 2.77   Graph of f: $f(x) = 2^x$
(Graph of $y = 2^x$)

x-axis. This is true of the graph of any exponential function, because for any real number *x* the expression $y = b^x$ is positive. In other words, the range of any exponential function is the set of all positive real numbers.

$$f(x) = b^x$$

$$\text{Range of } f = \{y \mid y > 0\}$$

The *y*-intercept of the graph of any exponential function is easy to find. For example, the *y*-intercept of the graph of $y = 2^x$ is (0, 1) because when $x = 0$, $y = 2^0 = 1$. The point (0, 1) is also the *y*-intercept of the graph of any exponential function because if $b > 0$, then $b^0 = 1$. In other words, the graph of any exponential function intersects the *y*-axis at (0, 1).

Note in Figure 2.77 that as *x* takes on larger and larger positive values, $y = 2^x$ also takes on larger and larger positive values. But as *x* takes on negative values that become larger and larger in absolute value, $y = 2^x$ takes on positive values that become nearer and nearer to zero. As we continue the sketch farther and farther to the left, the graph approaches the position of the left half of the *x*-axis. The *x*-axis is an asymptote. This is all true of the graph of any exponential function with base greater than the number 1. Tracing the graph of such a function from *left to right,* we see that as *x* increases, $y = b^x$ with $b > 1$ also increases so that the graph leaves its asymptote, the *x*-axis, goes through its intercept (0, 1), and rises off to the right.

If the base *b* of an exponential function is less than the number 1, if $0 < b < 1$, then the graph falls from left to right and approaches the position of the right half of the *x*-axis. This is demonstrated by the graph of $y = (\frac{1}{2})^x$ shown in Figure 2.78. Note that $(\frac{1}{2})^x = (2^{-1})^x = 2^{-x}$ so that the graph of $y = (\frac{1}{2})^x$ is precisely the graph of $y = 2^{-x}$.

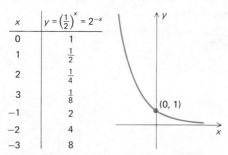

x	$y = \left(\frac{1}{2}\right)^x = 2^{-x}$
0	1
1	$\frac{1}{2}$
2	$\frac{1}{4}$
3	$\frac{1}{8}$
−1	2
−2	4
−3	8

FIGURE 2.78 Graph of f: $f(x) = (\frac{1}{2})^x$
(Graph of $y = (\frac{1}{2})^x$)

In calculus, as well as in many physical applications, it is frequently convenient to use the number *e* as the base of an exponential function. The number *e* was introduced in Section 1.1 and discussed in Example

8 of Section 1.2. Recall that e is an irrational number approximately equal to 2.718281828.

$$e \approx 2.718281828$$

Since e is greater than 1, the graph of $y = e^x$ will rise from left to right as did the graph of $y = 2^x$ in Figure 2.77. The graph of $y = e^x$ is shown in Figure 2.79.

Exercise 2.8, 1–10

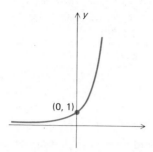

FIGURE 2.79   Graph of f: $f(x) = e^x$
(Graph of $y = e^x$)

Any exponential function is a one-to-one function. The sweeping horizontal line test verifies this, because the graph of any exponential function either looks similar to the graph of $y = 2^x$ in Figure 2.77 or similar to the graph of $y = (\frac{1}{2})^x$ in Figure 2.78. Thus any exponential function has an inverse that is itself a one-to-one function.

The inverse of any exponential function is called a *logarithmic function*. Let us find the inverse of the function f such that $f(x) = b^x$ where b is a positive real number other than the number 1. We know the domain, rule of assignment, and range of f.

Domain of $f = \{x | x$ is a real number$\}$

$f(x) = b^x$ where $b > 0$ and $b \neq 1$

Range of $f = \{y | y > 0\}$

In other words,

$f = \{(x, y) | y = b^x$ where $b > 0$ and $b \neq 1\}$.   Exponential function

To find f-inverse, we interchange the domain and range of f and "reverse" the rule of f. That is, to find f-inverse we interchange the members in each of the ordered pairs of f so that

$f^{-1} = \{(x, y) | x = b^y$ where $b > 0$ and $b \neq 1\}$. Logarithmic function

The domain of f^{-1} is the range of f, and the range of f^{-1} is the domain of f.

Domain of $f^{-1} = \{x | x > 0\}$

Range of $f^{-1} = \{y | y$ is a real number$\}$

If $x = b^y$, the number y is called the *logarithm of x to the base b* and is written $y = \log_b x$.

$$y = \log_b x \text{ provided that } x = b^y$$

Using this notation, we may write

$$f^{-1} = \{(x, y)|y = \log_b x \text{ where } b > 0 \text{ and } b \neq 1\}.$$

This function f^{-1} is called the *logarithm function to the base b*. For convenience, let us denote it by the letter g. The logarithm function to the base b is the function g such that

$$\text{Domain of } g = \{x|x > 0\},$$

$$\text{Range of } g = \{y|y \text{ is a real number}\},$$

and $$g(x) = \log_b x.$$

The graph of g is the graph of $y = \log_b x$.

EXAMPLE 1

Give the domain, range, and rule of assignment of the inverse of function f such that $f(x) = 10^x$. Sketch the graphs of f and its inverse on the same coordinate axes.

Solution

f

Domain of $f = \{x|x \text{ is a real number}\}$

Range of $f = \{y|y > 0\}$

Rule of f: $y = 10^x$

f^{-1}

Domain of $f^{-1} = \{x|x > 0\}$

Range of $f^{-1} = \{y|y \text{ is a real number}\}$

Rule of f^{-1}: $x = 10^y$ or $y = \log_{10} x$

The graph of f goes through (0, 1). The graph of f^{-1} goes through (1, 0). The graph of f^{-1} is the reflection of the graph of f in the line $y = x$. Both graphs are shown in Figure 2.80.

Exercise 2.8, 11–15

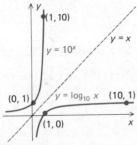

FIGURE 2.80 Graphs of $y = 10^x$ and $y = \log_{10} x$.

PART TWO | ELEMENTARY FUNCTIONS AND ANALYTIC GEOMETRY

The graph of any logarithm function *with base b greater than 1* looks similar to the graph of $y = \log_{10} x$ shown in Figure 2.80. The graph lies entirely to the right of the x-axis and has the y-axis as asymptote. The graph rises from left to right, intersecting the x-axis at $(1, 0)$. It is the reflection of the graph of $y = b^x$ in the line $y = x$.

The graph of $y = \log_b x$ for $0 < b < 1$ is again the reflection of the graph of $y = b^x$ in the line $y = x$. For example, the graph of $y = \log_{1/2} x$ is the reflection in the line $y = x$ of the graph of $y = (\frac{1}{2})^x$ shown in Figure 2.78. Logarithms with bases less than 1 are not often encountered in calculus.

EXAMPLE 2

Find the values of $\log_2 8$, $\log_2 \frac{1}{8}$, $\log_{10} 100$, $\log_{10} 1000$, $\log_{10} 1$, and $\log_e e$.

Solution

By definition,

$\log_b x = y$	provided that	$b^y = x.$
$\log_2 8 = 3$	because	$2^3 = 8$
$\log_2 \dfrac{1}{8} = -3$	because	$2^{-3} = \dfrac{1}{2^3} = \dfrac{1}{8}$
$\log_{10} 100 = 2$	because	$10^2 = 100$
$\log_{10} 1000 = 3$	because	$10^3 = 1000$
$\log_{10} 1 = 0$	because	$10^0 = 1$
$\log_e e = 1$	because	$e^1 = e$

Most of the logarithms encountered in calculus are to the base e, because using e as a base considerably simplifies many formulas. The logarithm function to the base e is sketched in the next example.

Exercise 2.8, 16–30

EXAMPLE 3

Sketch the graph of $y = \log_e x$.

Solution

The graph of $y = \log_e x$ is the reflection of the graph of $y = e^x$ in the line $y = x$. Since $\log_e x$ is only defined for positive real numbers x, the graph lies entirely to the right of the y-axis. The x-intercept is $(1, 0)$. (Log$_e 1 = 0$ because $e^0 = 1$.) The graph is shown in Figure 2.81.

Exercise 2.8, 31–40

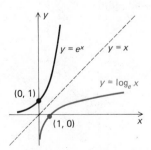

FIGURE 2.81 Graphs of $y = e^x$ and $y = \log_e x$

▶ Sketch the graph of each of the following.

EXAMPLE

$y = 3^x$

Solution

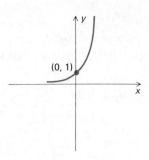

(0, 1)

1. $y = (\frac{1}{3})^x$ **2.** $y = 5^x$ **3.** $y = (\frac{1}{5})^x$

4. Function f: domain of $f = \{x \mid x \geq -1\}$, $f(x) = 2^x$

5. Function f: domain of $f = \{x \mid x \geq -1\}$, $f(x) = 2^{-x}$

6. Function f: domain of $f = \{x \mid x \geq -1\}$, $f(x) = e^x$

7. $y = e^{-x}$ **8.** $y = (\frac{1}{10})^x$

9. Function f: domain of $f = \{x \mid x \geq 0\}$, $f(x) = 10^x$

10. Function f: domain of $f = \{x \mid x \geq 0\}$, $f(x) = e^{-x}$

▶ In each of the following, give the domain, range, and rule of assignment of the inverse of f. Sketch the graphs of f and its inverse on the same coordinate axes.

EXAMPLE

$f(x) = (\frac{1}{2})^x$
Domain of $f = \{x \mid x \text{ is a real number}\}$
Range of $f = \{y \mid y > 0\}$
Rule of f: $y = (\frac{1}{2})^x$

Solution

Domain of $f^{-1} = \{x \mid x > 0\}$
Range of $f^{-1} = \{y \mid y \text{ is a real number}\}$
Rule of f^{-1}: $y = \log_{1/2} x$

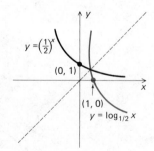

$y = (\frac{1}{2})^x$

(0, 1)

(1, 0)

$y = \log_{1/2} x$

11. $f(x) = 2^x$ **12.** $f(x) = 3^x$ **13.** $f(x) = (\frac{1}{3})^x$ **14.** $f(x) = e^{-x}$

15. $f(x) = 7^x$

▶ Find the value of each of the following.

EXAMPLES

(a) $\log_3 9$ (b) $\log_e 1$

Solutions

(a) $\log_3 9 = 2$ (because $3^2 = 9$)   (b) $\log_e 1 = 0$ (because $e^0 = 1$)

16. $\log_5 25$ **17.** $\log_5 1$ **18.** $\log_{25} 25$ **19.** $\log_e e^4$

20. $\log_{10} \frac{1}{10}$ **21.** $\log_{10} 0.01$ **22.** $\log_e \frac{1}{e}$ **23.** $\log_7 \frac{1}{7}$

24. $\log_4 64$ **25.** $\log_4 4$ **26.** $\log_{10} 10$ **27.** $\log_7 49$

28. $\log_{10} 1000$ **29.** $\log_3 81$ **30.** $\log_9 81$

▶ Sketch the graph of each of the following and label its intercept.

EXAMPLE

$y = \log_3 x$

Solution

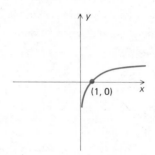
(1, 0)

31. $y = \log_4 x$ **32.** $y = \log_5 x$ **33.** $y = \log_{1/4} x$

34. Function $f: f(x) = \log_{10} x$, domain of $f = \{x \mid x \geq 1\}$

35. Function $f: f(x) = \log_e x$, domain of $f = \{x \mid x \geq 1\}$

36. Function $f: f(x) = \log_{10} x$, domain of $f = \{x \mid 0 < x \leq 1\}$

37. Function $f: f(x) = \log_2 x$, domain of $f = \{x \mid x \geq 1\}$

38. Function $f: f(x) = \log_e x$, domain of $f = \{x \mid 0 < x \leq 1\}$

39. $y = \log_7 x$ **40.** $y = \log_{1/7} x$

CHECK LIST FOR THE STUDENT

If you have learned the material in Section 2.8, you should be able to do the following:

1. Define an exponential function and give its domain.

2. Tell what numbers may be used as bases for exponential functions.

3. Give the range of an exponential function.

4. Sketch the graph of an exponential function with base $b > 1$ and label the intercept of the graph.

5. Sketch the graph of an exponential function with base $0 < b < 1$ and label the intercept of the graph.

6. Tell whether or not all exponential functions are one-to-one functions.

7. Sketch the graph of $y = e^x$.

8. Find the inverse of a given exponential function.

9. Sketch the graph of an exponential function and its inverse on the same coordinate system.

10. Complete the statement: $y = \log_b x$ provided that _____.

11. Find the values of $\log_b x$ for some given values of x and b.

12. Define the logarithm function to the base b and give its domain.

13. Give the range of the logarithm function to the base b.

14. Sketch the graph of $y = \log_b x$ with $b > 1$.

15. Sketch the graph of $y = \log_b x$ with $0 < b < 1$.

16. Sketch the graph of $y = \log_e x$.

SECTION QUIZ 2.8

Sketch the graph of each of the following.

1. $y = 6^x$

2. $y = (\frac{1}{6})^x$

3. Let $f(x) = 3^x$. Give the domain, range, and rule of assignment of the inverse of f. Sketch the graphs of f and its inverse on the same coordinate axes.

4. Find the value of each of the following.
   (a) $\log_2 64$ (b) $\log_3 1$ (c) $\log_7 7$ (d) $\log_2 \frac{1}{2}$ (e) $\log_{10} 0.001$

5. Sketch the graph of $y = \log_7 x$ and label its intercept.

2.9 Logarithms; Common and Natural Logarithms

The purpose of this section is to reinforce the definition of a logarithm, to provide manipulative practice with logarithmic notation used in beginning calculus, to present essential properties of logarithms, to show how to find values of common and natural logarithms, to show how common logarithms may be used in computations, and to illustrate one way in which properties of logarithms are utilized in calculus.

We have seen that if b is a positive real number other than 1, then for any positive real number x,

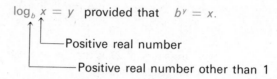

$$\log_b x = y \quad \text{provided that} \quad b^y = x.$$

In other words, the logarithm of positive number x to the base b is the exponent to which the base must be raised in order to produce the number x.

In Examples 2 and 3 of Section 2.8, we saw that $\log_{10} 1 = 0$ and that $\log_e 1 = 0$. In fact, if b is any acceptable base (b is any positive number other than 1), then

$$\log_b 1 = 0.$$

This is true because $b^0 = 1$.

In Example 2 of Section 2.8 we saw that $\log_e e = 1$. It is likewise true that if b is *any* acceptable base, then

$$\log_b b = 1.$$

This is true because $b^1 = b$.

In Example 2 of Section 2.8 we also saw that $\log_2 2^3 = 3$ and $\log_{10} 10^2 = 2$. In general, if b is any acceptable base and k is any real number,

$$\log_b b^k = k.$$

This is true because k is the exponent to which the base b must be raised in order to produce b^k.

If the base of a logarithm is the number 10, then the logarithm is called a **common logarithm**. It is customary to omit the subscript 10 from the notation for a common logarithm. For example, $\log_{10} x$ is customarily written $\log x$.

$$\log x \quad \text{means} \quad \log_{10} x$$

If the base of a logarithm is the number e, then the logarithm is called a **natural logarithm**. It is customary to write the symbols ln in place of $\log_e$ in the notation for a natural logarithm.

$$\ln x \quad \text{means} \quad \log_e x$$

EXAMPLE 1

Write $\log 10^{-1}$ and $\ln e^{x^2}$ in equivalent forms without using the symbols log or ln.

Solution

$$\log 10^{-1} = -1$$

(This means that -1 is the exponent to which the base 10 must be raised in order to produce the number 10^{-1}.)

$$\ln e^{x^2} = x^2$$

(This means that x^2 is the exponent to which the base e must be raised in order to produce e^{x^2}.)

Most calculus texts use the solution of the next example in deriving the formula known as "the derivative of e^x."

EXAMPLE 2

Write $y = e^x$ in logarithmic notation.

Solution

$y = e^x$ is equivalent to $x = \ln y$.

Exercise 2.9, 1–40

We have said that for positive number x, $\log_b x$ is the exponent to which the base must be raised to produce x. That is,

$$x = b^{\log_b x}.$$

This form of the definition is utilized in the next three examples. The solutions of these examples appear as the first steps in the derivations of basic formulas in beginning calculus. Learning to handle this notation with ease will make the related parts of beginning calculus much easier to follow.

EXAMPLE 3

Write 10 as a power of e. Write 10^x as a power of e.

Solution

$$10 = e^{\ln 10}$$
$$10^x = (e^{\ln 10})^x$$
$$\quad = e^{x \ln 10} \qquad \text{Exponent Law 3: } (a^r)^s = a^{rs}$$
$$10^x = e^{x \ln 10}$$

EXAMPLE 4

If a is any positive real number and x is any real number, write a^x as a power of e.

Solution

$$a = e^{\ln a}$$
$$a^x = (e^{\ln a})^x$$
$$\quad = e^{x \ln a} \qquad \text{Law 3: } (a^r)^s = a^{rs}$$
$$a^x = e^{x \ln a}$$

EXAMPLE 5

Let x be any positive real number, and let n be any real number. Write x as a power of e. Write x^n as a power of e.

Solution

$$x = e^{\ln x}$$
$$x^n = (e^{\ln x})^n$$

Exercise 2.9, 41–50

$$= e^{n \ln x} \qquad \text{Law 3: } (a^r)^s = a^{rs}$$

Familiarity with the meaning of $\ln |x|$ is needed for manipulations with many calculus formulas. From the definition of absolute value, we know that

$$\ln |x| = \ln x, \text{ if } x \text{ is positive}$$
$$\ln |x| = \ln(-x), \text{ if } x \text{ is negative.}$$

Use of this notation is demonstrated in the next two examples.

EXAMPLE 6

Find the value of $\ln |x|$ for $x = e^3$ and also for $x = -e^3$.

Solution

$$\ln |e^3| = \ln e^3 \qquad (e^3 \text{ is positive})$$
$$= 3$$
$$\ln |-e^3| = \ln (-(-e^3)) \qquad (-e^3 \text{ is negative})$$
$$= \ln e^3$$
$$= 3$$

EXAMPLE 7

Where x is any real number, why is it true that $\ln |1 + x^2| = \ln(1 + x^2)$? Is $\ln |x^3|$ equal to $\ln x^3$ for all real numbers x? Why or why not?

Solution

If x is any real number, then $1 + x^2$ is positive. If $1 + x^2$ is positive, then $|1 + x^2| = 1 + x^2$. It follows that

$$\ln |1 + x^2| = \ln(1 + x^2).$$

It is not true that $\ln |x^3|$ is equal to $\ln x^3$ for all real numbers x, because for some values of x the expression x^3 is nonpositive so that $\ln x^3$ is not even defined. If x^3 is *positive*, however, $|x^3| = x^3$, and then

Exercise 2.9, 51–55 $\ln |x^3| = \ln x^3$.

Three basic properties of logarithms to any acceptable base b may be derived directly from the exponent laws and the definition of a logarithm. These three properties are:

Property 1 $\log_b MN = \log_b M + \log_b N$

Property 2 $\log_b \dfrac{M}{N} = \log_b M - \log_b N$

Property 3 $\log_b M^k = k \log_b M.$

Here M and N are understood to be positive real numbers, and k is any real number. In Property 3, $\log_b M^k$ means $\log_b (M^k)$. Note that this is not the same as $(\log_b M)^k$.

To illustrate how these three properties follow from the exponent laws and the definition of a logarithm, let us prove Property 1 for the case in which the base is the number e.

Property 1 $\ln MN = \ln M + \ln N$

Proof   Let $x = \ln M$.   Then $M = e^x$.   Definition of $\ln M$

Let $y = \ln N$.   Then $N = e^y$.   Definition of $\ln N$

$$MN = e^x e^y$$
$$= e^{x+y} \qquad \text{Law 1: } a^r a^s = a^{r+s}$$

But $MN = e^{x+y}$ implies that

$$\ln MN = x + y. \qquad \text{Definition of } \ln MN$$

Since $x = \ln M$ and $y = \ln N$, we have

$$\ln MN = \ln M + \ln N.$$

EXAMPLE 8   Use Property 3 ($\log_b M^k = k \log_b M$) and prove that $\ln e^u = u$ for any real number u.

Solution $\ln e^u = u \ln e$ Property 3: $\log_b M^k = k \log_b M$

Exercise 2.9, 56–60 $= u$ $\ln e = 1$

Thus far we have only found the numerical values of the logarithms of numbers that are certain powers of the base. For example, we found $\log_2 8 = \log_2 2^3 = 3$, we found $\log 100 = \log 10^2 = 2$, and we found $\ln e^3 = 3$. Let us now see how to find the common logarithm of any positive number.

The common logarithm of a number that is any integral power of 10 may be found by direct application of the definition of a logarithm as an exponent. Some examples follow.

$$
\begin{array}{llll}
\log 10 = 1 & \text{because} & 10^1 = 10 \\
\log 100 = 2 & & 10^2 = 100 \\
\log 1000 = 3 & & 10^3 = 1000 \\
\log 10000 = 4 & & 10^4 = 10000 \\
\log 10^0 = \log 1 = 0 & & 10^0 = 1 \\
\log 0.1 = -1 & & 10^{-1} = 0.1 \\
\log 0.01 = -2 & & 10^{-2} = 0.01 \\
\log 0.001 = -3 & & 10^{-3} = 0.001 \\
\log 0.0001 = -4 & & 10^{-4} = 0.0001 \\
\end{array}
$$

The common logarithm of a number between 1 and 10 can be found approximately with a hand calculator or in a table of common logarithms. For example, Table I in Appendix B gives four-decimal-place approximations to the common logarithms of numbers between 1 and 10. In particular, in Table I we read that

$$\log 5 = 0.6990 \quad \text{and} \quad \log 5.23 = 0.7185.$$

With a hand calculator we would find that to six decimal places

$$\log 5 = 0.698970 \quad \text{and} \quad \log 5.23 = 0.718502.$$

Even though these are approximate values of log 5 and log 5.23, we adopt common convention and use "equals" signs instead of "approximately equals" signs.

Values of logarithms are found much more easily and to more decimal places with a hand calculator than with a table, but not all students have calculators, so we shall explain how to use logarithm tables.

The common logarithm of any positive number x that is not an integral power of 10 and that is not between 1 and 10 can be found approximately by first expressing x as the product of a number between 1 and 10 and an integral power of 10 and then by using Property 1 ($\log_b MN = \log_b M + \log_b N$), the definition of a logarithm as an exponent, and a table of common logarithms. This is demonstrated in the next example.

EXAMPLE 9

Use Table I to find log 523 to four decimal places.

Solution

We shall first express 523 as the product of a number between 1 and 10 and an integral power of 10. Then we will use Property 1 ($\log_b MN = \log_b M + \log_b N$).

$$523 = 5.23 \times 10^2$$

$$\begin{aligned}
\log 523 &= \log (5.23 \times 10^2) \\
&= \log 5.23 + \log 10^2 && \text{Property 1:} \\
& && \log_b MN = \log_b M + \log_b N \\
&= 0.7185 + \log 10^2 && \text{From Table I} \\
&= 0.7185 + 2 && \text{Definition of logarithm} \\
\log 523 &= 2.7185
\end{aligned}$$

On a calculator, to six decimal places, log 523 = 2.718502.

EXAMPLE 10

Use Table I to find log 0.00523 to four decimal places.

Solution

$$\begin{aligned}
0.00523 &= 5.23 \times 10^{-3} \\
\log 0.00523 &= \log (5.23 \times 10^{-3}) \\
&= \log 5.23 + \log 10^{-3} \\
&= 0.7185 + \log 10^{-3} \\
&= 0.7185 + (-3) \\
\log 0.00523 &= 0.7185 - 3 \\
&= -2.2815
\end{aligned}$$

On a calculator, to six decimal places, log $0.00523 = -2.281498$.

We have seen in Examples 9 and 10 that log $523 = 0.7185 + 2$ and log $0.00523 = 0.7185 + (-3)$. Any common logarithm can be expressed as the sum of a nonnegative decimal less than 1 and an integer. The nonnegative decimal is called the **mantissa**. The integer is called the **characteristic**. For a given positive number x, the mantissa of log x can be determined from a table of common logarithms. The characteristic of log x can be found by first expressing x as the product of a number between 1 and 10 and an integral power of 10 and then reading off the power of 10.

Our Table I only permits us to read off directly the common logarithms of numbers having three so-called significant digits. For example, we can read log 5.23, but we cannot read log 5.234. If the logarithm of a number with more than three significant digits is required, a technique called *interpolation* could be used, but it would be better to use a more

Exercise 2.9, 61–75

extensive table or a hand calculator.

Table I may also be used to approximate **antilogarithms,** numbers whose logarithms are known. For example, if we know that log $x = 2.6749$, we can find an approximation to x by looking in the table for mantissa 0.6749, reading off the associated three digits, and then placing the decimal point so that the characteristic of log x will be 2. In Table I we read

$$\log 4.73 = 0.6749.$$

It follows that

$$\log 473 = 2.6749,$$

so x is approximately 473. On a calculator, $x = 10^{2.6749}$, or approximately 473.0423.

If log x is known but the mantissa of log x is not given in Table I, we can approximate x by finding the two tabular values nearest the mantissa of log x. This will be demonstrated in Examples 13, 14, and 15. Better approximations can be obtained, however, by using a more extensive

Exercise 2.9, 81–85

table or a calculator.

To find the value of the natural logarithm of some given positive number x, we may use a table of natural logarithms such as Table II in Appendix B, we may use a calculator, or we may use the fact that the natural logarithm of x is the common logarithm of x divided by log e.

$$\ln x = \frac{\log x}{\log e}$$

We can prove this by first writing x as a power of e,

$$x = e^{\ln x},$$

and then taking the common logarithm of x:

$$\log x = \log (e^{\ln x})$$
$$= (\ln x)(\log e). \quad \text{Property 3:}$$
$$\log_b M^k = k \log_b M$$

Solving for $\ln x$:

$$\log x = (\ln x)(\log e)$$
$$\ln x = \frac{\log x}{\log e}.$$

This formula enables us to express a natural logarithm in terms of common logarithms. In other words, it enables us to change from base 10 to base e.

The formula $\ln x = \log x / \log e$ also enables us to change from base e to base 10, because it may be written $\log x = (\ln x)(\log e)$. In this form it simply says that the common logarithm of x is the product of the number $\log e$ and the natural logarithm of x.

Using general bases b and a instead of 10 and e, we may similarly derive a formula that enables us to change from base b to any other base a.

First writing x as a power of a:

$$x = a^{\log_a x}.$$

Then taking the logarithm to the base b of x:

$$\log_b x = \log_b (a^{\log_a x})$$
$$= (\log_a x)(\log_b a) \quad \text{Property 3:}$$
$$\log_b M^k = k \log_b M$$

Solving for $\log_a x$:

$$\log_b x = (\log_a x)(\log_b a)$$
$$\log_a x = \frac{\log_b x}{\log_b a} \qquad \begin{array}{l} \text{Change} \\ \text{of base} \\ \text{formula} \end{array}$$

This is called the *change of base formula*. As an example, take a to be 2 and b to be 10. The change of base formula then says:

$$\log_2 x = \frac{\log x}{\log 2}.$$

If x is 5,

$$\log_2 5 = \frac{\log 5}{\log 2}.$$

The number $\log e$ is approximately equal to 0.4343, so the formula for changing from base e to base 10 or from base 10 to base e may be written approximately as

$$\ln x = \frac{\log x}{0.4343}$$

or

$$\log x = (0.4343)(\ln x)$$

EXAMPLE 11

Use Table I to find $\ln 7$ to four decimal places.

Solution

$$\ln 7 = \frac{\log 7}{\log e}$$

In Table I we read $\log 7 = 0.8451$. Using $\log e = 0.4343$, we have

$$\ln 7 = \frac{0.8451}{0.4343}.$$

To four decimal places,

$$\ln 7 = 1.9459.$$

On a hand calculator having natural logarithms, we could read directly that $\ln 7 = 1.945910$ to six decimal places. In Table II, $\ln 7 = 1.9459$.

Exercise 2.9, 76–80

Common logarithms may be used to facilitate computations involving products, quotients, powers, and roots in the decimal system. But due to the advent of inexpensive hand calculators and high-speed computers, logarithms are no longer the essential computational tool that they once were. The next three examples show how logarithms may be used for computations.

EXAMPLE 12

Use logarithms and Table I to find an approximation to $\sqrt{76.8}$.

Solution

$$\sqrt{76.8} = (76.8)^{1/2}$$
$$\log \sqrt{76.8} = \tfrac{1}{2} \log (76.8) \quad \text{Property 3: } \log_b M^k = k \log_b M$$
$$= \tfrac{1}{2}(1.8854) \quad \text{Using Table I}$$
$$= 0.9427$$

Now to find an approximation to the number $\sqrt{76.8}$, whose logarithm is 0.9427, we look in Table I for mantissa 0.9427. In Table I we can find 0.9425 and 0.9430, but we cannot find 0.9427. We have

$$\log 8.76 = 0.9425$$
$$\log \sqrt{76.8} = 0.9427$$
$$\log 8.77 = 0.9430.$$

We see that $\sqrt{76.8}$ must be between 8.76 and 8.77. Since 0.9427 is closer to 0.9425 than it is to 0.9430, we say that

$$\sqrt{76.8} \text{ is approximately } 8.76.$$

A better approximation can be obtained from Table I by using a process called *linear interpolation.* A still better approximation can be obtained by using a more extensive table or a hand calculator. It would be much more efficient to perform such computations with a calculator. On a calculator, $\sqrt{76.8}$ is approximately 8.763561.

EXAMPLE 13   Use logarithms and Table I to find an approximation to $(53.2)^{10}$.

Solution

$$\log (53.2)^{10} = 10 \log (53.2)$$
$$= 10(1.7259)$$
$$= 17.2590$$

Using Table I, we have

$$\log (1.81 \times 10^{17}) = 17.2577$$
$$\log (53.2)^{10} = 17.2590$$
$$\log (1.82 \times 10^{17}) = 17.2601.$$

We see that $(53.2)^{10}$ is between 1.81×10^{17} and 1.82×10^{17}. Since 0.2590 is closer to 0.2601 than it is to 0.2577, we say that

$$(53.2)^{10} \text{ is approximately } 1.82 \times 10^{17}.$$

On a calculator, $(53.2)^{10}$ is approximately 1.8160×10^{17}.

EXAMPLE 14   Use logarithms and Table I to find an approximation to

$$\frac{762 \times 0.0342}{83.9}.$$

Solution

$$\log \frac{762 \times 0.0342}{83.9} = \log (762 \times 0.0342) - \log 83.9$$

Property 2: $\log_b \dfrac{M}{N} = \log_b M - \log_b N$

$$= \log 762 + \log 0.0342 - \log 83.9$$

Property 1: $\log_b MN = \log_b M + \log_b N$

$$= (2.8820) + (0.5340 - 2) - (1.9238)$$
$$= 1.4160 - (1.9238)$$
$$= -0.5078$$
$$= -0.5078 + 1 - 1$$
$$= 0.4922 - 1$$

We added and subtracted 1 from -0.5078 in order to express the logarithm of the quotient we are computing as the sum of an integer and a *nonnegative* four-place decimal so that we can use Table I. Using Table I, we have

$$\log 0.310 = 0.4914 - 1$$

$$\log \frac{762 \times 0.0342}{83.9} = 0.4922 - 1$$

$$\log 0.311 = 0.4928 - 1.$$

Since 0.4922 is closer to 0.4928 than to 0.4914, we say that

$$\frac{762 \times 0.0342}{83.9} \text{ is approximately } 0.311.$$

Exercise 2.9, 86–95 On a calculator, the given quotient is approximately 0.31061.

The properties of logarithms that we have considered in this section have many uses other than for computations. In calculus, for example, properties of logarithms may be used to advantage in a technique called "logarithmic differentiation." The next example illustrates the way they are applied in this technique.

EXAMPLE 15   Let

$$y = \frac{x^7 \sqrt{x^3 - 2x + 4}}{\sqrt[3]{x^2 + 5}}.$$

Simplify ln y as much as possible by using properties of logarithms. (Assume that y is positive and that the values of x are such that the various logarithms involved are all defined.)

Solution

$$\ln y = \ln\left[\frac{x^7 \sqrt{x^3 - 2x + 4}}{\sqrt[3]{x^2 + 5}}\right]$$

$$\ln y = \ln x^7 \sqrt{x^3 - 2x + 4} - \ln \sqrt[3]{x^2 + 5}$$

Property 2: $\log_b \dfrac{M}{N} = \log_b M - \log_b N$

$$= \ln x^7 + \ln \sqrt{x^3 - 2x + 4} - \ln \sqrt[3]{x^2 + 5}$$

Property 1: $\log_b MN = \log_b M + \log_b N$

$$= 7 \ln x + \tfrac{1}{2} \ln (x^3 - 2x + 4) - \tfrac{1}{3} \ln (x^2 + 5)$$

Property 3: $\log_b M^k = k \log_b M$

Believe it or not, this mechanical application of these properties of logarithms makes it much easier for a calculus student to find what is known as the "derivative of y with respect to x." Note that we had to assume that y is positive and that x is such that all expressions whose logarithms are involved are positive, because logarithms of nonpositive numbers are not defined.

Exercise 2.9, 96–100

EXERCISE 2.9 LOGARITHMS; COMMON AND NATURAL LOGARITHMS

▶ Express each of the following in the exponential form.

EXAMPLE $\qquad \log_{10} x = y$

Solution $\qquad 10^y = x$

1. $\log_3 9 = 2$ **2.** $\log_4 64 = 3$ **3.** $\log_5 25 = 2$

4. $\log x = \frac{1}{2}$ **5.** $\log x = 5$ **6.** $\ln x = y$

7. $\ln x = 5$ **8.** $\ln e^7 = 7$ **9.** $\ln x^2 = 3$

10. $\log_2 32 = 5$

▶ Express each of the following in logarithmic notation.

EXAMPLE $\qquad y = 3^4$

Solution $\qquad \log_3 y = 4$

11. $y = 2^6$ **12.** $y = 10^{-7}$ **13.** $y = e^{x^2}$ **14.** $y = e^u$

15. $2^3 = 8$ **16.** $y = e^{5x}$ **17.** $y = e^{x+h}$ **18.** $49 = 7^2$

19. $y = e^{m+n}$ **20.** $y = e^{2x}$

▶ Find the value of each of the following.

EXAMPLE $\qquad \ln e^{x^3}$

Solution $\qquad \ln e^{x^3} = x^3$

21. $\ln e^u$ **22.** $\ln e$ **23.** $\log 10$

24. $\log_2 2$ **25.** $\log_2 1$ **26.** $\log 10^x$

27. $\log_2 2^x$ **28.** $\log_3 3^{x^2}$ **29.** $\ln e^x$

30. $\ln e^{-x}$ **31.** $e^{\ln x}$ **32.** $10^{\log x}$

33. $2^{\log_2 x}$ **34.** $e^{\ln x^2}$ **35.** $e^{\ln 2x}$

36. $\log 100000$ **37.** $\log 0.00001$ **38.** $\ln e^5$

39. $\ln e^{-5}$ **40.** $\log 10^{-8}$

▶ Write each of the following as a power of e.

EXAMPLES

(a) 2^x (b) x^3

Solutions

(a) $2 = e^{\ln 2}$ (b) $x = e^{\ln x}$
$2^x = e^{x \ln 2}$ $x^3 = e^{3 \ln x}$

41. 5^x **42.** 2^{3x} **43.** 2^{x^2} **44.** $(\frac{1}{2})^x$ **45.** $(\sqrt{2})^x$

46. x^2 **47.** x^5 **48.** $x^{1/2}$ **49.** $x^{1/3}$ **50.** $x^{\sqrt{2}}$

▶ In each of the following, tell whether or not $\ln |P(x)|$ is equal to $\ln (P(x))$ for all real numbers x.

EXAMPLE

$P(x) = x - 1$

Solution

Here $\ln |x - 1| \neq \ln (x - 1)$ for all real numbers x because $(x - 1)$ is nonpositive when $x \leq 1$, and for such values of x, $\ln (x - 1)$ is not defined.

51. $P(x) = x^4$ **52.** $P(x) = x^5$ **53.** $P(x) = x^4 + 1$

54. $P(x) = x - 3$ **55.** $P(x) = x + 2$

▶ Where M and N are positive real numbers, prove each of the following.
56. $\log MN = \log M + \log N$

57. $\ln \dfrac{M}{N} = \ln M - \ln N$

58. $\ln M^k = k \ln M$ for any real number k

59. $\log M^k = k \log M$ for any real number k

60. $\log_b \dfrac{M}{N} = \log_b M - \log_b N$ for positive number $b \neq 1$

▶ Use Table I in Appendix B to find an approximation to each of the following.

EXAMPLES

(a) $\log 92.3$ (b) $\ln |-8|$

Solutions

(a) $\log 92.3 = 1.9652$ (b) $\ln |-8| = \ln 8$

$$= \frac{\log 8}{\log e}$$

$$= \frac{0.9031}{0.4343}$$

$$= 2.0794$$

61. log 923	62. log 0.0923	63. log 0.000578
64. log 5780	65. log 62500	66. log 0.00625
67. ln 347	68. ln $|-73.4|$	69. ln 27
70. ln $|-10|$	71. ln 0.134	72. ln $|-260|$
73. $\log_2 10$	74. $\log_5 21$	75. $\log_3 21$

► Express each of the following in terms of a natural logarithm.

EXAMPLES

(a) log 273   (b) log $(x^2 + 7)$

Solutions

(a) log 273 = (log e)(ln 273)   (b) log $(x^2 + 7)$ = (log e)(ln $(x^2 + 7)$)

76. log 35.9	77. log $(3x + 1)$	78. log 2
79. log 38	80. log $(x^2 - 9)$	

► Use Table I to find an approximation to x in each of the following.

EXAMPLE

log x = 1.4032

Solution

log 25.3 = 1.4031
    log x = 1.4032
log 25.4 = 1.4048
x is approximately 25.3

81. log x = 3.9167	82. log x = 0.8175 − 2
83. log x = 4.5530	84. log x = 0.3931 − 4
85. log x = 5.2718	

► Use logarithms and Table I to find an approximation to each of the following.

EXAMPLE

$\sqrt[3]{768}$

Solution

$\log \sqrt[3]{768} = \frac{1}{3} \log 768$
$= \frac{1}{3}(2.8854)$
$= 0.9618$
log 9.15 = 0.9614
$\log \sqrt[3]{768}$ = 0.9618
log 9.16 = 0.9619
$\sqrt[3]{768}$ is approximately 9.16.

86. $\sqrt{827}$ **87.** $(25.2)^6$ **88.** $\dfrac{359 \times \sqrt{2}}{21.7}$

89. $\sqrt[5]{4060}$ **90.** $\dfrac{32.8}{2.1 \times 7.6}$ **91.** $(0.0054)^{-1}$

92. $\dfrac{2370 \times 350}{5\sqrt{3}}$ **93.** $\sqrt[3]{4670}$ **94.** $\dfrac{3\sqrt{7}}{52}$

95. $(0.0121)^5$

▶ In each of the following, simplify ln y as much as possible by using properties of logarithms. (Assume that y is positive and that the values of x are such that the various logarithms involved are all defined.)

EXAMPLE

$$y = \frac{\sqrt[3]{x + 1}}{(x + 4)\sqrt{x + 2}}$$

Solution

$$\ln y = \tfrac{1}{3} \ln (x + 1) - \ln (x + 4) - \tfrac{1}{2} \ln (x + 2)$$

96. $y = \dfrac{x^3\sqrt{x + 2}}{\sqrt[3]{x - 2}}$ **97.** $y = \dfrac{(x + 2)^{2/3}}{\sqrt{1 - x^4}}$

98. $y = (3x - 7)(x^3 + 8)(2x^2 - 5)$

99. $y = \dfrac{x^2 - 3x + 2}{\sqrt{(x - 1)(x + 4)}}$ **100.** $y = \dfrac{x^{2/3}\sqrt[4]{x - 7}}{\sqrt[3]{x + 2}}$

CHECK LIST FOR THE STUDENT

If you have learned the material in Section 2.9, you should be able to do the following:

1. Define the logarithm of positive number x to the base b (b is positive and not equal to 1) as an exponent.

2. Find $\log_b 1$ for any acceptable base b.

3. Find $\log_b b$ for any acceptable base b.

4. Find $\log_b b^k$ for any acceptable base b and any real number k.

5. Define a common logarithm.

6. Write the customary notation for the common logarithm of x.

7. Define a natural logarithm.

8. Write the customary notation for the natural logarithm of x.

9. Write $y = e^x$ in logarithmic notation.

10. Write any positive real number a as a power of e.

11. Write x^n as a power of e.

12. Find the value of $\ln |x|$ for certain given values of x.

13. Tell whether or not $\ln |P(x)|$ is equal to $\ln (P(x))$ for certain given polynomials $P(x)$.

14. State, use, and derive the following basic properties of logarithms:

    1. $\log_b MN = \log_b M + \log_b N$

    2. $\log_b \dfrac{M}{N} = \log_b M - \log_b N$

    3. $\log_b M^k = k \log_b M$

15. Find the common logarithm of a number that is an integral power of 10.

16. Use a table of logarithms to find an approximation to the common logarithm of a number between 1 and 10.

17. Use a table of logarithms and find an approximation to the common logarithm of any positive number.

18. Express a given common logarithm as the sum of a nonnegative decimal less than 1 and an integer.

19. Give the meanings of the words *mantissa* and *characteristic*.

20. Use a table of logarithms to find an approximation to a number whose common logarithm is given.

21. Derive the formula $\ln x = \log x / \log e$.

22. Express a given natural logarithm in terms of common logarithms.

23. Express a given common logarithm in terms of natural logarithms.

24. State and derive the formula for expressing a logarithm to the base a in terms of logarithms to the base b.

25. Use common logarithms in computations involving products, quotients, powers, and roots in the decimal system.

26. Use properties of logarithms to simplify ln y for certain given expressions $y = f(x)$.

SECTION QUIZ 2.9

1. (a) Express $\log_2 a = b$ in exponential notation.
   (b) Express $y = 3^5$ in logarithmic notation.
   (c) Write 3 as a power of e.
   (d) Express $\log_{10} 37$ in terms of natural logarithms.

2. Find the value of each of the following.
   (a) ln 1   (b) $\log_{10} 10$ (c) ln e^5
   (d) $e^{\ln 5}$ (e) $10^{\log_{10} 2}$

Suppose that $\log_{10} 7 = 0.8451$ and $\log_{10} 5 = 0.6990$. Find the value of each of the following.

3. (a) $\log_{10} 700$ (b) $\log_{10} 0.7$

4. (a) $\log_{10} (5 \times 7)$ (b) $\log_{10} \frac{7}{5}$

5. (a) $\log_{10} 7^3$ (b) $\log_{10} \sqrt[3]{5}$

2.10 Operations on Functions

The purpose of this section is to define addition, subtraction, multiplication, division, and composition of functions and to provide practice in identifying composite functions as well as practice with the notation.

Let f and g be functions. Then the **sum** of f and g is denoted by $f + g$ and is defined to be the function whose rule of assignment assigns the number $f(x) + g(x)$ to x and whose domain is the set of all numbers common to the domains of f and g. That is, $f + g$ is the function such that

$$(f + g)(x) = f(x) + g(x),$$

$$\text{Domain of } f + g = \left\{ x \,\middle|\, \begin{array}{l} x \text{ is in the domain of } f \\ \text{and also in the domain of } g \end{array} \right\}.$$

To illustrate, let f be the function such that

$$f(x) = \ln x,$$

Domain of f is the set of all positive real numbers.

Let g be the function such that

$$g(x) = \sqrt{-x + 2},$$
$$\text{Domain of } g = \{x \,|\, x \leq 2\}.$$

Compare the domains of f and g graphically in Figure 2.82. The only numbers that are in the domain of f and also in the domain of g are those numbers x such that $0 < x \leq 2$. (Note that 2 is included in this set, but 0 is not.) For this f and g, $f + g$ is the function such that

$$(f + g)(x) = \ln x + \sqrt{-x + 2},$$
$$\text{Domain of } f + g = \{x \,|\, 0 < x \leq 2\}.$$

The **difference** of functions f and g is denoted by $f - g$ and is defined to be the function whose rule of assignment assigns the number $f(x) - g(x)$ to x and whose domain is the set of all numbers common to the domains of f and g. That is, $f - g$ is the function such that

$$(f - g)(x) = f(x) - g(x),$$

$$\text{Domain of } f - g = \left\{ x \,\middle|\, \begin{array}{l} x \text{ is in the domain of } f \\ \text{and also in the domain of } g \end{array} \right\}.$$

Domain of $f = \{x \,|\, x > 0\}$

```
————————————————————————
       0     Positive real numbers
```

```
————————————————————————
       0   2
```

Domain of $g = \{x \,|\, x \leq 2\}$

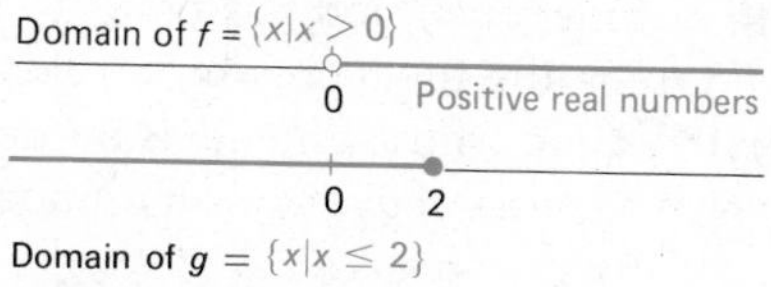

FIGURE 2.82

As an illustration, again consider f and g such that $f(x) = \ln x$ and $g(x) = \sqrt{-x + 2}$. For this f and g, $f - g$ is the function such that

$$(f - g)(x) = \ln x - \sqrt{-x + 2},$$
$$\text{Domain of } f - g = \{x \mid 0 < x \le 2\}.$$

The **product** of functions f and g is denoted by fg and is defined to be the function whose rule of assignment assigns the number $f(x) \cdot g(x)$ to x and whose domain is the set of all numbers common to the domains of f and g. That is, fg is the function such that

$$(fg)(x) = f(x) \cdot g(x),$$
$$\text{Domain of } fg = \left\{ x \;\middle|\; \begin{array}{l} x \text{ is in the domain of } f \\ \text{and also in the domain of } g \end{array} \right\}.$$

For example, if we again let f and g be such that $f(x) = \ln x$ and $g(x) = \sqrt{-x + 2}$, fg is the function such that

$$(fg)(x) = \ln x \cdot \sqrt{-x + 2}$$
$$\text{Domain of } fg = \{x \mid 0 < x \le 2\}.$$

The **quotient** of functions f and g is denoted by f/g and is defined to be the function whose rule of assignment assigns the number $f(x)/g(x)$ to x and whose domain is the set of all numbers common to the domains of f and g and for which $g(x)$ is not zero. That is, f/g is the function such that

$$\left(\frac{f}{g} \right)(x) = \frac{f(x)}{g(x)},$$
$$\text{Domain of } \frac{f}{g} = \left\{ x \;\middle|\; \begin{array}{l} x \text{ is in the domain of } f \\ \text{and also in the domain of } g, \\ \text{and } g(x) \ne 0 \end{array} \right\}.$$

If we once more let f and g be such that $f(x) = \ln x$ and $g(x) = \sqrt{-x + 2}$, the function f/g is such that

$$\left(\frac{f}{g} \right)(x) = \frac{\ln x}{\sqrt{-x + 2}},$$
$$\text{Domain of } \frac{f}{g} = \{x \mid 0 < x < 2\}.$$

The number 2 is not in the domain of f/g because $g(2) = 0$.

The **composite** of functions f and g is denoted by $f \circ g$ (read "f circle g") and is defined to be the function whose rule of assignment assigns the number $f(g(x))$ to x and whose domain is the set of all numbers in the domain of g to which the rule of g assigns a number that is in the domain of f. That is, $f \circ g$ is the function such that

$$(f \circ g)(x) = f(g(x)),$$

$$\text{Domain of } f \circ g = \left\{ x \;\middle|\; \begin{array}{l} x \text{ is in the domain of } g \\ \text{and } g(x) \text{ is in the} \\ \text{domain of } f \end{array} \right\}.$$

The process of finding the composite of two functions is sometimes called *composition*. To find $(f \circ g)(x) = f(g(x))$, we may think of working from right to left. First we take x from the domain of g. The rule of g assigns the number $g(x)$ to the x. Finally, if x is such that the number $g(x)$ is in the domain of f, the rule of f assigns the number $(f \circ g)(x)$ to the $g(x)$. If we let $y = f(u)$ and $u = g(x)$, this procedure may be described schematically as follows:

$$x \xrightarrow{\;g\;} u \xrightarrow{\;f\;} y$$
$$\quad \| \qquad \|$$
$$\quad g(x) \quad f(u)$$
$$\qquad\qquad \|$$
$$\qquad\quad f(g(x))$$

$$x \xrightarrow{\;\;f \circ g\;\;} y$$
$$\qquad\qquad \|$$
$$\qquad\quad f(g(x))$$

To illustrate the composition of functions, let functions f and g be such that

$$f(x) = \sqrt{x},$$

Domain of f is the set of all nonnegative real numbers;

$$g(x) = x - 3,$$

Domain of g is the set of all real numbers.

The only numbers in the domain of g for which $g(x) = x - 3$ is nonnegative (is in the domain of f) are those numbers x for which $x \geq 3$. For this f and g, $f \circ g$ is the function such that

$$(f \circ g)(x) = f(g(x))$$
$$= f(x - 3)$$
$$= \sqrt{x - 3},$$

Domain of $f \circ g = \{x \mid x \geq 3\}.$

We may describe the composition of this f and g schematically as follows:

$$x \xrightarrow{\;g\;} (x - 3) \xrightarrow{\;f\;} \sqrt{x - 3}$$
$$x \xrightarrow{\;\;f \circ g\;\;} \sqrt{x - 3}$$

As an additional illustration, let f and g be such that $f(x) = \ln x$ and $g(x) = \sqrt{-x + 2}$, as they were earlier in this section. In this case the domain of f is the set of all positive real numbers and the domain of g is the set of all numbers less than or equal to 2. The only numbers in the domain of g for which $g(x) = \sqrt{-x + 2}$ is positive (is in the domain of f) are the numbers x for which $x < 2$.

$$(f \circ g)(x) = f(g(x))$$
$$= f(\sqrt{-x + 2})$$
$$= \ln \sqrt{-x + 2},$$

Domain of $f \circ g = \{x \,|\, x < 2\}$.

Note that $x = 2$ is not in the domain of $f \circ g$ because $g(2) = 0$ is not in the domain of f; $f(x) = \ln x$ is not defined for $x = 0$.

EXAMPLE 1   Let functions f and g be such that $f(x) = \sqrt{x}$ and $g(x) = x - 3$. Find $(g \circ f)(x)$.

Solution
$$(g \circ f)(x) = g(f(x))$$
$$= g(\sqrt{x})$$
$$= \sqrt{x} - 3$$

In Example 1 we found that $(g \circ f)(x) = \sqrt{x} - 3$. For the same f and g, we found earlier that $(f \circ g)(x) = \sqrt{x} - 3$. This proves that $f \circ g$ is not

Exercise 2.10, 1–20   necessarily the same as $g \circ f$.

In calculus, it is frequently necessary to recognize a given function as the composite of two functions. For example, consider the function h such that $h(x) = e^{x^2}$. We may think of h as being the composite of f and g where $f(x) = e^x$ and $g(x) = x^2$. Indeed, for such f and g,

$$(f \circ g)(x) = f(g(x))$$
$$= f(x^2)$$
$$= e^{x^2}.$$

In identifying a given function as the composite $f \circ g$ of two functions, it sometimes clarifies matters to use the letter u for any element in the domain of f and to continue to use x for any element in the domain of g. If we let $y = f(u)$ and $u = g(x)$, then $y = f(u)$ is the rule of assignment of the composite $f \circ g$.

$$y = f(u) \qquad u = g(x)$$
$$y = f(g(x))$$

For example,

$$y = e^{x^2}$$

is the rule of assignment of composite $f \circ g$ where

$$f(u) = e^u \quad \text{and} \quad u = g(x) = x^2.$$

EXAMPLE 2   Find $f(u)$ and $u = g(x)$ so that $y = \ln (x^4 + 7)$ is the rule of assignment of composite function $f \circ g$.

Solution   Let $f(u) = \ln u$ where $u = g(x) = x^4 + 7$. Then

$$(f \circ g)(x) = f(g(x))$$
$$= f(x^4 + 7)$$
$$= \ln (x^4 + 7).$$

That is, $y = \ln (x^4 + 7)$ is the rule of assignment of $f \circ g$. Notice that this choice of $f(u)$ and u is not the only choice possible here. Another possibility is as follows:

Let $f(u) = \ln (u + 2)$ where $u = g(x) = x^4 + 5$.
Then $(f \circ g)(x) = f(g(x))$
$$= f(x^4 + 5)$$
$$= \ln (x^4 + 5 + 2)$$
$$= \ln (x^4 + 7).$$

There are actually infinitely many choices of functions f and g for which $y = \ln (x^4 + 7)$ is the rule of assignment of the composition of f and g. In general, there are infinitely many ways in which a function can be regarded as the composite of two functions.

Exercise 2.10, 21–30

EXERCISE 2.10 OPERATIONS ON FUNCTIONS

▶   In each of the following, find the rule of assignment and state the domain of $f + g$, $f - g$, fg, f/g, $f \circ g$, and $g \circ f$. If the domain of f or g is not explicitly stated, use the domain convention.

EXAMPLE $f(x) = e^x$
$g(x) = \sqrt{x}$

Solution   Domain of f is set of all real numbers.
Domain of g is set of all nonnegative real numbers.

$(f + g)(x) = e^x + \sqrt{x}$ Domain of $f + g = \{x \mid x \geq 0\}$
$(f - g)(x) = e^x - \sqrt{x}$ Domain of $f - g = \{x \mid x \geq 0\}$
$(fg)(x) = e^x \sqrt{x}$ Domain of $fg = \{x \mid x \geq 0\}$
$\left(\dfrac{f}{g}\right)(x) = \dfrac{e^x}{\sqrt{x}}$ Domain of $\dfrac{f}{g} = \{x \mid x > 0\}$

$$(f \circ g)(x) = f(g(x))$$
$$= e^{\sqrt{x}} \qquad \text{Domain of } f \circ g = \{x \mid x \geq 0\}$$

$$(g \circ f)(x) = g(f(x))$$
$$= \sqrt{e^x} \qquad \text{Domain of } g \circ f \text{ is the set of all real numbers}$$
(e^x is positive for all real numbers x)

1. $f(x) = e^{3x}$, $g(x) = \sqrt{-x}$

2. $f(x) = x^3$, $g(x) = 5x^2 + 2$

3. $f(x) = \ln |x|$, $g(x) = x^3$

4. $f(x) = \ln x$, $g(x) = \sqrt[3]{x}$

5. $f(x) = \ln x$; $g(x) = 2x$, domain of $g = \{x \mid -1 \leq x \leq 1\}$

6. $f(x) = 3x^2 + 2x$, $g(x) = e^{-x}$

7. $f(x) = \sqrt{x + 3}$, $g(x) = x^2$

8. $f(x) = -\sqrt{x}$, domain of $f = \{x \mid 0 \leq x \leq 9\}$; $g(x) = 2x$

9. $f(x) = x^2$, domain of $f = \{x \mid x \leq 0\}$; $g(x) = x + 5$

10. $f(x) = \log x$, domain of $f = \{x \mid x \geq 1\}$; $g(x) = x + 2$, domain of $g = \{x \mid x \geq 0\}$.

EXAMPLE

Let $f(x) = \sqrt[3]{x}$ and $g(x) = 2x + 3$. Find
(a) $(f + g)(8)$ (b) $(f - g)(8)$

(c) $(fg)(8)$ (d) $(f/g)(8)$

(e) $(f \circ g)(8)$ (f) $(g \circ f)(8)$

Solution

(a) $(f + g)(8) = \sqrt[3]{8} + 2(8) + 3$ (b) $(f - g)(8) = \sqrt[3]{8} - (16 + 3)$
$$= 2 + 16 + 3 \qquad\qquad\qquad\qquad = 2 - 19$$
$$= 21 \qquad\qquad\qquad\qquad\qquad\quad = -17$$

(c) $(fg)(8) = \sqrt[3]{8}(16 + 3)$ (d) $\left(\dfrac{f}{g}\right)(8) = \dfrac{\sqrt[3]{8}}{16 + 3}$
$$= 2(19) \qquad\qquad\qquad\qquad\qquad = \tfrac{2}{19}$$
$$= 38$$

(e) $(f \circ g)(8) = f(g(8))$ (f) $(g \circ f)(8) = g(f(8))$
$$= \sqrt[3]{16 + 3} \qquad\qquad\qquad = 2(\sqrt[3]{8}) + 3$$
$$= \sqrt[3]{19} \qquad\qquad\qquad\qquad = 4 + 3$$
$$\qquad\qquad\qquad\qquad\qquad\qquad = 7$$

Let $f(x) = \sqrt{x + 1}$ and $g(x) = x^2$. Find the value of each of the following.

11. $(f + g)(3)$ 12. $(f - g)(3)$ 13. $(fg)(3)$ 14. $\left(\dfrac{f}{g}\right)(3)$

15. $(f \circ g)(3)$ **16.** $(g \circ f)(3)$ **17.** $(f \circ g)(8)$ **18.** $(g \circ f)(8)$

19. $(f \circ g)(0)$ **20.** $(g \circ f)(0)$

▶ In each of the following, find one choice of $f(u)$ and $u = g(x)$ so that the given expression is the rule of assignment of composite function $f \circ g$.

EXAMPLE

$y = \sqrt{x^3 + 2}$

Solution

One choice of $f(u)$ and $u = g(x)$ so that $(f \circ g)(x) = \sqrt{x^3 + 2}$ is

$$f(u) = \sqrt{u} \qquad u = g(x) = x^3 + 2.$$
$$\quad\; = u^{1/2}$$

For this f and g, $(f \circ g)(x) = f(g(x))$
$$= \sqrt{x^3 + 2}.$$

21. $y = \sqrt[3]{x^2 - 5}$

22. $y = \ln x^5$

23. $y = e^{x^3}$

24. $y = (x^5 - 4x^2 + 2)^{2/3}$

25. $y = \sqrt{x^3 - 2x + 1}$

26. $y = \ln (2x^3 - 3x + 1)$

27. $y = 10^{-x}$

28. $y = e^{6x+1}$

29. $y = (x^2 + 3x - 5)^{-4/5}$

30. $y = 2^{3x^2}$

CHECK LIST FOR THE STUDENT

If you have learned the material in Section 2.10, you should be able to do the following:

1. Define the sum of two functions.

2. Find the rule of assignment and the domain of $f + g$ for given functions f and g.

3. Define the difference of two functions.

4. Find the rule of assignment and the domain of $f - g$ for given functions f and g.

5. Define the product of two functions.

6. Find the rule of assignment and the domain of fg for given functions f and g.

7. Define the quotient of two functions.

8. Find the rule of assignment and the domain of f/g for given functions f and g.

9. Define the composite of two functions.

10. Find the rule of assignment and the domain of $f \circ g$ for given functions f and g.

11. Prove that $f \circ g$ is not necessarily the same as $g \circ f$.

12. Find $f(u)$ and $u = g(x)$ so that a given $y = h(x)$ is the rule of assignment of composite function $f \circ g$.

SECTION QUIZ 2.10

1. Let $f(x) = \sqrt{x}$, and let the domain of f be $\{x \mid x \geq 0\}$. Let $g(x) = \sqrt{-x + 3}$, and let the domain of g be $\{x \mid x \leq 3\}$.
  (a) What is the domain of $f + g$?
  (b) What is the domain of f/g?

Let $f(x) = \sqrt[3]{x + 1}$, and let $g(x) = 5x - 4$. Find the value of each of the following.
2. (a) $(f + g)(7)$ (b) $(f - g)(7)$ (c) $(fg)(7)$

3. (a) $\left(\dfrac{f}{g}\right)(7)$ (b) $f(g(7))$ (c) $g(f(7))$

Let $f(x) = \sqrt{x}$, and let the domain of f be $\{x \mid x \geq 0\}$. Let $g(x) = x^2 + 9$, and let the domain of g be the set of all real numbers. Find the rule of assignment and state the domain of each of the following.
4. (a) $f + g$ (b) $f - g$ (c) fg

5. (a) $\dfrac{f}{g}$ (b) $f \circ g$ (c) $g \circ f$

Part 2 Final Test

1. Let f be the function whose domain is $\{-2, 0, 1, 3\}$ and which is such that $f(x) = x + 3$.
   (a) Find the range of f.   (b) Sketch the graph of f.

2. According to the domain convention, what is the domain of function f in each of the following?
   (a) $f(x) = \sqrt{x - 4}$ (b) $f(x) = 1/x^2$

3. Let f be the function whose domain is $\{x \mid x \geq -1\}$ and which is such that $f(x) = -x^2$. Sketch the graph of f.

4. Find the point of intersection of the following lines, sketch the graphs of the lines, and label the point of intersection.
$$x - 2y - 5 = 0 \quad \text{and} \quad 3x + 2y - 7 = 0$$

5. Let A and B be the points: $A(-3, 7)$, $B(2, 4)$.
   (a) Find the slope of the line through A and B.
   (b) Find the slope of a line perpendicular to the line through A and B.
   (c) Find the standard form of an equation of the line through A and B.

6. (a) Find the standard form of an equation of the line through the point $(2, 5)$ and with slope 3.
   (b) Find the slope and y-intercept of the line $5x - 3y - 7 = 0$.
   (c) Sketch the graph of the linear function f such that $f(x) = x - 3$, and label both its x- and y-intercepts.

7. In each of the following, find the intercepts of the graph, determine whether or not the graph is symmetric with respect to the x-axis, with respect to the y-axis, or with respect to the origin, find any vertical asymptote(s) that the graph has, and find any horizontal asymptote(s) that the graph has.
   (a) $4x^2 - y^2 = 4$ (b) $y = \dfrac{5x}{x^2 - 4}$

8. Sketch the graph of
$$y = \frac{3}{x - 4}.$$
   Label any intercept(s) or vertical or horizontal asymptote(s) that the graph has.

9. Let A and B be the points: $A(-2, 3)$, $B(1, 7)$.
   (a) Find the distance between A and B.
   (b) Find the midpoint of the segment joining A and B.

10. (a) Find the standard form of an equation of the circle with center at $(3, -7)$ and radius 6.
    (b) Find the center and radius of the circle $x^2 + y^2 - 10x + 4y - 52 = 0$.

11. Transform the equation $(x + 4)^2 + (y + 2)^2 = 4$ by translating axes so that the new origin is at $(-4, -2)$. Draw and label both sets of axes. Sketch the graph.

12. Sketch the graph of each of the following. Plot and label the focus, and draw and label the directrix.
(a) $x^2 = 24y$ (b) $y^2 = -24x$

13. Write the following equation in standard form. Label the vertex and sketch the graph.

$$y^2 - 6y - 24x - 39 = 0$$

14. If $f(x) = -3x^2 + 6x - 8$, sketch the graph of the function f.

Sketch the graph of each of the following. If it is an ellipse, label its center and give the lengths of its axes. If it is a hyperbola, give the equations of its asymptotes and draw its asymptotes.

15. $4(x + 5)^2 + 9(y - 4)^2 = 36$

16. $\dfrac{(x - 5)^2}{4} - \dfrac{y^2}{9} = 1$

17. In each of the following, if the equation has a graph, identify it. If the equation has no graph, state "No graph."
(a) $y = 2x$ (b) $y = 2x^2$
(c) $2x^2 + 2y^2 = -1$ (d) $2x^2 + 2y^2 = 1$
(e) $2x^2 + 2y^2 = 0$

18. In each of the following, state whether or not function f is one-to-one.
(a) $f(x) = 3x^2$ (b) $f(x) = 3x$

19. Let f be the function with domain $\{x \mid 0 \le x \le 2\}$, and let $f(x) = 3x - 1$. Where f^{-1} denotes the inverse of f, find the domain of f^{-1}, and find $f^{-1}(x)$.

20. Sketch the graph of $y = 8^x$.

21. Sketch the graph of $y = \log_8 x$.

22. Find the value of each of the following.
(a) $\log_7 1$ (b) $\log_7 7$ (c) $\ln e^9$ (d) $\log_{10} 0.0001$ (e) $\log_3 \frac{1}{9}$

23. Suppose that $\log_{10} 3 = 0.4771$ and $\log_{10} 4 = 0.6021$. Find the value of each of the following.
(a) $\log_{10} 300$ (b) $\log_{10} 0.004$ (c) $\log_{10} 12$
(d) $\log_{10} \frac{4}{3}$ (e) $\log_{10} \sqrt[3]{4}$ (f) $\log_{10} 3^4$

In each of the following, let $f(x) = \sqrt{x}$, let the domain of f be $\{x \mid x \ge 0\}$, let $g(x) = x - 3$, and let the domain of g be the set of all real numbers.

24. (a) Find $(f \circ g)(x) = f(g(x))$.
(b) Find $(g \circ f)(x) = g(f(x))$.

25. (a) What is the domain of $f + g$?
(b) What is the domain of f/g?
(c) What is the domain of $f \circ g$?
(d) What is the domain of $g \circ f$?

Part 3 Precalculus Trigonometry

Sometimes it is necessary to consider the so-called trigonometric functions to be functions whose domains are sets of angles, and sometimes it is necessary to consider them to be functions whose domains are sets of real numbers. In fact, some authors only call them *trigonometric functions* when they are considered to be functions whose domains are sets of angles; these authors call them *circular* functions when they are considered to be functions whose domains are sets of real numbers. But any angle may be associated with a real number, and any real number may be associated with an angle, so these functions may be defined so that they can be interpreted and used either way. In this text we shall consider both situations, and we shall call the functions trigonometric functions whether they are functions of angles or functions of real numbers. The purpose of Part 3 is to present the trigonometry essential for the study of calculus courses utilizing trigonometric functions.

3.1 Angles, Real Numbers, and Unit Circle Points

The purpose of this section is to lay the groundwork for the definitions of the trigonometric functions by introducing some terminology concerning angles, by considering angular measure, and by showing how angles, real numbers, and unit circle points can be associated with one another.

An **angle** may be defined as a pair of rays (half-lines) with a common endpoint called the **vertex**. Some angles are shown in Figure 3.1. The rays are called the **sides** of an angle. If one side is designated the *initial side* and the other side is designated the *terminal side,* the angle is a **directed angle** and may be thought of as being determined by rotating the initial side about the vertex to the position of the terminal side. If the rotation is counterclockwise, the angle is said to be *positive* and its measure is positive. If the rotation is clockwise, the angle is said to be *negative* and its measure is negative. Unless otherwise specified, by *angle* we mean *directed angle.* Some directed angles are shown in Figure 3.2.

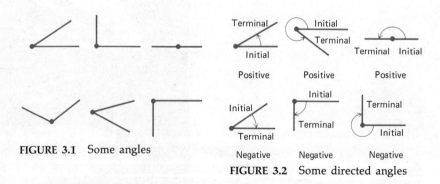

FIGURE 3.1 Some angles

FIGURE 3.2 Some directed angles

If an angle is thought of as being determined by rotating the initial side to the position of the terminal side, we must specify the sense of rotation as well as the initial and terminal sides, because two different angles may have the same initial and terminal sides. Two different angles, one positive and one negative, with the same initial and terminal sides are shown in Figure 3.3.

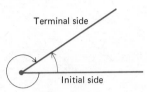

Terminal side

Initial side

FIGURE 3.3 Different angles with same sides

An angle may be referred to by naming a point on its initial side, then naming the vertex, and then naming a point on the terminal side. Angle *BAC* is shown in Figure 3.4. An angle may also be referred to simply by naming its vertex. Angle *BAC* in Figure 3.4 may simply be called angle *A*.

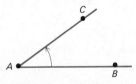

C

A *B*

FIGURE 3.4 Angle *BAC*

An angle is said to be in **standard position** provided that its vertex is at the origin of a Cartesian coordinate system and that its initial side is along the positive side of the *x*-axis. We sometimes refer to an angle in standard position as a *standard-position* angle. Several angles are shown in standard position in Figure 3.5. If the terminal side is in the first quadrant, the angle is said to be a *first-quadrant* angle or is said to be *in* the first quadrant; if the terminal side is in the second quadrant, the angle is said to be a *second-quadrant* angle or is said to be *in* the second quadrant, etc. If the terminal side lies on one of the axes, the angle is said to be a *quadrantal* angle.

Any given angle may be "placed" in standard position by choosing a Cartesian coordinate system with its origin at the vertex of the angle and with the positive half of its *x*-axis along the initial side of the angle. In this manner, any given angle may be identified with an angle in standard position.

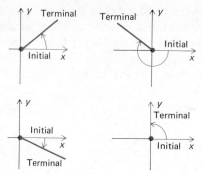

FIGURE 3.5 Angles in standard position

In many applications it is convenient to measure angles in *degrees*. One complete counterclockwise revolution of an initial side back to its own initial position is said to be 360 degrees. The symbol ° is used for degrees.

$$360° = 1 \text{ revolution}$$

When angles are measured in degrees, it is sometimes convenient to think of each degree as being subdivided into 60 smaller units of measure called *minutes*. The symbol ′ is used for minutes.

$$60' = 1°$$

An angle whose measure is 90° is said to be a **right angle**. A positive angle whose measure is less than 90° is said to be an **acute angle**. An angle whose measure is more than 90° and less than 180° is called an **obtuse angle**.

In calculus, angles are frequently measured in **radians**. One complete counterclockwise revolution is said to be 2π radians.

$$2\pi \text{ radians} = 1 \text{ revolution}$$

Thus we may convert from degree measure to radian measure or from radian measure to degree measure by using the fact that 2π radians equal 360 degrees.

$$2\pi \text{ radians} = 360°$$
or
$$\pi \text{ radians} = 180°$$

It is customary not to use a specific symbol for radians; "an angle of measure 2π" is understood to mean an angle of 2π radians.

EXAMPLE 1

An angle is determined by rotating its initial side clockwise through one-fourth revolution. What is its degree measure? What is its radian measure?

Solution	Since the rotation is clockwise, the measure of the angle is negative.

$$\text{The degree measure} = -\frac{1}{4}(360°)$$

$$= -90°$$

$$\text{The radian measure} = -\frac{1}{4}(2\pi)$$

$$= -\frac{\pi}{2}$$

EXAMPLE 2

Convert 30°, 45°, and 60° to radians. Convert $3\pi/4$, $3\pi/2$, and 5 radians to degrees.

Solution

π radians $= 180°$

$\dfrac{\pi}{180}$ radians $= 1°$ Dividing both sides by 180

1 radian $= \dfrac{180°}{\pi}$ Dividing both sides by π

$1° = \dfrac{\pi}{180}$ radians 1 radian $= \dfrac{180°}{\pi}$

$30° = \dfrac{30\pi}{180} = \dfrac{\pi}{6}$ $\dfrac{3\pi}{4} = \dfrac{3\pi}{4} \times \dfrac{180°}{\pi} = 135°$

$45° = \dfrac{45\pi}{180} = \dfrac{\pi}{4}$ $\dfrac{3\pi}{2} = \dfrac{3\pi}{2} \times \dfrac{180°}{\pi} = 270°$

$60° = \dfrac{60\pi}{180} = \dfrac{\pi}{3}$ $5 = 5 \times \dfrac{180°}{\pi} = \dfrac{900°}{\pi}$

If an angle is determined by rotating the initial side either counterclockwise or clockwise just until it coincides with the position of the terminal side, the absolute value of the degree measure must be less than or equal to 360°. Angles with degree measure greater than 360° or less than −360° (that is, angles with radian measure greater than 2π or less than -2π) are determined by rotating the initial side either counterclockwise or clockwise through more than one complete revolution. For example, an angle with degree measure 390° is determined by rotating the initial side counterclockwise through one complete revolution (360°) and then counterclockwise through an additional 30°. An angle with degree measure −780° is determined by rotating the initial side clockwise through two complete revolutions (−720°) and then clockwise through an additional 60°. As a further example, an angle with radian measure 3π is determined by rotating the initial side counterclockwise through one complete revolution (2π radians) and then counterclockwise through an additional π radians. Angles with measure 390° $-780°$, and 3π radians are shown in Figure 3.6.

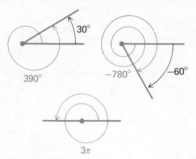

Exercise 3.1, 1–20   **FIGURE 3.6**   Angles with measure 390°, $-780°$, and 3π

An angle in standard position that is not a quadrantal angle may be associated with an acute angle θ called its *reference angle,* as shown in Figure 3.7. Note that in each case the reference angle of standard-position angle A is the acute angle whose sides are the terminal side of A (shown in color) and half the x-axis.

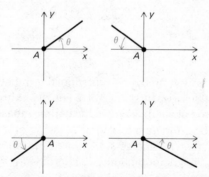

FIGURE 3.7   Reference angle θ for first-, second-, third-, or fourth-quadrant angle A

EXAMPLE 3

Find the measure in radians of the reference angle of A if the measure of A is 30°, 110°, $-135°$, or 300°.

Solution

In each case it is best to make a sketch so that we can easily see the quadrant in which the terminal side of A lies and so that we can then locate the reference angle. The sketches appear as Figures 3.8 through 3.11.

In Figure 3.8, the measure of A is 30°. The reference angle of A has measure 30° = π/6 radians. (Note that if A is an acute angle, then A is its own reference angle.)

In Figure 3.9, the measure of A is 110°. The reference angle of A has measure 70° = 70π/180 = 7π/18 radians.

In Figure 3.10, the measure of A is −135°. The reference angle of A has measure 45° = π/4 radians.

In Figure 3.11, the measure of A is 300°. The reference angle of A has measure 60° = π/3 radians.

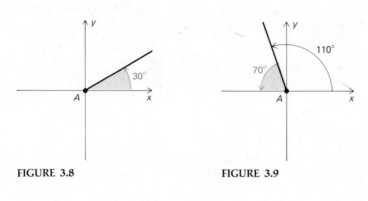

FIGURE 3.8   **FIGURE 3.9**

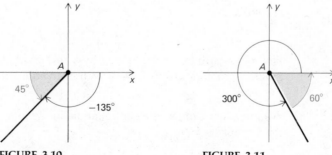

Exercise 3.1, 21–30 **FIGURE 3.10**   **FIGURE 3.11**

One of the reasons it is sometimes convenient to use radian measure is that if the central angle of a circle subtended by an arc of length s is of measure θ radians, then $s = r\theta$ where r is the radius of the circle. (See Figure 3.12.) This is true because the ratio of arc length s to the circumference $2\pi r$ is equal to the ratio of the measure θ radians of the angle subtended by the arc of length s to the measure 2π radians of the angle subtended by the entire circle. That is,

$$\frac{s}{2\pi r} = \frac{\theta}{2\pi}.$$

Solving for s, we have

$$s = 2\pi r \times \frac{\theta}{2\pi}$$

$$s = r\theta.$$

If the radius of the circle is 1 unit, the circle is called the **unit circle** and the length of arc s is simply equal to the number of radians in the central angle subtended by the arc.

If $r = 1$, $s = \theta$.

Length of arc of unit circle ⟶ ⟵ Central angle of θ *radians*

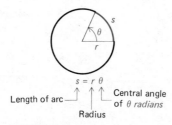

$s = r\theta$

Length of arc ⟶ ↑ ↑ Central angle
of θ *radians*
Radius

FIGURE 3.12

EXAMPLE 4

Find the length of arc subtended by a central angle of 60° in a circle of radius 12 meters. Using 3.14 for π, find an approximation to the length of this arc.

Solution

We must first express the central angle in radians, and then we may use the formula $s = r\theta$.

$$\theta = 60° = \frac{\pi}{3} \text{ radians}$$

$$s = r\theta$$

$$= 12 \times \frac{\pi}{3}$$

$$= 4\pi \text{ meters}$$

Using $\pi = 3.14$,

$$s = 4 \times 3.14$$

$$= 12.56 \text{ meters.}$$

EXAMPLE 5

Find the measure of the central angle θ subtending an arc s of length 2 meters on the unit circle. Use $\pi = 3.14$ to obtain a one-decimal-place approximation to the degree measure of θ.

Solution

Here $s = r\theta$ with $r = 1$. In other words, since the circle in this example is the unit circle,

$$s = \theta.$$

$$2 = \theta$$

$$\theta = 2 \text{ radians}$$

$$= 2 \times \frac{180}{\pi} \text{ degrees}$$

$$= \frac{360}{\pi} \text{ degrees.}$$

Using $\pi = 3.14$, to one decimal place,

$$\theta = \frac{360}{3.14}$$

$$= 114.6 \text{ degrees.}$$

Since 60 minutes equal one degree, and since $0.6 \times 60 = 36$, we may say that θ is approximately

Exercise 3.1, 31–50

$$114°36'.$$

Any angle A of radian measure α may be associated with a real number by assigning to A the number α of radians in the measure of A. Likewise any real number α may be associated with an angle by assigning to α the angle A in standard position with radian measure α. For example, an angle A of measure $30°$ is associated with the real number $\pi/6$, because the radian measure of A is $\pi/6$.

Angle	Real number
$A \longrightarrow$ (of measure $30°$)	$\dfrac{\pi}{6}$

The real number $\pi/6$ is associated with the standard-position angle A of measure $\pi/6$ radians. The angle A associated with the real number $\pi/6$ in this manner is shown in Figure 3.13.

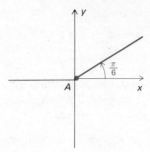

FIGURE 3.13   Angle A associated with real number $\pi/6$

Any angle A may be associated with a point on the unit circle $x^2 + y^2 = 1$ by assigning to A the point of intersection of the circle with the terminal side of the standard-position angle congruent to A. Likewise any real number α may be associated with a point on the circle $x^2 + y^2 = 1$ by assigning to α the point of intersection of the circle with the terminal side of the standard-position angle of radian measure α. For example, an angle A of measure $\pi/6$ is associated with the point P on $x^2 + y^2 = 1$ shown in Figure 3.14. Likewise the real number $\pi/6$ is associated with the same point P on $x^2 + y^2 = 1$ shown in Figure 3.14.

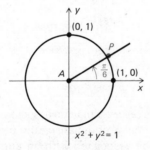

FIGURE 3.14   Unit circle point P associated with angle of measure $\pi/6$ and with real number $\pi/6$

To associate any real number α or any angle of measure α radians with a point on the unit circle, we do the following:

1. Consider the standard-position angle A of α radians.

2. Assign to α (or to the given angle of measure α radians) the point P of intersection of the terminal side of A with the unit circle.

Since the number of units in the arc subtending a central angle of the unit circle is precisely the number of radians in the central angle, the point P in Figure 3.15 associated with real number α and with any angle of α radians could also be located by starting at $(1, 0)$ and measuring off α units counterclockwise on the unit circle. The association of an angle or a real number with a unit circle point is shown schematically in Figure 3.15. Note that to any angle or to any real number there is assigned a *single* unit circle point. Any unit circle point, however, is assigned to infinitely many different angles or numbers. For example, the point $(1, 0)$ is assigned to 0, 2π, -2π, 4π, -4π, etc.

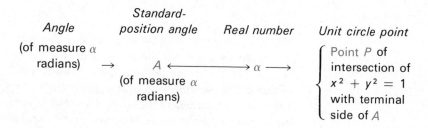

Angle	*Standard-position angle*	*Real number*	*Unit circle point*
(of measure α radians) $\rightarrow$	$A \longleftrightarrow$ (of measure α radians)	$\longleftrightarrow \alpha \longrightarrow$	$\left\{ \begin{array}{l} \text{Point } P \text{ of} \\ \text{intersection of} \\ x^2 + y^2 = 1 \\ \text{with terminal} \\ \text{side of } A \end{array} \right.$

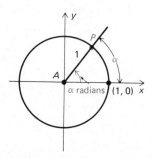

FIGURE 3.15   Unit circle point associated with real number α and with angle of measure α radians

EXAMPLE 6

Using the association process of the text, find the real number and the coordinates of the unit circle point associated with each of the standard-position angles of measure $0°$, $90°$, $180°$, $270°$, $360°$, and $-90°$, respectively.

Solution

Using the association process of the text, the real number associated with an angle is the number of radians in the radian measure of the angle. Accordingly, the real number associated with each of the given angles is

$$0° \rightarrow 0$$

$$90° \rightarrow \frac{\pi}{2}$$

$$180° \rightarrow \pi$$

$$270° \rightarrow \frac{3\pi}{2}$$

$$360° \rightarrow 2\pi$$

$$-90° \rightarrow -\frac{\pi}{2}$$

Again using the association process of the text, the unit circle point associated with each of the given angles is the point of intersection of $x^2 + y^2 = 1$ with the terminal side of the angle. These points of intersection are shown for each of the given angles in Figure 3.16. In each case, the terminal side of the angle is labeled with its degree measure and also with its radian measure. As a summary, there follows a list of the real numbers and unit circle points associated with the given angles.

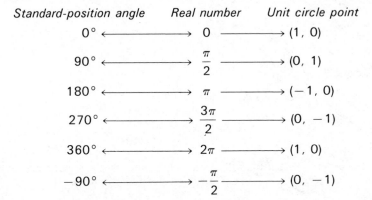

Standard-position angle	Real number	Unit circle point
$0°$ ⟷	0 ⟶	$(1, 0)$
$90°$ ⟷	$\dfrac{\pi}{2}$ ⟶	$(0, 1)$
$180°$ ⟷	π ⟶	$(-1, 0)$
$270°$ ⟷	$\dfrac{3\pi}{2}$ ⟶	$(0, -1)$
$360°$ ⟷	2π ⟶	$(1, 0)$
$-90°$ ⟷	$-\dfrac{\pi}{2}$ ⟶	$(0, -1)$

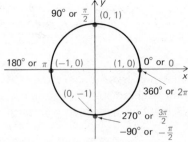

FIGURE 3.16 Unit circle points associated with 0, $\pi/2$, π, $3\pi/2$, 2π, and $-\pi/2$

As Example 6 suggests, it is easy to find the real number and the coordinates of the unit circle point associated with any of the quadrantal angles. The real number is some integral multiple of $\pi/2$, and the unit circle point is either $(1, 0)$, $(0, 1)$, $(-1, 0)$, or $(0, -1)$. In Section 3.3 we shall find the coordinates of the unit circle points associated with some other frequently encountered angles.

Exercise 3.1, 51–75

EXERCISE 3.1 ANGLES, REAL NUMBERS, AND UNIT CIRCLE POINTS

▶ In each of the following, find the degree measure and the radian measure of the angle determined by rotating its initial side as indicated.

EXAMPLE

Counterclockwise through one-third revolution.

Solution

Degree measure $= \dfrac{1}{3}(360°) = 120°$

Radian measure $= \dfrac{1}{3}(2\pi) = \dfrac{2\pi}{3}$

1. Clockwise through one-third revolution.
2. Counterclockwise through one-fourth revolution.
3. Counterclockwise through one-half revolution.
4. Clockwise through one-half revolution.
5. Clockwise through three-fourths revolution.
6. Counterclockwise through three-fourths revolution.
7. Clockwise through two-thirds revolution.
8. Counterclockwise through three revolutions.
9. Clockwise through one and one-half revolutions.
10. Counterclockwise through two and one-fourth revolutions.

▶ Convert each of the following to radians or to degrees.

EXAMPLES

(a) $150°$ (b) $2\pi/3$ (c) 7

Solutions

(a) $150° = \dfrac{150\pi}{180} = \dfrac{5\pi}{6}$

(b) $\dfrac{2\pi}{3} = \dfrac{2(180°)}{3} = 120°$

(c) 7 radians $= 7 \times \dfrac{180°}{\pi} = \dfrac{1260°}{\pi}$

11. 210° **12.** 240° **13.** 300° **14.** 330° **15.** 420°

16. $\dfrac{5\pi}{4}$ **17.** $\dfrac{7\pi}{4}$ **18.** $\dfrac{7\pi}{6}$ **19.** 4 **20.** 11

▶ In each of the following, the degree measure of an angle in standard position is given. Find the measure in radians of its reference angle. Draw and label the reference angle.

EXAMPLE 230°

Solution 230° − 180° = 50°

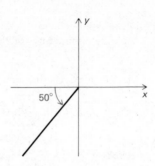

21. −230° **22.** 15° **23.** 310° **24.** 80°

25. 125° **26.** −125° **27.** −30° **28.** 260°

29. −300° **30.** 280°

▶ In each of the following, find the length of arc subtended by a central angle with the given measure in a circle with the given radius.

EXAMPLE 30°, radius 4 cm

Solution $30° = \dfrac{\pi}{6}$

$s = r\theta$

$= 4 \times \dfrac{\pi}{6} = \dfrac{2\pi}{3}$ cm

31. 60°, radius 4 ft **32.** 45°, radius 8 ft

33. 115°, radius 16 cm **34.** 270°, radius 6 mm

35. 80°, radius 10 in. **36.** 300°, radius 1 ft

37. 315°, radius 18 cm **38.** 100°, radius 12 m

39. $\dfrac{3\pi}{4}$, radius 1 cm **40.** $\dfrac{\pi}{8}$, radius 24 cm

▶ In each of the following, find the measure of the central angle subtending an arc of the given length on the unit circle. Use $\pi = 3.14$ to obtain a one-decimal-place approximation to the degree measure of the angle.

EXAMPLE   5 cm

Solution

$$5 \text{ radians} = \frac{5 \times 180°}{\pi}$$

$$\simeq \frac{900°}{3.14}$$

$$\simeq 286.6°$$

41. 4 ft   **42.** 3 cm   **43.** 1 in.   **44.** $\frac{7}{2}$ ft   **45.** $\frac{9}{2}$ cm

46. 6 cm   **47.** $\frac{5}{2}$ in.   **48.** 3.7 cm   **49.** 5.2 ft   **50.** 4.6 in.

▶ Using the association process of the text, in each of the following find the real number and the coordinates of the unit circle point associated with the standard-position angle whose measure is given.

EXAMPLE $-180°$

Solution $-180° = -\pi$ radians.
Angle of measure $-180° \rightarrow -\pi \rightarrow (-1, 0)$

51. 540°   **52.** $-270°$ **53.** 5π **54.** $-\dfrac{3\pi}{2}$

55. $\dfrac{9\pi}{2}$ **56.** $\dfrac{5\pi}{2}$ **57.** $-\pi$ **58.** $-630°$

59. $\dfrac{7\pi}{2}$ **60.** 630°

▶ Using the association process of the text, find the degree measure of the standard-position angle and find the coordinates of the unit circle point associated with the given real number.

EXAMPLE 3π

Solution

$$3\pi \text{ radians} = 3\pi \times \frac{180°}{\pi}$$

$$= 540°$$

$$3\pi \rightarrow (-1, 0)$$

61. 0 **62.** $\dfrac{\pi}{2}$ **63.** π **64.** $\dfrac{3\pi}{2}$ **65.** 2π

66. $\dfrac{5\pi}{2}$ **67.** $\dfrac{7\pi}{2}$ **68.** $\dfrac{-3\pi}{2}$ **69.** 7π **70.** -5π

71. -2π **72.** $-\pi$ **73.** 5π **74.** 4π **75.** -9π

CHECK LIST FOR THE STUDENT

If you have learned the material in Section 3.1, you should be able to do the following:

1. Define positive and negative angles.

2. Tell what must be given in order to specify a directed angle.

3. Tell how to refer to a directed angle.

4. Define standard position of an angle.

5. Define a first-, second-, third-, or fourth-quadrant angle.

6. Define a quadrantal angle.

7. Define a right angle.

8. Define an acute angle and define an obtuse angle.

9. Convert from radian measure to degree measure or from degree measure to radian measure.

10. Tell how to determine angles with degree measure greater than 360° or less than −360°.

11. Find the reference angle for any given standard-position angle that is not a quadrantal angle.

12. State and use the formula giving the length s of the arc subtended by a central angle of θ radians in a circle of radius r.

13. Define the unit circle.

14. State and use the formula giving the length s of the arc subtended by a central angle of θ radians in the unit circle.

15. Associate any angle with given degree or radian measure with a real number.

16. Associate any given real number with an angle.

17. Associate any given real number or any angle of known radian measure with a point on the unit circle.

18. Find more than one real number associated with each of the points $(1, 0)$, $(0, 1)$, $(−1, 0)$, and $(0, −1)$.

19. Find the real number and the coordinates of the unit circle point associated with any given quadrantal angle.

SECTION QUIZ 3.1

1. (a) Convert 400° to radians.
   (b) Convert $5\pi/3$ radians to degrees.
   (c) Convert 2 radians to degrees.

2. In each of the following, the degree measure of an angle in standard position is given. Find the degree measure and then the radian measure of its reference angle. Draw and label the reference angle.
   (a) 155°   (b) 198°   (c) 292°   (d) −68°   (e) −343°

3. Find the length of arc subtended by a central angle of measure 70° in a circle with radius 8 cm.

4. Find the radian measure of the central angle subtending an arc of length 2 cm on the unit circle.

5. Use the association process of the text in each of the following.
   (a) Find the real number and the coordinates of the unit circle point associated with the standard-position angle whose measure is 90°.
   (b) Find the degree measure of the standard-position angle and find the coordinates of the unit circle point associated with the real number -3π.

3.2 Trigonometric Functions of Angles and Numbers; Right Triangle Trigonometry

The purpose of this section is to define the trigonometric functions as functions of angles or numbers and to show that the values of the trigonometric functions of an acute angle may be interpreted as ratios of the lengths of the sides of a right triangle.

The trigonometric functions are called the cosine, sine, tangent, cotangent, secant, and cosecant. Before actually defining these six functions, let us familiarize ourselves with the notation we will be using for them.

We have become accustomed to using a single letter such as f to denote a function. We have been using the notation $f(x)$ to denote the number that the rule of f assigns to x. The function called the cosine, however, is customarily denoted by the three letters cos, and the notation *cos x* denotes the number that the rule of the cosine function assigns to x. The symbols *cos x* are read "cosine x." Do not make the mistake of thinking that the symbols *cos x* mean some sort of multiplication. The symbols *cos x* together represent the single number assigned to x by the cosine function—that is, by the *rule of assignment* of the cosine function. Similar notation is employed for all the six trigonometric functions and is summarized as follows:

cos x Number assigned to x by cosine function

sin x Number assigned to x by sine function

tan x Number assigned to x by tangent function

cot x Number assigned to x by cotangent function

sec x Number assigned to x by secant function

csc x Number assigned to x by cosecant function

If x is an angle, in the notation just introduced it is customary to replace x by its measure. For example, it is customary to write cos 90° for the number assigned to an angle of measure 90° by the cosine function. If the measure is negative, we sometimes insert parentheses for clarity. For example, we may write cos $(-90°)$ for the number assigned to an angle of measure $-90°$ by the cosine function.

In order to define any function, we must specify its domain, either explicitly or by the domain convention, and we must in some way give its rule of assignment. Table 3.1 specifies the domain and rule for each of the trigonometric functions. In each case, the rule is stated in terms of the point on the unit circle $x^2 + y^2 = 1$ that is associated with each domain element.

TABLE 3.1 DEFINITIONS OF THE TRIGONOMETRIC FUNCTIONS*

Function	Domain	Rule
Cosine	All angles A or All real numbers α	$\left. \begin{array}{l} \cos A \\ \text{or} \\ \cos \alpha \end{array} \right\} = x$
Sine	All angles A or All real numbers α	$\left. \begin{array}{l} \sin A \\ \text{or} \\ \sin \alpha \end{array} \right\} = y$
Tangent	All angles A or all real numbers α except those for which $\cos A = \cos \alpha = 0$ (that is, except those for which $x = 0$)	$\left. \begin{array}{l} \tan A \\ \text{or} \\ \tan \alpha \end{array} \right\} = \dfrac{y}{x}$
Cotangent	All angles A or all real numbers α except those for which $\sin A = \sin \alpha = 0$ (that is, except those for which $y = 0$)	$\left. \begin{array}{l} \cot A \\ \text{or} \\ \cot \alpha \end{array} \right\} = \dfrac{x}{y}$
Secant	All angles A or all real numbers α except those for which $\cos A = \cos \alpha = 0$ (that is, except those for which $x = 0$)	$\left. \begin{array}{l} \sec A \\ \text{or} \\ \sec \alpha \end{array} \right\} = \dfrac{1}{x}$
Cosecant	All angles A or all real numbers α except those for which $\sin A = \sin \alpha = 0$ (that is, except those for which $y = 0$)	$\left. \begin{array}{l} \csc A \\ \text{or} \\ \csc \alpha \end{array} \right\} = \dfrac{1}{y}$

*Here (x, y) is the point on $x^2 + y^2 = 1$ associated with angle A or real number α. See Figure 3.17.

Notice that, as shown in Figure 3.17, $\cos A$ and $\sin A$ (or $\cos \alpha$ and $\sin \alpha$) are simply the coordinates (x, y) of the unit circle point associated with angle A or real number α. That is why some authors call these functions *circular* functions.

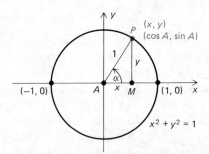

FIGURE 3.17 Definitions of cosine and sine
$\cos A = \cos \alpha = x$
$\sin A = \sin \alpha = y$
(Measure of $A = \alpha$ radians)

Exercise 3.2, 1–10

EXAMPLE 1   Find the values of the trigonometric functions for angle A of measure $90°$.

Solution   We see in Figure 3.18 that the point P on $x^2 + y^2 = 1$ associated with an angle of measure $90°$ is $(0, 1)$. That is, the x-coordinate of P is 0 and the y-coordinate of P is 1. It follows from the definitions of the trigonometric functions that

$$\cos 90° = 0 \qquad\qquad \sin 90° = 1$$

$$\tan 90° \text{ is not defined} \qquad \cot 90° = \frac{0}{1} = 0$$

$$\sec 90° \text{ is not defined} \qquad \csc 90° = \frac{1}{1} = 1.$$

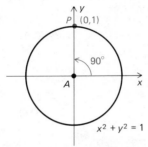

FIGURE 3.18   Point $(0, 1)$ associated with angle of measure $90°$

EXAMPLE 2   Find the values of the trigonometric functions of real number $\alpha = \pi$.

Solution   In Figure 3.19 the point P on $x^2 + y^2 = 1$ associated with real number π is $(-1, 0)$.

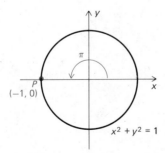

FIGURE 3.19   Point $(-1, 0)$ associated with angle α of measure π

It follows from the definitions of the trigonometric functions that

$$\cos \pi = -1 \qquad\qquad \sin \pi = 0$$

$$\tan \pi = \frac{0}{-1} = 0 \qquad \cot \pi \text{ is not defined}$$

$$\sec \pi = \frac{1}{-1} = -1 \qquad \csc \pi \text{ is not defined.}$$

EXAMPLE 3

Find the values of the trigonometric functions at 0 and also at $3\pi/2$.

Solution

In Figure 3.20 we see that on the unit circle the point associated with 0 is (1, 0), and the point associated with $3\pi/2$ is (0, −1). It follows that

$$\cos 0 = 1 \qquad\qquad \sin 0 = 0$$

$$\tan 0 = \frac{0}{1} = 0 \qquad \cot 0 \text{ is not defined}$$

$$\sec 0 = \frac{1}{1} = 1 \qquad \csc 0 \text{ is not defined.}$$

Likewise,

$$\cos \frac{3\pi}{2} = 0 \qquad\qquad \sin \frac{3\pi}{2} = -1$$

$$\tan \frac{3\pi}{2} \text{ is not defined} \qquad \cot \frac{3\pi}{2} = \frac{0}{-1} = 0$$

$$\sec \frac{3\pi}{2} \text{ is not defined} \qquad \csc \frac{3\pi}{2} = \frac{1}{-1} = -1.$$

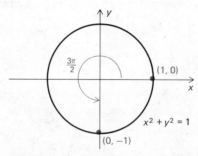

FIGURE 3.20   Point (1, 0) associated with 0
Point (0, −1) associated with $3\pi/2$

As Examples 1—3 suggest, it is relatively easy to find the values of the trigonometric functions of quadrantal angles or at real numbers that are integral multiples of $\pi/2$. For reference, the values of some of these are tabulated in Table 3.2. In the next section we shall see how to find values of the trigonometric functions of other angles or of other numbers.

TABLE 3.2 VALUES OF TRIGONOMETRIC FUNCTIONS OF SOME ANGLES AND NUMBERS

Degree measure of A	Radian measure α of A	Unit circle point associated with A or α	$\cos A$ or $\cos \alpha$	$\sin A$ or $\sin \alpha$	$\tan A$ or $\tan \alpha$
0°	0	(1, 0)	1	0	0
90°	$\dfrac{\pi}{2}$	(0, 1)	0	1	Not defined
180°	π	(−1, 0)	−1	0	0
270°	$\dfrac{3\pi}{2}$	(0, −1)	0	−1	Not defined
360°	2π	(1, 0)	1	0	0
−90°	$-\dfrac{\pi}{2}$	(0, −1)	0	−1	Not defined
−180°	$-\pi$	(−1, 0)	−1	0	0

Exercise 3.2, 11–30

If angle A is an *acute* angle, then the values of the trigonometric functions of A are certain ratios of the lengths of the sides of a right triangle having angle A as an acute angle. For example, the value of $\cos A$ is the ratio of the length of the side adjacent to angle A to the length of the hypotenuse. To see this, first consider right triangle ABC in Figure 3.21. Then introduce a Cartesian coordinate system so that A is at the origin. Draw the unit circle. Let P be the point of intersection of the unit circle with hypotenuse $\overline{AB}$. Note that P is the unit circle point associated with angle A, so the coordinates of P are ($\cos A$, $\sin A$). The lengths of

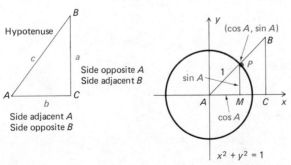

FIGURE 3.21 Right triangle with acute angle A; $\cos A = b/c$, $\sin A = a/c$

the sides of triangle *APM* are therefore cos *A* and sin *A*, as shown in the figure. Now compare right triangles *APM* and *ABC*. They are similar. The ratios of the lengths of their corresponding sides are equal. In particular,

$$\frac{\cos A}{1} = \frac{b}{c}$$

and
$$\frac{\sin A}{1} = \frac{a}{c}.$$

In other words, the coordinates of *P* are (*b/c*, *a/c*). It follows that tan *A* = *a/b*, cot *A* = *b/a*, sec *A* = *c/b*, and csc *A* = *c/a*.

$$\cos A = \frac{b}{c} \qquad \sin A = \frac{a}{c}$$

$$\tan A = \frac{a}{b} \qquad \cot A = \frac{b}{a}$$

$$\sec A = \frac{c}{b} \qquad \csc A = \frac{c}{a}$$

We can indicate these facts with words as follows:

$$\cos A = \frac{\text{side adjacent } A}{\text{hypotenuse}} \qquad \sin A = \frac{\text{side opposite } A}{\text{hypotenuse}}$$

$$\tan A = \frac{\text{side opposite } A}{\text{side adjacent } A} \qquad \cot A = \frac{\text{side adjacent } A}{\text{side opposite } A}$$

$$\sec A = \frac{\text{hypotenuse}}{\text{side adjacent } A} \qquad \csc A = \frac{\text{hypotenuse}}{\text{side opposite } A}$$

In Figure 3.21, angle *B* is also an acute angle, so we can write the trigonometric functions of *B* as ratios.

$$\cos B = \frac{a}{c} \qquad \sin B = \frac{b}{c}$$

$$\tan B = \frac{b}{a} \qquad \cot B = \frac{a}{b}$$

$$\sec B = \frac{c}{a} \qquad \csc B = \frac{c}{b}$$

Notice that

$$\sin A = \cos B \qquad \tan A = \cot B \qquad \sec A = \csc B$$
$$\cos A = \sin B \qquad \cot A = \tan B \qquad \csc A = \sec B.$$

Here A and B are *complementary* angles—that is, the sum of their measures is $90°$; these relationships are true for any pair of complementary angles A and B.

EXAMPLE 4

If the lengths of the sides and hypotenuse of right triangle ABC are 3, 4, and 5 units, as shown in Figure 3.22, find $\sin A$, $\cos A$, $\tan A$, $\sin B$, $\cos B$, and $\cot B$.

Solution

$$\sin A = \frac{\text{side opposite } A}{\text{hypotenuse}} = \frac{3}{5} \qquad \sin B = \frac{\text{side opposite } B}{\text{hypotenuse}} = \frac{4}{5}$$

$$\cos A = \frac{\text{side adjacent } A}{\text{hypotenuse}} = \frac{4}{5} \qquad \cos B = \frac{\text{side adjacent } B}{\text{hypotenuse}} = \frac{3}{5}$$

$$\tan A = \frac{\text{side opposite } A}{\text{side adjacent } A} = \frac{3}{4} \qquad \cot B = \frac{\text{side adjacent } B}{\text{side opposite } B} = \frac{3}{4}$$

Notice that $\sin A = \cos B$, $\cos A = \sin B$, and $\tan A = \cot B$.

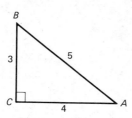

FIGURE 3.22

EXAMPLE 5

A demolition expert needs to estimate the height of a vertical building. He stands at point P on the ground 50 m from the base B of the building and determines that the angle up from the ground to a line from P to the top T of the building measures $60°$. He knows that $\tan 60° = \sqrt{3}$. What is the height of the building?

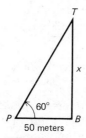

FIGURE 3.23

Solution

In Figure 3.23 we see that PTB is a right triangle with acute angle P. Let x be the height of the building. Since

$$\tan P = \frac{\text{side opposite } P}{\text{side adjacent } P},$$

$$\tan 60° = \frac{x}{50}.$$

$$\sqrt{3} = \frac{x}{50}$$

$$x = 50\sqrt{3} \text{ m}$$

(The building is approximately 86.6 m high.)

It will be easier to remember the definitions of the trigonometric functions if we make particular note of the following basic relationships. These are true for any angle or any number for which the functions involved are defined. As is so often done in the literature, instead of using A or α, we shall use the Greek letter θ (theta) in stating these basic relationships.

$$\tan \theta = \frac{\sin \theta}{\cos \theta} \qquad \cot \theta = \frac{\cos \theta}{\sin \theta}$$

$$\sec \theta = \frac{1}{\cos \theta} \qquad \csc \theta = \frac{1}{\sin \theta}$$

$$\cot \theta = \frac{1}{\tan \theta}$$

EXAMPLE 6

Verify that in right triangle ABC of Example 4, $\tan A = \sin A / \cos A$.

Solution

In Example 4 we saw that $\tan A = \frac{3}{4}$ and that $\sin A = \frac{3}{5}$ and $\cos A = \frac{4}{5}$. It is true that

$$\frac{3}{4} = \frac{\frac{3}{5}}{\frac{4}{5}}.$$

In integral calculus, use is made of a device known as *trigonometric substitution*. The next example illustrates the way in which the definitions of the trigonometric functions are utilized in the application of this device.

Exercise 3.2, 31–45

EXAMPLE 7

Let $x = 3 \sin \theta$. Assume that θ is an acute angle. Find $\cos \theta$ and $\tan \theta$ in terms of x.

Solution

It is helpful to construct a so-called reference triangle, as shown in Figure 3.24. Since $\sin \theta = x/3$, we call x the length of the side opposite θ, and we acknowledge that 3 is the length of the hypotenuse. Using the Pythagorean theorem, we find that the side adjacent θ is of length

$$\sqrt{9 - x^2}.$$

Now, since

$$\cos\theta = \frac{\text{side adjacent }\theta}{\text{hypotenuse}} \quad \text{and} \quad \tan\theta = \frac{\text{side opposite }\theta}{\text{side adjacent }\theta},$$

we see from the reference triangle that

Exercise 3.2, 46–50

$$\cos\theta = \frac{\sqrt{9 - x^2}}{3} \quad \text{and} \quad \tan\theta = \frac{x}{\sqrt{9 - x^2}}.$$

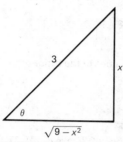

FIGURE 3.24 $\sin\theta = x/3$

EXERCISE 3.2 TRIGONOMETRIC FUNCTIONS OF ANGLES AND NUMBERS; RIGHT TRIANGLE TRIGONOMETRY

▶ In each of the following, the given point is the point of intersection of the unit circle with the terminal side of standard-position angle A. Find the trigonometric functions of A.

EXAMPLE

$$\left(\frac{1}{\sqrt{5}}, \frac{2}{\sqrt{5}}\right)$$

Solution

$$\cos A = \frac{1}{\sqrt{5}}, \quad \sin A = \frac{2}{\sqrt{5}}, \quad \tan A = 2$$

$$\cot A = \frac{1}{2}, \quad \sec A = \sqrt{5}, \quad \csc A = \frac{\sqrt{5}}{2}$$

1. $\left(\dfrac{\sqrt{2}}{2}, \dfrac{\sqrt{2}}{2}\right)$ 2. $\left(\dfrac{\sqrt{3}}{2}, \dfrac{1}{2}\right)$ 3. $\left(\dfrac{1}{2}, \dfrac{\sqrt{3}}{2}\right)$
4. $\left(\dfrac{-\sqrt{2}}{2}, \dfrac{\sqrt{2}}{2}\right)$ 5. $\left(\dfrac{-1}{2}, \dfrac{-\sqrt{3}}{2}\right)$

▶ In each of the following, the given point is the point on the unit circle associated with real number α. Find the trigonometric functions of α.

EXAMPLE $\left(\dfrac{-1}{2}, \dfrac{\sqrt{3}}{2}\right)$

Solution $\cos \alpha = \dfrac{-1}{2}, \quad \sin \alpha = \dfrac{\sqrt{3}}{2}, \quad \tan \alpha = -\sqrt{3}$

$\cot \alpha = \dfrac{-1}{\sqrt{3}}, \quad \sec \alpha = -2, \quad \csc \alpha = \dfrac{2}{\sqrt{3}}$

6. $\left(\dfrac{1}{2}, \dfrac{-\sqrt{3}}{2}\right)$ 7. $\left(\dfrac{-\sqrt{3}}{2}, \dfrac{-1}{2}\right)$ 8. $\left(\dfrac{\sqrt{2}}{2}, \dfrac{-\sqrt{2}}{2}\right)$

9. $\left(\dfrac{1}{5}, \dfrac{2\sqrt{6}}{5}\right)$ 10. $\left(\dfrac{-2\sqrt{6}}{5}, \dfrac{1}{5}\right)$

▶ In each of the following, find the values of the trigonometric functions of an angle with the given measure.

EXAMPLE $-270°$

Solution The unit circle point associated with $-270°$ is $(0, 1)$. Therefore $\cos(-270°) = 0$, $\sin(-270°) = 1$, $\tan(-270°)$ is undefined, $\cot(-270°) = 0$, $\sec(-270°)$ is undefined, and $\csc(-270°) = 1$.

11. $450°$ 12. $-180°$ 13. $\dfrac{5\pi}{2}$ 14. -3π

15. $720°$ 16. $-630°$ 17. $-\dfrac{9\pi}{2}$ 18. $540°$

19. $810°$ 20. 10π

▶ Find the values of the trigonometric functions of each of the following real numbers.

EXAMPLE 3π

Solution The unit circle point associated with 3π is $(-1, 0)$. Therefore $\cos 3\pi = -1$, $\sin 3\pi = 0$, $\tan 3\pi = 0$, $\cot 3\pi$ is undefined, $\sec 3\pi = -1$, and $\csc 3\pi$ is undefined.

21. 5π 22. -4π 23. $\dfrac{7\pi}{2}$ 24. 8π

25. $\dfrac{9\pi}{2}$ 26. $\dfrac{-11\pi}{2}$ 27. 6π 28. -9π

29. $\dfrac{13\pi}{2}$ 30. -23π

In right triangle ABC, let a, b, and c represent the lengths of the sides opposite angles A, B, and C, respectively. Let angle C be the right angle. In each of the following, draw the triangle, label the sides, and find the trigonometric functions of A and of B.

EXAMPLE

$a = 12$ cm, $b = 5$ cm

Solution

By the Pythagorean theorem,

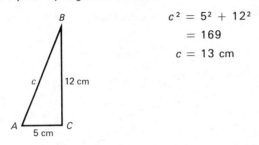

$$c^2 = 5^2 + 12^2$$
$$= 169$$
$$c = 13 \text{ cm}$$

$\sin A = \frac{12}{13}$, $\cos A = \frac{5}{13}$, $\tan A = \frac{12}{5}$, $\cot A = \frac{5}{12}$, $\sec A = \frac{13}{5}$, $\csc A = \frac{13}{12}$

$\sin B = \frac{5}{13}$, $\cos B = \frac{12}{13}$, $\tan B = \frac{5}{12}$, $\cot B = \frac{12}{5}$, $\sec B = \frac{13}{12}$, $\csc B = \frac{13}{5}$

31. $a = 6$ ft, $b = 8$ ft   **32.** $a = 10$ m, $c = 26$ m

33. $a = 9$ mm, $c = 15$ mm   **34.** $a = 2\frac{1}{2}$ in., $b = 6$ in.

35. $a = 12$ cm, $b = 16$ cm

36. An archeologist needs to estimate the height of a cliff. She stands at a point P on the ground 100 m from the base B of the cliff and determines that the angle up from the ground to a line from P to the top T of the cliff measures $80°$. Using the value 5.67 for $\tan 80°$, and assuming that the cliff is perpendicular to the ground, find the height of the cliff.

37. An alpine climber is at point A on a level with the base B of a certain cliff. He wants to know if he has enough rope to reach directly from A to B, but there is a deep ravine between A and B so he cannot walk there to find out. The top T of the cliff is reported to be 400 ft high. If the climber uses a pocket transit and determines that angle BAT is $58°$, if the cliff is perpendicular to AB, and if $\tan 58° = 1.6$, what is the distance along a straight line from A to B?

38. A forest ranger wants to estimate the height of a certain sugar pine tree. A trimmer has already climbed to the top T and has thrown down one end of a 500-ft rope. The ranger stands on the ground holding the end of the rope at R. There is no slack in the rope. The tree is perpendicular to the ground at B. Angle BRT is $21°$. If $\sin 21° = 0.36$, how high is the tree?

39. A house painter wants to know how long a ladder he should buy so that he can paint the gutter on a certain house. He must place the foot of the ladder at F on a strip of pavement exactly 22 ft from point H on the foundation of the house, and the top of the ladder must rest on the gutter at G. If GH is perpendicular to the ground line FH, if angle HFG is $60°$, and if $\cos 60° = \frac{1}{2}$, what is the distance FG?

40. A marina operator is stringing a safety net across a river starting at point P, exactly 100 m upstream from a dam D, and ending at Q. If the net PQ is perpendicular to PD, if angle PDQ is $75°$, and if $\tan 75° = 3.7$, what is the length of PQ?

▶ Verify that each of the following is true in right triangle ABC of Figure 3.22.

41. $\cot A = \dfrac{\cos A}{\sin A}$ 42. $\sec A = \dfrac{1}{\cos A}$ 43. $\csc A = \dfrac{1}{\sin A}$

44. $\tan B = \dfrac{\sin B}{\cos B}$ 45. $\sec B = \dfrac{1}{\cos B}$

▶ In each of the following, assume that θ is an acute angle. Draw a "reference triangle" and express the trigonometric functions of θ in terms of x.

46. $x = 3 \cos \theta$ 47. $x = 3 \tan \theta$ 48. $x = 3 \sec \theta$

49. $x = 2 \sin \theta$ 50. $x = 2 \tan \theta$

CHECK LIST FOR THE STUDENT

If you have learned the material in Section 3.2, you should be able to do the following:

1. Define the trigonometric functions of angles or numbers by referring to the unit circle.

2. Find the values of the trigonometric functions of quadrantal angles or at real numbers that are integral multiples of $\pi/2$.

3. Interpret the trigonometric functions of an acute angle as certain ratios of the lengths of the sides and hypotenuse of a right triangle.

4. Find the trigonometric functions of the acute angles of a right triangle if the lengths of the sides and hypotenuse are known.

5. Use the trigonometric functions as ratios to find the length of a side of a right triangle if an appropriate trigonometric function of one of the acute angles and the length of one side are known.

6. State and use the following:

$$\tan \theta = \frac{\sin \theta}{\cos \theta}, \quad \cot \theta = \frac{\cos \theta}{\sin \theta}, \quad \sec \theta = \frac{1}{\cos \theta}, \quad \csc \theta = \frac{1}{\sin \theta},$$

$$\cot \theta = \frac{1}{\tan \theta}.$$

7. Construct a reference triangle and express the trigonometric functions of θ in terms of x, given that $x = a \sin \theta$, $x = a \cos \theta$, $x = a \sec \theta$, or $x = a \tan \theta$ for some real number a and some acute angle θ.

SECTION QUIZ 3.2

1. Let the measure of angle A be nonnegative and less than 2π. Let A be in standard position, and let

$$\left(\frac{1}{3}, \frac{-\sqrt{8}}{3} \right)$$

be the point on $x^2 + y^2 = 1$ associated with A. Sketch the circle, plot the given point, draw angle A, and find $\cos A$, $\sin A$, $\tan A$, $\cot A$, $\sec A$, and $\csc A$.

2. Let

$$\left(\frac{-1}{\sqrt{5}}, \frac{2}{\sqrt{5}} \right)$$

be the unit circle point associated with real number α. Find $\cos \alpha$, $\sin \alpha$, $\tan \alpha$, $\cot \alpha$, $\sec \alpha$, and $\csc \alpha$.

3. Find the unit circle point associated with each of the following.

(a) $90°$ (b) π (c) $\dfrac{3\pi}{2}$ (d) $360°$

(e) 0 (f) $-\pi$ (g) $\dfrac{-3\pi}{2}$ (h) $\dfrac{-\pi}{2}$

4. Find the value of each of the following.

(a) $\cos \dfrac{\pi}{2}$ (b) $\sin \dfrac{\pi}{2}$ (c) $\cos \pi$

(d) $\sin \pi$ (e) $\tan \pi$ (f) $\sin 270°$

(g) $\csc 270°$ (h) $\cos 0°$ (i) $\sec 0°$

5. In triangle ABC the right angle is at C, the hypotenuse is of length 10 m, and the lengths of the sides opposite A and B are $a = 6$ m and $b = 8$ m, respectively. Find $\sin A$, $\cos A$, $\tan A$, $\cot A$, $\sec A$, and $\csc A$. Find $\sin B$ and $\cos B$.

3.3 Trigonometric Functional Values

The purpose of this section is to show how to find values of the trigonometric functions of angles other than the quadrantal angles or of numbers other than integral multiples of $\pi/2$.

We noted in Section 3.2 that the values of the trigonometric functions of an angle A or a real number α may be found from the coordinates of the unit circle point associated with A or α. We saw that it is easy to find the values of the trigonometric functions at quadrantal angles or at real numbers that are integral multiples of $\pi/2$, because it is easy to find the coordinates of the unit circle points associated with such angles or numbers. It is also relatively easy to find the values of the trigonometric functions of angles with measures $45°$, $30°$, or $60°$ or of the numbers $\pi/4$, $\pi/6$, or $\pi/3$.

First let us consider an angle whose measure is $45°$. Choose a coordinate system so that this angle with measure $45°$ is in standard position, and call this angle A. Since

$$45° = \frac{\pi}{4} \text{ radians,}$$

angle A is the standard-position angle associated with the real number $\pi/4$. The trigonometric functional values of $\pi/4$ or of an angle of measure $45°$ can thus be found from the coordinates of the unit circle point associated with angle A. The unit circle point P associated with angle A is shown in Figure 3.25. It can be proved that the coordinates of P are $(\sqrt{2}/2, \ \sqrt{2}/2)$. That proof follows.

The unit circle point P associated with an angle of $45°$ or with the real number $\pi/4$ is $(\sqrt{2}/2, \ \sqrt{2}/2)$.

Sketch of Proof

Step 1   Drop a perpendicular from P to the point M on the x-axis. (See Figure 3.25.)

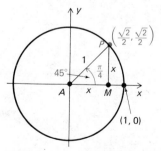

FIGURE 3.25   Unit circle point associated with angle of $45°$ or with $\pi/4$

Step 2 Note that triangle *AMP* is a right triangle and that it is isoceles (length of side $\overline{AM}$ equals length of side $\overline{MP}$), because the angles at *A* and *P* both measure $45°$. (The sum of the angles of a triangle equals $180°$; the angle at *M* is $90°$ and the angle at *A* is $45°$, so the angle at *P* must be $180° - (90° + 45°) = 45°$.)

Step 3 Note that since the length of $\overline{AM}$ equals the length of $\overline{MP}$, the coordinates of *P* are equal.

Step 4 Let the coordinates of *P* be (x, x).

Step 5 Note that the hypotenuse $\overline{AP}$ of triangle *AMP* is of length 1 because $\overline{AP}$ is a radius of the unit circle.

Step 6 Apply the Pythagorean theorem to triangle *AMP* and solve for *x*.

$$x^2 + x^2 = 1$$
$$2x^2 = 1$$
$$x^2 = \tfrac{1}{2}$$
$$x = \frac{1}{\sqrt{2}}$$
$$= \frac{1}{\sqrt{2}} \cdot \frac{\sqrt{2}}{\sqrt{2}}$$
$$= \frac{\sqrt{2}}{2}$$

The coordinates of *P* are $\left(\dfrac{\sqrt{2}}{2}, \dfrac{\sqrt{2}}{2}\right)$.

Since the coordinates of the unit circle point associated with an angle of $45°$ or with $\pi/4$ are $(\sqrt{2}/2, \sqrt{2}/2)$, it follows from the definitions of the trigonometric functions that

$$\cos 45° = \cos \frac{\pi}{4} = \frac{\sqrt{2}}{2} \qquad \sin 45° = \sin \frac{\pi}{4} = \frac{\sqrt{2}}{2}$$

$$\tan 45° = \tan \frac{\pi}{4} = 1 \qquad \cot 45° = \cot \frac{\pi}{4} = 1$$

$$\sec 45° = \sec \frac{\pi}{4} = \frac{2}{\sqrt{2}} \qquad \csc 45° = \csc \frac{\pi}{4} = \frac{2}{\sqrt{2}}.$$

Note that

$$\frac{2}{\sqrt{2}} = \frac{2 \times \sqrt{2}}{\sqrt{2} \times \sqrt{2}} = \sqrt{2},$$

so we can also write

$$\sec 45° = \sec \frac{\pi}{4} = \sqrt{2} \quad \text{and} \quad \csc 45° = \csc \frac{\pi}{4} = \sqrt{2}.$$

Now let us find the trigonometric functional values of an angle of 30° or of the number $\pi/6$ by finding the coordinates of the unit circle point associated with such an angle or with $\pi/6$. Consider an angle of 30°. Choose a coordinate system so that this angle is in standard position, and call this angle A. Since

$$30° = \frac{\pi}{6} \text{ radians,}$$

angle A is the standard-position angle associated with real number $\pi/6$. The unit circle point P associated with A or with $\pi/6$ is shown in Figure 3.26. It can be shown that the coordinates of this P are $(\sqrt{3}/2, \frac{1}{2})$. The proof follows.

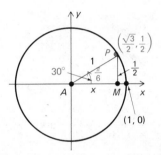

FIGURE 3.26 Unit circle point associated with angle of 30° or with $\pi/6$

The unit circle point P associated with an angle of 30° or with the real number $\pi/6$ is $(\sqrt{3}/2, \frac{1}{2})$.

Sketch of Proof

Step 1 Drop a perpendicular from P to the point M on the x-axis. (See Figure 3.26.)

Step 2 Note that the hypotenuse $\overline{AP}$ of right triangle AMP is of length 1 because $\overline{AP}$ is a radius of the unit circle.

Step 3 Use the fact usually developed in the plane geometry that in a "30°-60° right triangle" the length of the side opposite the angle of measure 30° is one-half the length of the hypotenuse, and conclude that the length of $\overline{MP}$ (the y-coordinate of P) is $\frac{1}{2}$.

Step 4 Apply the Pythagorean theorem to triangle *AMP* to find the length *x* of $\overline{AM}$ (the *x*-coordinate of *P*).

$$x^2 + (\tfrac{1}{2})^2 = 1^2$$
$$x^2 = 1 - (\tfrac{1}{2})^2$$
$$= 1 - \tfrac{1}{4}$$
$$= \tfrac{3}{4}$$
$$x = \frac{\sqrt{3}}{2}$$

The coordinates of *P* are $(\sqrt{3}/2, \tfrac{1}{2})$.

Since the coordinates of the unit circle point associated with an angle of 30° or with $\pi/6$ are $(\sqrt{3}/2, \tfrac{1}{2})$, it follows from the definitions of the trigonometric functions that

$$\cos 30° = \cos \frac{\pi}{6} = \frac{\sqrt{3}}{2} \qquad \sin 30° = \sin \frac{\pi}{6} = \frac{1}{2}$$

$$\tan 30° = \tan \frac{\pi}{6} = \frac{1}{\sqrt{3}} \qquad \cot 30° = \cot \frac{\pi}{6} = \sqrt{3}$$

$$\sec 30° = \sec \frac{\pi}{6} = \frac{2}{\sqrt{3}} \qquad \csc 30° = \csc \frac{\pi}{6} = 2.$$

Note that

$$\frac{1}{\sqrt{3}} = \frac{1 \times \sqrt{3}}{\sqrt{3} \times \sqrt{3}} = \frac{\sqrt{3}}{3} \quad \text{and} \quad \frac{2}{\sqrt{3}} = \frac{2\sqrt{3}}{3},$$

so equivalent ways of writing tan 30° and sec 30° are

$$\tan 30° = \tan \frac{\pi}{6} = \frac{\sqrt{3}}{3} \quad \text{and} \quad \sec 30° = \sec \frac{\pi}{6} = \frac{2}{\sqrt{3}}.$$

Finally, let us find the trigonometric functional values of an angle of 60° or of the real number $\pi/3$. Consider an angle of 60°. Choose a coordinate system so that this angle is in standard position, and call this angle *A*. Since

$$60° = \frac{\pi}{3} \text{ radians,}$$

this angle *A* is the standard-position angle associated with the real number $\pi/3$. The unit circle point *P* associated with *A* and hence with $\pi/3$ is shown in Figure 3.27. It can be shown that the coordinates of this *P* are $(\tfrac{1}{2}, \sqrt{3}/2)$. The proof is very similar to the proof just sketched for the case in which *A* has measure 30°, because in the triangle *AMP* of Figure 3.27, the angle at *P* must be 30° (180° − (90° + 60°) = 30°), so triangle *AMP* of Figure 3.27 is also a "30°-60° right triangle."

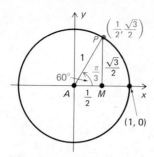

FIGURE 3.27 Unit circle point associated with angle of 60° or with $\pi/3$

Using the fact that the coordinates of the unit circle point associated with an angle of 60° or with real number $\pi/3$ are $(\frac{1}{2}, \sqrt{3}/2)$, the trigonometric functional values of such an angle or of $\pi/3$ can be deduced from the definitions of the trigonometric functions.

$$\cos 60° = \cos \frac{\pi}{3} = \frac{1}{2} \qquad \sin 60° = \sin \frac{\pi}{3} = \frac{\sqrt{3}}{2}$$

$$\tan 60° = \tan \frac{\pi}{3} = \sqrt{3} \qquad \cot 60° = \cot \frac{\pi}{3} = \frac{1}{\sqrt{3}}$$

$$\sec 60° = \sec \frac{\pi}{3} = 2 \qquad \csc 60° = \csc \frac{\pi}{3} = \frac{2}{\sqrt{3}}$$

Note that equivalent ways of writing cot 60° and csc 60° are

$$\cot 60° = \cot \frac{\pi}{3} = \frac{\sqrt{3}}{3} \quad \text{and} \quad \csc 60° = \csc \frac{\pi}{3} = \frac{2\sqrt{3}}{3}.$$

For easy reference, the trigonometric functional values of 45°, 30°, and 60° are summarized in Table 3.3.

TABLE 3.3 TRIGONOMETRIC FUNCTIONAL VALUES OF ANGLES OF MEASURE 45°, 30°, AND 60° (OF NUMBERS $\pi/4$, $\pi/6$, AND $\pi/3$)

Degree measure of A	Radian measure α of A	Unit circle point associated with A or α	$\cos A$ or $\cos \alpha$	$\sin A$ or $\sin \alpha$	$\tan A$ or $\tan \alpha$	$\cot A$ or $\cot \alpha$	$\sec A$ or $\sec \alpha$	$\csc A$ or $\csc \alpha$
45°	$\dfrac{\pi}{4}$	$\left(\dfrac{\sqrt{2}}{2}, \dfrac{\sqrt{2}}{2}\right)$	$\dfrac{\sqrt{2}}{2}$	$\dfrac{\sqrt{2}}{2}$	1	1	$\dfrac{2}{\sqrt{2}} = \sqrt{2}$	$\dfrac{2}{\sqrt{2}} = \sqrt{2}$
30°	$\dfrac{\pi}{6}$	$\left(\dfrac{\sqrt{3}}{2}, \dfrac{1}{2}\right)$	$\dfrac{\sqrt{3}}{2}$	$\dfrac{1}{2}$	$\dfrac{1}{\sqrt{3}} = \dfrac{\sqrt{3}}{3}$	$\sqrt{3}$	$\dfrac{2}{\sqrt{3}} = \dfrac{2\sqrt{3}}{3}$	2
60°	$\dfrac{\pi}{3}$	$\left(\dfrac{1}{2}, \dfrac{\sqrt{3}}{2}\right)$	$\dfrac{1}{2}$	$\dfrac{\sqrt{3}}{2}$	$\sqrt{3}$	$\dfrac{1}{\sqrt{3}} = \dfrac{\sqrt{3}}{3}$	2	$\dfrac{2}{\sqrt{3}} = \dfrac{2\sqrt{3}}{3}$

We now know the exact values of the trigonometric functions at the quadrantal angles (or at numbers that are integral multiples of $\pi/2$) and also at the three acute angles $45°$, $30°$, and $60°$ (or at the numbers $\pi/4$, $\pi/6$, and $\pi/3$). We can find approximations to the trigonometric functional values of any acute angle (or of any positive number less than $\pi/2$) in a table such as Table III in Appendix B. A still easier way to find such approximations is to use a hand calculator. But not every student has a hand calculator, so we shall use Table III in this text.

In Table III, headings at the top of the page are used with entries in the degrees column and the radians column at the left; headings at the bottom of the page are used with entries in the degrees column and the radians column at the right. The entries in Table III can be arranged in this manner because $\cos A = \sin B$, $\sin A = \cos B$, $\tan A = \cot B$, and $\cot A = \tan B$ whenever A and B are complementary angles.

Notice that there are no columns in Table III for $\sec \theta$ and $\csc \theta$. If necessary, these can be derived from Table III by using $\sec \theta = 1/\cos \theta$ or $\csc \theta = 1/\sin \theta$. Again it would be more efficient to use a more extensive table or a calculator.

EXAMPLE 1

Solution

Use Table III to find an approximation to $\sin 12°$.

In Table III we read that $\sin 12°$ is approximately 0.2079. As in using the logarithm tables, it is customary to use an equals sign and write

$$\sin 12° = 0.2079,$$

even though 0.2079 is an approximation to $\sin 12°$. (On a calculator, $\sin 12° = 0.207912$ correct to six decimal places.)

EXAMPLE 2

Solution

Use Table III to find an approximation to $\cos 0.17$.

We look in the radians column for the real number 0.17. We cannot find 0.17. The entry nearest 0.17 is 0.1716, so we use that entry and say that $\cos 0.17$ is approximately 0.9853. Following the custom of using the equals sign, we write

$$\cos 0.17 = 0.9853.$$

We could get a better approximation from Table III by using the technique called linear interpolation, but use of a more extensive table or a calculator would be more efficient. (A more extensive table or a calculator would show that correct to four decimal places, $\cos 0.17$ is actually 0.9856.)

EXAMPLE 3

Solution

Use Table III to find an approximation to t, if $\tan t = 1.23$ and $0° < t < 90°$.

This time we look in the body of the table in the tangent columns. The entry in those columns nearest in value to 1.23 is 1.2276. We read that

$$\tan 50°50' = 1.2276$$

or $$\tan 0.8872 = 1.2276.$$

Accordingly we say that t is approximately 50°50', or 0.8872 radians. If a closer approximation is required, we should use a calculator or more extensive table. (On a calculator, t is approximately 50.8886°, or 0.888174 radians.)

EXAMPLE 4

Use Table III to find, to the nearest degree, an approximation to the measure of angle A in the triangle with sides of length 3 and 5 units, as shown in Figure 3.28.

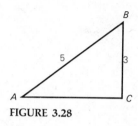

FIGURE 3.28

Solution

$$\sin A = \frac{\text{Side opposite } A}{\text{Hypotenuse}}$$

$$= \frac{3}{5}$$

$$= 0.6$$

In Table III, the entry in the sine columns nearest 0.6 is 0.5995. We read

$$\sin 36°50' = 0.5995.$$

Accordingly we say that the measure of angle A is approximately 37°. For brevity, it is customary to write

$$A = 37°.$$

(On a calculator we would find A to be about 36.87°.)

We have just seen how to find approximations to the values of the trigonometric functions of acute angles or of positive numbers less than $\pi/2$. Such angles or numbers are associated with unit circle points in the first quadrant. Approximations to the trigonometric functional values of angles or numbers associated with unit circle points in the other three quadrants can be found from the functional values of the acute angles we have called their reference angles (Figure 3.7). For example, we can show that if A is a second-quadrant angle with acute angle θ as reference angle, then

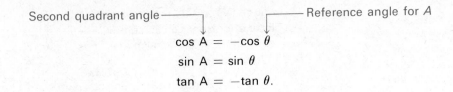

$$\cos A = -\cos \theta$$
$$\sin A = \sin \theta$$
$$\tan A = -\tan \theta.$$

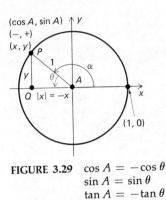

FIGURE 3.29 $\cos A = -\cos \theta$
$\sin A = \sin \theta$
$\tan A = -\tan \theta$

To show this, consider Figure 3.29. Let P be the unit circle point associated with angle A of radian measure α. Let $\overline{PQ}$ be perpendicular to the x-axis. The reference angle θ of A is an acute angle in triangle PAQ. Let (x, y) be the coordinates of P. Since P is in the second quadrant, x is negative and y is positive. In triangle PAQ, therefore, the length of $\overline{QA}$ is $|x|$ $= -x$ and the length of $\overline{QP}$ is y, so that

$$\cos \theta = \frac{\text{Side adjacent } \theta}{\text{Hypotenuse}} = \frac{-x}{1} = -x$$

$$\sin \theta = \frac{\text{Side opposite } \theta}{\text{Hypotenuse}} = \frac{y}{1} = y$$

$$\tan \theta = \frac{\text{Side opposite } \theta}{\text{Side adjacent } \theta} = \frac{y}{-x}.$$

In other words,

$$x = -\cos \theta$$
$$y = \sin \theta$$
$$\frac{y}{x} = -\tan \theta.$$

By the definitions of $\cos A$, $\sin A$, and $\tan A$, we also have

$$\cos A = \cos \alpha = x$$
$$\sin A = \sin \alpha = y$$
$$\tan A = \tan \alpha = \frac{y}{x}$$

It follows that

$$\cos A = \cos \alpha = -\cos \theta$$
$$\sin A = \sin \alpha = \sin \theta$$
$$\tan A = \tan \alpha = -\tan \theta$$

for A in the second quadrant. Since

$$\cot A = \frac{\cos A}{\sin A}, \quad \sec A = \frac{1}{\cos A}, \quad \text{and} \quad \csc A = \frac{1}{\sin A},$$

it is also true that for A in the second quadrant,

$$\cot A = \cot \alpha = -\cot \theta$$
$$\sec A = \sec \alpha = -\sec \theta$$
$$\csc A = \csc \alpha = \csc \theta.$$

We have just seen that if A is a second-quadrant angle of radian measure α, to find a trigonometric functional value of A or α we do the following:

1. Find the corresponding functional value of the reference angle of A.
2. Prefix the appropriate sign.

The same is true if A is in any other quadrant. The appropriate sign is determined by recalling that $\cos A$ and $\sin A$ are simply the coordinates x and y of the unit circle point (x, y) associated with A. Accordingly $\cos A$, $\sin A$, and all the other trigonometric functions will have the signs indicated in Fig. 3.30.

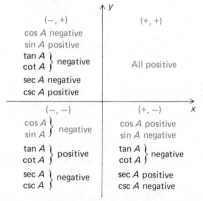

Exercise 3.3, 26–35   **FIGURE 3.30**   Signs of the trigonometric functional values

EXAMPLE 5   Find
(a) cos 150°   (b) tan (−3π/4)   (c) sin 300°   (d) sec (13π/6)

Solution   (a) The standard-position angle of measure 150° is in the second quadrant. For brevity, we simply say, "150° is in the second quadrant." (See Figure 3.31.)
In the second quadrant, cos A is negative.
The reference angle is 30°.

$$\cos 150° = -\cos 30°$$

$$= -\frac{\sqrt{3}}{2}$$

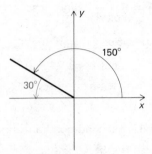

FIGURE 3.31

(b) −3π/4 is in the third quadrant.
In the third quadrant, tan A is positive. (See Figure 3.32.)
The reference angle is π/4.

$$\tan\left(\frac{-3\pi}{4}\right) = +\tan\frac{\pi}{4}$$

$$= 1$$

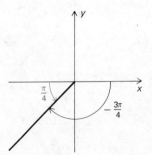

FIGURE 3.32

(c) 300° is in the fourth quadrant.

In the fourth quadrant, sin A is negative. (See Figure 3.33.)
The reference angle is 60°.

$$\sin 300° = -\sin 60°$$

$$= -\frac{\sqrt{3}}{2}$$

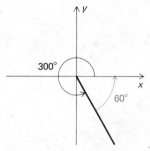

FIGURE 3.33

(d) $13\pi/6 = 2\pi + \pi/6$ is in the first quadrant.

In the first quadrant, sec A is positive. (See Figure 3.34.)
The reference angle is $\pi/6$.

$$\sec \frac{13\pi}{6} = +\sec \frac{\pi}{6}$$

$$= \frac{2}{\sqrt{3}}$$

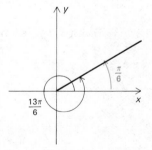

Exercise 3.3, 1–25   **FIGURE 3.34**

EXAMPLE 6

Use Table III to find an approximation to tan 830°.

Solution

$830° = 2 \times (360°) + 110°$ is in the second quadrant.

In the second quadrant, tan A is negative. (See Figure 3.35.)
The reference angle is $70°$.

tan $830° = -$tan $70°$

$\qquad = -2.7475$ (from Table III)

(On a calculator, to six decimal places, tan $830° = -2.747477$.)

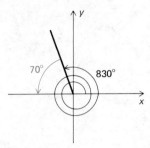

FIGURE 3.35

EXAMPLE 7

Solution

Use 3.14 for π, use Table III, and find an approximation to cos 2.19.

If we use 3.14 for π, then $\pi/2$ is approximately 1.57, and the number 2.19 is between $\pi/2$ and π.

2.19 is in the second quadrant. (See Figure 3.36.)

In the second quadrant, cos A is negative.

The reference angle, $\pi - 2.19$, is approximately $3.14 - 2.19 = 0.95$, so cos $2.19 \simeq -$cos 0.95.

The entry nearest 0.95 in the Radians columns in Table III is 0.9512. We read cos $0.9512 = 0.5807$. We say that cos 0.95 is approximately 0.5807, so that

$$\cos 2.19 = -0.5807 \text{ approximately.}$$

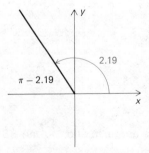

FIGURE 3.36

(As usual, a much better approximation could be obtained more efficiently with a calculator or more extensive table. In fact, a calculator would show that, correct to four decimal places, cos 2.19 is actually -0.5804.)

EXAMPLE 8

If $\tan \theta = -2.1445$ and $-90° < \theta < 90°$, use Table III and find an approximation to θ.

Solution

Since $-90° < \theta < 90°$, θ is in the fourth or first quadrant or else $\theta = 0°$. But $\tan \theta$ is negative, so θ must be in the fourth quadrant with reference angle t such that $\tan t = 2.1445$. From Table III, $t = 65°$ so $\theta = -65°$. (On a calculator, θ is $-64.9999°$ to four decimal places.)

EXAMPLE 9

If $\cos \theta = -0.9205$ and $0 < \theta < \pi$, use Table III and use 3.14 for π to find an approximation to θ.

Solution

Since $0 < \theta < \pi$, θ is in the first or second quadrant or else $\theta = \pi/2$. But $\cos \theta$ is negative, so θ must be in the second quadrant with reference angle t such that $\cos t = 0.9205$. (See Figure 3.37.) By Table III, $t = 0.4014$. Therefore θ is approximately $3.14 - 0.4014 = 2.7386$. (A calculator would show that, correct to four decimal places, $\theta = 2.7402$.)

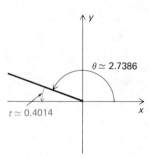

Exercise 3.3, 36–65
Applications to Right
Triangles
Exercise 3.3, 66–75

FIGURE 3.37

In calculus we sometimes have to substitute for one variable, say x, in terms of another, say θ, and then we may have to determine what values of the second variable correspond to given values of the first variable. For example, we might have to make the substitution $x = 2 \cos \theta$, and then we might have to determine which values of θ correspond to $x = 0$ and $x = 1$. Other types of trigonometric substitutions that we might have to make include $x = a \sin \theta$, $x = a \tan \theta$, and $x = a \sec \theta$ for some specific real number a. Some work with such substitutions is illustrated in the next two examples.

EXAMPLE 10

If $x = 2 \cos \theta$ and $0 \leq \theta \leq \pi$, find θ when $x = 0$; find θ when $x = 1$.

Solution

Let $x = 0$.

$$0 = 2 \cos \theta$$
$$\cos \theta = 0$$

The only number θ for which $0 \leq \theta \leq \pi$ and $\cos \theta = 0$ is $\pi/2$.

$$\theta = \frac{\pi}{2}$$

Let $x = 1$.

$$1 = 2 \cos \theta$$
$$\cos \theta = \tfrac{1}{2}$$

The only number θ for which $0 \leq \theta \leq \pi$ and $\cos \theta = \tfrac{1}{2}$ is $\pi/3$.

$$\theta = \frac{\pi}{3}$$

EXAMPLE 11

If $x = 3 \tan \theta$ and $-\pi/2 < \theta < \pi/2$, find θ when $x = 0$; find θ when $x = 3$.

Solution

Let $x = 0$.

$$0 = 3 \tan \theta$$
$$\tan \theta = 0$$
$$\theta = 0$$

(Note that 0 is the only number θ for which $-\pi/2 < \theta < \pi/2$ and $\tan \theta = 0$.)

Let $x = 3$.

$$3 = 3 \tan \theta$$
$$\tan \theta = 1$$

$$\theta = \frac{\pi}{4}$$

(Note that $\pi/4$ is the only number θ for which $-\pi/2 < \theta < \pi/2$ and $\tan \theta = 1$.)

Exercise 3.3, 76–80

EXERCISE 3.3 TRIGONOMETRIC FUNCTIONAL VALUES

▶ Find $\cos \theta$, $\sin \theta$, $\tan \theta$, $\cot \theta$, $\sec \theta$, and $\csc \theta$ for each of the given values of θ.

EXAMPLE

$210°$

Solution

$210°$ is in the third quadrant with reference angle $30°$.

$$\cos 210° = -\cos 30° = -\frac{\sqrt{3}}{2} \qquad \sin 210° = -\sin 30° = -\frac{1}{2}$$

$$\tan 210° = \tan 30° = \frac{1}{\sqrt{3}} \qquad \cot 210° = \cot 30° = \sqrt{3}$$

$$\sec 210° = -\sec 30° = -\frac{2}{\sqrt{3}} \qquad \csc 210° = -\csc 30° = -2$$

1. 120° 2. 135° 3. 150° 4. 225° 5. 240°

6. 300° 7. 315° 8. 330° 9. −45°

10. −120° 11. −390° 12. 765° 13. 1020° 14. $\dfrac{2\pi}{3}$

15. $\dfrac{3\pi}{4}$ 16. $\dfrac{5\pi}{6}$ 17. $\dfrac{7\pi}{6}$ 18. $\dfrac{5\pi}{4}$ 19. $\dfrac{4\pi}{3}$

20. $\dfrac{5\pi}{3}$ 21. $\dfrac{7\pi}{4}$ 22. $\dfrac{11\pi}{6}$ 23. $-\dfrac{11\pi}{4}$

24. $-\dfrac{17\pi}{6}$ 25. $-\dfrac{7\pi}{3}$

▶ In what quadrant is standard-position angle A in each of the following?

EXAMPLE

$\cos A$ is positive and $\tan A$ is negative

Solution

The following table shows that A is in the fourth quadrant.

	1st	2nd	3rd	4th
$\cos A$	+	−	−	+
$\tan A$	+	−	+	−

26. $\sin A$ is negative and $\tan A$ is positive

27. $\cos A$ is negative and $\sin A$ is positive

28. $\sin A$ is positive and $\sec A$ is negative

29. $\cos A$ is positive and $\tan A$ is positive

30. $\cot A$ is negative and $\sin A$ is positive

▶ In what quadrant is the standard-position angle associated with real number α in each of the following?

31. $\tan \alpha$ is negative and $\sin \alpha$ is negative

32. $\tan \alpha$ is negative and $\sin \alpha$ is positive

33. $\cos \alpha$ is negative and $\cot \alpha$ is positive

34. sec α is positive and tan α is negative

35. csc α is positive and cot α is negative

▶ Use Table III and find an approximation to each of the following. Use 3.14 for π, if necessary.

EXAMPLE

cos 159°

Solution

159° is in the second quadrant with reference angle 21°.

$$\cos 159° = -\cos 21° = -0.9336$$

36. sin 245°   **37.** tan 304°   **38.** cos 141°   **39.** cot 204°

40. sin 436°   **41.** cos 3.85   **42.** sin 2.73   **43.** tan 1.93

44. cot 5.32   **45.** cos 7.69   **46.** sin 2   **47.** cos 3

48. tan (-1) **49.** sin (-3) **50.** cos (-4)

▶ Use Table III and find an approximation to θ in each of the following. Use 3.14 for π, if necessary.

EXAMPLE

$\tan \theta = 2.14, \ 0° < \theta < 90°$

Solution

$\theta = 65°$

51. $\cot \theta = 0.404, \ 0° < \theta < 90°$

52. $\sin \theta = 0.731, \ 0 < \theta < \pi/2$

53. $\cos \theta = 0.956, \ 0° < \theta < 90°$

54. $\tan \theta = 0.158, \ 0 < \theta < \pi/2$

55. $\cot \theta = 0.123, \ 0° < \theta < 90°$

56. $\sin \theta = -0.9397, \ -90° < \theta < 90°$

57. $\tan \theta = 0.2679, \ -\pi/2 < \theta < \pi/2$

58. $\sin \theta = -0.1736, \ -90° < \theta < 90°$

59. $\tan \theta = -0.1051, \ -\pi/2 < \theta < \pi/2$

60. $\sin \theta = 0.7431, \ -90° < \theta < 90°$

61. $\cos \theta = -0.6820, \ 0° < \theta < 180°$

62. $\cot \theta = -5.1446, \ 0 < \theta < \pi$

63. $\cos \theta = -0.9511, \ 0° < \theta < 180°$

64. $\cot \theta = -0.6745, \ 0 < \theta < \pi$

65. $\cos \theta = -0.6293, \ 0 < \theta < \pi$

▶ Let a, b, and c be the lengths of the sides opposite angles A, B, and C in right triangle ABC with right angle at C. If necessary, use Table III and find an approximation to the required angle or side.

EXAMPLE

$a = \sqrt{3}$ cm, $c = 2$ cm   Find A.

Solution

$\sin A = \dfrac{\sqrt{3}}{2}$. The measure of A is $60°$. We write: $A = 60°$.

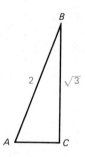

66. $a = \sqrt{2}$ m, $c = 2$ m   Find A.

67. $b = \sqrt{3}$ ft, $a = 3$ ft   Find B.

68. $a = 1$ cm, $b = 2$ cm   Find A.

69. $\tan A = \dfrac{\sqrt{3}}{3}$, $a = 1$ ft   Find c.

70. $\cot B = 2$, $a = 3$ cm   Find b.

71. $\tan A = 1$, $a = 7$ mm   Find b.

72. $\sec A = 2$, $c = 6$ km   Find a.

73. $a = 6$ ft, $b = 8$ ft   Find A and c.

74. $a = 5$ cm, $c = 13$ cm   Find b and A.

75. $\sin A = \frac{1}{2}$, $c = 12$ m   Find a and B.

▶ In each of the following, find the values of θ corresponding to the given values of x.

EXAMPLE

$x = 3 \sec \theta$; $0 \le \theta < \dfrac{\pi}{2}$; $x = \sqrt{3}$, $x = 3$.

Solution

$x = \sqrt{3}$ $x = 3$

$\sec \theta = \dfrac{\sqrt{3}}{3}$ $\sec \theta = 1$

$\theta = \dfrac{\pi}{6}$ $\theta = 0$

76. $x = 2 \tan \theta; \dfrac{-\pi}{2} < \theta < \dfrac{\pi}{2}; x = 0, x = 2$

77. $x = 2 \sin \theta; \dfrac{-\pi}{2} < \theta < \dfrac{\pi}{2}; x = \sqrt{2}, x = \sqrt{3}$

78. $x = 2 \cos \theta; 0 \leq \theta \leq \pi; x = -2, x = 2$

79. $x = 2 \sec \theta; 0 \leq \theta < \dfrac{\pi}{2}; x = 2\sqrt{2}, x = 4$

80. $x = 5 \tan \theta; \dfrac{-\pi}{2} < \theta < \dfrac{\pi}{2}; x = 5, x = 5\sqrt{3}$

CHECK LIST FOR THE STUDENT

If you have learned the material in Section 3.3, you should be able to do the following:

1. Give the coordinates of the unit circle point associated with an angle of 45° or with the real number $\pi/4$.

2. Give the coordinates of the unit circle point associated with an angle of 30° or with the real number $\pi/6$.

3. Give the coordinates of the unit circle point associated with an angle of 60° or with the real number $\pi/3$.

4. Give the six trigonometric functional values of 45°, 30°, and 60° (or of $\pi/4$, $\pi/6$, and $\pi/3$).

5. Use a table of trigonometric functions (Table III of Appendix B).

6. Find approximations to the trigonometric functions of A provided that A is an acute angle whose measure is given in degrees or radians.

7. Find approximations to the trigonometric functions of α where α is a positive real number less than $\pi/2$.

8. Use Table III to find an approximation to θ provided that $\cos \theta$, $\sin \theta$, $\tan \theta$, or $\cot \theta$ is given and $0° < \theta < 90°$ or $0 < \theta < \pi/2$.

9. Find approximations to the trigonometric functions of any angle or any number for which the functions are defined.

10. Give the signs of the functional values of each of the trigonometric functions in each of the four quadrants.

11. Use Table III and find an approximation to θ provided that $\sin \theta$ or $\tan \theta$ is given and $-90° < \theta < 90°$ or $-\pi/2 < \theta < \pi/2$, or provided that $\cos \theta$ or $\cot \theta$ is given and $0° < \theta < 180°$ or $0 < \theta < \pi$.

12. Find values of θ corresponding to given values of x provided that $x = a \cos \theta$, $x = a \sin \theta$, $x = a \tan \theta$, or $x = a \sec \theta$ where a is some specific real number.

SECTION QUIZ 3.3

Give the value of each of the following.

1. (a) $\cos \dfrac{\pi}{6}$ (b) $\sin 60°$ (c) $\tan \dfrac{\pi}{4}$

    (d) $\sec 45°$ (e) $\cot 30°$ (f) $\csc \dfrac{\pi}{3}$

2. (a) $\cos 150°$ (b) $\sin \dfrac{3\pi}{4}$ (c) $\tan \left(\dfrac{-\pi}{4} \right)$

    (d) $\cot 240°$ (e) $\sec \dfrac{5\pi}{4}$ (f) $\csc 330°$

3. Find an approximation to each of the following by using some or all of the approximate values given below. Use 3.14 for π.

$\cos 20° = 0.9397 \qquad \cos 0.512 = 0.8718$
$\sin 20° = 0.3420 \qquad \sin 0.512 = 0.4899$
$\cos 50° = 0.6428 \qquad \cos 0.256 = 0.9674$
$\sin 50° = 0.7660 \qquad \sin 0.256 = 0.2532$

   (a) $\cos 340°$ (b) $\cos 230°$ (c) $\sin 380°$
   (d) $\cos 3.652$ (e) $\sin (-3.396)$ (f) $\cos 6.024$

4. In each of the following, find an approximation to θ by using some of the approximate values given in the previous question.

   (a) $\sin \theta = -0.6428$ and $-90° \le \theta \le 90°$
   (b) $\cos \theta = -0.9397$ and $0 \le \theta \le \pi$

5. Let a, b, and c be the lengths of the sides opposite angles A, B, and C in right triangle ABC with right angle at C. Let $a = \frac{1}{2}$ cm and $c = 2$ cm. Find B.

3.4 Properties and Graphs of Sine and Cosine Functions

The purpose of this section is to note some basic properties of the sine and cosine functions and to use those properties in sketching the graphs of the sine and cosine, as well as in sketching the graphs of other functions with similar rules of assignment.

One of the basic properties of the sine function is that for any angle θ or any real number θ, $\sin \theta$ is never greater than 1 and never less than -1. Likewise, $\cos \theta$ is never greater than 1 and never less than -1. We can see that this property of the sine and cosine is true by recalling the definitions of the sine and cosine and referring to the unit circle. Figure 3.38 reminds us that for any θ in the domain of the sine or cosine, $\cos \theta$ and $\sin \theta$ are the coordinates of the unit circle point P associated with θ. Since the unit circle has radius 1, each coordinate of P must be between 1 and -1, inclusively. For all θ,

$$-1 \leq \sin \theta \leq 1$$

and

$$-1 \leq \cos \theta \leq 1.$$

In fact, the range of both the sine function and the cosine function is the set of all real numbers y such that $-1 \leq y \leq 1$.

$$\text{Range of sine } = \{y \mid -1 \leq y \leq 1\}$$
$$\text{Range of cosine } = \{y \mid -1 \leq y \leq 1\}$$

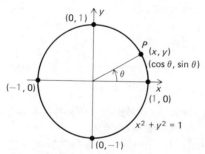

FIGURE 3.38 $\quad -1 \leq \cos \theta \leq 1$
$\qquad\qquad\qquad -1 \leq \sin \theta \leq 1$

It is also readily apparent from the definitions of the sine and cosine that for any θ in the domain of the sine and cosine,

$$\cos (-\theta) = \cos \theta$$
$$\sin (-\theta) = -\sin \theta.$$

Indeed, as we see in Figure 3.39, if the unit circle point P associated with θ has coordinates (x, y), then the unit circle point M associated with $-\theta$

has coordinates $(x, -y)$. That is, P and M have the same x-coordinates, and their y-coordinates differ only in sign. But the coordinates of P are $(\cos \theta, \sin \theta)$ and the coordinates of M are $(\cos (-\theta), \sin (-\theta))$, so

$$\cos (-\theta) = \cos \theta \quad \text{and} \quad \sin (-\theta) = -\sin \theta.$$

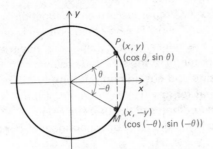

FIGURE 3.39 $\cos(-\theta) = \cos \theta$
$\sin(-\theta) = -\sin \theta$

Another basic property of the sine and cosine is that the values of $\sin \theta$ and $\cos \theta$ repeat themselves every time θ varies through $360°$, or 2π, and then through another $360°$, or 2π. In Figure 3.38, for example, we see that as θ varies from 0 to 2π, P moves completely around the circle once. Then as θ varies from 2π to 4π, P moves around the circle again, and the values of $\cos \theta$ and $\sin \theta$ repeat the same pattern they made as θ varied from 0 to 2π. We say that the sine and cosine functions are *periodic* functions with period 2π. In other words, for any θ in the domain of the sine or cosine,

$$\left.\begin{array}{l} \sin (\theta + 2\pi) = \sin \theta \\ \cos (\theta + 2\pi) = \cos \theta. \end{array}\right\} \quad \begin{array}{l} \text{Periodic with} \\ \text{period } 2\pi \end{array}$$

We shall verify these properties in Section 3.6 when we develop formulas for $\sin (a + b)$ and $\cos (a + b)$.

Any function f or the graph of f is said to be **periodic** with period k provided that $x + k$ is in the domain of f, $k \neq 0$, and

$$f(x + k) = f(x) \quad \text{Periodic with period } k$$

for all x in the domain of f. If there is a smallest such positive number k, it is called the **fundamental period**. Notice that 4π, 6π, 8π, -2π, -4π, or any nonzero integral multiple of 2π is also a period for the sine or cosine function. For example, $\sin (\theta + 4\pi) = \sin \theta$, $\cos (\theta + 6\pi) = \cos \theta$, and $\cos (\theta + (-2\pi)) = \cos \theta$. But 2π is the smallest positive number k for which $\sin (\theta + k) = \sin \theta$ or $\cos (\theta + k) = \cos \theta$.

The fundamental period of the sine and cosine functions is 2π.

Unless otherwise specified, by *the* period of a function we mean the *fundamental* period. The period of the sine and cosine functions is 2π.

Still another basic property of the sine and cosine is that they are *complementary* functions, or *cofunctions,* in that the cosine of an angle is equal to the sine of its complement. (Two angles are complementary provided that the sum of their measures is $90°$ or $\pi/2$.) More generally, for any θ in the domain of the cosine,

$$\cos \theta = \sin \left(\frac{\pi}{2} - \theta\right).$$

It is likewise true that for any θ in the domain of the cosine,

$$\cos \theta = \sin \left(\frac{\pi}{2} + \theta\right).$$

These properties will be verified in Section 3.6, where general formulas for $\sin (a + b)$ and $\sin (a - b)$ will be developed.

Let us now sketch the graph of the sine function. That is, let us sketch the graph of $y = \sin x$ for all real numbers x. Since we have just observed that the range is $\{y \mid -1 \leq y \leq 1\}$, we know that no part of the graph will be above the line $y = 1$ and no part of the graph will be below the line $y = -1$. In Figure 3.40, the lines $y = 1$ and $y = -1$ were drawn as guides to remind us of these facts. We have also just observed that the sine function is periodic with fundamental period 2π, so we know that whatever pattern the graph makes for $0 \leq x \leq 2\pi$, that pattern will be repeated every 2π units both to the right and to the left. To visualize the behavior of the graph for x between 0 and 2π, inclusively, consider the table in Figure 3.40. This table shows how $\sin x$ varies as x increases through values corresponding to each of the four quadrants. For example, as x increases, taking on all values from 0 to $\pi/2$, $\sin x$ increases continuously from 0 to 1 in the sense that it takes on all values from 0 to 1. If we plot the points $(0, 0)$, $(\pi/2, 1)$, $(\pi, 0)$, $(3\pi/2, -1)$, and $(2\pi, 0)$, they all appear to be strung together like beads on the part of the graph that is between 0 and 2π. This pattern repeats indefinitely in both directions.

Next let us sketch the graph of $y = \cos x$ for all real numbers x. Since we know that the range of the cosine function is $\{y \mid -1 \leq y \leq 1\}$, once again we know that the graph lies between or just touches the guide lines $y = 1$ and $y = -1$. Once again, we also know that the fundamental period is 2π, so the pattern the graph makes for $0 \leq x \leq 2\pi$ will be repeated every 2π units both to the right and to the left. The table in Figure 3.41 shows how $\cos x$ varies as x increases through values corresponding to each of the four quadrants. For instance, as x increases, taking on all values from 0 to $\pi/2$, $\cos x$ decreases, continuously taking on all values from 1 to 0. If we plot the points shown in the table in Figure

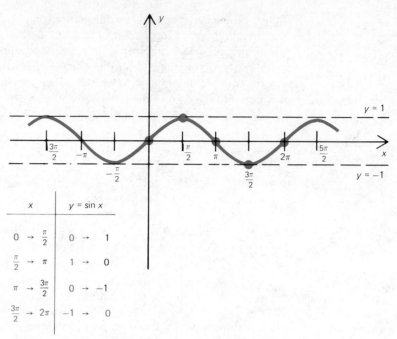

x			$y = \sin x$	
0	$\to$	$\dfrac{\pi}{2}$	$0 \to 1$	
$\dfrac{\pi}{2}$	$\to$	π	$1 \to 0$	
π	$\to$	$\dfrac{3\pi}{2}$	$0 \to -1$	
$\dfrac{3\pi}{2}$	$\to$	2π	$-1 \to 0$	

FIGURE 3.40   Graph of $y = \sin x$
(Graph of the sine function)

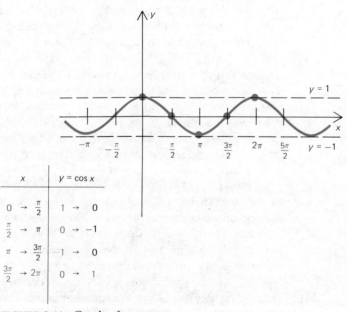

x			$y = \cos x$	
0	$\to$	$\dfrac{\pi}{2}$	$1 \to 0$	
$\dfrac{\pi}{2}$	$\to$	π	$0 \to -1$	
π	$\to$	$\dfrac{3\pi}{2}$	$-1 \to 0$	
$\dfrac{3\pi}{2}$	$\to$	2π	$0 \to 1$	

FIGURE 3.41   Graph of $y = \cos x$
(Graph of the cosine function)

3.41, together with intervening points, they all appear to be strung together like beads on the part of the graph that is between 0 and 2π. This pattern repeats indefinitely in both directions.

Figure 3.42 shows the graphs of both the sine function and the cosine function on the same coordinate axes. The cosine curve looks exactly like the sine curve shifted $\pi/2$ units to the left. This demonstrates the fact noted earlier that for all real numbers x,

$$\cos x = \sin\left(\frac{\pi}{2} + x\right).$$

In applications, we sometimes need to work with functions f and g such that $f(x) = a\sin(bx + c)$ and $g(x) = a\cos(bx + c)$ where a, b, and c are specific real numbers. The graphs of these functions, that is, the graphs of the equations

$$y = a\sin(bx + c)$$

and
$$y = a\cos(bx + c),$$

can be readily obtained by suitably modifying or shifting the graphs of the sine and cosine functions. We shall work up to a consideration of the graphs of these equations by first considering the graphs of $y = a\sin x$ and $y = a\cos x$, then considering the graphs of $y = a\sin bx$ and $y =$

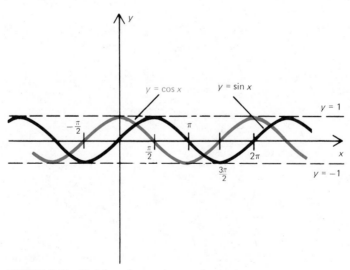

FIGURE 3.42   Graphs of $y = \sin x$ and $y = \cos x$
$\cos x = \sin(\pi/2 + x)$

$a \cos bx$, and finally considering the graphs of $y = a \sin (bx + c)$ and $y = a \cos (bx + c)$.

Consider the graphs of $y = a \sin x$ *and* $y = a \cos x$. If a is a positive real number, the graphs of $y = a \sin x$ and $y = a \cos x$ look like the graphs of the sine and cosine except that they lie between or just touch the guide lines $y = a$ and $y = -a$ instead of the guide lines $y = 1$ and $y = -1$. After all, since

$$-1 \leq \sin x \leq 1 \quad \text{and} \quad -1 \leq \cos x \leq 1,$$

it must be true that if a is positive,

$$-a \leq a \sin x \leq a \quad \text{and} \quad -a \leq \cos x \leq a.$$

If a is negative, the graphs of $y = a \sin x$ and $y = a \cos x$ look like the graphs of the sine and cosine reflected in the x-axis and lying between or just touching the guide lines $y = a$ and $y = -a$. The absolute value of a is the maximum distance any point on either graph is from the x-axis. The number $|a|$ is called the **amplitude** of the function defined by $y = a \sin x$ or by $y = a \cos x$. The number $|a|$ is also called the amplitude of the *graphs* of these functions. Specific cases in which $a = 3$ and $a = -3$ appear in the next two examples.

EXAMPLE 1

Find the amplitude and fundamental period and sketch the graph of $y = 3 \sin x$.

Solution

The amplitude is $|3| = 3$ and the fundamental period is 2π because the fundamental period of the sine function is 2π.

$$3 \sin (x + 2\pi) = 3 \sin x$$

The graph is shown in Figure 3.43.

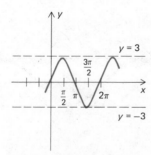

FIGURE 3.43 Graph of $y = 3 \sin x$

EXAMPLE 2

Find the amplitude and fundamental period and sketch the graph of $y = -3 \cos x$.

Solution   The amplitude is $|-3| = 3$ and the fundamental period is 2π. The graph is shown in Figure 3.44.

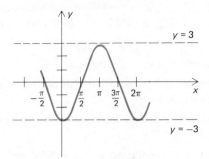

FIGURE 3.44   Graph of $y = -3 \cos x$

Consider the graphs of $y = a \sin bx$ *and* $y = a \cos bx$. For positive number b, the graph of $y = a \sin bx$ looks like the graph of $y = a \sin x$, except that the period is different. The expression $a \sin x$ repeats itself every time x varies through 2π and then through another 2π. But $a \sin bx$ repeats itself every time bx varies through 2π and then through another 2π. That is, $a \sin bx$ repeats itself every time x varies through $2\pi/b$ and then through another $2\pi/b$. Indeed, it must be true that

$$0 \le bx \le 2\pi$$

whenever $0 \le x \le \dfrac{2\pi}{b}.$ Dividing through by positive b

The fundamental period of the function defined by $y = a \sin bx$ must be $2\pi/b$.

Likewise, for positive number b, the graph of $y = a \cos bx$ looks like the graph of $y = a \cos x$, except that the fundamental period of the function defined by $y = a \cos bx$ is $2\pi/b$.

$$\left. \begin{array}{l} y = a \sin bx \\ y = a \cos bx \end{array} \right\} \qquad \begin{array}{l} \text{Amplitude } |a| \\ \text{Fundamental period } \dfrac{2\pi}{b} \end{array}$$

If b is negative, we can use the fact that $\sin(-\theta) = -\sin\theta$ to show that for negative b, the graph of $y = a \sin bx$ looks like the graph of $y = -a \sin(|b|x)$. For example, if b is -2, the graph of $y = a \sin(-2x)$ looks like the graph of $y = -a \sin 2x$, because

$$a \sin(-2x) = -a \sin 2x.$$

Similarly, if b is negative, the graph of $y = a \cos bx$ looks like the graph of $y = a \cos(|b|x)$, because $\cos(-\theta) = \cos\theta$. For example, if b is -2, the graph of $y = a \cos(-2x)$ looks like the graph of $y = a \cos 2x$, because

$$a \cos(-2x) = a \cos 2x.$$

EXAMPLE 3   Find the amplitude and fundamental period and sketch the graph of $y = 3 \sin 2x$.

Solution   The amplitude is $|3| = 3$ and the fundamental period is $2\pi/2 = \pi$. The graph is sketched in Figure 3.45. To sketch the graph quickly, we do the following:

1. Draw the guide lines $y = 3$ and $y = -3$.

2. Start at the origin and mark off one fundamental period on the x-axis by putting a tick mark for π.

3. Divide that marked period into quarters by putting tick marks for $\pi/4$, $\pi/2$, and $3\pi/4$.

4. Plot the points corresponding to the origin and the tick marks.

5. Recall the general configuration of the sine curve and sketch the graph.

x	$2x$	$y = 3 \sin 2x$
0	0	0
$\frac{\pi}{4}$	$\frac{\pi}{2}$	3
$\frac{\pi}{2}$	π	0
$\frac{3\pi}{4}$	$\frac{3\pi}{2}$	-3
π	2π	0

Amplitude 3
Fundamental period π

FIGURE 3.45   Graph of $y = 3 \sin 2x$

EXAMPLE 4   Find the amplitude and fundamental period and sketch the graph of $y = 3 \cos \frac{1}{2}x$.

Solution   The amplitude is $|3| = 3$ and the fundamental period is $2\pi/\frac{1}{2} = 4\pi$. The graph is sketched in Figure 3.46. To sketch the graph, we do the following:

1. Draw the guide lines $y = 3$ and $y = -3$.

2. Put a tick mark on the x-axis for 4π (the fundamental period).

3. Divide the interval from 0 to 4π into quarters by putting tick marks for π, 2π, and 3π.

4. Plot the points corresponding to the origin and the tick marks.

5. Recall the general configuration of the cosine curve and sketch the graph.

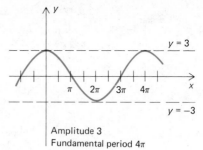

x	$\frac{1}{2}x$	$y = 3\cos\frac{1}{2}x$
0	0	3
π	$\frac{\pi}{2}$	0
2π	π	-3
3π	$\frac{3\pi}{2}$	0
4π	2π	3

Amplitude 3
Fundamental period 4π

FIGURE 3.46 Graph of $y = 3\cos\frac{1}{2}x$

EXAMPLE 5

Find the amplitude and fundamental period and sketch the graph of $y = \frac{1}{2}\sin(-4x)$.

Solution

Since $\sin(-4x) = -\sin 4x$, the graph of $y = \frac{1}{2}\sin(-4x)$ is the graph of $y = -\frac{1}{2}\sin 4x$. This graph looks like the graph of $y = \frac{1}{2}\sin 4x$ reflected in the x-axis.

$$y = -\frac{1}{2}\sin 4x$$

The amplitude is $\left|-\frac{1}{2}\right| = \frac{1}{2}$ and the fundamental period is $2\pi/4 = \pi/2$. The graph is sketched in Fig. 3.47.

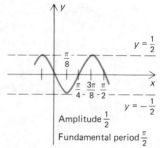

x	$4x$	$y = -\frac{1}{2}\sin 4x$
0	0	0
$\frac{\pi}{8}$	$\frac{\pi}{2}$	$-\frac{1}{2}$
$\frac{\pi}{4}$	π	0
$\frac{3\pi}{8}$	$\frac{3\pi}{2}$	$\frac{1}{2}$
$\frac{\pi}{2}$	2π	0

Amplitude $\frac{1}{2}$
Fundamental period $\frac{\pi}{2}$

FIGURE 3.47 Graph of $y = \frac{1}{2}\sin(-4x)$
(Graph of $y = \frac{-1}{2}\sin 4x$)

To sketch the graph of $y = a\sin bx$ or $y = a\cos bx$ where b is nonzero, we do the following:

1. Draw the guide lines $y = a$ and $y = -a$.

2. Start at the origin and mark off one fundamental period on the x-axis.

3. Divide the marked period into quarters.

4. Plot the points corresponding to the origin and the tick marks.

5. Recall the general configuration of the sine curve or the cosine curve and sketch the graph.

Exercise 3.4, 1–25

Consider the graphs of $y = a \sin (bx + c)$ and $y = a \cos (bx + c)$. For nonzero b, the graph of $y = a \sin (bx + c)$ looks like the graph of $y = a \sin bx$ shifted to the left or right on the x-axis. **To see this, write**

$$y = a \sin (bx + c)$$

in the form

$$y = a \sin \left[b \left(x + \frac{c}{b} \right) \right].$$

Now translate the coordinate axes so that the origin of the new x'-y' axes is at the old point $x = -c/b$, $y = 0$. That is, let

$$x' = x + \frac{c}{b}$$

$$y' = y.$$

Relative to the x'-y' axes, the equation $y = a \sin [b(x + c/b)]$ then becomes

$$y' = a \sin bx'.$$

In other words, the graph of $y = a \sin (bx + c)$ is the same as the graph of $y = a \sin bx$, except that it is shifted $|c/b|$ units to the left or right. If c/b is positive, the new origin is at the old point $(-c/b, 0)$ to the left of the old origin, and the graph of $y = a \sin (bx + c)$ looks like the graph of $y = a \sin bx$ shifted $|c/b|$ units to the left. (See Figure 3.48.) If c/b is negative, then $-c/b$ is positive so that the new origin is $|c/b|$ units to the right of the old origin and the graph of $y = a \sin (bx + c)$ looks like the graph of $y = a \sin bx$ shifted $|c/b|$ units to the right. Similar remarks apply to the graph of the function defined by $y = a \cos (bx + c)$. These

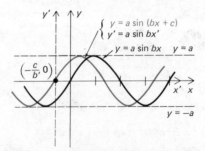

FIGURE 3.48 Graph of $y = a \sin(bx + c)$, c/b and a positive (Graph of $y' = a \sin bx'$)

remarks are summarized as follows and are illustrated in the next two examples.

$$y = a \sin (bx + c) \left.\begin{matrix} \\ \\ \end{matrix}\right\} \quad \text{Amplitude } |a|$$

$$y = a \cos (bx + c) \qquad \text{Fundamental period } \dfrac{2\pi}{|b|}$$

$y = a \sin (bx + c)$ is $y' = a \sin bx'$ ⎫ Origin of x'-y' axes at old point

$y = a \cos (bx + c)$ is $y' = a \cos bx'$ ⎭ $\left(-\dfrac{c}{b}, 0\right)$

To sketch the graph of $y = a \sin (bx + c)$ or of $y = a \cos (bx + c)$, we do the following:

1. Translate axes so that origin of new x'-y' axes is at old $(-c/b, 0)$.
2. Sketch the graph of $y' = a \sin bx'$ or $y' = a \cos bx'$.

EXAMPLE 6

Find the amplitude and fundamental period and sketch the graph of $y = 3 \sin (2x + \pi)$.

Solution

We can write $y = 3 \sin (2x + \pi)$ as

$$y = 3 \sin \left(2\left(x + \dfrac{\pi}{2}\right)\right).$$

Let $x' = x + \pi/2$ and $y' = y$ so that the new origin is at the old point $(-\pi/2, 0)$ to the left of the old origin. Relative to the x'-y' axes, the equation $y = 3 \sin (2x + \pi)$ becomes

$$y' = 3 \sin 2x'.$$

The graph is sketched in Figure 3.49. The tick marks are labeled relative to the x-y axes above the horizontal axis and relative to the x'-y' axes below the horizontal axis.

The graph of $y = 3 \sin 2x$ has also been sketched in Figure 3.49 so that the student can see that the graph of $y = 3 \sin (2x + \pi)$ does look like the graph of $y = 3 \sin 2x$ shifted $\pi/2$ units to the left.

EXAMPLE 7

Find the amplitude and fundamental period and sketch the graph of $y = 3 \cos (2x - \pi/2)$.

Solution

The amplitude is $|3| = 3$ and the fundamental period is $2\pi/2 = \pi$. We can write $y = 3 \cos (2x - \pi/2)$ as

$$y = 3 \cos \left(2\left(x - \dfrac{\pi}{4}\right)\right).$$

Let

$$x' = x - \dfrac{\pi}{4}$$

$$y' = y.$$

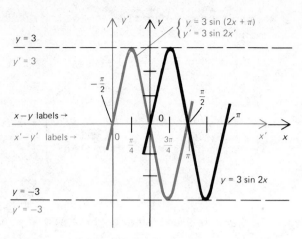

Amplitude 3
Fundamental period π

x	x'	$2x'$	$y' = 3 \sin 2x'$
$-\dfrac{\pi}{2}$	0	0	0
$-\dfrac{\pi}{4}$	$\dfrac{\pi}{4}$	$\dfrac{\pi}{2}$	3
0	$\dfrac{\pi}{2}$	π	0
$\dfrac{\pi}{4}$	$\dfrac{3\pi}{4}$	$\dfrac{3\pi}{2}$	-3
$\dfrac{\pi}{2}$	π	2π	0

FIGURE 3.49 Graph of $y = 3 \sin (2x + \pi)$
(Graph of $y' = 3 \sin 2x'$)

That is, translate axes so that the new origin is at the old point $(\pi/4, 0)$. The equation becomes

$$y' = 3 \cos 2x'.$$

The graph is sketched in Figure 3.50. The graph of $y = 3 \cos 2x$ has also been sketched in Figure 3.50 so that the student can see that the graph of $y = 3 \cos (2x - \pi/2)$ does look like the graph of $y = 3 \cos 2x$ shifted $\pi/4$ units to the right.

Exercise 3.4, 26–35

In Section 2.6, we noted that in calculus it is sometimes necessary to find an expression for the circular cross section of the solid generated by revolving a graph or part of a graph about a coordinate axis. This is demonstrated again in the next example.

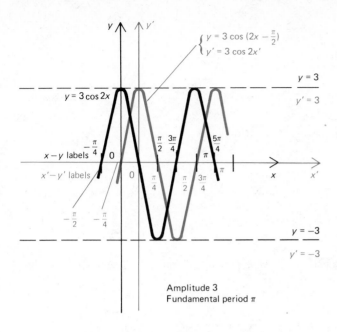

$$\begin{cases} y = 3 \cos \left(2x - \frac{\pi}{2}\right) \\ y' = 3 \cos 2x' \end{cases}$$

y = 3

y' = 3

y = 3 cos 2x

x − y labels

x' − y' labels

Amplitude 3
Fundamental period π

y = −3

y' = −3

x	x'	2x'	y' = 3 cos 2x'
0	0	0	3
$\frac{\pi}{2}$	$\frac{\pi}{4}$	$\frac{\pi}{2}$	0
$\frac{3\pi}{4}$	$\frac{\pi}{2}$	π	−3
π	$\frac{3\pi}{4}$	$\frac{3\pi}{2}$	0
$\frac{5\pi}{4}$	π	2π	3

FIGURE 3.50 Graph of $y = 3 \cos (2x - \pi/2)$
(Graph of $y' = 3 \cos 2x'$)

EXAMPLE 8

If that part of the graph of $y = \sin x$ for $0 \le x \le \pi$ is revolved about the x-axis, what is the cross-sectional area generated by the line segment from $(x, 0)$ to (x, y) on the graph? Express this cross-sectional area in terms of x.

Solution

The cross-sectional area is the area of a circle. As shown in Figure 3.51, the radius of the circle is the length of the line segment from $(x, 0)$ to (x, y). This length is y. But $y = \sin x$. In terms of x, the radius is therefore sin x.

$$\text{Cross-sectional area} = \pi \ (\text{radius})^2$$
$$= \pi \ (\sin x)^2$$

Exercise 3.4, 36–40

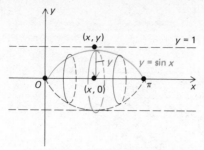

FIGURE 3.51 Cross-sectional area $= \pi (\sin x)^2$

EXERCISE 3.4 PROPERTIES AND GRAPHS OF SINE AND COSINE FUNCTIONS

▶ Find the amplitude and fundamental period and sketch the graph of each of the following.

1. $y = \cos(-x)$ **2.** $y = \sin(-x)$

3. $y = -\cos(-x)$ **4.** $y = -\sin(-x)$

5. $y = \sin(x + 2\pi)$ **6.** $y = \cos(x + 2\pi)$

7. $y = \sin\left(\dfrac{\pi}{2} - x\right)$ **8.** $y = \sin\left(\dfrac{\pi}{2} + x\right)$

9. $y = -\sin(x + 2\pi)$ **10.** $y = -\sin\left(\dfrac{\pi}{2} + x\right)$

11. $y = 4\sin x$ **12.** $y = 4\cos x$

13. $y = \frac{1}{4}\sin x$ **14.** $y = \frac{1}{4}\cos x$

15. $y = -4\sin x$ **16.** $y = 2\sin 4x$

17. $y = 2\cos 4x$ **18.** $y = 3\sin \pi x$

19. $y = 3\cos \pi x$ **20.** $y = -2\sin \frac{1}{2}x$

21. $y = -3\cos \frac{1}{2}x$ **22.** $y = \sin \dfrac{\pi}{2}x$

23. $y = \cos \dfrac{\pi}{2}x$ **24.** $y = -\sin \frac{1}{8}x$

25. $y = -\cos \frac{1}{8}x$ **26.** $y = 2\sin(4x + \pi)$

27. $y = 3\cos(4x - \pi)$ **28.** $y = \frac{1}{2}\sin(4x + 8\pi)$

29. $y = \dfrac{-1}{4}\cos(x - \pi)$ **30.** $y = 5\sin\left(x + \dfrac{\pi}{4}\right)$

31. $y = 2\sin(-2x + 8\pi)$ **32.** $y = \cos(-x + \pi)$

33. $y = -\sin(-x - \pi)$

34. $y = \frac{1}{2}\cos(8x - 4\pi)$

35. $y = \dfrac{-1}{2}\sin\left(\dfrac{1}{2}x + \dfrac{\pi}{2}\right)$

▶ In each of the following, sketch the graph. If the graph is revolved about the x-axis, find an expression for the cross-sectional area generated by the line segment from $(x, 0)$ to (x, y) on the graph. Express this cross-sectional area in terms of x.

36. $y = \cos x$ for $0 \le x \le \dfrac{\pi}{2}$

37. $y = -\cos x$ for $\dfrac{-\pi}{2} \le x \le \dfrac{\pi}{2}$

38. $y = \sin 2x$ for $\dfrac{\pi}{4} \le x \le \dfrac{\pi}{2}$

39. $y = \cos 2x$ for $0 \le x \le \dfrac{\pi}{4}$

40. $y = -\sin \frac{1}{4}x$ for $0 \le x \le 4\pi$

CHECK LIST FOR THE STUDENT

If you have learned the material in Section 3.4, you should be able to do the following:

1. State the range of the sine function and of the cosine function.
2. Express $\cos(-\theta)$ in terms of $\cos\theta$.
3. Express $\sin(-\theta)$ in terms of $\sin\theta$.
4. Define a periodic function f with period $k \neq 0$.
5. Define the fundamental period of a periodic function.
6. State the fundamental period of the sine function and of the cosine function.
7. Express $\sin(\pi/2 - \theta)$ and $\sin(\pi/2 + \theta)$ in terms of $\cos\theta$.
8. Sketch the graph of the sine function.
9. Sketch the graph of the cosine function.
10. Sketch the graph of $y = a \sin x$ and $y = a \cos x$ for a positive or negative.
11. Define the amplitude of the function defined by $y = a \sin x$ or $y = a \cos x$.
12. Find the amplitude and fundamental period and sketch the graph of $y = a \sin bx$ or $y = a \cos bx$ for a positive or negative and for b positive or negative.
13. Find the amplitude and fundamental period and sketch the graph of $y = a \sin(bx + c)$ or $y = a \cos(bx + c)$ for real numbers $a \neq 0$, $b \neq 0$, and c.

SECTION QUIZ 3.4

1. Express each of the following in terms of $\cos\theta$ or $\sin\theta$.
   (a) $\sin(-\theta)$ (b) $\cos(-\theta)$ (c) $\sin(\theta + 2\pi)$

   (d) $\cos(\theta + 2\pi)$ (e) $\sin\left(\dfrac{\pi}{2} - \theta\right)$ (f) $\cos\left(\dfrac{\pi}{2} - \theta\right)$

Find the amplitude and fundamental period and sketch the graph of each of the following. Label any intercept(s) shown on your graph.

2. $y = 5 \sin 2x$

3. $y = -\cos \frac{1}{2} x$

4. $y = 5 \sin(2x + 3\pi)$

5. $y = 3 \cos(4x - \pi)$

3.5 Properties and Graphs of Other Trigonometric Functions

The purpose of this section is to present basic properties of the tangent, cotangent, secant, and cosecant functions, and to use these properties in sketching the graphs of the functions.

THE TANGENT FUNCTION

The range of the tangent function is the set of all real numbers. To see this, consider the definition of the tangent function. Figure 3.52 reminds us that $\tan \theta = y/x$ and shows us that $\tan \theta$ can be interpreted as the slope of the line through the origin and the unit circle point P associated with θ. (By the slope formula, the slope of the line through the two points $(0, 0)$ and (x, y) is $(y - 0)/(x - 0) = y/x$.) The slope of a vertical line is not defined, and if θ is such that P is at N or at M (that is, if θ is any odd multiple of $\pi/2$ or of $90°$), then $\tan \theta$ is not defined. For example, $\tan \theta$ is not defined for $\theta = -\pi/2$ or for $\theta = \pi/2$. (The tangent function is not defined for any number θ where $\cos \theta = 0$.) Now let us think of what happens to the values of $\tan \theta$ (the slope of $\overline{OP}$) as θ increases continuously from $-\pi/2$ to 0 and then from 0 to $\pi/2$. As θ increases continuously from $-\pi/2$ to 0, P swings from N to Q, and the slope of $\overline{OP}$ ($\tan \theta$) increases continuously through all the *negative* real numbers to zero. Then as θ increases continuously from 0 up to $\pi/2$, P swings from Q up to M, and the slope of $\overline{OP}$ ($\tan \theta$) increases from zero through all the *positive* real numbers. It is true, then, that as θ varies from $-\pi/2$ to $\pi/2$, the values of $\tan \theta$ range over all the real numbers.

Range of tangent = set of all real numbers

The tangent function is periodic with period π. In other words,

$$\tan (\theta + \pi) = \tan \theta \qquad \text{Periodic with period } \pi$$

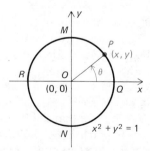

FIGURE 3.52 $\tan \theta = y/x$, for $x \neq 0$
$\tan \theta = $ Slope of $\overline{OP}$

for all θ in the domain of the tangent. We shall verify this in Section 3.6 when we develop a general formula for tan $(a + b)$. In the meantime, we can intuitively see the truth of this property by again considering the definition of the tangent function. In Figure 3.52 we see that as θ varies through the period π from $-\pi/2$ to $\pi/2$, the values of tan θ increase through all the negative real numbers to zero and then continue to increase through all the positive real numbers. Then as θ varies through another period π, from $\pi/2$ to $3\pi/2$, P swings from M through R to N, and the values of the slope of $\overline{OP}$ (tan θ) repeat the same pattern they made when θ varied from $-\pi/2$ to $\pi/2$. Every time θ varies continuously for 180° or π through values in the domain of the tangent and then varies continuously for another such 180° or π, the values of tan θ repeat themselves in the same pattern; the number π is a period of the tangent. Furthermore, π is the smallest positive number k for which tan $(\theta + k)$ = tan θ, so π is the *fundamental* period of the tangent.

The fundamental period of the tangent function is π.

For any θ in the domain of the tangent.

$$\tan (-\theta) = -\tan \theta.$$

This follows from the fact that

$$\tan (-\theta) = \frac{\sin (-\theta)}{\cos (-\theta)},$$

while sin $(-\theta) = -\sin \theta$ and cos $(-\theta) = \cos \theta$.

$$\tan (-\theta) = \frac{-\sin \theta}{\cos \theta} = -\tan \theta$$

Now let us sketch the graph of the tangent function. The graph of the tangent function is the graph of the equation $y = \tan x$. In Figure 3.53, the dotted vertical lines have been drawn at odd integral multiples of $\pi/2$ to remind us that the tangent function is not defined at such numbers, so no points of the graph are on these lines. These dotted lines are actually vertical asymptotes of the graph. If we plot the points in the table shown in Figure 3.53, together with intervening points, they all appear to be strung together like beads on the part of the graph shown between $-\pi/2$ and $\pi/2$. This same pattern is repeated for $\pi/2 < x < 3\pi/2$, for $-3\pi/2 < x < -\pi/2$, etc. The graph consists of an unlimited number of separate pieces each of which looks exactly like any other one. The graph displays the fact that the domain of the tangent is the set of all real numbers except odd integral multiples of $\pi/2$, the range is the set of all real numbers, the fundamental period is π, and tan $(-\theta)$ = $-\tan \theta$.

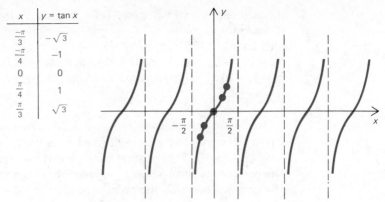

x	$y = \tan x$
$\dfrac{-\pi}{3}$	$-\sqrt{3}$
$\dfrac{-\pi}{4}$	-1
0	0
$\dfrac{\pi}{4}$	1
$\dfrac{\pi}{3}$	$\sqrt{3}$

Exercise 3.5, 1–5

FIGURE 3.53 Graph of $y = \tan x$
(Graph of the tangent function)

THE COTANGENT FUNCTION

The range of the cotangent function is also the set of all real numbers. This may be seen from its graph in Figure 3.54.

Range of cotangent = set of all real numbers

Again, like the tangent, the cotangent function is periodic with period π. That is,

$$\cot (\theta + \pi) = \cot \theta \qquad \text{Periodic with period } \pi$$

for all θ in the domain of the cotangent. In Figure 3.54 we see that as x varies through the period π from 0 to π, cot x decreases continuously

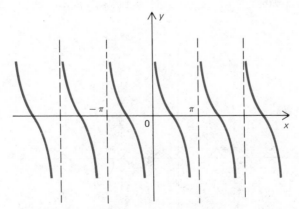

FIGURE 3.54 Graph of $y = \cot x$
(Graph of the cotangent function)

through all the real numbers. Then as x varies through another period π, for example from π to 2π, the values of cot x repeat the same pattern. This same pattern is repeated for $-\pi < x < 0$, for $-\pi < x < -2\pi$, for $2\pi < x < 3\pi$, etc. Since π is the smallest positive number k for which cot $(\theta + k) = $ cot θ, π is the *fundamental* period of the cotangent.

The fundamental period of the cotangent function is π.

The dotted vertical lines in Figure 3.54 drawn at integral multiples of π remind us that the cotangent function is not defined at any number θ where sin $\theta = 0$. These dotted lines are vertical asymptotes of the graph. The graph consists of an unlimited number of separate congruent pieces. The graph shows that the domain of the cotangent is the set of all real numbers except integral multiples of π, the range is the set of all real numbers, and the fundamental period is π.

Exercise 3.5, 6–10

THE COSECANT FUNCTION

The graph of the cosecant function can be readily deduced from the graph of the sine by using the fact that

$$\csc \theta = \frac{1}{\sin \theta}$$

for all values of θ for which csc θ is defined. The graph of $y = \sin x$ has been dotted in Figure 3.55 for reference. The vertical dotted lines drawn at numbers where sin $x = 0$ remind us that csc x is not defined at such numbers. These dotted lines are vertical asymptotes of the graph of the cosecant function.

The dotted horizontal lines $y = 1$ and $y = -1$ in Figure 3.55 remind us that $-1 \leq \sin x \leq 1$. Since csc x is the *reciprocal* of sin x, the values

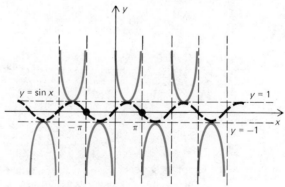

FIGURE 3.55 Graph of $y = \csc x$
(Graph of the cosecant function)

of csc x must be greater than or equal to 1 or less than or equal to -1; the graph of $y = \csc x$ must be above, or below, or just touching the guide lines $y = 1$ and $y = -1$. In other words,

$$\text{Range of cosecant} = \{y \mid y \geq 1 \text{ or } y \leq 1\}.$$

Furthermore, csc x is positive whenever sin x is positive, and csc x is negative whenever sin x is negative.

We can see in Figure 3.55 that the cosecant is periodic with fundamental period 2π. This is to be expected, because the fundamental period of the sine function is 2π. That is,

$$\csc (\theta + 2\pi) = \csc \theta \qquad \text{Periodic with period } 2\pi$$

for all θ in the domain of the cosecant.

Exercise 3.5, 11–15   The fundamental period of the cosecant function is 2π.

THE SECANT FUNCTION

The graph of the secant function is shown in Figure 3.56. Since

$$\sec \theta = \frac{1}{\cos \theta}$$

for all values of θ for which sec θ is defined, the graph of the secant can be deduced from the graph of the cosine, just as the graph of the cosecant was deduced from the graph of the sine. In Figure 3.56 we observe that

$$\text{Range of secant} = \{y \mid y \geq 1 \quad \text{or} \quad y \leq -1\}.$$

Like the cosine, the secant is periodic with fundamental period 2π. That is,

$$\sec (\theta + 2\pi) = \sec \theta \qquad \text{Periodic with period } 2\pi$$

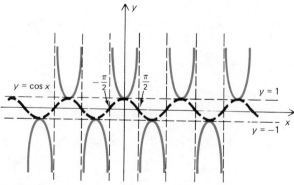

FIGURE 3.56 Graph of $y = \sec x$
(Graph of the secant function)

for all θ in the domain of the secant.

The fundamental period of the secant function is 2π.

The dotted vertical lines in Figure 3.56 remind us that sec x is not defined at values of x such that cos $x = 0$. These dotted vertical lines are vertical asymptotes of the graph of the secant function.

Compare Figures 3.55 and 3.56, and note that the graph of the secant looks exactly like the graph of the cosecant shifted $\pi/2$ units to the left. This is to be expected, because we noted in Section 3.4 that cos $x =$ sin $(\pi/2 + x)$, and the graph of the cosine looks like the graph of the sine shifted $\pi/2$ units to the left.

Exercise 3.5, 16–20
Miscellaneous
Exercise 3.5, 21–30

EXERCISE 3.5 PROPERTIES AND GRAPHS OF OTHER TRIGONOMETRIC FUNCTIONS

▶ Sketch the graph and give the range of the function defined by each of the following.

1. $y = \tan x$, $\dfrac{-\pi}{4} \leq x \leq \dfrac{\pi}{4}$ **2.** $y = \tan x$, $\dfrac{-\pi}{2} < x < \dfrac{\pi}{2}$

3. $y = \tan x$, $-\pi \leq x \leq \pi$ and $x \neq \dfrac{-\pi}{2}$ or $\dfrac{\pi}{2}$

4. $y = \tan x$, $\dfrac{\pi}{2} < x < \dfrac{3\pi}{2}$

5. $y = \tan x$, $0 \leq x \leq \pi$ and $x \neq \dfrac{\pi}{2}$

6. $y = \cot x$, $\dfrac{\pi}{4} \leq x \leq \dfrac{3\pi}{4}$ **7.** $y = \cot x$, $0 < x < \pi$

8. $y = \cot x$, $-\pi < x < \pi$ and $x \neq 0$

9. $y = \cot x$, $\dfrac{-\pi}{2} \leq x < 0$

10. $y = \cot x$, $0 < x \leq \dfrac{7\pi}{4}$ and $x \neq \pi$

11. $y = \csc x$, $\dfrac{\pi}{4} \leq x \leq \dfrac{3\pi}{4}$ **12.** $y = \csc x$, $0 < x < \pi$

13. $y = \csc x$, $\pi < x < 2\pi$

14. $y = \csc x$, $-\pi < x < 2\pi$ and $x \neq 0$ or π

15. $y = \csc x$, $\dfrac{\pi}{6} \leq x \leq \dfrac{3\pi}{2}$ and $x \neq \pi$

16. $y = \sec x$, $\dfrac{-\pi}{4} \leq x \leq \dfrac{\pi}{4}$ **17.** $y = \sec x$, $\dfrac{-\pi}{2} < x < \dfrac{\pi}{2}$

18. $y = \sec x$, $0 \le x \le \pi$ and $x \ne \dfrac{\pi}{2}$

19. $y = \sec x$, $\dfrac{-3\pi}{2} < x \le \dfrac{\pi}{3}$ and $x \ne \dfrac{-\pi}{2}$

20. $y = \sec x$, $0 \le x \le 2\pi$ and $x \ne \dfrac{\pi}{2}$ or $\dfrac{3\pi}{2}$

21. Prove that $\cot(-\theta) = -\cot\theta$.   **22.** Prove that $\sec(-\theta) = \sec\theta$.

23. Prove that $\csc(-\theta) = -\csc\theta$.

24. Sketch the graph of $y = \tan(-x)$

25. Sketch the graph of $y = \cot(-x)$

CHECK LIST FOR THE STUDENT

If you have learned the material in Section 3.5, you should be able to do the following:

1. State the range of the tangent function.

2. State the fundamental period of the tangent function.

3. Express $\tan(-\theta)$ in terms of $\tan\theta$.

4. Sketch and label intercepts of vertical asymptotes of the graph of the tangent function.

5. Sketch the graph of the tangent function.

6. State the range and fundamental period of the cotangent function.

7. Sketch and label intercepts of vertical asymptotes of the graph of the cotangent function.

8. Sketch the graph of the cotangent function.

9. Obtain the graph of the cosecant function from the graph of the sine.

10. State the range and fundamental period of the cosecant function.

11. Obtain the graph of the secant function from the graph of the cosine.

12. State the range and fundamental period of the secant function.

SECTION QUIZ 3.5

1. Express each of the following in terms of $\tan\theta$ or $\cot\theta$.
   (a) $\tan(-\theta)$ (b) $\tan(\theta + \pi)$ (c) $\cot(\theta + \pi)$

Sketch the graph and give the range of the function defined by each of the following.

2. $y = \tan x$, $\dfrac{-\pi}{2} < x < \dfrac{3\pi}{2}$ and $x \neq \dfrac{\pi}{2}$

3. $y = \cot x$, $0 < x < 2\pi$ and $x \neq \pi$

4. $y = \sec x$, $\dfrac{-\pi}{2} < x < \dfrac{3\pi}{2}$ and $x \neq \dfrac{\pi}{2}$

5. $y = \csc x$, $0 < x < 2\pi$ and $x \neq \pi$

3.6 Basic Identities

The purpose of this section is to present and apply some identities that are used very frequently in calculus.

An **identity** is an equation that is true for all values of the variable(s) for which its terms are defined. Several identities were noted in Section 3.2.

$$\tan \theta = \frac{\sin \theta}{\cos \theta} \qquad \cot \theta = \frac{\cos \theta}{\sin \theta}$$

$$\sec \theta = \frac{1}{\cos \theta} \qquad \csc \theta = \frac{1}{\sin \theta} \qquad \cot \theta = \frac{1}{\tan \theta}$$

These follow directly from the definitions of the trigonometric functions.

Three other identities noted in Sections 3.4 and 3.5 are

$$\cos(-\theta) = \cos \theta$$
$$\sin(-\theta) = -\sin \theta$$
$$\tan(-\theta) = -\tan \theta.$$

One of the most commonly used identities is

$$(\cos \theta)^2 + (\sin \theta)^2 = 1.$$

Using the notation $\cos^2 \theta$ for $(\cos \theta)^2$ and $\sin^2 \theta$ for $(\sin \theta)^2$, this is usually written

$$\cos^2 \theta + \sin^2 \theta = 1.$$

This identity follows directly from the fact that $\cos \theta$ and $\sin \theta$ are defined respectively as the *x*- and *y*-coordinates of a point on the unit circle (see Figure 3.38). The identity is simply the distance-formula assertion that the distance from the point $(\cos \theta, \sin \theta)$ to the origin is 1.

Another very commonly used identity is

$$1 + \tan^2 \theta = \sec^2 \theta.$$

The notation $\tan^2 \theta$ is used for $(\tan \theta)^2$; the notation $\sec^2 \theta$ is used for $(\sec \theta)^2$. This identity can be obtained from $\cos^2 \theta + \sin^2 \theta = 1$ by first dividing both sides through by $\cos^2 \theta$ and then using the fact that $(\sin \theta)/(\cos \theta) = \tan \theta$ and $1/(\cos \theta) = \sec \theta$.

$$\cos^2 \theta + \sin^2 \theta = 1$$

$$1 + \frac{\sin^2 \theta}{\cos^2 \theta} = \frac{1}{\cos^2 \theta}$$

$$1 + \left(\frac{\sin \theta}{\cos \theta}\right)^2 = \left(\frac{1}{\cos \theta}\right)^2$$

$$1 + \tan^2 \theta = \sec^2 \theta$$

Still another commonly used identity,

$$\cot^2 \theta + 1 = \csc^2 \theta,$$

is obtained from $\cos^2 \theta + \sin^2 \theta = 1$ by first dividing both sides through by $\sin^2 \theta$. The notation $\cot^2 \theta$ is used for $(\cot \theta)^2$; the notation $\csc^2 \theta$ is used for $(\csc \theta)^2$.

$$\cos^2 \theta + \sin^2 \theta = 1$$

$$\frac{\cos^2 \theta}{\sin^2 \theta} + 1 = \frac{1}{\sin^2 \theta}$$

$$\left(\frac{\cos \theta}{\sin \theta}\right)^2 + 1 = \left(\frac{1}{\sin \theta}\right)^2$$

$$\cot^2 \theta + 1 = \csc^2 \theta$$

EXAMPLE 1

If $\sin \theta = \frac{3}{5}$ and $\tan \theta < 0$, use basic identities to find the values of the other trigonometric functions of θ.

Solution

$$\cos^2 \theta + \sin^2 \theta = 1$$

$$\cos^2 \theta = 1 - \sin^2 \theta$$

If $\sin \theta = \frac{3}{5}$,

$$\cos^2 \theta = 1 - \frac{9}{25}$$

$$= \frac{16}{25}.$$

Since $\sin \theta$ is positive and $\tan \theta$ is negative, θ must be in the second quadrant, and so $\cos \theta$ must be negative.

$$\cos \theta = -\frac{4}{5}$$

$$\tan \theta = \frac{\sin \theta}{\cos \theta} \qquad\qquad \cot \theta = \frac{1}{\tan \theta}$$

$$= \frac{3}{5} \div -\frac{4}{5} \qquad\qquad = -\frac{4}{3}$$

$$= -\frac{3}{4}$$

$$\sec \theta = \frac{1}{\cos \theta} \qquad\qquad \csc \theta = \frac{1}{\sin \theta}$$

Exercise 3.6, 1–10

$$= -\frac{5}{4} \qquad\qquad\qquad = \frac{5}{3}$$

The so-called **addition formulas** are very useful identities.

$$\begin{cases} \cos (a + b) = \cos a \cos b - \sin a \sin b \\ \cos (a - b) = \cos a \cos b + \sin a \sin b \end{cases}$$

$$\begin{cases} \sin (a + b) = \sin a \cos b + \cos a \sin b \\ \sin (a - b) = \sin a \cos b - \cos a \sin b \end{cases}$$

$$\begin{cases} \tan(a + b) = \dfrac{\tan a + \tan b}{1 - \tan a \tan b} \\[3mm] \tan(a - b) = \dfrac{\tan a - \tan b}{1 + \tan a \tan b} \end{cases}$$

These are true for all values of a and b for which their terms are defined. The three pairs of addition formulas can be condensed and written as three single formulas by using the double signs, $\pm$ or $\mp$, with the understanding that choice of all the upper signs in an equation produces an identity and that choice of all the lower signs produces an identity.

Addition formulas

$$\cos(a \pm b) = \cos a \cos b \mp \sin a \sin b$$

$$\sin(a \pm b) = \sin a \cos b \pm \cos a \sin b$$

$$\tan(a \pm b) = \frac{\tan a \pm \tan b}{1 \mp \tan a \tan b}$$

Let us derive the formula for $\cos(a - b)$ for the case in which a, b, and $a - b$ are all positive and less than 2π. Let P, Q, and R be the points on the unit circle associated with a, b, and $a - b$, respectively. One possible arrangement of these points is shown in Figure 3.57. The coordinates of P are $(\cos a, \sin a)$, the coordinates of Q are $(\cos b, \sin b)$, and the coordinates of R are $(\cos(a - b), \sin(a - b))$. In the figure, arc SP is of length a, arc SQ is of length b, arc QP is of length $a - b$, and arc SR is also of length $a - b$. Since arc SR is the same length as arc QP, chord SR is the same length as chord QP.

$$\overline{SR} = \overline{QP}$$

By the distance formula, we then have

$$\sqrt{[\cos(a - b) - 1]^2 + [\sin(a - b) - 0]^2}$$

$$= \sqrt{(\cos a - \cos b)^2 + (\sin a - \sin b)^2}.$$

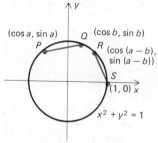

FIGURE 3.57 $\overline{SR} = \overline{QP}$

Squaring both sides and performing the operations indicated under the radical signs, we obtain

$$\cos^2 (a - b) - 2 \cos (a - b) + 1 + \sin^2 (a - b)$$
$$= \cos^2 a - 2 \cos a \cos b + \cos^2 b + \sin^2 a - 2 \sin a \sin b + \sin^2 b.$$

Now, using the identities $\cos^2 (a - b) + \sin^2 (a - b) = 1$, $\cos^2 a + \sin^2 a = 1$, and $\cos^2 b + \sin^2 b = 1$, we have

$$2 - 2 \cos (a - b) = 2 - 2 \cos a \cos b - 2 \sin a \sin b.$$

Simplifying, we finally have

$$\cos (a - b) = \cos a \cos b + \sin a \sin b$$

This is the addition formula for $\cos (a - b)$. Using similar methods, it can be derived for all real numbers or all angles a and b.

EXAMPLE 2

Use the formula for $\cos (a - b)$ to show that $\sin \theta = \cos (\pi/2 - \theta)$.

Solution

$$\cos \left(\frac{\pi}{2} - \theta \right) = \cos \frac{\pi}{2} \cos \theta + \sin \frac{\pi}{2} \sin \theta$$

But $\cos \pi/2 = 0$ and $\sin \pi/2 = 1$, so

$$\cos \left(\frac{\pi}{2} - \theta \right) = 0 \times \cos \theta + 1 \times \sin \theta$$
$$= \sin \theta.$$

EXAMPLE 3

Use Example 2 to show that $\sin (\pi/2 - \theta) = \cos \theta$.

Solution

$$\sin \left(\frac{\pi}{2} - \theta \right) = \cos \left(\frac{\pi}{2} - \left(\frac{\pi}{2} - \theta \right) \right) \qquad \text{By Example 2}$$
$$= \cos \left(\frac{\pi}{2} - \frac{\pi}{2} + \theta \right)$$
$$= \cos \theta$$

To derive the addition formula for $\sin (a + b)$, we use Example 2, rewrite, use the formula for the cosine of a difference, and then use Examples 2 and 3.

$$\sin (a + b) = \cos \left(\frac{\pi}{2} - (a + b) \right) \qquad \text{By Example 2}$$
$$= \cos \left(\left(\frac{\pi}{2} - a \right) - b \right)$$
$$= \cos \left(\frac{\pi}{2} - a \right) \cos b + \sin \left(\frac{\pi}{2} - a \right) \sin b$$

But by Example 2, $\cos(\pi/2 - a) = \sin a$, and by Example 3, $\sin(\pi/2 - a)$ $= \cos a$, so

$$\sin(a + b) = \sin a \cos b + \cos a \sin b.$$

The addition formula for $\sin(a - b)$ can be obtained from the formula for $\sin(a + b)$ by first writing $\sin(a - b)$ as $\sin(a + (-b))$.

$$\sin(a - b) = \sin(a + (-b))$$
$$= \sin a \cos(-b) + \cos a \sin(-b)$$

But $\cos(-b) = \cos b$, and $\sin(-b) = -\sin b$, so

$$\sin(a - b) = \sin a \cos b - \cos a \sin b.$$

EXAMPLE 4

Use the formula for $\sin(a + b)$ to show that $\sin(\pi/2 + \theta) = \cos\theta$.

Solution

$$\sin\left(\frac{\pi}{2} + \theta\right) = \sin\frac{\pi}{2}\cos\theta + \cos\frac{\pi}{2}\sin\theta$$

But $\sin\pi/2 = 1$ and $\cos\pi/2 = 0$, so

$$\sin\left(\frac{\pi}{2} + \theta\right) = 1 \times \cos\theta + 0 \times \sin\theta$$
$$= \cos\theta.$$

EXAMPLE 5

Use the addition formulas to prove that the cosine and sine functions are periodic with period 2π.

Solution

We want to prove that $\sin(\theta + 2\pi) = \sin\theta$ and $\cos(\theta + 2\pi) = \cos\theta$. Using the addition formulas, we have

$$\sin(\theta + 2\pi) = \sin\theta\cos 2\pi + \cos\theta\sin 2\pi$$
$$= \sin\theta \times 1 \quad + \cos\theta \times 0$$
$$= \sin\theta$$

$$\cos(\theta + 2\pi) = \cos\theta\cos 2\pi - \sin\theta\sin 2\pi$$
$$= \cos\theta \times 1 \quad - \sin\theta \times 0$$
$$= \cos\theta$$

The addition formulas for $\tan(a \pm b)$ can be obtained from the formulas for $\sin(a \pm b)$ and $\cos(a \pm b)$.

$$\tan(a \pm b) = \frac{\sin(a \pm b)}{\cos(a \pm b)}$$

$$= \frac{\sin a \cos b \pm \cos a \sin b}{\cos a \cos b \mp \sin a \sin b}$$

Dividing numerator and denominator through by cos a cos b, we have

$$\tan(a \pm b) = \frac{\dfrac{\sin a \cancel{\cos b}}{\cos a \cancel{\cos b}} \pm \dfrac{\cancel{\cos a} \sin b}{\cancel{\cos a} \cos b}}{1 \mp \dfrac{\sin a \sin b}{\cos a \cos b}}$$

$$= \frac{\tan a \pm \tan b}{1 \mp \tan a \tan b}$$

EXAMPLE 6

Use the addition formula for the tangent to prove that the tangent function is periodic with period π.

Solution

We want to prove that $\tan(\theta + \pi) = \tan \theta$.

$$\tan(\theta + \pi) = \frac{\tan \theta + \tan \pi}{1 - \tan \theta \tan \pi}$$

But $\tan \pi = 0$, so

$$\tan(\theta + \pi) = \frac{\tan \theta}{1}$$

$$= \tan \theta.$$

EXAMPLE 7

Show that the cotangent function is periodic with period π.

Solution

We want to show that $\cot(\theta + \pi) = \cot \theta$.

$$\cot(\theta + \pi) = \frac{1}{\tan(\theta + \pi)}$$

$$= \frac{1}{\tan \theta} \qquad \text{By Example 6}$$

Exercise 3.6, 11–20

$$= \cot \theta$$

Identities known as the *double-angle formulas* can be derived from the addition formulas. The **double-angle formulas** are

$$\sin 2\theta = 2 \sin \theta \cos \theta$$
$$\cos 2\theta = \cos^2 \theta - \sin^2 \theta \qquad \text{Double-angle formulas}$$
$$\tan 2\theta = \frac{2 \tan \theta}{1 - \tan^2 \theta}.$$

They follow directly from the addition formulas.

$$\sin 2\theta = \sin(\theta + \theta)$$
$$= \sin \theta \cos \theta + \cos \theta \sin \theta$$
$$= 2 \sin \theta \cos \theta$$

$$\cos 2\theta = \cos (\theta + \theta)$$
$$= \cos \theta \cos \theta - \sin \theta \sin \theta$$
$$= \cos^2 \theta - \sin^2 \theta$$
$$\tan 2\theta = \tan (\theta + \theta)$$
$$= \frac{\tan \theta + \tan \theta}{1 - \tan \theta \tan \theta}$$
$$= \frac{2 \tan \theta}{1 - \tan^2 \theta}$$

We have seen that a formula for $\cos 2\theta$ is

$$\cos 2\theta = \cos^2 \theta - \sin^2 \theta.$$

Two alternative formulas for $\cos 2\theta$ are

$$\cos 2\theta = 1 - 2 \sin^2 \theta$$

and
$$\cos 2\theta = 2 \cos^2 \theta - 1.$$

These alternative formulas for $\cos 2\theta$ can be obtained from the first by using the identity $\cos^2 \theta + \sin^2 \theta = 1$. In particular, using $\cos^2 \theta = 1 - \sin^2 \theta$, we have

$$\cos 2\theta = \cos^2 \theta - \sin^2 \theta$$
$$= (1 - \sin^2 \theta) - \sin^2 \theta$$
$$= 1 - 2 \sin^2 \theta.$$

Or using $\sin^2 \theta = 1 - \cos^2 \theta$, we have

$$\cos 2\theta = \cos^2 \theta - \sin^2 \theta$$
$$= \cos^2 \theta - (1 - \cos^2 \theta)$$
$$= \cos^2 \theta - 1 + \cos^2 \theta$$
$$= 2 \cos^2 \theta - 1.$$

Exercise 3.6, 21–25

It is sometimes necessary to express $\sin^2 \theta$ or $\cos^2 \theta$ in terms of $\cos 2\theta$. This can be done by solving the two alternative formulas for $\cos 2\theta$ for $\sin^2 \theta$ or $\cos^2 \theta$, respectively.

$$\cos 2\theta = 1 - 2 \sin^2 \theta \qquad \cos 2\theta = 2 \cos^2 \theta - 1$$
$$2 \sin^2 \theta = 1 - \cos 2\theta \qquad 2 \cos^2 \theta = 1 + \cos 2\theta$$
$$\sin^2 \theta = \frac{1 - \cos 2\theta}{2} \qquad \cos^2 \theta = \frac{1 + \cos 2\theta}{2}$$

If we let $2\theta = u$ so that $\theta = u/2$, we can write these last two identities in forms called the **half-angle formulas**.

$$\sin^2 \frac{u}{2} = \frac{1 - \cos u}{2} \qquad \cos^2 \frac{u}{2} = \frac{1 + \cos u}{2}$$

These identities are true for all values of u. If the symbol θ is used in place of u, the half-angle formulas are written

$$\sin^2 \frac{\theta}{2} = \frac{1 - \cos \theta}{2} \qquad \cos^2 \frac{\theta}{2} = \frac{1 + \cos \theta}{2}.$$

EXAMPLE 8

Use the half-angle formulas to find $\sin 112.5°$ and $\cos 112.5°$.

Solution

By the half-angle formulas,

$$\sin^2 112.5° = \frac{1 - \cos 225°}{2}, \text{ and}$$

$$\cos^2 112.5° = \frac{1 + \cos 225°}{2}.$$

The angle $225°$ is in the third quadrant with reference angle $45°$. Since $\cos \theta$ is negative in the third quadrant,

$$\cos 225° = -\cos 45°$$

$$= -\frac{\sqrt{2}}{2}.$$

It follows that

$$\sin^2 112.5° = \frac{1 - \left(\dfrac{-\sqrt{2}}{2}\right)}{2}$$

and

$$\cos^2 112.5° = \frac{1 + \left(\dfrac{-\sqrt{2}}{2}\right)}{2}.$$

Simplifying, we have

$$\sin^2 112.5° = \frac{2 + \sqrt{2}}{4}$$

and

$$\cos^2 112.5° = \frac{2 - \sqrt{2}}{4}.$$

In solving these equations for $\sin 112.5°$ and $\cos 112.5°$, we must note that $112.5°$ is in the second quadrant, so that $\sin 112.5°$ is positive and $\cos 112.5°$ is negative.

$$\sin 112.5° = +\sqrt{\frac{2 + \sqrt{2}}{4}} \qquad \cos 112.5° = -\sqrt{\frac{2 - \sqrt{2}}{4}}$$

$$= \frac{\sqrt{2 + \sqrt{2}}}{2} \qquad\qquad = -\frac{\sqrt{2 - \sqrt{2}}}{2}$$

A half-angle formula for the tangent can be deduced from the half-angle formulas for the sine and cosine.

$$\tan^2 \frac{\theta}{2} = \frac{\sin^2 \frac{\theta}{2}}{\cos^2 \frac{\theta}{2}}$$

Exercise 3.6, 26–50

$$\tan^2 \frac{\theta}{2} = \frac{1 - \cos \theta}{1 + \cos \theta}$$

The basic identities we have considered in this section may be used to prove that other given equations are identities. One way to do this is to choose one side of the given equation and utilize the basic identities to rewrite or transform the chosen side until it is exactly the same as the other side. This method is demonstrated in the next example.

In proving an equation is an identity, we must bear in mind that the basic identities are valid only for all those values of the variable(s) for which the various functions and terms are defined.

EXAMPLE 9

Prove that the following equation is an identity by transforming the left side to the right side.

$$\tan \theta + \cot \theta = \sec \theta \csc \theta$$

Solution

$$\tan \theta + \cot \theta = \frac{\sin \theta}{\cos \theta} + \frac{\cos \theta}{\sin \theta}$$

$$= \frac{\sin^2 \theta + \cos^2 \theta}{\cos \theta \sin \theta}$$

$$= \frac{1}{\cos \theta \sin \theta}$$

$$= \frac{1}{\cos \theta} \cdot \frac{1}{\sin \theta}$$

$$= \sec \theta \csc \theta$$

The given equation is an identity. It is true for all values of θ for which the tangent, cotangent, secant, and cosecant are defined.

Another way to prove that a given equation is an identity is to work independently with the two sides and transform each of them to the same expression. The identity of Example 9 is proved in this manner in the next example.

EXAMPLE 10

Prove that the following equation is an identity by working independently with the two sides.

$$\tan \theta + \cot \theta = \sec \theta \csc \theta$$

We shall work with the two sides independently, expressing each in terms of sines and cosines.

Left Side $\tan \theta + \cot \theta = \dfrac{\sin \theta}{\cos \theta} + \dfrac{\cos \theta}{\sin \theta}$

$$= \dfrac{\sin^2 \theta + \cos^2 \theta}{\cos \theta \sin \theta}$$

$$= \dfrac{1}{\cos \theta \sin \theta}$$

Right Side $\sec \theta \csc \theta = \dfrac{1}{\cos \theta} \cdot \dfrac{1}{\sin \theta}$

$$= \dfrac{1}{\cos \theta \sin \theta}$$

Since the left side and the right side are each equal to the same expression, $1 / (\cos \theta \sin \theta)$, we conclude that the left side and the right side are equal to each other.

Sometimes the easiest way to prove that a given equation is an identity is to express each side independently in terms of sines and/or cosines, as we did in Example 10. Sometimes we may use a bit of ingenuity in addition to the basic identities to prove that a given equation is an identity. This is illustrated in the next example.

EXAMPLE 11

Prove that the following equation is an identity.

$$\frac{1 - \cos \theta}{\sin \theta} = \frac{\sin \theta}{1 + \cos \theta}$$

Solution

We choose to work with the left side, and we begin to transform it by multiplying numerator and denominator by $1 + \cos \theta$.

$$\frac{1 - \cos \theta}{\sin \theta} = \frac{(1 - \cos \theta)(1 + \cos \theta)}{\sin \theta (1 + \cos \theta)}$$

$$= \frac{1 - \cos^2 \theta}{\sin \theta (1 + \cos \theta)}$$

$$= \frac{\sin^2 \theta}{\sin \theta (1 + \cos \theta)} \qquad \begin{aligned} \sin^2 \theta + \cos^2 \theta &= 1 \\ \sin^2 \theta &= 1 - \cos^2 \theta \end{aligned}$$

Exercise 3.6, 51–65

$$= \frac{\sin \theta}{1 + \cos \theta}$$

Developing the ability to manipulate readily with basic trigonometric identities is essential preparation for calculus. The next example illustrates the way identities are sometimes used in calculus to replace algebraic expressions involving radicals with trigonometric expressions not involving radicals.

EXAMPLE 12

Let $\tan \theta = x/3$, and let θ be in the first quadrant. Express $\sqrt{9 + x^2}$ in terms of θ, and write it in a form without a radical sign.

Solution

$$x = 3 \tan \theta$$
$$\sqrt{9 + x^2} = \sqrt{9 + 9 \tan^2 \theta}$$
$$= 3\sqrt{1 + \tan^2 \theta}$$
$$= 3\sqrt{\sec^2 \theta}$$
$$= 3 \sec \theta$$

In Example 12, the radical $\sqrt{9 + x^2}$ was replaced by a nonradical expression by making the substitution $x = 3 \tan \theta$. Any radical of the form $\sqrt{a^2 + x^2}$ where a is a positive real number can be replaced with a nonradical expression by making the substitution $x = a \tan \theta$ with $0 < \theta < \pi/2$.

$$\text{If } x = a \tan \theta, \quad \sqrt{a^2 + x^2} = \sqrt{a^2 + a^2 \tan^2 \theta}$$
$$= a\sqrt{1 + \tan^2 \theta}$$
$$= a\sqrt{\sec^2 \theta}$$
$$= a \sec \theta.$$

Likewise, any radical of the form $\sqrt{a^2 - x^2}$ where a is positive can be replaced by a nonradical by making the substitution $x = a \cos \theta$ or the substitution $x = a \sin \theta$ with $0 < \theta < \pi/2$.

$$\text{If } x = a \cos \theta, \quad \sqrt{a^2 - x^2} = \sqrt{a^2 - a^2 \cos^2 \theta}$$
$$= a\sqrt{1 - \cos^2 \theta}$$
$$= a\sqrt{\sin^2 \theta}$$
$$= a \sin \theta.$$

$$\text{If } x = a \sin \theta, \quad \sqrt{a^2 - x^2} = \sqrt{a^2 - a^2 \sin^2 \theta}$$
$$= a\sqrt{1 - \sin^2 \theta}$$
$$= a\sqrt{\cos^2 \theta}$$
$$= a \cos \theta.$$

Also, any radical of the form $\sqrt{x^2 - a^2}$ where a is positive can be replaced by a nonradical by making the substitution $x = a \sec \theta$ with $0 < \theta < \pi/2$.

$$\text{If } x = a \sec \theta, \quad \sqrt{x^2 - a^2} = \sqrt{a^2 \sec^2 \theta - a^2}$$
$$= a\sqrt{\sec^2 \theta - 1}$$
$$= a\sqrt{\tan^2 \theta}$$
$$= a \tan \theta.$$

Substitutions such as these made to replace radicals of the form $\sqrt{a^2 + x^2}$, $\sqrt{a^2 - x^2}$, or $\sqrt{x^2 - a^2}$ by $a \sec \theta$, $a \sin \theta$, $a \cos \theta$, or $a \tan \theta$ form the basis of the integral calculus technique called trigonometric substitution.

Exercise 3.6, 66–70

In calculus it is sometimes necessary to work with fractions such as

$$\frac{\sin^3 \theta \cos \theta}{1 + \cos \theta},$$

whose numerators and denominators are products, sums, or differences of positive integral powers of $\sin \theta$ and $\cos \theta$. It is sometimes necessary to replace such a fraction by a rational expression whose numerator and denominator are polynomials in a single variable, say the variable u. Such a replacement can be effected by making the substitution

$$u = \tan \frac{\theta}{2}.$$

This is demonstrated in the next example.

EXAMPLE 13

Replace

$$\frac{\sin^3 \theta \cos \theta}{1 + \cos \theta}$$

by a rational expression whose numerator and denominator are polynomials in a single variable.

Solution

Let $u = \tan \theta/2$. Our main problem is to express $\sin \theta/2$ and $\cos \theta/2$ and then $\sin \theta$ and $\cos \theta$ in terms of u. For reference, Figure 3.58 shows a right triangle with acute angle $\theta/2$. Since $\tan \theta/2 = u/1$, the side opposite $\theta/2$ is labeled u and the side adjacent $\theta/2$ is labeled 1.

From the reference triangle, we read

$$\sin \frac{\theta}{2} = \frac{u}{\sqrt{1 + u^2}} \quad \text{and} \quad \cos \frac{\theta}{2} = \frac{1}{\sqrt{1 + u^2}}.$$

Now, to find $\sin \theta$ and $\cos \theta$ in terms of u, we need only find $\sin \theta$ and

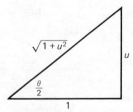

FIGURE 3.58 $\tan (\theta/2) = u/1$

cos θ in terms of sin $\theta/2$ and cos $\theta/2$. Since θ is *twice* $\theta/2$, we can use the double-angle formulas and write

$$\sin \theta = 2 \sin \frac{\theta}{2} \cos \frac{\theta}{2}$$

$$\cos \theta = \cos^2 \frac{\theta}{2} - \sin^2 \frac{\theta}{2}.$$

Substituting for sin $\theta/2$ and cos $\theta/2$, we have

$$\sin \theta = 2\left(\frac{u}{\sqrt{1 + u^2}}\right)\left(\frac{1}{\sqrt{1 + u^2}}\right)$$

$$= \frac{2u}{1 + u^2}$$

and

$$\cos \theta = \left(\frac{1}{\sqrt{1 + u^2}}\right)^2 - \left(\frac{u}{\sqrt{1 + u^2}}\right)^2$$

$$= \frac{1}{1 + u^2} - \frac{u^2}{1 + u^2}$$

$$= \frac{1 - u^2}{1 + u^2}.$$

It follows that

$$\frac{\sin^3 \theta \cos \theta}{1 + \cos \theta} = \frac{\left(\dfrac{2u}{1 + u^2}\right)^3\left(\dfrac{1 - u^2}{1 + u^2}\right)}{1 + \dfrac{1 - u^2}{1 + u^2}}$$

$$= \frac{8u^3(1 - u^2)}{(1 + u^2)^4} \times \frac{(1 + u^2)}{2}$$

$$= \frac{4u^3 - 4u^5}{(1 + u^2)^3}.$$

This illustrates the type of manipulations that are sometimes necessary in calculus.

Exercise 3.6, 71–75

LIST OF BASIC IDENTITIES

The basic identities presented in this section are listed here for reference.

$$\tan \theta = \frac{\sin \theta}{\cos \theta} \qquad\qquad \cot \theta = \frac{\cos \theta}{\sin \theta}$$

$$\sec \theta = \frac{1}{\cos \theta} \qquad\qquad \csc \theta = \frac{1}{\sin \theta}$$

$$\cot \theta = \frac{1}{\tan \theta}$$

$$\cos (-\theta) = \cos \theta \qquad\qquad \cos^2 \theta + \sin^2 \theta = 1$$
$$\sin (-\theta) = -\sin \theta \qquad\qquad 1 + \tan^2 \theta = \sec^2 \theta$$
$$\tan (-\theta) = -\tan \theta \qquad\qquad \cot^2 \theta + \quad 1 \quad = \csc^2 \theta$$

$$\left.\begin{array}{l} \cos (a \pm b) = \cos a \cos b \mp \sin a \sin b \\[4pt] \sin (a \pm b) = \sin a \cos b \pm \cos a \sin b \\[4pt] \tan (a \pm b) = \dfrac{\tan a \pm \tan b}{1 \mp \tan a \tan b} \end{array}\right\} \text{ Addition formulas}$$

$$\left.\begin{array}{l} \sin 2\theta = 2 \sin \theta \cos \theta \\[4pt] \cos 2\theta = \cos^2 \theta - \sin^2 \theta \\[4pt] \qquad = 1 - 2 \sin^2 \theta \\[4pt] \qquad = 2 \cos^2 \theta - 1 \end{array}\right\} \text{ Double-angle formulas}$$

$$\sin^2 \theta = \frac{1 - \cos 2\theta}{2}$$

$$\cos^2 \theta = \frac{1 + \cos 2\theta}{2}$$

$$\left.\begin{array}{l} \sin^2 \dfrac{\theta}{2} = \dfrac{1 - \cos \theta}{2} \\[10pt] \cos^2 \dfrac{\theta}{2} = \dfrac{1 + \cos \theta}{2} \\[10pt] \tan^2 \dfrac{\theta}{2} = \dfrac{1 - \cos \theta}{1 + \cos \theta} \end{array}\right\} \text{ Half-angle formulas}$$

EXERCISE 3.6 BASIC IDENTITIES

▶ In each of the following, use the given information and the basic identities to find the values of the other trigonometric functions of θ.

EXAMPLE

$\sec \theta = \frac{13}{12}$, $\sin \theta < 0$.

Solution

Since $\sin \theta < 0$ and $\sec \theta > 0$, θ is in the fourth quadrant and $\tan \theta < 0$.

$$\cos \theta = \tfrac{12}{13} \qquad \tan \theta = -\sqrt{\tfrac{169}{144} - 1} = -\tfrac{5}{12} \qquad \cot \theta = -\tfrac{12}{5}$$
$$\sin \theta = -\sqrt{1 - \tfrac{144}{169}} = -\tfrac{5}{13} \qquad \csc \theta = -\tfrac{13}{5}$$

1. $\cos\theta = \frac{-3}{5}$, $\tan\theta > 0$ $\qquad$ **2.** $\cot\theta = \frac{4}{3}$, $\sin\theta < 0$

3. $\sec\theta = \frac{-5}{3}$, $\sin\theta > 0$ $\qquad$ **4.** $\sin\theta = \frac{1}{7}$, $\cot\theta < 0$

5. $\tan\theta = \frac{-8}{15}$, $\sin\theta > 0$ $\qquad$ **6.** $\cos\theta = \frac{1}{3}$, $\csc\theta < 0$

7. $\sin\theta = \frac{-15}{17}$, $\cos\theta > 0$ $\qquad$ **8.** $\sec\theta = \frac{3}{2}$, $\sin\theta > 0$

9. $\cot\theta = \frac{-5}{12}$, $\cos\theta < 0$ $\qquad$ **10.** $\csc\theta = \frac{13}{2}$, $\cos\theta > 0$

▶ Use the addition formulas to find the value of each of the following.

EXAMPLE

$$\cos\left(\frac{\pi}{4} + \frac{\pi}{3}\right) = \cos\frac{\pi}{4}\cos\frac{\pi}{3} - \sin\frac{\pi}{4}\sin\frac{\pi}{3}$$

$$= \left(\frac{\sqrt{2}}{2}\right)\left(\frac{1}{2}\right) - \left(\frac{\sqrt{2}}{2}\right)\left(\frac{\sqrt{3}}{2}\right)$$

$$= \frac{\sqrt{2} - \sqrt{6}}{4}$$

11. $\sin\left(\frac{\pi}{4} + \frac{\pi}{3}\right)$ $\qquad$ **12.** $\sin\left(\frac{\pi}{4} - \frac{\pi}{3}\right)$ $\qquad$ **13.** $\cos\left(\frac{\pi}{4} - \frac{\pi}{3}\right)$

14. $\sin\left(\frac{\pi}{2} + \frac{\pi}{6}\right)$ $\qquad$ **15.** $\tan\left(\frac{\pi}{4} - \frac{\pi}{6}\right)$ $\qquad$ **16.** $\sin\,(30° + 45°)$

17. $\tan\,(30° + 45°)$ $\qquad$ **18.** $\tan\,(60° - 45°)$ $\qquad$ **19.** $\cos\,(45° - 30°)$

20. $\sin\,(90° + 30°)$

▶ In each of the following, use the given information and basic identities to find $\sin 2\theta$, $\cos 2\theta$, and $\tan 2\theta$.

EXAMPLE

$$\tan\theta = \frac{3}{4}, \quad \pi < \theta < \frac{3\pi}{2}$$

Solution

$$\sec^2\theta = 1 + \frac{9}{16} = \frac{25}{16} \qquad\qquad \sin 2\theta = 2\left(\frac{-3}{5}\right)\left(\frac{-4}{5}\right) = \frac{24}{25}$$

$$\cos^2\theta = \frac{16}{25} \qquad\qquad\qquad \cos 2\theta = \left(\frac{-4}{5}\right)^2 - \left(\frac{-3}{5}\right)^2$$

$$\pi < \theta < \frac{3\pi}{2}; \quad \cos\theta < 0, \sin\theta < 0 \qquad\qquad = \frac{16}{25} - \frac{9}{25} = \frac{7}{25}$$

$$\cos\theta = \frac{-4}{5}, \quad \sin\theta = -\sqrt{1 - \frac{16}{25}} = \frac{-3}{5} \quad \tan 2\theta = \frac{24}{25} \div \frac{7}{25} = \frac{24}{7}$$

21. $\sin\theta = \frac{3}{5}$, $\frac{\pi}{2} < \theta < \pi$ $\qquad\qquad$ **22.** $\cos\theta = \frac{12}{13}$, $\sin\theta < 0$

23. $\tan \theta = \dfrac{-8}{15}$, $\dfrac{3\pi}{2} < \theta < 2\pi$ **24.** $\sin \theta = \dfrac{3}{5}$, $\cos \theta > 0$

25. $\tan \theta = \dfrac{5}{12}$, $\sin \theta < 0$

▶ Use the information given in each of the following to find $\sin \theta$ and $\cos \theta$.

EXAMPLE

$\cos 2\theta = \dfrac{-15}{17}$, $\pi < 2\theta < \dfrac{3\pi}{2}$

Solution

$\dfrac{\pi}{2} < \theta < \dfrac{3\pi}{4}$ implies that $\cos \theta < 0$ and $\sin \theta > 0$.

$$\sin \theta = \sqrt{\dfrac{1 - \left(\dfrac{-15}{17}\right)}{2}} \qquad \cos \theta = -\sqrt{\dfrac{1 + \left(\dfrac{-15}{17}\right)}{2}}$$

$$= \sqrt{\dfrac{32}{34}} \qquad\qquad\qquad = -\sqrt{\dfrac{2}{34}}$$

26. $\cos 2\theta = \dfrac{12}{13}$, $0 < \theta < \dfrac{\pi}{4}$ **27.** $\cos 2\theta = \dfrac{-3}{5}$, $\cos \theta < 0$

28. $\cos 2\theta = \dfrac{4}{5}$, $\cos \theta < 0$ **29.** $\cos 2\theta = \dfrac{1}{2}$, $\dfrac{3\pi}{2} < 2\theta < 2\pi$

30. $\cos 2\theta = 1$, $\cos \theta < 0$

▶ Use the information given in each of the following to find $\sin \theta/2$, $\cos \theta/2$, and $\tan \theta/2$.

EXAMPLE

$\tan \theta = \dfrac{-4}{3}$, $\dfrac{3\pi}{2} < \theta < 2\pi$

Solution

$\sec^2 \theta = 1 + \dfrac{16}{9} = \dfrac{25}{9}$; $\cos^2 \theta = \dfrac{9}{25}$

$\dfrac{3\pi}{2} < \theta < 2\pi$ implies $\cos \theta > 0$, $\cos \theta = \dfrac{3}{5}$

$\dfrac{3\pi}{4} < \dfrac{\theta}{2} < \pi$ implies $\sin \dfrac{\theta}{2} > 0$, $\cos \dfrac{\theta}{2} < 0$.

$$\sin \frac{\theta}{2} = \sqrt{\frac{1 - \frac{3}{5}}{2}} \qquad\qquad \cos \frac{\theta}{2} = -\sqrt{\frac{1 + \frac{3}{5}}{2}} \qquad\qquad \tan \frac{\theta}{2} = \frac{\sin \dfrac{\theta}{2}}{\cos \dfrac{\theta}{2}}$$

$$= \sqrt{\frac{2}{10}} = \sqrt{\frac{1}{5}} = \frac{1}{\sqrt{5}} \qquad\qquad = -\sqrt{\frac{8}{10}} \qquad\qquad = -\sqrt{\frac{2}{8}}$$

$$\qquad\qquad = -\sqrt{\frac{4}{5}} = -\frac{2}{\sqrt{5}} \qquad\qquad = -\frac{1}{2}$$

31. $\tan \theta = \dfrac{-4}{3}, \; \dfrac{\pi}{2} < \theta < \pi$ **32.** $\tan \theta = \dfrac{3}{4}, \; 0 < \theta < \dfrac{\pi}{2}$

33. $\sin \theta = \dfrac{-8}{17}, \; \dfrac{3\pi}{2} < \theta < 2\pi$ **34.** $\cos \theta = \dfrac{12}{13}, \; \dfrac{3\pi}{2} < \theta < 2\pi$

35. $\sec \theta = \dfrac{-5}{3}, \; \pi < \theta < \dfrac{3\pi}{2}$

▶ Use the half-angle formulas to find the exact value of each of the following.

36. $\sin 22.5°$ **37.** $\cos 22.5°$ **38.** $\sin 157.5°$ **39.** $\cos 157.5°$

40. $\cos 67.5°$ **41.** $\sin 67.5°$ **42.** $\tan 67.5°$ **43.** $\tan 22.5°$

44. $\tan 157.5°$ **45.** $\tan 30°$

▶ Express each of the following in terms of $\cos \theta$ and/or $\sin \theta$.

EXAMPLE $\sin 3\theta$

Solution
$$\begin{aligned}
\sin 3\theta &= \sin (\theta + 2\theta) \\
&= \sin \theta \cos 2\theta + \cos \theta \sin 2\theta \\
&= \sin \theta \,(\cos^2 \theta - \sin^2 \theta) + \cos \theta \,(2 \sin \theta \cos \theta) \\
&= 3 \sin \theta \cos^2 \theta - \sin^3 \theta
\end{aligned}$$

46. $\cos 3\theta$ **47.** $\sin 4\theta$ **48.** $\cos 4\theta$ **49.** $\sin 6\theta$ **50.** $\cos 5\theta$

▶ Prove that each of the following equations is an identity.

51. $\sec \theta + \tan \theta = \dfrac{1 + \sin \theta}{\cos \theta}$

52. $\sec^2 \theta + \cos^2 \theta = \dfrac{\sec^4 \theta + 1}{\sec^2 \theta}$

53. $\tan \theta \csc^2 \theta - \cot \theta = \tan \theta$

54. $\dfrac{\cos \theta}{\sin \theta \cot^2 \theta} = \tan \theta$

55. $\cos \theta + \tan \theta \sin \theta = \sec \theta$

56. $\sin^2 \theta - \tan^2 \theta = -\tan^2 \theta \sin^2 \theta$

57. $\dfrac{1 + \cos \theta}{\sin \theta} + \dfrac{\sin \theta}{1 + \cos \theta} = 2 \csc \theta$

58. $\sin 2\theta + \dfrac{(1 - \tan \theta)^2}{\sec^2 \theta} = 1$

59. $2 \cos^2 \theta \csc^2 \theta + 1 = \csc^2 \theta + \cot^2 \theta$

60. $\dfrac{\sin^2 \theta}{1 - \cos \theta} = 1 + \cos \theta$

61. $\dfrac{1 - \cos^2 \theta}{\tan \theta + \cot \theta} = \dfrac{\cos \theta}{\csc^3 \theta}$

62. $\dfrac{\sin^2 \theta \sec \theta}{1 + \sec \theta} = 1 - \cos \theta$

63. $\dfrac{1 + \sec \theta}{-\tan^2 \theta} = \dfrac{1}{1 - \sec \theta}$

64. $\dfrac{1 + \cot^2 \theta}{\sec^2 \theta} = \cot^2 \theta$

65. $\dfrac{(1 - \cot \theta)^2}{\sec^2 \theta} + \dfrac{\cos 2\theta}{\tan^2 \theta} = \cot^2 \theta \, (2 \cos^2 \theta - \sin 2\theta)$

▶ Make an appropriate substitution and replace each of the following by $a \sec \theta$, $a \sin \theta$, $a \cos \theta$, or $a \tan \theta$ where a is a real number.

EXAMPLE

$\sqrt{25 + x^2}$

Solution

Let $x = 5 \tan \theta$ where $0 < \theta < \dfrac{\pi}{2}$.

$\sqrt{25 + x^2} = \sqrt{25 + 25 \tan^2 \theta} = 5 \sec \theta$

66. $\sqrt{9 - x^2}$ **67.** $\sqrt{x^2 - 9}$ **68.** $\sqrt{16 + x^2}$

69. $\sqrt{16 - x^2}$ **70.** $\sqrt{x^2 - 16}$

▶ Make an appropriate substitution and replace each of the following by a rational expression whose numerator and denominator are polynomials in u.

EXAMPLE

$\dfrac{\sin \theta}{\cos \theta}$

Solution   Let $u = \tan \theta/2$. Then

$$\sin \theta = \frac{2u}{1 + u^2} \quad \text{and} \quad \cos \theta = \frac{1 - u^2}{1 + u^2}.$$

(See Example 13.)

$$\frac{\sin \theta}{\cos \theta} = \frac{2u}{1 + u^2} \div \frac{1 - u^2}{1 + u^2} = \frac{2u}{1 - u^2}$$

71. $\dfrac{1}{\cos \theta}$ **72.** $\dfrac{\sin^2 \theta}{\cos \theta}$ **73.** $\dfrac{3 + \cos \theta}{\sin^2 \theta}$

74. $\dfrac{1 + 4 \sin \theta}{1 - 3 \cos \theta}$ **75.** $\dfrac{\sin \theta + \cos \theta}{\cos^3 \theta - 2 \sin \theta}$

CHECK LIST FOR THE STUDENT

If you have learned the material in Section 3.6, you should be able to do the following:

1. State and use the basic identities. (See the list at the end of Section 3.6.)

2. Derive the identities $1 + \tan^2 \theta = \sec^2 \theta$ and $\cot^2 \theta + 1 = \csc^2 \theta$ from the identity $\cos^2 \theta + \sin^2 \theta = 1$.

3. Use the addition formulas to prove that the cosine and sine functions are periodic with period 2π.

4. Use the addition formulas to prove that the tangent and cotangent functions are periodic with period π.

5. Express $\sin^2 \theta$ and $\cos^2 \theta$ in terms of $\cos 2\theta$.

6. Use the basic identities to prove that certain given equations are identities.

7. Make an appropriate substitution and replace a radical of the form $\sqrt{a^2 + x^2}$, $\sqrt{a^2 - x^2}$, or $\sqrt{x^2 - a^2}$ by $a \sec \theta$, $a \sin \theta$, $a \cos \theta$, or $a \tan \theta$.

8. Make an appropriate substitution and replace a fraction whose numerator and denominator are products, sums, or differences of positive integral powers of $\sin \theta$ and $\cos \theta$ by a rational expression whose numerator and denominator are polynomials in a single variable.

SECTION QUIZ 3.6

In each of the following, use the given information and basic identities to find the values of the other trigonometric functions of θ.

1. $\sec \theta = \frac{-13}{12}$, $\sin \theta > 0$

2. $\sin \theta = \frac{1}{5}$, $\tan \theta < 0$

3. Use the addition formulas to find the exact value of $\cos (\pi/6 + \pi/4)$.

4. Use the fact that $\tan \theta = \frac{-12}{5}$ and $\pi/2 < \theta < \pi$ to find $\sin 2\theta$.

5. Use the fact that $\cos 2\theta = \frac{-2}{5}$ and $2\pi < 2\theta < 3\pi$ to find $\cos \theta$.

6. Use the fact that $\sin \theta = \frac{-1}{3}$ and $\pi < \theta < 3\pi/2$ to find $\cos \theta/2$.

7. Use a half-angle formula to find the exact value of $\sin 165°$.

8. Express $\sec^2 \theta \cot^2 \theta + \csc \theta \tan \theta$ in terms of $\cos \theta$ and/or $\sin \theta$.

Prove that each of the following is an identity.

9. $\cos^2 \theta - \cot^2 \theta = -\cot^2 \theta \cos^2 \theta$

10. $\dfrac{\sin 2\theta \cos 2\theta}{\tan^2 \theta} = 2 \cot \theta \cos^4 \theta - \sin 2\theta \cos^2 \theta$

3.7 The Laws of Sines and Cosines

The purpose of this section is to present, use, and prove the law of sines and the law of cosines.

The law of sines and the law of cosines may be used to determine the sides or angles of any triangle, whether it is a right triangle or not, provided that certain sides or angles are given. These laws are needed from time to time in calculus and its applications, just as the Pythagorean theorem is needed from time to time. In fact, the law of cosines is a generalization of the Pythagorean theorem.

THE LAW OF SINES

If a, b, and c represent the lengths of the sides of triangle ABC with angles α, β, and γ as shown in Figure 3.59, the law of sines says that

$$\frac{a}{\sin \alpha} = \frac{b}{\sin \beta},$$

$$\frac{a}{\sin \alpha} = \frac{c}{\sin \gamma},$$

and

$$\frac{b}{\sin \beta} = \frac{c}{\sin \gamma}.$$

These three equations may be condensed by writing the **law of sines** as follows:

$$\frac{a}{\sin \alpha} = \frac{b}{\sin \beta} = \frac{c}{\sin \gamma} \qquad \text{Law of sines}$$

A proof is presented after the next example.

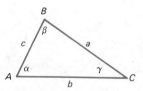

FIGURE 3.59 Triangle ABC

EXAMPLE 1

If triangle ABC is such that the measure of angle A is $30°$, the measure of angle C is $15°$, and the length b of the side opposite B is 20 cm, find angle B and find the lengths of the other two sides correct to two decimal places.

Triangle ABC is shown in Figure 3.60. Since the sum of the angles of any triangle is $180°$, the measure of angle B is

$$180° - (30° + 15°) = 180° - 45° = 135°.$$

By the law of sines,

$$\frac{a}{\sin 30°} = \frac{20}{\sin 135°}.$$

Solving for a and using $\sin 135° = \sqrt{2}/2$, we have

$$a = \sin 30° \times \frac{20}{\sin 135°}$$

$$= \frac{1}{2} \times \frac{20}{\frac{\sqrt{2}}{2}}$$

$$= 10\sqrt{2} \text{ cm.}$$

Correct to two decimal places, $a = 14.14$ cm.

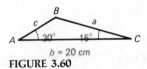

$b = 20$ cm

FIGURE 3.60

Again by the law of sines, and using $a = 10\sqrt{2}$,

$$\frac{c}{\sin 15°} = \frac{10\sqrt{2}}{\sin 30°}.$$

Solving for c, we have

$$c = \sin 15° \times \frac{10\sqrt{2}}{\sin 30°}.$$

By Table III in Appendix B, $\sin 15° = 0.2588$, so

$$c = \frac{2.588\sqrt{2}}{\frac{1}{2}}$$

$$= (5.176)\sqrt{2}.$$

Correct to two decimal places, $c = 7.32$ cm.

A proof of the law of sines follows for triangle ABC, shown in Figure 3.61. In that figure, α and γ are acute angles. The proof given may be slightly modified to establish that the law of sines holds even if α or γ is not acute.

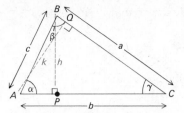

FIGURE 3.61 $\sin \gamma = h/a$; $\sin \alpha = h/c$
$\sin \gamma = k/b$; $\sin \beta = k/c$

A Proof of the Law of Sines

Drop a perpendicular *BP* from *B* to the opposite side, and let its length be denoted by *h*. Then triangles *BCP* and *BAP* are right triangles, so that

$$\frac{h}{a} = \sin \gamma \quad \text{and} \quad \frac{h}{c} = \sin \alpha.$$

In other words,

$$h = a \sin \gamma \quad \text{and} \quad h = c \sin \alpha.$$

It follows that

$$a \sin \gamma = c \sin \alpha.$$

This may be written

$$\frac{a}{\sin \alpha} = \frac{c}{\sin \gamma}.$$

Now, to show that each of these ratios is also equal to $b/(\sin \beta)$, drop a perpendicular *AQ* from *A* to the opposite side, and let its length be denoted by *k*. From right triangles *ACQ* and *ABQ*, we read

$$\frac{k}{b} = \sin \gamma \quad \text{and} \quad \frac{k}{c} = \sin \beta.$$

That is,

$$k = b \sin \gamma \quad \text{and} \quad k = c \sin \beta$$

so that

$$b \sin \gamma = c \sin \beta.$$

This may be written

$$\frac{b}{\sin \beta} = \frac{c}{\sin \gamma}.$$

But we have already shown that $a/(\sin \alpha) = c/(\sin \gamma)$, so it is also true that

$$\frac{a}{\sin \alpha} = \frac{b}{\sin \beta}.$$

Combining our results, we have

$$\frac{a}{\sin \alpha} = \frac{b}{\sin \beta} = \frac{c}{\sin \gamma},$$

Exercise 3.7, 1–5 which is the law of sines.

In Example 1 we were given *two angles and a side*. We found the third angle, so then we had *two angles and an opposite side*. We can always determine exactly one triangle by using the law of sines with those specifications. If we are given *two sides and the angle opposite one of them*, we can use the law of sines, but we may not be able to determine exactly one triangle with the given specifications. In fact, if we are given two sides and the angle opposite one of them, there are three distinct possibilities.

1. No triangle satisfies the specifications.

2. Two triangles satisfy the specifications.

3. Exactly one triangle satisfies the specifications.

The three possibilities are illustrated in Figure 3.62. As shown in that figure, the three cases depend on the relative lengths of the given sides and the perpendicular dropped to the third side. In Figure 3.62, the given angle is acute. If the given angle is obtuse (has measure greater than $90°$), there are only the two possibilities illustrated in Figure 3.63.

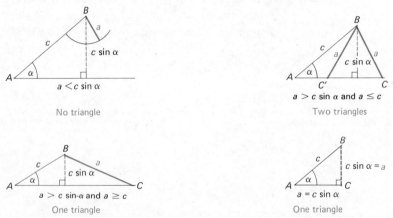

FIGURE 3.62 Given two sides and opposite acute angle: no triangle, two triangles, or one triangle

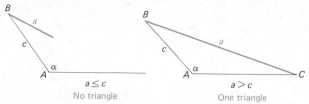

FIGURE 3.63   Given two sides and opposite obtuse angle: no triangle or one triangle

Note   In each of the following examples, it is assumed that all the given lengths are in the same unit.

EXAMPLE 2   Find the angle at B in each triangle ABC such that $\alpha = 30°$, $a = 4$, and $c = 10$. If no such triangle exists, say "No triangle."

Solution   We can tell in advance that there is no triangle that satisfies the given specifications, because $c \sin \alpha = 10(\frac{1}{2}) = 5$ so that $a < c \sin \alpha$, as shown in Figure 3.64. But even if we did not notice in advance that there is no triangle with the given a, c, and α, we would discover that no such triangle exists when we tried to apply the law of sines. To demonstrate this, let us apply the law of sines with the given a, c, and α.

$$\frac{a}{\sin \alpha} = \frac{c}{\sin \gamma}$$

$$\frac{4}{\frac{1}{2}} = \frac{10}{\sin \gamma}$$

$$8 \sin \gamma = 10$$

$$\sin \gamma = 1.2$$

But there is no γ for which $\sin \gamma = 1.2$, because there is no γ for which $\sin \gamma > 1$. (The range of the sine function is the set of all real numbers *between* 1 and -1, inclusive.) Thus there is no triangle satisfying the given specifications.

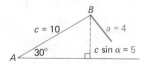

FIGURE 3.64   No triangle

EXAMPLE 3   Find, to the nearest ten minutes, the angle at B in each triangle ABC such that $\alpha = 30°$, $a = 8$, and $c = 10$.

Solution

Here $a > c \sin \alpha$ and $a < c$. As shown in Figure 3.65, the two triangles ABC' and ABC both satisfy the given specifications. In triangles ABC' and ABC we can determine the angles at C' and C by the law of sines.

$$\frac{a}{\sin \alpha} = \frac{c}{\sin \gamma}$$

$$\frac{8}{\frac{1}{2}} = \frac{10}{\sin \gamma}$$

$$16 \sin \gamma = 10$$

$$\sin \gamma = 0.625$$

From Table III in Appendix B, the acute angle whose sine is 0.625 is about $38°40'$. Thus the angle γ at C in triangle ABC is about $38°40'$. Since $BC'C$ is an isosceles triangle, angle $BC'C$ has the same measure as γ. And in triangle ABC', the angle γ' at C' is such that

$$\gamma' = 180° - \gamma.$$

Thus, to the nearest ten minutes,

$$\gamma' = 180° - 38°40'$$
$$= 179°60' - 38°40'$$
$$= 141°20'$$

Now, using the fact that the sum of the angles of a triangle is $180°$,

the angle at B in triangle $ABC = 180° - (30° + \gamma)$
$$= 180° - (68°40')$$
$$= 111°20', \text{ and}$$

the angle at B in triangle $ABC' = 180° - (30° + \gamma')$
$$= 180° - (171°20')$$
$$= 8°40'.$$

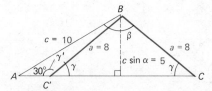

FIGURE 3.65 Two triangles, ABC' and ABC

EXAMPLE 4

Find, to the nearest ten minutes, the angle at B in each triangle ABC such that $\alpha = 30°$, $a = 20$, and $c = 10$.

Since $a > c \sin \alpha$ and also $a > c$, the triangle shown in Figure 3.66 is the only triangle with the given specifications. By the law of sines,

$$\frac{a}{\sin \alpha} = \frac{c}{\sin \gamma}$$

$$\frac{20}{\frac{1}{2}} = \frac{10}{\sin \gamma}$$

$$40 \sin \gamma = 10$$

$$\sin \gamma = 0.25.$$

From Table III, γ is about $14°30'$. Thus, to the nearest ten minutes and using Table III,

$$\beta = 180° - (30° + 14°30')$$
$$= 135°30'.$$

(On a calculator also, γ is $14°30'$ to the nearest ten mintues.)

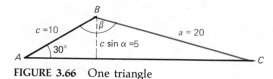

Exercise 3.7, 6–15   **FIGURE 3.66**   One triangle

THE LAW OF COSINES

If we are given two sides and the angle included between them, or if we are given three sides, the law of sines alone will not enable us to determine the other parts. In these cases, however, the law of cosines will enable us to determine the parts not given.

The law of cosines expresses the length (actually, the *square* of the length) of one side of a triangle in terms of the lengths of the other two sides and the cosine of their included angle. In particular, referring to triangle *ABC* in Figure 3.59, the **law of cosines** says

$$a^2 = b^2 + c^2 - 2bc \cos \alpha,$$
$$b^2 = a^2 + c^2 - 2ac \cos \beta, \qquad \text{Law of cosines}$$
and $\qquad c^2 = a^2 + b^2 - 2ab \cos \gamma.$

In words, the law of cosines says that *the square of the length of a side of a triangle equals the sum of the squares of the lengths of the other two sides minus twice the product of the lengths of the other two sides multiplied by the cosine of the angle included between them.* A proof will be presented after the next two examples.

EXAMPLE 5   In triangle *ABC*, if $c = 10$, $b = 6$, and the angle at *A* is 60°, find *a* and find the other angles correct to the nearest degree.

Solution   By the law of cosines,

$$a^2 = b^2 + c^2 - 2bc \cos 60°$$
$$= 36 + 100 - 120 \times \tfrac{1}{2}$$
$$= 136 - 60$$
$$= 76$$
$$a = \sqrt{76}$$

We can now use the law of cosines again to find the angle at *B* or at *C*. Let us find the angle β at *B*.

$$b^2 = a^2 + c^2 - 2ac \cos \beta$$
$$36 = 76 + 100 - 2\sqrt{76}(10) \cos \beta$$
$$36 = 176 - 20\sqrt{76} \cos \beta$$
$$\cos \beta = \frac{-140}{-20\sqrt{76}}$$
$$= \frac{7}{\sqrt{76}}$$

Correct to four decimal places,

$$\cos \beta = 0.8030.$$

By Table III, to the nearest degree,

$$\beta = 37°.$$

Using $\beta = 37°$,

$$\gamma = 180° - (60° + 37°)$$
$$= 83°.$$

Once we found *a*, we could have used the law of sines to find the angle at *B* or at *C*. Let us use the law of sines to find β.

$$\frac{a}{\sin \alpha} = \frac{b}{\sin \beta}$$
$$\frac{\sqrt{76}}{\sin 60°} = \frac{6}{\sin \beta}$$
$$\sqrt{76} \sin \beta = 6 \sin 60°$$
$$\sin \beta = \frac{6}{\sqrt{76}} \frac{\sqrt{3}}{2} = \frac{3\sqrt{3}}{\sqrt{76}}$$

Correct to four decimal places,

$$\sin \beta = 0.5960.$$

From Table III, we get (once again)

$$\beta = 37°$$

to the nearest degree.

EXAMPLE 6

In triangle ABC, if $c = 10$, $b = 9$, and $a = 8$, find each of the angles to the nearest degree.

Solution

Let us first use the law of cosines to find α.

$$a^2 = b^2 + c^2 - 2bc \cos \alpha$$

Solving for $\cos \alpha$, we have

$$\cos \alpha = \frac{a^2 - b^2 - c^2}{-2bc}$$

$$= \frac{64 - 81 - 100}{-2(9)(10)}$$

$$= \frac{64 - 181}{-180}$$

$$= \frac{117}{180}$$

$$= 0.65.$$

From Table III, to the nearest degree,

$$\alpha = 49°.$$

We may now use the law of cosines to find β or γ. (Now that we know α, we may also use the law of sines to find β or γ. But since we only found α to the nearest degree, it is best to use the given data and find β or γ with the law of cosines.) Let us use the law of cosines to find β.

$$b^2 = a^2 + c^2 - 2ac \cos \beta$$

$$\cos \beta = \frac{b^2 - a^2 - c^2}{-2ac}$$

$$= \frac{81 - 64 - 100}{-2(8)(10)}$$

$$= \frac{83}{160}$$

Correct to four decimal places,

$$\cos \beta = 0.5188.$$

From Table III, to the nearest degree,

$$\beta = 59°.$$

Using $\alpha = 49°$ and $\beta = 59°$, we get

$$\gamma = 180° - (49° + 59°)$$
$$= 72°.$$

A Proof of the Law of Cosines

Place triangle ABC of Figure 3.59 on a Cartesian coordinate system as shown in Figure 3.67. The coordinates of C are $(b, 0)$. The x-coordinate of B is $c \cos \alpha$ (because in right triangle ABC, $x/c = \cos \alpha$). The y-coordinate of B is $c \sin \alpha$ (because $y/c = \sin \alpha$). Since a is the distance between B and C, by the distance formula

$$a^2 = (c \cos \alpha - b)^2 + (c \sin \alpha - 0)^2$$
$$= c^2 \cos^2 \alpha - 2bc \cos \alpha + b^2 + c^2 \sin^2 \alpha$$
$$= c^2 (\cos^2 \alpha + \sin^2 \alpha) - 2bc \cos \alpha + b^2.$$

But $\cos^2 \alpha + \sin^2 \alpha = 1$, so

$$a^2 = c^2 - 2bc \cos \alpha + b^2,$$

or

$$a^2 = b^2 + c^2 - 2bc \cos \alpha.$$

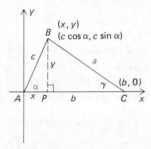

FIGURE 3.67 $a^2 = b^2 + c^2 - 2bc \cos \alpha$

By placing the triangle with B at the origin or with C at the origin, we may similarly show that

$$b^2 = a^2 + c^2 - 2ac \cos \beta$$

or

$$c^2 = a^2 + b^2 - 2ab \cos \gamma.$$

In Figure 3.67, angles α and γ are acute. The proof may be modified to establish that the law of cosines is true even if α or γ is not acute.

Notice that if triangle ABC is a right triangle, one of the three expressions of the law of cosines reduces to the Pythagorean theorem. Indeed, suppose triangle ABC is a right triangle with right angle at α. Then a is the length of the hypotenuse and $\cos \alpha = \cos 90° = 0$, so that the expression

$$a^2 = b^2 + c^2 - 2bc \cos \alpha$$

reduces to

Exercise 3.7, 16–30
Verbal Problems
Exercise 3.7, 31–35

$$a^2 = b^2 + c^2,$$

which is the Pythagorean theorem.

EXERCISE 3.7 THE LAWS OF SINES AND COSINES

▶ In each of the following, find the other angle and find, correct to two decimal places, the lengths of the remaining sides of triangle ABC with the given specifications. Use Table III in Appendix B if necessary. Assume that all the given lengths are in meters.

1. $\alpha = 60°$, $\gamma = 20°$, $b = 10$ 2. $\beta = 30°$, $\gamma = 63°$, $a = 5$

3. $\alpha = 50°$, $\beta = 75°$, $a = 100$ 4. $\beta = 30°$, $\gamma = 45°$, $c = 10$

5. $\alpha = 10°$, $\gamma = 120°$, $b = 8$

▶ In each of the following, find the remaining angles of each triangle ABC with the given specifications. If no such triangle exists, say "No triangle." Express answers to the nearest ten minutes. Use Table III if necessary. Assume that all the given lengths are in centimeters.

6. $\alpha = 60°$, $a = 6$, $c = 10$ 7. $\alpha = 60°$, $a = 5\sqrt{3}$, $c = 10$

8. $\alpha = 60°$, $a = 12$, $c = 10$ 9. $\alpha = 60°$, $a = 9$, $c = 10$

10. $\beta = 45°$, $b = 4$, $c = 10$ 11. $\beta = 45°$, $b = 5\sqrt{2}$, $c = 10$

12. $\beta = 45°$, $b = 9$, $c = 10$ 13. $\beta = 45°$, $b = 12$, $c = 10$

14. $\gamma = 135°$, $a = 10$, $c = 6$ 15. $\gamma = 150°$, $b = 12$, $c = 4$

▶ In each of 16–25, find the remaining side, and in 26–30 find the angles of the triangle ABC with the given specifications. If no such triangle exists, say "No triangle." Express angles to the nearest degree and lengths to the nearest two decimal places. Use Table III if necessary. Assume that all the given lengths are in meters.

16. $c = 12$, $b = 8$, $\alpha = 30°$ 17. $c = 15$, $a = 6$, $\beta = 45°$

18. $a = 9$, $b = 4$, $\gamma = 60°$ 19. $a = 11$, $b = 7$, $\gamma = 120°$

20. $c = 3$, $b = 7$, $\alpha = 37°$ 21. $a = 8$, $c = 12$, $\beta = 53°$

22. $b = 10$, $a = 12$, $\gamma = 90°$ 23. $a = 9$, $b = 7$, $\gamma = 62°$

24. $c = 8$, $b = 6$, $\alpha = 150°$ 25. $a = 4$, $c = 13$, $\beta = 135°$

26. $a = 9$, $b = 7$, $c = 5$ 27. $a = 11$, $b = 8$, $c = 5$

28. $a = 13$, $b = 10$, $c = 9$ 29. $a = 8$, $b = 7$, $c = 11$

30. $a = 5$, $b = 7$, $c = 10$

31. A ferry ramp must be constructed so that at low tide it has one end at a point B on the beach and makes an angle of $20°$ with the horizontal beach. The other end of the ramp must be at point D on the deck of the ferry. If a taut cable from D to point C on the beach is 15 m long, and if the distance along the horizontal beach from C to B is 21 m, what must be the length of the ramp to the nearest meter?

32. In order to find the distance between two points A and B on opposite sides of a building, a surveyor measures the distance from B to a third point C. The surveyor then measures the distance from A to C, and also measures angle ACB. If $\overline{BC}$ is 10 m, $\overline{AC}$ is 23 m, and angle ACB is $38°$, what is the distance between A and B to the nearest meter?

33. A forester needs to know the length of a particular sugar pine tree. The tree is leaning so that it makes an angle of $85°$ with the horizontal ground. The forester stands, with the tree leaning toward him, at a point P on the ground 30 m from the base of the tree and measures the angle up from the ground to the line of sight PT to the top T of the tree. If that angle is $74°$, what is the length of the tree to the nearest meter?

34. A sailing course is laid out so that there are buoys at points A, B, and C. If $\overline{AB}$ is 2 km, $\overline{BC}$ is 2 km, and $\overline{CA}$ is 1 km, what are the angles of triangle ABC to the nearest degree?

35. To reach the treasure at T, according to a pirate map, one should start at a rock R, walk x meters to point S, make a turn so that angle RST is $40°$, and then walk 100 m to T. If the map also shows that $\overline{TR}$ is 80 m, what is x to the nearest meter?

CHECK LIST FOR THE STUDENT

If you have learned the material in Section 3.7, you should be able to do the following:

1. State the law of sines.

2. Use the law of sines and find the remaining parts of a triangle, if two angles and a side are given.

3. Give a proof of the law of sines for triangle ABC with angles α and γ acute.

4. Determine whether or not there exists a triangle having two given sides and a given angle opposite one of them.

5. Find the remaining angles of each triangle having two sides and an opposite angle given.

6. State the law of cosines.

7. Use the law of cosines and find the remaining parts of a triangle, if two sides and the included angle are given or if three sides are given.

8. Give a proof of the law of cosines for triangle ABC with angles α and γ acute.

SECTION QUIZ 3.7

Let triangle ABC have angles of measure α, β, and γ at A, B, and C, respectively, and let a, b, and c represent the lengths of the sides opposite A, B, and C.

1. If $\alpha = 60°$, $\beta = 45°$, and $b = 10$ cm, find a.

2. In each of the following, tell how many triangles there are with the given specifications.
   (a) $\alpha = 30°$, $c = 12$ m, $a = 2$ m
   (b) $\alpha = 30°$, $c = 12$ m, $a = 6$ m
   (c) $\alpha = 30°$, $c = 12$ m, $a = 9$ m
   (d) $\alpha = 30°$, $c = 12$ m, $a = 15$ m
   (e) $\alpha = 30°$, $c = 12$ m, $a = 12$ m

3. If $\gamma = 45°$, $c = 5\sqrt{2}$ m, and $a = 5\sqrt{3}$ m, find α.

4. If $c = 10$ cm, $b = 6$ cm, and $\alpha = \pi/6$, find a.

5. If $a = 10$ m, $b = 8$ m, and $c = 6$ m, find $\cos \beta$.

3.8 Inverse Trigonometric Functions

The purpose of this section is to review some facts about functions and to define and present the graphs of the six inverse trigonometric functions. Before proceeding to define the inverse trigonometric functions, let us first review some pertinent facts that were introduced in Sections 2.1 and 2.7.

1. A real-valued function is defined provided that the following two things are given.

(a) A set called the domain

(b) A rule assigning a single real number to each element in the domain

The set of all numbers assigned to the domain elements by the rule of assignment is called the *range* of the function. By *function* we mean *real-valued* function.

2. By the *domain convention,* if the domain of a function is not explicitly stated, the domain is understood to be the set of all real numbers except those to which the rule of assignment would not assign a real number.

3. A function f may be "defined" by an equation $y = f(x)$. In this case, the rule of assignment assigns the number $f(x)$ to each element x in the domain, and the domain is determined by the domain convention.

Example: The equation $y = \sqrt{x}$ defines the function whose rule of assignment assigns the number $\sqrt{x}$ to each element x in its domain and whose domain is the set of all nonnegative real numbers x.

4. There is a set of ordered pairs associated with any function f; the first member in each pair is a domain element, and the second member is the number assigned to the first member by the rule of assignment of the function. Some authors define a function as a set of ordered pairs; the function f defined by $y = f(x)$ may be defined as the set of ordered pairs $\{(x, y) | y = f(x)\}$.

Example: The function defined by $y = \sqrt{x}$ may be defined as the set of ordered pairs $\{(x, y) | y = \sqrt{x}\}$. Since the domain is not explicitly stated, it is understood to be the set of all nonnegative real numbers.

Example: The function f whose domain is the set of all real numbers greater than 4 and whose rule of assignment is written $f(x) = \sqrt{x}$ may be defined as the set of ordered pairs $\{(x, y) | y = \sqrt{x} \text{ and } x > 4\}$.

5. The graph of a function f whose domain and range are sets of real numbers is the set of all points (x, y) on a Cartesian coordinate system such that x is in the domain of f and $y = f(x)$.

6. The sweeping vertical line test can be used to determine whether or not a graph represents a function. If no vertical line intersects a graph in more than one point, the graph represents a function.

7. A one-to-one function is a function in which there is a one-to-one correspondence between the domain and the range.

8. The sweeping *horizontal* line test can be used to determine whether or not a graph is the graph of a *one-to-one* function. If no horizontal line intersects the graph of a function in more than one point, then the function is one-to-one.

Example: Applying the sweeping horizontal line test to the graph of the sine function (Figure 3.40), we see that the sine function is not one-to-one.

9. The inverse of a one-to-one function f defined by the equation $y = f(x)$ can be found by simply interchanging x and y in the equation. Such interchanging is equivalent to interchanging the domain and range and "reversing" the rule of assignment. If f is defined as $\{(x, y)|y = f(x)\}$, then f-inverse is $\{(x, y)|x = f(y)\}$.

Example: If f is $\{(x, y)|y = \sqrt{x} \text{ and } x > 4\}$, then f-inverse is $\{(x, y)|x = \sqrt{y} \text{ and } y > 4\}$. The domain of f, and hence the range of f-inverse, is the set of all real numbers greater than 4. The range of f, and hence the domain of f-inverse, is the set of all real numbers greater than $\sqrt{4} = 2$. In other words, f-inverse is $\{(x, y)|y = x^2 \text{ and } x > 2\}$.

10. The inverse of a one-to-one function is itself a one-to-one function.

11. The inverse of a function that is not one-to-one is not itself a function. (Its rule of assignment would fail to assign just a single number to each element in its domain.)

Example: If f is $\{(x, y)|y = x^2\}$, then f is a function (see Figure 2.7), but the sweeping horizontal line test shows that f is not one-to-one. The inverse of f is $\{(x, y)|x = y^2\}$; Figure 2.7 shows that this f-inverse is not itself a function.

THE INVERSE-SINE FUNCTION

Now let us prepare to define the *inverse-sine function.* We have seen that the sweeping horizontal line test shows that the sine function is *not* one-to-one. Therefore, the inverse of the sine function is not a function. For this reason, the so-called inverse sine is *not* the inverse of the sine function itself, but is the inverse of a one-to-one function having the same rule of assignment as the sine. In particular, the inverse sine is the inverse of the function whose domain is $\{x|-\pi/2 \leq x \leq \pi/2\}$ and whose rule of assignment is $y = \sin x$. In other words, the inverse-sine function is the *inverse* of the function

$$\left\{(x, y) \,\middle|\, y = \sin x \text{ and } -\frac{\pi}{2} \leq x \leq \frac{\pi}{2}\right\} \qquad \text{The inverse sine is the inverse of this function.}$$

whose graph is just that part of the graph of the sine that lies between $-\pi/2$ and $\pi/2$, inclusive. The graph of this function is shown in Figure 3.68. Note that the sweeping horizontal line test shows that it is indeed a one-to-one function. Its range is the set of all real numbers between -1 and 1, inclusive.

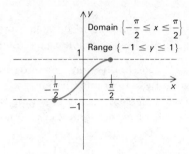

FIGURE 3.68 A one-to-one function

$$\left\{(x, y)\,|\,y = \sin x \text{ and } -\frac{\pi}{2} \le x \le \frac{\pi}{2}\right\}$$

(The inverse of this is called the inverse sine.)

We have said that the inverse sine is the inverse of the one-to-one function shown in Figure 3.68. Specifically, the **inverse-sine function** is defined as follows:

$$\text{Inverse sine} = \left\{(x, y)\,|\,x = \sin y \text{ and } -\frac{\pi}{2} \le y \le \frac{\pi}{2}\right\}$$

Its domain is the range of the function shown in Figure 3.68; its range is the domain of the function shown in Figure 3.68. The graph of the inverse sine is shown in Figure 3.69.

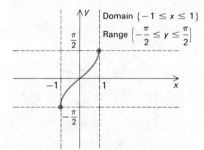

FIGURE 3.69 Inverse-sine function

$$\left\{(x, y)\,|\,x = \sin y \text{ and } -\frac{\pi}{2} \le y \le \frac{\pi}{2}\right\}$$

$$\{(x, y)\,|\,y = \text{Arcsin } x\}$$

The statement $x = \sin y$ and $-\pi/2 \leq y \leq \pi/2$ is frequently written more concisely as $y = $ Arcsin x.

$y = $ Arcsin x provided that $x = \sin y$ and $-\pi/2 \leq y \leq \pi/2$

Sometimes the notation $\text{Sin}^{-1}\, x$ is used for Arcsin x.

$$\text{Sin}^{-1}\, x \text{ means Arcsin } x$$

In other words, if $y = $ Arcsin x or if $y = \text{Sin}^{-1}\, x$, then y is the number between $-\pi/2$ and $\pi/2$, inclusive, whose sine is equal to x. With this new notation, the inverse-sine function can be written as

$$\{(x,\, y)\,|\, y = \text{Arcsin } x\} \quad \text{or as} \quad \{(x,\, y)\,|\, y = \text{Sin}^{-1}\, x\}.$$

EXAMPLE 1   Find the values of

$$\text{Arcsin } \frac{\sqrt{2}}{2} \quad \text{Arcsin} \left(\frac{-\sqrt{2}}{2} \right) \quad \text{Sin}^{-1}\, \frac{1}{2} \quad \text{and} \quad \text{Sin}^{-1}\!\left(\frac{-1}{2} \right).$$

Solution $\text{Arcsin } \dfrac{\sqrt{2}}{2} = \dfrac{\pi}{4}$ because $\sin \dfrac{\pi}{4} = \dfrac{\sqrt{2}}{2}$ and $-\dfrac{\pi}{2} \leq \dfrac{\pi}{4} \leq \dfrac{\pi}{2}$.

$\text{Arcsin} \left(\dfrac{-\sqrt{2}}{2} \right) = -\dfrac{\pi}{4}$ because $\sin \left(-\dfrac{\pi}{4} \right) = \dfrac{-\sqrt{2}}{2}$ and $-\dfrac{\pi}{2} \leq -\dfrac{\pi}{4} \leq \dfrac{\pi}{2}$.

$\text{Sin}^{-1}\, \dfrac{1}{2} = \dfrac{\pi}{6}$ because $\sin \dfrac{\pi}{6} = \dfrac{1}{2}$ and $-\dfrac{\pi}{2} \leq \dfrac{\pi}{6} \leq \dfrac{\pi}{2}$.

$\text{Sin}^{-1} \left(\dfrac{-1}{2} \right) = -\dfrac{\pi}{6}$ because $\sin \left(-\dfrac{\pi}{6} \right) = \dfrac{-1}{2}$ and $-\dfrac{\pi}{2} \leq -\dfrac{\pi}{6} \leq \dfrac{\pi}{2}$.

If it is necessary for us to think of the domain of the sine function as being a set of *angles* rather than *numbers,* then $y = $ Arcsin x provided that $x = \sin y$ and y is the angle whose measure is between $-\pi/2$ and $\pi/2$ radians, inclusive.

THE INVERSE-COSINE FUNCTION

The inverse-cosine function is not defined as the inverse of the cosine itself, because the cosine function is not one-to-one (apply the sweeping horizontal line test to its graph in Figure 3.41), and hence its inverse would not even be a function. Instead, the inverse cosine is defined as the inverse of that one-to-one function whose rule of assignment is the same as the rule of assignment of the cosine, but whose domain is only the set of real numbers between 0 and π, inclusive. The graph of this one-to-one function whose inverse is defined as the inverse cosine is shown in Figure 3.70. It may be defined as

$\{(x,\, y)\,|\, y = \cos x \text{ and } 0 \leq x \leq \pi\}$ The inverse cosine is the
inverse of this function.

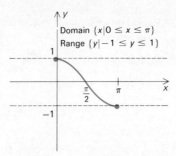

FIGURE 3.70   A one-to-one function
$$\{(x, y) \,|\, y = \cos x \text{ and } 0 \le x \le \pi\}$$
(The inverse of this is called the inverse cosine.)

Its graph is just that part of the graph of the cosine that lies between 0 and π, inclusive.

The inverse cosine is the inverse of the function shown in Figure 3.70. The **inverse-cosine function** may be defined as follows:

$$\text{Inverse cosine} = \{(x, y) \,|\, x = \cos y \text{ and } 0 \le y \le \pi\}$$

Its domain is the range of the function shown in Figure 3.70; its range is the domain of the function shown in Figure 3.70. The graph of the inverse cosine is shown in Figure 3.71.

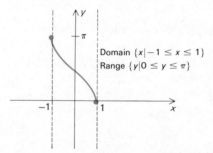

FIGURE 3.71   Inverse-cosine function
$$\{(x, y) \,|\, x = \cos y \text{ and } 0 \le y \le \pi\}$$
$$\{(x, y) \,|\, y = \text{Arccos } x\}$$

The notation $y = \text{Arccos } x$ is used for the statement $x = \cos y$ and $0 \le y \le \pi$. In other words, if $y = \text{Arccos } x$, then y is the number between 0 and π, inclusive, such that $\cos y = x$.

$$y = \text{Arccos } x \text{ provided that } x = \cos y \text{ and } 0 \le y \le \pi$$

Sometimes the notation $\text{Cos}^{-1} x$ is used for $\text{Arccos } x$.

$$\text{Cos}^{-1} x \text{ means Arccos } x$$

Suppose it is necessary to think of the cosine function as being a function whose domain is a set of *angles* rather than numbers. Then if $y = $ Arccos x, y is the angle whose radian measure is between 0 and π, inclusive, and which is such that cos $y = x$.

EXAMPLE 2

Find the values of

$$\text{Arccos } \frac{\sqrt{3}}{2} \quad \text{Arccos} \left(\frac{-\sqrt{3}}{2} \right) \quad \text{Cos}^{-1} 1 \quad \text{and} \quad \text{Cos}^{-1}(-1).$$

Solution

$\text{Arccos } \dfrac{\sqrt{3}}{2} = \dfrac{\pi}{6}$ because $\cos \dfrac{\pi}{6} = \dfrac{\sqrt{3}}{2}$ and $0 \leq \dfrac{\pi}{6} \leq \pi$.

$\text{Arccos} \left(\dfrac{-\sqrt{3}}{2} \right) = \dfrac{5\pi}{6}$ because $\cos \dfrac{5\pi}{6} = \dfrac{-\sqrt{3}}{2}$ and $0 \leq \dfrac{5\pi}{6} \leq \pi$.

$\text{Cos}^{-1} 1 = 0$ because $\cos 0 = 1$ and $0 \leq 0 \leq \pi$.

Exercise 3.8, 1–10, 31–36 $\text{Cos}^{-1} (-1) = \pi$ because $\cos \pi = -1$ and $0 \leq \pi \leq \pi$.

THE INVERSE-TANGENT FUNCTION

Like the sine function and the cosine function, the tangent function is not one-to-one. (Apply the sweeping horizontal line test to its graph in Figure 3.53.) The inverse-tangent function is therefore not defined as the inverse of the tangent itself. Instead, the inverse tangent is defined as the *inverse* of the one-to-one function whose rule of assignment is the same as the rule of assignment of the tangent function, but whose domain is only the set of real numbers strictly between $-\pi/2$ and $\pi/2$. The graph of this one-to-one function is shown in Figure 3.72. It is the function

$$\left\{ (x, y) \,\middle|\, y = \tan x \text{ and } -\frac{\pi}{2} < x < \frac{\pi}{2} \right\}. \qquad \begin{array}{l} \text{The inverse tangent is the} \\ \text{inverse of this function.} \end{array}$$

The graph consists of just that branch of the graph of the tangent that lies between $-\pi/2$ and $\pi/2$. As we have said, the inverse tangent is the inverse of the one-to-one function shown in Figure 3.72. The **inverse-tangent function** may be defined as follows:

$$\text{Inverse tangent} = \left\{ (x, y) \,\middle|\, x = \tan y \text{ and } -\frac{\pi}{2} < y < \frac{\pi}{2} \right\}$$

Its domain is the set of all real numbers x; its range is the set of all numbers y strictly between $-\pi/2$ and $\pi/2$. The graph of the inverse tangent is shown in Figure 3.73.

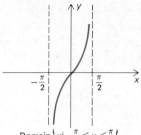

Domain $\left\{x | -\dfrac{\pi}{2} < x < \dfrac{\pi}{2}\right\}$
Range $\{$ All real numbers $\}$

FIGURE 3.72 A one-to-one function

$$\left\{(x, y) \,\middle|\, y = \tan x \text{ and } -\frac{\pi}{2} < x < \frac{\pi}{2}\right\}$$

(The inverse of this is called the inverse tangent.)

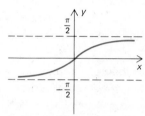

Domain $\{$ All real numbers $\}$
Range $\left\{y | -\dfrac{\pi}{2} < y < \dfrac{\pi}{2}\right\}$

FIGURE 3.73 Inverse-tangent function

$$\left\{(x, y) \,\middle|\, x = \tan y \text{ and } -\frac{\pi}{2} < y < \frac{\pi}{2}\right\}$$

$$\{(x, y) \,|\, y = \text{Arctan } x\}$$

The notation $y = \text{Arctan } x$ means that $x = \tan y$ and $-\pi/2 < y < \pi/2$.

$$y = \text{Arctan } x \text{ provided that } x = \tan y \text{ and } -\frac{\pi}{2} < y < \frac{\pi}{2}$$

The symbols $\text{Tan}^{-1} x$ are sometimes used for Arctan x.

$$\text{Tan}^{-1} x \text{ means } \text{Arctan } x$$

If the domain of the tangent is considered to be a set of angles, then $y = \text{Arctan } x$ provided that y is the angle with radian measure strictly between $-\pi/2$ and $\pi/2$ such that $\tan y = x$.

EXAMPLE 3   Find the values of Arctan 1, Arctan(-1), $\text{Tan}^{-1} \sqrt{3}$, and $\text{Tan}^{-1}(-\sqrt{3})$.

Solution

$$\text{Arctan } 1 = \frac{\pi}{4} \qquad \text{because} \quad \tan \frac{\pi}{4} = 1 \qquad \text{and} \quad -\frac{\pi}{2} < \frac{\pi}{4} < \frac{\pi}{2}.$$

$$\text{Arctan } (-1) = -\frac{\pi}{4} \qquad \text{because} \quad \tan \left(-\frac{\pi}{4}\right) = -1 \quad \text{and} \quad -\frac{\pi}{2} < -\frac{\pi}{4} < \frac{\pi}{2}.$$

$$\text{Tan}^{-1} \sqrt{3} = \frac{\pi}{3} \qquad \text{because} \quad \tan \frac{\pi}{3} = \sqrt{3} \qquad \text{and} \quad -\frac{\pi}{2} < \frac{\pi}{3} < \frac{\pi}{2}.$$

$$\text{Tan}^{-1} (-\sqrt{3}) = -\frac{\pi}{3} \quad \text{because} \quad \tan \left(-\frac{\pi}{3}\right) = -\sqrt{3} \text{ and } -\frac{\pi}{2} < -\frac{\pi}{3} < \frac{\pi}{2}.$$

Exercise 3.8, 11–13, 37–39

THE OTHER INVERSE TRIGONOMETRIC FUNCTIONS

We have seen that the inverse-sine, inverse-cosine, and inverse-tangent functions are not inverses of the sine, cosine, and tangent functions themselves but rather are inverses of certain one-to-one functions with the same rules of assignment as the sine, cosine, and tangent functions. Similarly, the inverse-cotangent, inverse-secant, and inverse-cosecant functions are not defined as inverses of the cotangent, secant, and cosecant functions. Instead they are defined as inverses of particular one-to-one functions whose rules of assignment are the same as the rules of assignment of the cotangent, secant, and cosecant. The one-to-one function whose inverse is defined as the inverse cotangent is shown, together with the inverse cotangent, in Figure 3.74. In Figure 3.75, the one-to-one function whose inverse is defined as the inverse secant is shown, together with the inverse secant. Finally, Figure 3.76 shows the one-to-one function whose inverse is defined as the inverse cosecant, and it also shows the inverse cosecant. In Figures 3.74–3.76 we see that

$y = \text{Arccot } x$ provided that $x = \cot y$ and $0 < y < \pi$

$y = \text{Arcsec } x$ provided that $x = \sec y$ and $0 \leq y \leq \pi$ with $y \neq \dfrac{\pi}{2}$

$y = \text{Arccsc } x$ provided that $x = \csc y$ and $-\dfrac{\pi}{2} \leq y \leq \dfrac{\pi}{2}$ with $y \neq 0$.

The symbols Cot^{-1}, Sec^{-1}, and Csc^{-1} are used for Arccot, Arcsec, and Arccsc.

$$\text{Cot}^{-1} x \text{ means Arccot } x$$
$$\text{Sec}^{-1} x \text{ means Arcsec } x$$
$$\text{Csc}^{-1} x \text{ means Arccsc } x$$

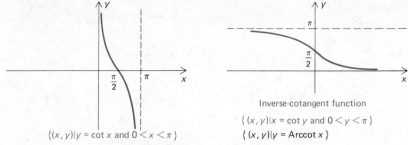

$\{(x, y)|y = \cot x \text{ and } 0 < x < \pi\}$

Inverse-cotangent function

$\{(x, y)|x = \cot y \text{ and } 0 < y < \pi\}$

$\{(x, y)|y = \text{Arccot } x\}$

FIGURE 3.74 Inverse cotangent

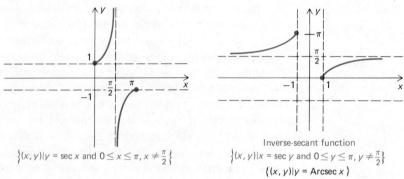

$\left\{(x, y)|y = \sec x \text{ and } 0 \leq x \leq \pi, x \neq \dfrac{\pi}{2}\right\}$

Inverse-secant function

$\left\{(x, y)|x = \sec y \text{ and } 0 \leq y \leq \pi, y \neq \dfrac{\pi}{2}\right\}$

$\{(x, y)|y = \text{Arcsec } x\}$

FIGURE 3.75 Inverse secant

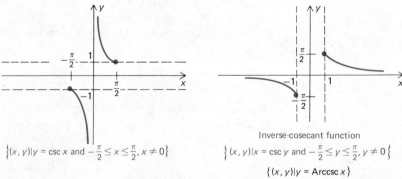

$\left\{(x, y)|y = \csc x \text{ and } -\dfrac{\pi}{2} \leq x \leq \dfrac{\pi}{2}, x \neq 0\right\}$

Inverse-cosecant function

$\left\{(x, y)|x = \csc y \text{ and } -\dfrac{\pi}{2} \leq y \leq \dfrac{\pi}{2}, y \neq 0\right\}$

$\{(x, y)|y = \text{Arccsc } x\}$

FIGURE 3.76 Inverse cosecant

EXAMPLE 4   Find the values of Arccot (-1), Cot^{-1} 1, Arcsec 2, Sec^{-1} (-2), Arccsc (-2), and Csc^{-1} 2.

Solution

Arccot $(-1) = \dfrac{3\pi}{4}$ because $\cot \dfrac{3\pi}{4} = -1$ and $0 < \dfrac{3\pi}{4} < \pi$.

Cot^{-1} 1 $= \dfrac{\pi}{4}$ because $\cot \dfrac{\pi}{4} = 1$ and $0 < \dfrac{\pi}{4} < \pi$.

Arcsec 2 $= \dfrac{\pi}{3}$ because $\sec \dfrac{\pi}{3} = 2$ and $0 \le \dfrac{\pi}{3} \le \pi$ with $\dfrac{\pi}{3} \ne \dfrac{\pi}{2}$.

Sec^{-1} $(-2) = \dfrac{2\pi}{3}$ because $\sec \dfrac{2\pi}{3} = -2$ and $0 \le \dfrac{2\pi}{3} \le \pi$ with $\dfrac{2\pi}{3} \ne \dfrac{\pi}{2}$.

Arccsc $(-2) = -\dfrac{\pi}{6}$ because $\csc\left(-\dfrac{\pi}{6}\right) = -2$ and $-\dfrac{\pi}{2} \le -\dfrac{\pi}{6} \le \dfrac{\pi}{2}$ with $-\dfrac{\pi}{6} \ne 0$.

Csc^{-1} 2 $= \dfrac{\pi}{6}$ because $\csc \dfrac{\pi}{6} = 2$ and $-\dfrac{\pi}{2} \le \dfrac{\pi}{6} \le \dfrac{\pi}{2}$ with $\dfrac{\pi}{6} \ne 0$.

Exercise 3.8, 14–30, 40–50

EXAMPLE 5   Use Table III in Appendix B and 3.14 for π to find an approximation to Arccot (-1.2497).

Solution   Let $\theta =$ Arccot (-1.2497). Then $\cot \theta = -1.2497$ and $0 < \theta < \pi$. But $\cot \theta$ is negative, so θ must be in the second quadrant with reference angle t such that $\cot t = 1.2497$. (See Examples 8 and 9 in Section 3.3.) By Table III, $t = 0.6749$. Since $\theta = \pi - t$, θ is approximately $3.14 - 0.6749 = 2.4651$. That is, Arccot (-1.2497) is approximately 2.4651.

If a hand calculator with inverse trigonometric functions is available, the computation of such values can be accomplished much more efficiently and with greater accuracy. For example, if we used a calculator instead of using Table III and 3.14 for π, we would find that Arccot $(-1.2497) = \pi -$Arctan $(1/1.2497) = 2.466735$ instead of 2.4651.

Exercise 3.8, 51–60

EXAMPLE 6   Find the exact value of sin (Cos^{-1} $(\frac{-4}{5})$).

Solution   Let $\theta =$ Cos^{-1} $(\frac{-4}{5})$. We want to find $\sin \theta$. First consider $\cos \theta$. Note that $\cos \theta = \frac{-4}{5}$ and $0 \le \theta \le \pi$. Since $\cos \theta$ is negative, we must have $\pi/2 < \theta < \pi$. For convenience, let us interpret the reference angle t as an angle in the right triangle drawn in Figure 3.77. Since $\cos \theta = -\cos t$ and $\cos \theta = \frac{-4}{5}$, it follows that

$$\cos t = \tfrac{4}{5},$$

so the side adjacent to t and the hypotenuse have been labeled 4 and 5,

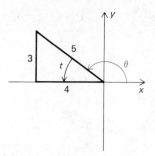

FIGURE 3.77 $\theta = \text{Cos}^{-1}(\frac{-4}{5})$

respectively. By the Pythagorean theorem, the side opposite t is of length 3. Since

$$\sin \theta = \sin t,$$

we can read directly from the reference triangle that

$$\sin \theta = \tfrac{3}{5}.$$

That is,

Exercise 3.8, 61–65

$$\sin \left(\text{Cos}^{-1}\left(\tfrac{-4}{5}\right)\right) = \tfrac{3}{5}.$$

In performing various derivations and manipulations in calculus, we sometimes need to find a trigonometric function of a given inverse trigonometric function. In such cases it is often helpful to construct a reference triangle as was first illustrated in Example 7 of Section 3.2. This is demonstrated in the next example.

EXAMPLE 7

Find an algebraic expression for $\sin (\text{Arctan } u)$ where u is negative.

Solution

Let $\theta = \text{Arctan } u$. Then $\tan \theta = u$ and $-\pi/2 < \theta < \pi/2$. But u is negative, so we must have $-\pi/2 < \theta < 0$.

For convenience, let us interpret the reference angle $t = -\theta$ as an angle in the right triangle drawn in Figure 3.78. Since $\tan \theta = -\tan t$ and $\tan \theta = u/1$, it follows that $\tan t = -u/1$, so the sides opposite and adjacent to t have been labeled $-u$ and 1, respectively. By the Pythagorean theorem, the hypotenuse is $\sqrt{1 + u^2}$. We can now read directly from the reference triangle that

$$\sin \theta = -\sin t$$

$$= -\frac{\text{Opposite}}{\text{Hypotenuse}}$$

$$= -\frac{-u}{\sqrt{1 + u^2}} = \frac{u}{\sqrt{1 + u^2}}.$$

That is,

$$\sin (\text{Arctan } u) = \frac{u}{\sqrt{1 + u^2}}.$$

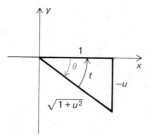

FIGURE 3.78 $\theta = \text{Arctan } u$ and $u < 0$

Exercise 3.8, 66–75

$$\sin \theta = -\sin t = \frac{u}{\sqrt{1 + u^2}}$$

EXERCISE 3.8 INVERSE TRIGONOMETRIC FUNCTIONS

▶ Find the value of each of the following.

1. $\text{Arcsin } \dfrac{\sqrt{3}}{2}$

2. $\text{Arcsin } \left(\dfrac{-\sqrt{3}}{2}\right)$

3. $\text{Sin}^{-1} \, 0$

4. $\text{Sin}^{-1} \, 1$

5. $\text{Arcsin } (-1)$

6. $\text{Arccos } \frac{1}{2}$

7. $\text{Arccos } \left(\frac{-1}{2}\right)$

8. $\text{Cos}^{-1} \, 0$

9. $\text{Arccos } \dfrac{\sqrt{2}}{2}$

10. $\text{Arccos } \left(\dfrac{-\sqrt{2}}{2}\right)$

11. $\text{Arctan } \dfrac{\sqrt{3}}{3}$

12. $\text{Arctan } \left(\dfrac{-\sqrt{3}}{3}\right)$

13. $\text{Tan}^{-1} \, 0$

14. $\text{Arccot } \sqrt{3}$

15. $\text{Arccot } (-\sqrt{3})$

16. $\text{Cot}^{-1} \dfrac{\sqrt{3}}{3}$

17. $\text{Cot}^{-1} \left(-\dfrac{\sqrt{3}}{3}\right)$

18. $\text{Cot}^{-1} \, 0$

19. $\text{Arcsec } \dfrac{2}{\sqrt{3}}$

20. $\text{Sec}^{-1} \left(\dfrac{-2}{\sqrt{3}}\right)$

21. $\text{Arcsec } \dfrac{2}{\sqrt{2}}$

22. $\text{Arcsec } \left(\dfrac{-2}{\sqrt{2}}\right)$

23. $\text{Sec}^{-1} \, (-1)$

24. $\text{Sec}^{-1} \, 1$

25. $\text{Arccsc } \dfrac{2}{\sqrt{2}}$

26. $\text{Arccsc } \left(\dfrac{-2}{\sqrt{2}}\right)$

27. $\text{Csc}^{-1} \dfrac{2}{\sqrt{3}}$

28. $\text{Csc}^{-1} \left(\dfrac{-2}{\sqrt{3}}\right)$

29. $\text{Arccsc } (-1)$

30. $\text{Arccsc } 1$

► Sketch the graph of each of the following.

31. $\left\{(x, y) \mid x = \sin y \text{ and } 0 \leq y \leq \dfrac{\pi}{2}\right\}$

32. $y = \text{Arcsin } x, \ 0 \leq x \leq 1$ **33.** $y = \text{Sin}^{-1} x, \ -1 \leq x \leq 0$

34. $\left\{(x, y) \mid x = \cos y \text{ and } 0 \leq y \leq \dfrac{\pi}{2}\right\}$

35. $y = \text{Arccos } x, \ 0 \leq x \leq 1$ **36.** $y = \text{Cos}^{-1} x, \ -1 \leq x \leq 0$

37. $\left\{(x, y) \mid x = \tan y \text{ and } 0 \leq y \leq \dfrac{\pi}{2}\right\}$

38. $y = \text{Arctan } x, \ x \geq 0$ **39.** $y = \text{Tan}^{-1} x, \ x \leq 0$

40. $\left\{(x, y) \mid x = \cot y \text{ and } 0 < y \leq \dfrac{\pi}{2}\right\}$

41. $y = \text{Arccot } x, \ x \geq 0$ **42.** $y = \text{Cot}^{-1} x, \ x \leq 0$

43. $\left\{(x, y) \mid x = \sec y \text{ and } 0 \leq y < \dfrac{\pi}{2}\right\}$

44. $y = \text{Arcsec } x, \ x \geq 1$ **45.** $y = \text{Sec}^{-1} x, \ x \leq -1$

46. $\left\{(x, y) \mid x = \csc y \text{ and } 0 < y \leq \dfrac{\pi}{2}\right\}$

47. $y = \text{Arccsc } x, \ x \geq 1$ **48.** $y = \text{Csc}^{-1} x, \ x \leq -1$

49. $y = \text{Arcsin } \dfrac{\sqrt{2}}{2}$ **50.** $y = \text{Arccos } 1$

► Use Table III in Appendix B and 3.14 for π to find an approximation to each of the following.

51. Arcsin 0.1536 **52.** Arcsin (-0.2079) **53.** Cos^{-1} 0.9976

54. Arccos (-0.9976) **55.** Arctan 0.2339 **56.** Tan^{-1} (-0.4074)

57. Sin^{-1} 0.3934 **58.** Arccot 2.3945 **59.** Cot^{-1} (-3.7321)

60. Arccos (-0.8021)

► Find the exact value of each of the following.

61. $\cos \left(\text{Sin}^{-1} \frac{3}{5}\right)$ **62.** $\tan \left(\text{Cos}^{-1} \left(\frac{-3}{5}\right)\right)$ **63.** $\sec \left(\text{Tan}^{-1} \frac{4}{3}\right)$

64. $\sin \left(\text{Cos}^{-1} \left(\frac{-12}{13}\right)\right)$ **65.** $\tan \left(\text{Sec}^{-1} \frac{13}{5}\right)$

► Find an algebraic expression for each of the following.

66. $\cos (\text{Arctan } u), \ u < 0$ **67.** $\tan (\text{Arccos } u), \ u > 0$

68. $\tan (\text{Arcsec } u), \ u > 0$ **69.** $\sin (\text{Arccos } u), \ u < 0$

70. $\sin (\text{Arcsin } u), \ u < 0$ **71.** $\tan (\text{Arcsin } u), \ u > 0$

72. $\sec (\text{Arctan } u), \ u < 0$ **73.** $\csc (\text{Arcsin } u), \ u > 0$

74. $\cos (\text{Arcsin } u), \ u < 0$ **75.** $\cos (\text{Arcsec } u), \ u > 0$

CHECK LIST FOR THE STUDENT

If you have learned the material in Section 3.8, you should be able to do the following:

1. Define and sketch the graph of the one-to-one function whose inverse is defined as the inverse sine; define and sketch the graph of the inverse-sine function.

2. Define and sketch the graph of the one-to-one function whose inverse is defined as the inverse cosine; define and sketch the graph of the inverse-cosine function.

3. Define and sketch the graph of the one-to-one function whose inverse is defined as the inverse tangent; define and sketch the graph of the inverse-tangent function.

4. Define and sketch the graph of the one-to-one function whose inverse is defined as the inverse cotangent; define and sketch the graph of the inverse-cotangent function.

5. Define and sketch the graph of the one-to-one function whose inverse is defined as the inverse secant; define and sketch the graph of the inverse-secant function.

6. Define and sketch the graph of the one-to-one function whose inverse is defined as the inverse cosecant; define and sketch the graph of the inverse-cosecant function.

7. Find the values of Arcsin x, Arccos x, Arctan x, Arccot x, Arcsec x, and Arccsc x for certain given values of x.

8. Use the notation Sin^{-1}, Cos^{-1}, Tan^{-1}, Cot^{-1}, Sec^{-1}, and Csc^{-1}.

9. Find trigonometric functions of a given inverse trigonometric function by constructing a reference triangle.

SECTION QUIZ 3.8

1. (a) Sketch the graph of the function f whose domain is $\{x \mid -\pi/2 \leq x \leq \pi/2\}$ and which is such that $f(x) = \sin x$. What is its range?
   (b) Sketch the graph of the inverse-sine function. Give its domain and range.

2. (a) Sketch the graph of the function f whose domain is $\{x \mid 0 \leq x \leq \pi\}$ and which is such that $f(x) = \cos x$. What is its range?
   (b) Sketch the graph of the inverse-cosine function. Give its domain and range.

3. (a) Sketch the graph of the function f whose domain is $\{x \mid -\pi/2 < x < \pi/2\}$ and which is such that $f(x) = \tan x$. What is its range?
   (b) Sketch the graph of the inverse-tangent function. Give its domain and range.

4. Find the value of each of the following.
   (a) Arccos $\frac{1}{2}$ (b) Arcsin (-1) (c) Arccot 0
   (d) Arctan 0 (e) Arcsec $\sqrt{2}$ (f) Arccsc $(-\sqrt{2})$

5. Find the exact value of each of the following.
   (a) $\sin (\text{Sin}^{-1} 0.78)$ (b) $\cos (\text{Sin}^{-1} (\frac{-3}{5}))$

3.9 Conditional Equations

The purpose of this section is to show how to solve some trigonometric equations in one variable that are true for some but not all values of the variable.

In Section 3.6 we considered *identities,* trigonometric equations that are true for all values of the variable(s) for which their terms are defined. Some trigonometric equations, however, are not identities. For example, the equation $\sin \theta = 2$ is not true for *any* values of θ because $-1 \leq \sin \theta \leq 1$. Also, the equation $\sin \theta = 0$ is true only for those values of θ that are integral multiples of π (for angles θ with radian measures that are integral multiples of π). An equation that is true for some but not all values of its variable(s) is called a **conditional equation**. The solutions of a conditional trigonometric equation may be interpreted as numbers or as angles having those numbers as radian measures.

EXAMPLE 1

Solve the equation $\cos \theta = \frac{1}{2}$ for $0 \leq \theta \leq \pi$.

Solution

We know that $\cos \pi/3 = \frac{1}{2}$ and that $\pi/3$ is the only number between 0 and π, inclusive, for which this is true. Using notation introduced for the inverse cosine in Section 3.8, we can write the solution of

$$\cos \theta = \frac{1}{2} \text{ for } 0 \leq \theta \leq \pi$$

as

$$\theta = \text{Arccos } \frac{1}{2}$$

$$= \frac{\pi}{3}.$$

If the domain of the cosine is considered to be a set of numbers, the solution of $\cos \theta = \frac{1}{2}$ is the *number* $\pi/3$; if the domain of the cosine is considered to be a set of angles, the solution of $\cos \theta = \frac{1}{2}$ is an *angle* with radian measure $\pi/3$ or degree measure $60°$.

Exercise 3.9, 1–10

EXAMPLE 2

Solve the equation $\cos \theta = \frac{1}{2}$ for $0 \leq \theta \leq 2\pi$. Write an expression for the set of *all* solutions of $\cos \theta = \frac{1}{2}$.

Solution

We know that the cosine is positive in the first and fourth quadrants and that there are two values of θ between 0 and 2π, inclusive, for which $\cos \theta = \frac{1}{2}$. We know that $\cos \pi/3 = \frac{1}{2}$. Thus $\pi/3$ is the first-quadrant solution. The fourth-quadrant angle with reference angle $\pi/3$ has radian measure $2\pi - \pi/3 = 5\pi/3$. Thus $\cos 5\pi/3 = \frac{1}{2}$, and $5\pi/3$ is the fourth-quadrant solution. The solutions of $\cos \theta = \frac{1}{2}$ with $0 \leq \theta \leq 2\pi$ are $\pi/3$, $5\pi/3$.

Since the cosine function is periodic with period 2π, an unlimited number of solutions of $\cos \theta = \frac{1}{2}$ can be found simply by adding integral multiples of 2π to $\pi/3$ or to $5\pi/3$. If k represents any integer, the set of all solutions is

$$\left\{\theta \,\middle|\, \theta = \frac{\pi}{3} + 2k\pi \text{ or } \theta = \frac{5\pi}{3} + 2k\pi\right\}.$$

EXAMPLE 3

Solve the equation $\tan \theta = 1$ for $0 \le \theta \le \pi$. Write an expression for the set of *all* solutions of $\tan \theta = 1$.

Solution

We know that $\pi/4$ is the only value of θ between 0 and π, inclusive, for which $\tan \theta = 1$. The solution of $\tan \theta = 1$ for $0 \le \theta \le \pi$ is therefore $\pi/4$.

Since the tangent function is periodic with period π, an unlimited number of solutions of $\tan \theta = 1$ can be found by adding integral multiples of π to $\pi/4$. If k represents any integer, the set of all solutions is

Exercise 3.9, 11–20

$$\left\{\theta \,\middle|\, \theta = \frac{\pi}{4} + k\pi\right\}.$$

Algebraic methods may be used to solve more complicated conditional trigonometric equations than those presented in the previous three examples. In each case, the overall plan is to solve the given equation for one particular trigonometric function or to reduce the problem to that of solving one or more equations similar to those of Examples 1–3. In some cases, identities may be used to advantage. The following examples demonstrate some methods of solution.

EXAMPLE 4

Solve the equation $4 \sin \theta - 1 = 3 \sin \theta$.

Solution

The given equation is a linear equation in $\sin \theta$. (If we substituted x for $\sin \theta$, we would have the linear equation $4x - 1 = 3x$.) Solving for $\sin \theta$, we have

$$4 \sin \theta - 3 \sin \theta = 1$$
$$\sin \theta = 1.$$

Since the sine function is periodic with period 2π, let us first find the solutions θ for which $0 \le \theta \le 2\pi$. We find that the only value of θ for which $\sin \theta = 1$ and $0 \le \theta \le 2\pi$ is $\pi/2$.

Other solutions can be found by adding integral multiples of 2π to $\pi/2$. The set of all solutions is

$$\left\{\theta \,\middle|\, \theta = \frac{\pi}{2} + 2k\pi\right\}$$

where k represents any integer.

EXAMPLE 5

Solve the equation $\sin^2 \theta - \frac{1}{2} \sin \theta = 0$.

Solution	This equation is quadratic in $\sin \theta$. (If we substituted x for $\sin \theta$, we would have $x^2 - \frac{1}{2}x = 0$.) It can be solved by factoring.

$$\sin^2 \theta - \tfrac{1}{2}\sin \theta = 0$$
$$\sin \theta \,(\sin \theta - \tfrac{1}{2}) = 0$$
$$\sin \theta = 0 \quad \text{or} \quad \sin \theta - \tfrac{1}{2} = 0$$
$$\sin \theta = \tfrac{1}{2}$$

If $\sin \theta = 0$, θ is some integral multiple of π. If $\sin \theta = \frac{1}{2}$, θ is $\pi/6$ or $5\pi/6$ or any integral multiple of 2π *plus* $\pi/6$ or $5\pi/6$. The set of all solutions of the given equation is

$$\left\{\theta \,\middle|\, \theta = k\pi, \text{ or } \theta = \frac{\pi}{6} + 2k\pi, \text{ or } \theta = \frac{5\pi}{6} + 2k\pi\right\}$$

where k represents any integer. (Notice that we did *not* begin the task of solving the given equation by dividing through by $\sin \theta$. Such a division would have caused us to "lose" all the solutions that are solutions of $\sin \theta = 0$.)

EXAMPLE 6

Solve the equation $\tan \theta \cos \theta + \tan \theta = 0$.

Solution

Two different trigonometric functions appear in this equation, but this particular equation can still be solved by factoring.

$$\tan \theta \,(\cos \theta + 1) = 0$$
$$\tan \theta = 0 \quad \text{or} \quad \cos \theta + 1 = 0$$
$$\cos \theta = -1$$

If $\tan \theta = 0$, θ is some integral multiple of π. If $\cos \theta = -1$, θ is π or θ is any integral multiple of 2π *plus* π. The set of all solutions of the given equation is

$$\{\theta \,|\, \theta = k\pi \quad \text{or} \quad \theta = \pi + 2k\pi\}$$

where k represents any integer.

Notice that $\pi + 2k\pi = (2k + 1)\pi$ represents any *odd* integral multiple of π, whereas $k\pi$ represents *any* integral multiple of π. Thus every solution of $\cos \theta = -1$ is also a solution of $\tan \theta = 0$, and the set of all solutions of the given equation can be written in a more condensed form as

$$\{\theta \,|\, \theta = k\pi\}.$$

EXAMPLE 7

Solve the equation $2\tan^2 \theta - \sec^2 \theta = 0$.

Solution

This equation contains two different trigonometric functions. One way to solve it is to first use the identity $\sec^2 \theta = 1 + \tan^2 \theta$ and then solve for $\tan \theta$.

$$2 \tan^2 \theta - (1 + \tan^2 \theta) = 0$$

$$2 \tan^2 \theta - 1 - \tan^2 \theta = 0$$

$$\tan^2 \theta = 1$$

$$\tan \theta = 1 \text{ or } -1$$

If $\tan \theta = 1$, $\theta = \pi/4$ or any integral multiple of π plus $\pi/4$. If $\tan \theta = -1$, $\theta = 3\pi/4$ or any integral multiple of π plus $3\pi/4$. The set of all solutions of the given equation is

$$\left\{ \theta \,\middle|\, \theta = \frac{\pi}{4} + k\pi \quad \text{or} \quad \theta = \frac{3\pi}{4} + k\pi \right\}$$

where k represents any integer.

EXAMPLE 8

Solve the equation $\sin^2 \theta + \cos 2\theta = 0$.

Solution

If we first use the identity

$$\sin^2 \theta = \frac{1 - \cos 2\theta}{2},$$

the only trigonometric function in the equation will then be $\cos 2\theta$.

$$\frac{1 - \cos 2\theta}{2} + \cos 2\theta = 0$$

$$1 - \cos 2\theta + 2 \cos 2\theta = 0$$

$$\cos 2\theta = -1$$

$$2\theta = \pi + 2k\pi \text{ where } k \text{ is any integer}$$

$$\theta = \frac{\pi}{2} + k\pi \quad \text{where } k \text{ is any integer}$$

We could also have done this problem by first using the identity $\cos 2\theta = 1 - 2 \sin^2 \theta$, but this would have been more complicated, because we would then have had to solve a second-degree equation for $\sin \theta$ rather than a first-degree equation for $\cos 2\theta$.

Exercise 3.9, 21–45

EXAMPLE 9

Use Table III in Appendix B and 3.14 for π to find approximations to the solutions of the equation $4 \sin^2 \theta + \sin \theta - 3 = 0$.

Solution

The given equation is quadratic in $\sin \theta$. (If we substituted x for $\sin \theta$, we would have the quadratic equation $4x^2 + x - 3 = 0$.) It can be solved by factoring.

$$(\sin \theta + 1)(4 \sin \theta - 3) = 0$$

$$\sin \theta + 1 = 0 \quad \text{or} \quad 4 \sin \theta - 3 = 0$$

$$\sin \theta = -1 \qquad\qquad \sin \theta = \tfrac{3}{4}$$

If $\sin \theta = -1$, θ is $3\pi/2$ or θ is some integral multiple of 2π plus $3\pi/2$. That is, θ is approximately

$$\frac{3(3.14)}{2} + 2k(3.14)$$

or $\qquad\qquad 4.71 + (6.28)k$

where k represents any integer. If $\sin \theta = \frac{3}{4}$, then $\sin \theta = 0.75$, and by Table III, θ is approximately

$$0.8494.$$

But this entry in Table III approximates only the positive value of θ less than $\pi/2$ for which $\sin \theta = 0.75$. There is another positive value of θ less than 2π that satisfies the equation, namely the second-quadrant value having reference angle with measure approximately 0.8494 radians. This is the value

$$\pi - 0.8494,$$

which is approximately

$$3.14 - 0.8494 = 2.2906.$$

Thus the values of θ between 0 and 2π for which $\sin \theta = 0.75$ are approximately

$$0.8494 \text{ and } 2.2906.$$

Since the sine function is periodic with period 2π, each of these approximate solutions plus any integral multiple of 2π is an approximate solution of the equation. It follows that if $\sin \theta = \frac{3}{4}$, then θ is approximately

$$0.8494 + 2k(3.14) \quad \text{or} \quad 2.2906 + 2k(3.14)$$

where k represents any integer.

Combining the solutions of $\sin \theta = -1$ and $\sin \theta = \frac{3}{4}$, we see that approximate solutions of the original equation $4 \sin^2 \theta + \sin \theta - 3 = 0$ are

$$\{\theta \mid \theta = 4.71 + (6.28)k \quad \text{or} \quad \theta = 0.8494 + (6.28)k \quad \text{or} \quad \theta = 2.2906 + (6.28)k\}.$$

If a hand calculator were used instead of using Table III and 3.14 for π to find approximations to the solutions of $\sin \theta = -1$, we would find that one solution θ is approximately -1.5708, because this is approximately Arcsin (-1) where Arcsin (-1) lies between $-\pi/2$ and $\pi/2$, inclusive. And since the sine is periodic with period 2π, any integral multiple of 2π plus -1.5708 is also an approximate solution. In particular, $2\pi + (-1.5708)$ is also an approximate solution. But $2\pi + (-1.5708)$ is approximately

$$2(3.14) + (-1.5708) = 4.7092,$$

which is approximately 4.71. In other words, 2π plus the solution of $\sin\theta = -1$ given by a calculator is approximately the first solution we found in Example 9.

Likewise, if a hand calculator were used instead of using Table III and 3.14 for π to find approximations to the solutions of $\sin\theta = \frac{3}{4}$, we would find that one solution θ is approximately 0.8481. Using Table III without interpolating, we found θ to be approximately 0.8494. As usual, a calculator is more efficient and produces more accurate results than a table.

Exercise 3.9, 46–50

Examples 10 and 11 of Section 3.3 illustrated some conditional equations that arise when it is necessary to make trigonometric substitutions in calculus. The next example is a further illustration.

EXAMPLE 10   If $x = 4\sin\theta$ and $-\pi/2 \le \theta \le \pi/2$, find θ when $x = 0$; find θ when $x = 2$.

Solution   Let $x = 0$.

$$4\sin\theta = 0$$
$$\sin\theta = 0$$

The only value of θ for which $-\pi/2 \le \theta \le \pi/2$ and $\sin\theta = 0$ is 0. In the notation of Section 3.8,

$$\theta = \text{Arcsin } 0$$
$$= 0.$$

Let $x = 2$.

$$4\sin\theta = 2$$
$$\sin\theta = \tfrac{1}{2}$$

The only value of θ for which $-\pi/2 \le \theta \le \pi/2$ and $\sin\theta = \frac{1}{2}$ is $\pi/6$. In the notation of Section 3.8,

$$\theta = \text{Arcsin } \tfrac{1}{2}$$
$$= \frac{\pi}{6}.$$

Exercise 3.9, 51–55

EXERCISE 3.9 CONDITIONAL EQUATIONS

▶ Solve each of the following equations for θ as prescribed.

1. $\cos\theta = \dfrac{\sqrt{3}}{2}$, $0 \le \theta \le \pi$

2. $\sin\theta = \dfrac{\sqrt{3}}{2}$, $\dfrac{-\pi}{2} \le \theta \le \dfrac{\pi}{2}$

3. $\tan\theta = \sqrt{3}$, $\dfrac{-\pi}{2} < \theta < \dfrac{\pi}{2}$

4. $\cot\theta = \sqrt{3}$, $0 < \theta < \pi$

3.9 CONDITIONAL EQUATIONS

453

5. $\sec \theta = 2$, $0 \leq \theta \leq \pi$ and $\theta \neq \dfrac{\pi}{2}$

6. $\csc \theta = 2$, $\dfrac{-\pi}{2} \leq \theta \leq \dfrac{\pi}{2}$ and $\theta \neq 0$

7. $\cos \theta = \dfrac{-\sqrt{3}}{2}$, $0 \leq \theta \leq \pi$ **8.** $\sin \theta = \dfrac{-\sqrt{3}}{2}$, $\dfrac{-\pi}{2} \leq \theta \leq \dfrac{\pi}{2}$

9. $\tan \theta = -\sqrt{3}$, $\dfrac{-\pi}{2} < \theta < \dfrac{\pi}{2}$ **10.** $\cot \theta = -\sqrt{3}$, $0 < \theta < \pi$

▶   Solve each of the following equations for θ as prescribed, and then write an expression for the set of *all* solutions.

11. $\cos \theta = \dfrac{\sqrt{3}}{2}$, $0 \leq \theta \leq 2\pi$ **12.** $\sin \theta = \dfrac{\sqrt{3}}{2}$, $0 \leq \theta \leq 2\pi$

13. $\tan \theta = \sqrt{3}$, $0 < \theta < \pi$ **14.** $\cot \theta = 1$, $0 < \theta < \pi$

15. $\sec \theta = 1$, $0 \leq \theta \leq \pi$ **16.** $\csc \theta = 1$, $\dfrac{-\pi}{2} \leq \theta \leq \dfrac{\pi}{2}$

17. $\cos \theta = \dfrac{-\sqrt{3}}{2}$, $0 \leq \theta \leq 2\pi$ **18.** $\sin \theta = \dfrac{-\sqrt{3}}{2}$, $0 \leq \theta \leq 2\pi$

19. $\tan \theta = -\sqrt{3}$, $0 \leq \theta \leq \pi$ **20.** $\cot \theta = -1$, $0 < \theta < \pi$

▶   Solve each of the following equations.

21. $5 \tan \theta - 2 \tan \theta = \sqrt{3}$ **22.** $2 \cot \theta - 5 = \cot \theta - 4$

23. $\sec \theta = 6 - 2 \sec \theta$ **24.** $5 \csc \theta = 2 \csc \theta - 2\sqrt{3}$

25. $5 \cos \theta - 1 = 7 \cos \theta$ **26.** $\sin^3 \theta + \sin^2 \theta = 0$

27. $2 \cos^2 \theta + \cos \theta = 0$ **28.** $\tan^2 \theta - \tan \theta = 0$

29. $\cos^2 \theta - \cos \theta - 2 = 0$ **30.** $\sec^2 \theta - \sec \theta - 2 = 0$

31. $\tan \theta \sin \theta - \sin \theta = 0$ **32.** $\cot \theta \cos \theta + \cos \theta = 0$

33. $\tan \theta \sec \theta + \sqrt{3} \sec \theta = 0$ **34.** $\tan^2 \theta - 3 = 0$

35. $\sin^2 \theta - \frac{1}{4} = 0$ **36.** $3 \cos^2 \theta + 2 \sin^2 \theta = 3$

37. $2 \sec^2 \theta - \tan^2 \theta = 4$ **38.** $16 \sin^2 \theta \cos^2 \theta = 1$

39. $\tan \theta - 3 \cot \theta = 0$ **40.** $\cot^2 \theta + 2 \csc^2 \theta = 3$

41. $3 \cos 2\theta - 4 \cos^2 \theta + 2 = 0$

42. $16 \sin^2 \theta \cos^2 \theta - 3 \sin^2 2\theta = 1$

43. $\cos 2\theta - 2 \sin^2 \theta + 1 = \sqrt{3}$

44. $2 \cos 2\theta \sin^2 2\theta + 2 \cos^3 2\theta + 1 = 0$

45. $\cos^2 2\theta + \sin^2 2\theta = \cos^2 2\theta$

► Use Table III and 3.14 for π to find approximations to the solutions of each of the following equations.

46. $4 \cos^2 \theta - \cos \theta - 3 = 0$ **47.** $8 \sin^2 \theta - 15 \sin \theta + 7 = 0$

48. $\tan^2 \theta - 2 \tan \theta - 3 = 0$ **49.** $\sec^2 \theta - 6 \sec \theta + 8 = 0$

50. $\cot^2 \theta + 4 \cot \theta - 5 = 0$

► In each of the following, find θ as prescribed for each of the given values of x.

51. $x = 3 \cos \theta;\ 0 \leq \theta \leq \pi;\ x = 0,\ x = 3$

52. $x = 7 \tan \theta;\ \dfrac{-\pi}{2} < \theta < \dfrac{\pi}{2};\ x = 0,\ x = 7$

53. $x = 3 \sin \theta;\ \dfrac{-\pi}{2} \leq \theta \leq \dfrac{\pi}{2};\ x = \sqrt{3},\ x = 3$

54. $x = 3 \sec \theta;\ 0 \leq \theta < \dfrac{\pi}{2};\ x = 3,\ x = 2\sqrt{3}$

55. $x = 7 \sin \theta;\ \dfrac{-\pi}{2} \leq \theta \leq \dfrac{\pi}{2};\ x = \dfrac{7}{2},\ x = 7$

CHECK LIST FOR THE STUDENT

If you have learned the material in Section 3.9, you should be able to do the following:

1. Solve conditional equations of the form $\sin \theta = a$, $\cos \theta = a$, $\tan \theta = a$, $\cot \theta = a$, $\sec \theta = a$, or $\csc \theta = a$ for certain real numbers a.

2. Use algebraic methods such as solution by factoring to solve certain conditional trigonometric equations.

3. Use identities and algebraic methods to solve certain conditional trigonometric equations.

SECTION QUIZ 3.9

1. (a) Solve the equation $\cos \theta = \frac{-1}{2}$ for $0 \leq \theta \leq \pi$.

    (b) Solve the equation $\cos \theta = \frac{-1}{2}$ for $0 \leq \theta \leq 2\pi$. Write an expression for the set of *all* solutions of $\cos \theta = \frac{-1}{2}$.

2. Solve the equation $\tan \theta = -1$ for $-\pi/2 < \theta < \pi/2$. Write an expression for the set of *all* solutions of $\tan \theta = -1$.

3. Solve the equation $\cos^2 \theta - \frac{1}{2} \cos \theta = 0$.

4. Solve the equation $\sin \theta \tan \theta - \sin \theta = 0$.

5. Solve the equation $2 \sec^2 \theta + \tan^2 \theta = 3$.

3.10 Polar Coordinates

The purpose of this section is to define polar coordinates, to show the relationship between Cartesian coordinates and polar coordinates, and to demonstrate some graphs in polar coordinates.

Using Cartesian coordinates, we locate a point in a plane by referring to two mutually perpendicular lines called the coordinate axes. Sometimes it is more convenient to locate a point in a plane by using *polar coordinates* and referring to a fixed point and a fixed ray (half-line) emanating from the fixed point. The fixed point is called the **pole** and is usually labeled O. The fixed ray is called the **polar axis** and is usually shown along the horizontal like the positive half of the x-axis. For convenience, we shall let A represent any point other than O on the polar axis, and we shall denote the polar axis by $\overrightarrow{OA}$. A polar coordinate system is shown in Figure 3.79.

FIGURE 3.79   A polar coordinate system

The members r and θ of an ordered pair (r, θ) are called **polar coordinates** of a point P provided that the first coordinate r specifies the distance from P to the pole, and that the second coordinate θ specifies how much the polar axis would have to be rotated to make it coincide with $\overrightarrow{OP}$. In particular, P is located a distance of $|r|$ units from the pole, and if the polar axis is rotated about the pole through the angle θ to position $\overrightarrow{OA}\,'$, then P is located either on $\overrightarrow{OA}\,'$ (if $r > 0$) or on $\overrightarrow{OA}\,'$ extended (if $r < 0$). To locate point P with polar coordinates (r, θ) we proceed as follows:

1. Rotate polar axis $\overrightarrow{OA}$ through θ to the position $\overrightarrow{OA}\,'$. (Rotate counterclockwise if $\theta > 0$; rotate clockwise if $\theta < 0$.)

2. If r is positive, P is r units from O on $\overrightarrow{OA}\,'$.

3. If r is negative, P is $|r|$ units from O on $\overrightarrow{OA}\,'$ extended. ($\overrightarrow{OA}\,'$ extended is shown as a dashed half-line in Figure 3.80.)

4. If r is 0, P is at O.

Point P having polar coordinates (r, θ) with r and θ both positive is shown in Figure 3.80.

The second member of a pair of polar coordinates may be specified by giving either its degree measure or its radian measure. For example, the point $(2, 30°)$ is the same as the point with polar coordinates $(2, \pi/6)$.

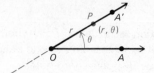

FIGURE 3.80 (r, θ) with r and θ positive

If radian measure is used, polar coordinates simply look like an ordered pair of real numbers. In this case, it may be necessary to specify that they are actually polar coordinates so that the ordered pair will not be mistaken for Cartesian coordinates.

EXAMPLE 1   Plot the points with polar coordinates $(2, 30°)$, $(-2, 30°)$, $(2, -30°)$, and $(-2, -30°)$.

Solution   The points are plotted in Figure 3.81.

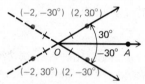

Exercise 3.10, 1–10   FIGURE 3.81   Some points on a polar coordinate system

In Cartesian coordinates, each point in the plane has exactly one pair of coordinates. There is a one-to-one correspondence between Cartesian coordinates and points in the plane. In polar coordinates, however, this is not the case. It is true that any given pair of polar coordinates locates exactly one point, but any one point has an unlimited number of different pairs of polar coordinates. For example, the point P with polar coordinates $(2, 30°)$ also has polar coordinates $(-2, 180° + 30°)$ or $(-2, 210°)$. Furthermore, the same point can be located by first rotating through $30°$ *plus* any integral multiple of $360°$ or by first rotating through $210°$ *plus* any integral multiple of $360°$. Several pairs of polar coordinates of this point P are shown in Figure 3.82. In general, the point P with polar coordinates $(2, 30°)$ and $(-2, 210°)$ also has polar coordinates $(2, 30° + k \cdot 360°)$ and $(-2, 210° + k \cdot 360°)$ where k represents any integer. Generalizing still further, and using radian measure, we may say that

(r, θ) also has polar coordinates $(r, \theta + 2k\pi)$ and $(-r, \theta + \pi + 2k\pi)$

where k represents any integer.

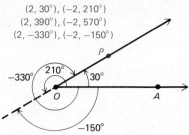

(2, 30°), (−2, 210°)
(2, 390°), (−2, 570°)
(2, −330°), (−2, −150°)

FIGURE 3.82 Some pairs of polar coordinates of (2, 30°)

EXAMPLE 2

Find three other pairs of polar coordinates of the point (−2, 60°) such that the second coordinate of each pair is between −360° and 360°.

Solution

The point (−2, 60°) is plotted in Figure 3.83. Other polar coordinates of (−2, 60°) are (2, 60° + 180°) or (2, 240°), (2, 60° − 180°) or (2, −120°), and (−2, 60° − 360°) or (−2, −300°).

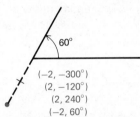

60°

(−2, −300°)
(2, −120°)
(2, 240°)
(−2, 60°)

Exercise 3.10, 11–20   **FIGURE 3.83** The point (−2, 60°)

If a Cartesian coordinate system is superimposed on a polar coordinate system with the origin at the pole and the positive *x*-axis along the polar axis, any particular point *P* in the plane can be located with either Cartesian coordinates or polar coordinates. The relationship between the Cartesian coordinates (*x*, *y*) and polar coordinates (*r*, θ) of a point *P* can be read off from the right triangle *OPM* shown in Figure 3.84. Referring to triangle *OPM*, we see that

$$\frac{x}{r} = \cos \theta \quad \text{and} \quad \frac{y}{r} = \sin \theta.$$

In other words,

$$x = r \cos \theta$$
$$y = r \sin \theta$$

Relationship between Cartesian coordinates (*x*, *y*) and polar coordinates (*r*, θ) of a point *P*.

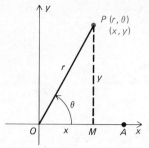

FIGURE 3.84 $x = r \cos \theta$
$y = r \sin \theta$

These two equations are sometimes called *transformation equations.* They may be used to transform from polar coordinates to Cartesian coordinates. The point P in Figure 3.84 is shown with r positive. If r is negative or zero, the transformation equations still hold.

Two relationships that will enable us to transform from Cartesian coordinates to polar coordinates are readily derivable from the transformation equations $x = r \cos \theta$, $y = r \sin \theta$. These two relationships are

$$r^2 = x^2 + y^2$$

and $$\tan \theta = \frac{y}{x}.$$

The first of these relationships may be obtained by squaring and adding the equations $x = r \cos \theta$, $y = r \sin \theta$ and then using the identity $\cos^2 \theta + \sin^2 \theta = 1$.

$$x^2 = r^2 \cos^2 \theta$$
$$y^2 = r^2 \sin^2 \theta$$
$$x^2 + y^2 = r^2 \cos^2 \theta + r^2 \sin^2 \theta$$
$$= r^2(\cos^2 \theta + \sin^2 \theta)$$
$$= r^2$$

It follows that

$$r^2 = x^2 + y^2.$$

This relationship may also be obtained by noting that $|r|$ is the distance between $(0, 0)$ and (x, y) so that $r^2 = (x - 0)^2 + (y - 0)^2$.

The relationship $\tan \theta = y/x$ may be obtained by dividing $y = r \sin \theta$ by $x = r \cos \theta$ and using the identity $\tan \theta = (\sin \theta)/(\cos \theta)$.

$$y = r \sin \theta$$

$$\frac{y}{x} = \frac{r \sin \theta}{r \cos \theta} = \tan \theta$$

$$\tan \theta = \frac{y}{x}.$$

EXAMPLE 3   Find the Cartesian coordinates of the point with polar coordinates $(2, 30°)$.

Solution $x = r \cos \theta$ and $y = r \sin \theta$

$x = 2 \cos 30°$ $y = 2 \sin 30°$

$$x = 2\left(\frac{\sqrt{3}}{2}\right) \qquad y = 2\left(\frac{1}{2}\right)$$

$\quad = \sqrt{3}$ $\quad = 1$

Exercise 3.10, 21–30   The Cartesian coordinates are $(\sqrt{3}, 1)$.

EXAMPLE 4   Find a pair of polar coordinates of the point with Cartesian coordinates $(-3, 3)$.

Solution   The point $(-3, 3)$ is in the second quadrant, and we must choose r and θ accordingly.

$$r^2 = x^2 + y^2$$
$$= 9 + 9$$
$$= 18$$
$$r = \pm 3\sqrt{2}$$

We may use either $3\sqrt{2}$ or $-3\sqrt{2}$ for r, but we must choose a combination of r and θ so that the point (r, θ) will be in the second quadrant.

$$\tan \theta = \frac{y}{x}$$

$$= \frac{3}{-3}$$

$$= -1$$

$$\tan \theta = -1$$

The set of all solutions of this equation is

$$\left\{\theta \,\middle|\, \theta = \frac{3\pi}{4} + k\pi \text{ or } \theta = \frac{7\pi}{4} + k\pi\right\}$$

where k represents any integer. We must use one of these values of θ. If we use $3\pi/4$, we must use $r = +3\sqrt{2}$, and one pair of polar coordinates of the point $(-3, 3)$ is $(3\sqrt{2}, 3\pi/4)$. Using $7\pi/4$, another pair of polar coordinates of $(-3, 3)$ is $(-3\sqrt{2}, 7\pi/4)$. There is an unlimited number
Exercise 3.10, 31–40   of pairs of polar coordinates of $(-3, 3)$.

If an equation is written in polar coordinates, it is said to be in **polar form**; if it is written in Cartesian coordinates, it is said to be in **Cartesian form**. The relationships between Cartesian and polar coordinates that we have developed may be used to transform equations from one of these forms to the other.

EXAMPLE 5

Write the equation $x^2 + y^2 - 6y + 5 = 0$ in polar form.

Solution

In the given equation, let $x = r \cos \theta$ and $y = r \sin \theta$.

$$r^2 \cos^2 \theta + r^2 \sin^2 \theta - 6r \sin \theta + 5 = 0$$

This is a polar form of the given equation. It may be simplified by using the identity $\cos^2 \theta + \sin^2 \theta = 1$.

$$r^2(\cos^2 \theta + \sin^2 \theta) - 6r \sin \theta + 5 = 0$$
$$r^2 - 6r \sin \theta + 5 = 0$$

Exercise 3.10, 41–50

EXAMPLE 6

Write the equation $r - r \cos \theta = 2$ in Cartesian form.

Solution

We have seen that

$$r = \pm \sqrt{x^2 + y^2}$$

and

$$x = r \cos \theta.$$

Making these substitutions in the given equation, we have

$$\pm \sqrt{x^2 + y^2} - x = 2.$$

This is a Cartesian form of the given equation. It may be simplified by first isolating the radical and squaring both sides.

$$\pm \sqrt{x^2 + y^2} = x + 2$$
$$x^2 + y^2 = x^2 + 4x + 4$$
$$y^2 = 4x + 4$$

Exercise 3.10, 51–60

The graph of an equation in polar coordinates is the set of all points having a pair of polar coordinates that satisfy the equation. Some relatively common graphs are illustrated in the following examples.

EXAMPLE 7

Sketch the graph of $r = 3$.

Solution

The variable θ is not present in the given equation. Any point with $r = 3$, regardless of its second coordinate θ, will satisfy the equation. The points on the graph must all be located at a distance of 3 units from the pole. That is, the graph must be the circle with center at the pole and radius 3. Any equation of the form $r = c$ where c is a specific real number represents a circle with center at the pole and radius $|c|$.

The graph of $r = 3$ is shown in Figure 3.85. It could have been sketched by first transforming the given equation to the Cartesian form $x^2 + y^2 = 9$.

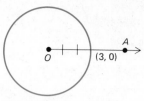

Exercise 3.10, 61–62 **FIGURE 3.85** $r = 3$

EXAMPLE 8

Sketch the graph of $\theta = \pi/6$.

Solution

The variable r is not present in this equation. Any point with $\theta = \pi/6$, regardless of the value of r, will satisfy the equation. The graph must be the line through the pole making an angle of $\pi/6$ with the polar axis. Any equation of the form $\theta = c$ where c is a specific real number represents a line through the pole making an angle of c radians with the polar axis. The graph of $\theta = \pi/6$ is shown in Figure 3.86.

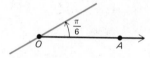

Exercise 3.10, 63–66 **FIGURE 3.86** $\theta = \pi/6$

EXAMPLE 9

Sketch the graph of $r = 6 \sin \theta$.

Solution

A table of values showing how r varies as θ varies from 0 to 2π will help us picture the graph. The entries in the table take into account variations of r as θ varies through each quadrant.

$r = 6 \sin \theta$	θ
$0 \rightarrow 6$	$0 \rightarrow \dfrac{\pi}{2}$
$6 \rightarrow 0$	$\dfrac{\pi}{2} \rightarrow \pi$
$0 \rightarrow -6$	$\pi \rightarrow \dfrac{3\pi}{2}$
$-6 \rightarrow 0$	$\dfrac{3\pi}{2} \rightarrow 2\pi$

The graph of $r = 6 \sin \theta$ is shown in Figure 3.87. It is a circle with center at $(3, \pi/2)$ and radius 3. Any equation of the form $r = 2a \sin \theta$ where a is a specific real number represents a circle with center at $(a, \pi/2)$ and radius $|a|$.

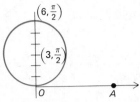

FIGURE 3.87 $r = 6 \sin \theta$

In Example 9 we could have sketched the graph by first transforming the given equation to Cartesian form, because the Cartesian form is a familiar one. The Cartesian form may be obtained as follows.

$$r = 6 \sin \theta$$

$$r = 6\frac{y}{r}$$

$$r^2 = 6y$$

$$x^2 + y^2 = 6y$$

Putting this in standard form by completing squares, we have

$$x^2 + (y - 3)^2 = 9.$$

This is the circle whose center has Cartesian coordinates $(0, 3)$ and whose radius is 3.

EXAMPLE 10   Sketch the graph of $r = 6 \cos \theta$.

Solution   A table of values showing how r varies as θ varies through each quadrant from 0 to 2π is as follows:

$r = 6 \cos \theta$	θ
$6 \to 0$	$0 \to \dfrac{\pi}{2}$
$0 \to -6$	$\dfrac{\pi}{2} \to \pi$
$-6 \to 0$	$\pi \to \dfrac{3\pi}{2}$
$0 \to 6$	$\dfrac{3\pi}{2} \to 2\pi$

The graph is shown in Figure 3.88. It is a circle with center at (3, 0) and radius 3. Any equation of the form $r = 2a \cos \theta$ where a is a specific real number is a circle with center at $(a, 0)$ and radius $|a|$.

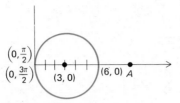

FIGURE 3.88 $r = 6 \cos \theta$

The graph in Example 10 could have been obtained by first transforming the given equation to Cartesian form and then recognizing the Cartesian form. It is left as an exercise for the student to show that the Cartesian form of $r = 6 \cos \theta$ is $x^2 + y^2 = 6x$, which may be written

Exercise 3.10, 67–70

$(x - 3)^2 + y^2 = 9$.

EXAMPLE 11 Sketch the graph of $r = \frac{1}{3}\theta$ for $\theta > 0$.

Solution In this case, as θ increases, r also increases. The graph is sketched in Figure 3.89. It is called a *spiral of Archimedes.* Any equation of the form $r = a\theta$ where a is a specific real number represents a spiral of Archimedes. It spirals counterclockwise, as shown in Figure 3.89, if $a\theta > 0$; it spirals clockwise if $a\theta < 0$.

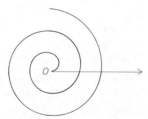

Exercise 3.10, 71–75 **FIGURE 3.89** $r = \frac{1}{3}\theta$; $\theta > 0$ (spiral of Archimedes)

EXAMPLE 12 Sketch the graph of $r = 3 + 2 \cos \theta$.

Solution A table of values showing how r and θ vary as θ varies through each quadrant from 0 to 2π is as follows:

$r = 3 + 2 \cos \theta$	θ
$5 \to 3$	$0 \to \dfrac{\pi}{2}$
$3 \to 1$	$\dfrac{\pi}{2} \to \pi$
$1 \to 3$	$\pi \to \dfrac{3\pi}{2}$
$3 \to 5$	$\dfrac{3\pi}{2} \to 2\pi$

Since the cosine is periodic with period 2π, continuing the table we would simply repeat the existing entries and hence effectively retrace the graph.

The graph of $r = 3 + 2 \cos \theta$ is shown in Figure 3.90. It is called a *limaçon*. The graph of any equation of the form $r = a + b \cos \theta$ where a and b are specific real numbers with $a \geq b$ looks similar to this graph. If $a = b$, the graph dips in sharply to the pole and is then called a *cardioid*. The graph of any equation of the form $r = a + b \sin \theta$ with $a \geq b$ also looks similar to this except that it is symmetric to the line $\theta = \pi/2$ rather than to the polar axis. In calculus one learns how to find the slope of the line tangent to a graph at any point on it, and with this information these graphs can be drawn more precisely.

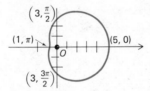

Exercise 3.10, 76–82   **FIGURE 3.90** $r = 3 + 2 \cos \theta$ (limaçon)

EXAMPLE 13

Solution

Sketch the graph of $r = 2 \cos 3\theta$.

In order to trace out the graph, we consider how r varies as θ varies from 0 to 2π. But as θ varies from 0 to 2π, 3θ varies from 0 to 6π, and the values of $\cos 3\theta$ will repeat each time 3θ varies through 2π. Table 3.4 is a table of values. The graph is shown in Figure 3.91. In this case, the graph is completely traced out as θ varies from 0 to π. Letting θ vary from π to 2π retraces the graph.

The graph in Figure 3.91 is called a *rose curve*. Any equation of the form $r = a \cos n\theta$ or $r = a \sin n\theta$ where a is a specific real number and n is a specific positive integer is a rose curve. It can be shown that if n is odd, Exercise 3.10, 83–90 there are exactly n leaves (loops); if n is even, there are exactly $2n$ leaves.

In calculus one learns to solve problems such as finding the area of a region bounded by graphs whose equations are given in polar form, find-

466   PART THREE | PRECALCULUS TRIGONOMETRY

TABLE 3.4   TABLE OF VALUES FOR $r = 2 \cos 3\theta$

r	θ	3θ	$\cos 3\theta$
$2 \to 0$	$0 \to \dfrac{\pi}{6}$	$0 \to \dfrac{\pi}{2}$	$1 \to 0$
$0 \to -2$	$\dfrac{\pi}{6} \to \dfrac{\pi}{3}$	$\dfrac{\pi}{2} \to \pi$	$0 \to -1$
$-2 \to 0$	$\dfrac{\pi}{3} \to \dfrac{\pi}{2}$	$\pi \to \dfrac{3\pi}{2}$	$-1 \to 0$
$0 \to 2$	$\dfrac{\pi}{2} \to \dfrac{2\pi}{3}$	$\dfrac{3\pi}{2} \to 2\pi$	$0 \to 1$
$2 \to 0$	$\dfrac{2\pi}{3} \to \dfrac{5\pi}{6}$	$2\pi \to \dfrac{5\pi}{2}$	$1 \to 0$
$0 \to -2$	$\dfrac{5\pi}{6} \to \pi$	$\dfrac{5\pi}{2} \to 3\pi$	$0 \to -1$
$-2 \to 0$	$\pi \to \dfrac{7\pi}{6}$	$3\pi \to \dfrac{7\pi}{2}$	$-1 \to 0$
$0 \to 2$	$\dfrac{7\pi}{6} \to \dfrac{4\pi}{3}$	$\dfrac{7\pi}{2} \to 4\pi$	$0 \to 1$
$2 \to 0$	$\dfrac{4\pi}{3} \to \dfrac{3\pi}{2}$	$4\pi \to \dfrac{9\pi}{2}$	$1 \to 0$
$0 \to -2$	$\dfrac{3\pi}{2} \to \dfrac{5\pi}{3}$	$\dfrac{9\pi}{2} \to 5\pi$	$0 \to -1$
$-2 \to 0$	$\dfrac{5\pi}{3} \to \dfrac{11\pi}{6}$	$5\pi \to \dfrac{11\pi}{2}$	$-1 \to 0$
$0 \to 2$	$\dfrac{11\pi}{6} \to 2\pi$	$\dfrac{11\pi}{2} \to 6\pi$	$0 \to 1$

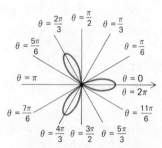

FIGURE 3.91 $r = 2 \cos 3\theta$

ing the volume generated by revolving such a region about a line, and finding the length of a segment of a graph whose equation is given in polar form.

EXERCISE 3.10 POLAR COORDINATES

▶ Plot and label the pairs of polar coordinates in each of the following.

1. $(3, 45°), (-3, 45°), (3, -45°), (-3, -45°)$

2. $\left(5, \dfrac{\pi}{4}\right), \left(-5, \dfrac{\pi}{4}\right), \left(5, \dfrac{-\pi}{4}\right), \left(-5, \dfrac{-\pi}{4}\right)$

3. $(1, 30°), (-1, 30°), \left(1, \dfrac{-\pi}{6}\right), \left(-1, \dfrac{-\pi}{6}\right)$

4. $\left(4, \dfrac{\pi}{3}\right), \left(-4, \dfrac{\pi}{3}\right), (4, -60°), (-4, -60°)$

5. $(3, 0), (-3, 0), \left(3, \dfrac{\pi}{2}\right), \left(-3, \dfrac{\pi}{2}\right)$

6. $(3, \pi), (-3, \pi), \left(3, \dfrac{3\pi}{2}\right), \left(-3, \dfrac{3\pi}{2}\right)$

7. $(0, 30°), \left(0, \dfrac{\pi}{3}\right), (0, 0), \left(0, \dfrac{\pi}{4}\right)$

8. $(1, 0), (2, 0), (3, 0), (4, 0)$

9. $\left(3, \dfrac{3\pi}{4}\right), \left(2, \dfrac{5\pi}{4}\right), \left(-1, \dfrac{7\pi}{4}\right), \left(4, \dfrac{11\pi}{4}\right)$

10. $(3, 10°), \left(2, \dfrac{\pi}{5}\right), (-4, 100°), \left(-2, \dfrac{2\pi}{5}\right)$

▶ In each of the following, find three other pairs of polar coordinates (r, θ) of the given point such that $-360° \leq \theta \leq 360°$.

11. $(3, 45°)$ **12.** $(-3, 150°)$ **13.** $(2, 135°)$

14. $(3, 90°)$ **15.** $(-3, 270°)$ **16.** $(8, 15°)$

17. $(-7, 80°)$ **18.** $(5, 250°)$ **19.** $(-3, -75°)$

20. $(2, -130°)$

▶ Find the Cartesian coordinates of the point given in each of the following.

21. $(3, 60°)$ **22.** $(-2, 45°)$ **23.** $(3, 270°)$

24. $(-5, 135°)$ **25.** $(3, 135°)$ **26.** $(2, 150°)$

27. $(5, 210°)$ **28.** $(6, 300°)$ **29.** $(-2, -60°)$

30. $(5, -135°)$

▶ Find a pair of polar coordinates (r, θ) of the point given in each of the following, with (r, θ) such that $0° \leq \theta \leq 360°$ and $r \geq 0$.

31. $(-3, -3)$ **32.** $(3, -3)$ **33.** $(2\sqrt{3}, 2)$

34. $(-2\sqrt{3}, 2)$ **35.** $(2\sqrt{3}, -2)$ **36.** $(-2\sqrt{3}, -2)$

37. $(0, 5)$ **38.** $(-5, 0)$ **39.** $(-3, 3\sqrt{3})$

40. $(3, -3\sqrt{3})$

Write each of the following equations in polar form.

41. $x^2 + y^2 - 4x - 5 = 0$ **42.** $x^2 + y^2 = 36$

43. $x^2 + y^2 - 6x - 2y + 6 = 0$ **44.** $y = 3x$

45. $2x - 3y + 5 = 0$

46. $(x^2 + y^2)^2 - 2x(x^2 - 3y^2) = 0$

47. $y^2 = 4x$ **48.** $x^2 = 4(y - 1)$

49. $4x^2 + 9y^2 = 36$ **50.** $4x^2 - 9y^2 = 36$

Write each of the following equations in Cartesian form.

51. $r - r \sin \theta = 2$ **52.** $r = 8 \sin \theta$

53. $r = 3 + 2 \cos \theta$ **54.** $\theta = 3$

55. $r = \dfrac{4}{2 + \cos \theta}$ **56.** $r = \dfrac{4}{1 + \cos \theta}$

57. $r = \dfrac{8}{1 + 2 \cos \theta}$ **58.** $r^2 + 9 - 6r \cos \theta = 16$

59. $r = 2 \cos \theta + 3 \sin \theta$ **60.** $r = \dfrac{6}{1 - 3 \sin \theta}$

Sketch the graph of each of the following.

61. $r = 5$ **62.** $r = -5$ **63.** $\theta = \dfrac{\pi}{3}$

64. $\theta = \dfrac{-\pi}{3}$ **65.** $\theta = \dfrac{\pi}{2}$ **66.** $\theta = 0$

67. $r = 8 \sin \theta$ **68.** $r = 8 \cos \theta$ **69.** $r = -10 \sin \theta$

70. $r = -10 \cos \theta$ **71.** $r = \theta$ **72.** $r = -\theta$

73. $r = -3\theta$ **74.** $r = 2\theta$ **75.** $r = -2\theta$

76. $r = 2 + \cos \theta$ **77.** $r = 3 + 3 \cos \theta$ **78.** $r = 3 + 2 \sin \theta$

79. $r = 3 + 3 \sin \theta$ **80.** $r = 2 + \sin \theta$ **81.** $r = 5 + 3 \cos \theta$

82. $r = 5 + 3 \sin \theta$ **83.** $r = 4 \cos 3\theta$ **84.** $r = 2 \sin 3\theta$

85. $r = 3 \cos 2\theta$ **86.** $r = 3 \sin 2\theta$ **87.** $r = 3 \cos 4\theta$

88. $r = 3 \sin 4\theta$ **89.** $r = 3 \sin 5\theta$ **90.** $r = 3 \cos 5\theta$

CHECK LIST FOR THE STUDENT

If you have learned the material in Section 3.10, you should be able to do the following:

1. Plot a point in a plane with polar coordinates.

2. Write expressions for an unlimited number of pairs of polar coordinates of the point with polar coordinates (r, θ).

3. Find three other pairs of polar coordinates of the point (r, θ) such that the second coordinate of each pair is between $-360°$ and $360°$, inclusive, or between -2π and 2π, inclusive.

4. Give the relationship between Cartesian coordinates (x, y) and polar coordinates (r, θ) of a point P provided that the Cartesian coordinate system is superimposed on the polar coordinate system with the origin at the pole and with the positive x-axis along the polar axis.

5. Find the Cartesian coordinates of a point with given polar coordinates.

6. Find a pair of polar coordinates of a point with given Cartesian coordinates.

7. Transform an equation from Cartesian form to polar form.

8. Transform an equation from polar form to Cartesian form.

9. Sketch the graphs of $r = c$, $\theta = c$, $r = 2a \sin \theta$, $r = 2a \cos \theta$, $r = a\theta$, $r = a + b \cos \theta$ with $a \geq b$, $r = a + b \sin \theta$ with $a \geq b$, $r = a \cos n\theta$, and $r = a \sin n\theta$ where c, a, and b are specific real numbers and n is a specific positive integer.

SECTION QUIZ 3.10

1. Find three other pairs of polar coordinates (r, θ) of the point $(5, 40°)$ such that $-360° \leq \theta \leq 360°$.

2. Find the Cartesian coordinates of the point $(3, -45°)$.

3. Find a pair of polar coordinates (r, θ) of the point $(-5, 5)$, with (r, θ) such that $0° \leq \theta \leq 360°$ and $r \geq 0$.

4. Write the equation $x^2 + y^2 - 6y - 16 = 0$ in polar form.

5. Write the equation $r = 4 \cos \theta - 6 \sin \theta$ in Cartesian form.

Sketch the graph of each of the following.

6. $r = 1$

7. $\theta = \dfrac{\pi}{4}$

8. $r = 12 \cos \theta$

9. $r = 2 + 2 \sin \theta$

10. $r = 3 \cos 3\theta$

3.11 Parametric Equations

The purpose of this section is to introduce parametric equations and to present parametric equations of circles, ellipses, hyperbolas, the path of a projectile, and cycloids.

Any pair of equations giving the two coordinates of each point on a graph in terms of a third variable is called a pair of *parametric equations* of the graph. The third variable is called the *parameter.* For example, if a particle moves so that its position at any time t is given on a Cartesian coordinate system by the equations

$$x = 2t$$
$$y = 6t,$$

then these two equations are called parametric equations of the graph representing the path of the particle. The variable t is the parameter. *An equation of the graph can be found by eliminating the parameter.* In this case, $t = x/2$ so

$$y = 6\left(\frac{x}{2}\right)$$

$$y = 3x.$$

Thus the pair of parametric equations $x = 2t$, $y = 6t$ represents the line $y = 3x$.

The same graph may be represented by an unlimited number of different pairs of parametric equations. For example, the pair

$$\begin{cases} x = 2t + 5 \\ y = 6t + 15 \end{cases}$$

also represents the line $y = 3x$. To see this, eliminate the parameter. Here

$$t = \frac{x - 5}{2}, \quad y = 6\left(\frac{x - 5}{2}\right) + 15$$

or $y = 3x - 15 + 15$; that is, $y = 3x$.

A pair of parametric equations may represent only part of the graph of the equation obtained by eliminating the parameter. For example, consider the pair

$$\begin{cases} x = 2t^2 \\ y = 6t^2. \end{cases}$$

If we eliminate t, we get $y = 3x$, which is an equation of a line through the origin. But the given parametric equations only represent that part of this line that has endpoint at the origin and lies in the first quadrant. The reason for this is that t^2 must be greater than or equal to zero, so both x

and y are greater than or equal to zero. The graph of $y = 3x$ and the graph of $x = 2t^2$, $y = 6t^2$ are shown in Figure 3.92.

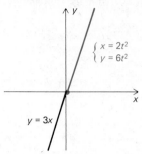

FIGURE 3.92   Graph of $x = 2t^2$, $y = 6t^2$

In applications, it is sometimes more convenient to represent a graph by a pair of parametric equations than it is to represent the graph by a single equation. The parameter may represent time, or it may represent something else. For instance, a convenient parametric representation of a circle has a parameter that represents a certain angle. This is illustrated in the next paragraph.

Consider the circle $x^2 + y^2 = 9$ shown in Figure 3.93. Let P be any point (x, y) on the graph, and let t be the smallest nonnegative standard-position angle with terminal side on $\overline{OP}$. In terms of t, the coordinates of P are

$$x = 3 \cos t$$
$$y = 3 \sin t.$$

This is a pair of parametric equations of the circle $x^2 + y^2 = 9$. In general,

$$\begin{cases} x = r \cos t \\ y = r \sin t \end{cases}$$

is a pair of parametric equations of the circle

$$x^2 + y^2 = r^2.$$

$\begin{cases} x = r \cos t \\ y = r \sin t \end{cases}$ represents the circle $x^2 + y^2 = r^2$

EXAMPLE 1

Eliminate the parameter and obtain an equation in x and y of the graph represented by

$$\begin{cases} x = 3 \cos t + 2 \\ y = 3 \sin t - 5. \end{cases}$$

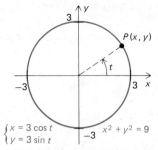

$$\begin{cases} x = 3 \cos t \\ y = 3 \sin t \end{cases} \qquad x^2 + y^2 = 9$$

FIGURE 3.93   Graph of $x = 3 \cos t$, $y = 3 \sin t$

Solution

We sometimes have to exercise a bit of ingenuity in eliminating parameters. In this case, if we write the given equations in the form

$$x - 2 = 3 \cos t$$
$$y + 5 = 3 \sin t,$$

and if we then square both sides of each of them and add, we will be able to utilize the identity $\cos^2 t + \sin^2 t = 1$ and hence eliminate t. Squaring both sides of each of the given equations, we have

$$(x - 2)^2 = 9 \cos^2 t$$
$$(y + 5)^2 = 9 \sin^2 t.$$

Now adding, we have

$$(x - 2)^2 + (y + 5)^2 = 9 \cos^2 t + 9 \sin^2 t$$
$$= 9(\cos^2 t + \sin^2 t)$$
$$= 9.$$

The given parametric equations represent the circle

$$(x - 2)^2 + (y + 5)^2 = 9$$

with center at $(2, -5)$ and radius 3.

We can use the same technique used in Example 1 to show that the pair of equations

$$\begin{cases} x = a \cos t \\ y = b \sin t \end{cases}$$

represents an ellipse. If we write the equations in the form

$$\frac{x}{a} = \cos t$$

$$\frac{y}{b} = \sin t,$$

and then if we square both sides of each equation and add, we have

$$\frac{x^2}{a^2} + \frac{y^2}{b^2} = \cos^2 t + \sin^2 t$$

$$\frac{x^2}{a^2} + \frac{y^2}{b^2} = 1.$$

This is the standard form of an equation of an ellipse (see Section 2.6).

$$\begin{cases} x = a \cos t \\ y = b \sin t \end{cases} \text{ represents the ellipse } \frac{x^2}{a^2} + \frac{y^2}{b^2} = 1.$$

EXAMPLE 2

Eliminate the parameter from the equations $x = 3 \sec t$, $y = 2 \tan t$. Identify the graph.

Solution

Recall that $1 + \tan^2 t = \sec^2 t$, so that $\sec^2 t - \tan^2 t = 1$. If we write the given equations in the form

$$\frac{x}{3} = \sec t$$

$$\frac{y}{2} = \tan t,$$

and if we then square both sides of each and subtract, we will eliminate the parameter. Squaring both sides of each, we have

$$\frac{x^2}{9} = \sec^2 t$$

$$\frac{y^2}{4} = \tan^2 t.$$

Subtracting, we have

$$\frac{x^2}{9} - \frac{y^2}{4} = \sec^2 t - \tan^2 t$$

$$\frac{x^2}{9} - \frac{y^2}{4} = 1.$$

This is an equation of a hyperbola (see Section 2.6).

The parametric equations presented in Example 2 represent a hyperbola. In general, the pair of equations

$$\begin{cases} x = a \sec t \\ y = b \tan t \end{cases}$$

represents a hyperbola. We can eliminate the parameter by using the same technique used in Example 2. Writing the equations in the form

$$\frac{x}{a} = \sec t$$

$$\frac{y}{b} = \tan t$$

and squaring both sides of each, we have

$$\frac{x^2}{a^2} = \sec^2 t$$

$$\frac{y^2}{b^2} = \tan^2 t.$$

Subtracting, we have

$$\frac{x^2}{a^2} - \frac{y^2}{b^2} = \sec^2 t - \tan^2 t$$

$$\frac{x^2}{a^2} - \frac{y^2}{b^2} = 1.$$

$$\begin{cases} x = a \sec t \\ y = b \tan t \end{cases} \quad \text{represents the hyperbola} \quad \frac{x^2}{a^2} - \frac{y^2}{b^2} = 1.$$

Similarly, it can be shown that

Exercise 3.11, 1–20

$$\begin{cases} x = a \tan t \\ y = b \sec t \end{cases} \quad \text{represents the hyperbola} \quad \frac{y^2}{b^2} - \frac{x^2}{a^2} = 1.$$

The graph representing the path of a projectile is part of a parabola that is usually represented by a pair of parametric equations. Such a graph is shown in Figure 3.94. The projectile was fired from the origin at an angle A with the positive x-axis. In physics it can be shown that if the

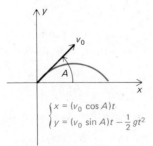

$$\begin{cases} x = (v_0 \cos A)t \\ y = (v_0 \sin A)t - \frac{1}{2} gt^2 \end{cases}$$

(t represents time after fire)

FIGURE 3.94   Path of a projectile

initial velocity is v_0 meters per second, if we neglect air resistance, and if acceleration due to gravity is g meters per second per second, then the position of the particle at any time t seconds after fire is given by the pair of parametric equations

$$\begin{cases} x = (v_0 \cos A)t \\ y = (v_0 \sin A)t - \tfrac{1}{2}gt^2. \end{cases} \qquad \text{Path of a projectile}$$

This pair of equations represents the parabolic path shown in Figure 3.94. (At any given locality, the g is a constant approximately equal to 9.8 meters per second per second.) The next example demonstrates the special case in which the initial velocity v_0 is ten meters per second and angle A is $\pi/3$.

EXAMPLE 3

The pair of parametric equations

$$x = \left(10 \cos \frac{\pi}{3}\right)t$$

$$y = \left(10 \sin \frac{\pi}{3}\right)t - \frac{1}{2}gt^2$$

represents the path of a projectile fired from the origin with an initial velocity of ten meters per second and at an angle of $\pi/3$ with the positive x-axis. The parameter t represents time in seconds after fire. Take g to be 9.8 meters per second per second, and eliminate the parameter. Identify the graph.

Solution

Taking g to be 9.8 meters per second per second, the equations become

$$x = \left(10 \cos \frac{\pi}{3}\right)t$$

$$y = \left(10 \sin \frac{\pi}{3}\right)t - \frac{9.8}{2}t^2$$

or

$$x = 5t$$
$$y = 5\sqrt{3}\,t - 4.9t^2.$$

We can eliminate t by solving the first equation for t and substituting in the second equation. Solving the first equation for t, we have

$$t = \frac{x}{5}.$$

Substituting in the second equation, we have

$$y = 5\sqrt{3}\left(\frac{x}{5}\right) - 4.9\left(\frac{x}{5}\right)^2$$

$$y = \sqrt{3}x - \frac{4.9}{25}x^2.$$

This equation can be written in the form

$$y = \frac{-4.9}{25}x^2 + \sqrt{3}x.$$

It is a special case of the equation

$$y = ax^2 + bx + c.$$

Its graph is a parabola with axis parallel to the y-axis. (See the discussion of quadratic functions in Section 2.5). It opens down. Since the x-axis represents the ground, the path of the projectile is that part of this parabola that lies above the x-axis.

Exercise 3.11, 21–24

One graph that appears in calculus applications and is almost always represented by parametric equations is the *cycloid.* Part of a cycloid is shown in Figure 3.95. A cycloid consists of all the points traced out by a fixed point on the circumference of a circle as it rolls along a fixed line. In Figure 3.95 the rolling circle has radius a. The point P on the circle went through the origin and traced out the graph shown as the circle rolled to the right along the x-axis. A pair of parametric equations of this cycloid is

$$\begin{cases} x = a(t - \sin t) \\ y = a(1 - \cos t). \end{cases} \qquad \text{Cycloid}$$

We can see that this pair does represent a cycloid by examining Figure 3.95. In that figure, the circle has rolled through angle t since P passed through the origin. Since passing through the origin, P has covered the distance OR, which is equal to the length of arc PR. But if t is measured in radians, the length of arc PR equals at (see Section 3.1), so

$$OR = at.$$

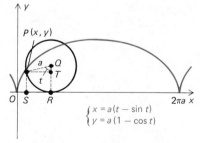

$$\begin{cases} x = a(t - \sin t) \\ y = a(1 - \cos t) \end{cases}$$

FIGURE 3.95 A cycloid

The x-coordinate of P is OS.

$$x = OS$$
$$= OR - SR$$
$$= OR - PT$$
$$= at - a \sin t$$
$$x = a(t - \sin t)$$

The y-coordinate of P is SP.

$$y = SP$$
$$= RT$$
$$= RQ - TQ$$
$$= a - a \cos t$$
$$y = a(1 - \cos t)$$

The parameter t can be eliminated from the given parametric equations of a cycloid, but the resulting equation in x and y is so complicated that the graph is usually represented by parametric equations.

EXAMPLE 4

The pair of equations

$$\begin{cases} x = 3(t - \sin t) \\ y = 3(1 - \cos t) \end{cases}$$

represents a cycloid. Find the point on the graph that corresponds to $t = \pi/2$.

Solution

$$x = 3\left(\frac{\pi}{2} - \sin \frac{\pi}{2}\right) \qquad y = 3\left(1 - \cos \frac{\pi}{2}\right)$$

$$= 3\left(\frac{\pi}{2} - 1\right) \qquad\qquad = 3(1 - 0)$$

$$= \frac{3\pi - 6}{2} \qquad\qquad\qquad = 3$$

Exercise 3.11, 25–30

EXERCISE 3.11 PARAMETRIC EQUATIONS

▶   In each of the following, eliminate the parameter and identify the graph.

EXAMPLES

(a) $x = 5 \sin t + 3$, $y = 5 \cos t - 1$
(b) $x = 3t$, $y = 9t^2 - 6t + 1$

Solutions

(a) $(x - 3) = 5 \sin t$
    $(y + 1) = 5 \cos t$

$$(x - 3)^2 + (y + 1)^2 = 25 \sin^2 t + 25 \cos^2 t$$
$$(x - 3)^2 + (y + 1)^2 = 25; \quad \text{A circle}$$

(b) $t = \dfrac{x}{3}, \quad y = 9\left(\dfrac{x^2}{9}\right) - 6\left(\dfrac{x}{3}\right) + 1$

$$y = x^2 - 2x + 1; \quad \text{A parabola}$$

1. $x = 4 \cos t - 2, \ y = 4 \sin t + 1$

2. $x = 5 \cos t, \ y = 5 \sin t$ 3. $x = 5 \cos t, \ y = 12 \sin t$

4. $x = 12 \sin t, \ y = 5 \cos t$ 5. $x = 3t + 1, \ y = t + 4$

6. $x = 2t - 5, \ y = t - 1$ 7. $x = 3t^2, \ y = 12t^2$

8. $x = 3t - 1, \ y = 12t - 4$ 9. $x = 3 \tan t, \ y = 4 \sec t$

10. $x = 5 \sec t, \ y = 2 \tan t$

11. $x = 3 \cos t - 1, \ y = 4 \sin t + 2$

12. $x = 3 \sec t - 1, \ y = 4 \tan t + 2$

13. $x = 2 \cot t, \ y = 3 \csc t$ 14. $x = 5 \csc t, \ y = 4 \cot t$

15. $x = 2 \cot t + 5, \ y = 3 \csc t - 1$

16. $x = 3t, \ y = 9t^2$ 17. $x = 9t^2, \ y = 3t$

18. $x = 2t, \ y = 16t^2 + 48t + 37$

19. $x = 2t, \ y = 12t^2 + 1$ 20. $x = t^2, \ y = t^4$

▶ Each of the following pairs of parametric equations represents the path of a projectile. Take g to be 9.8 meters per second per second, eliminate the parameter, and identify the graph.

EXAMPLE

$$x = \left(8 \cos \dfrac{\pi}{4}\right)t, \ y = \left(8 \sin \dfrac{\pi}{4}\right)t - \dfrac{1}{2}gt^2$$

Solution

$$x = 4\sqrt{2}t, \ y = 4\sqrt{2}t - 4.9t^2;$$

$$t = \dfrac{x}{4\sqrt{2}}, \ y = 4\sqrt{2}\,\dfrac{x}{4\sqrt{2}} - 4.9\left(\dfrac{x^2}{32}\right),$$

$$y = x - \dfrac{4.9}{32}x^2; \quad \text{A parabola}$$

(The path is that part of the parabola lying above the x-axis.)

21. $x = \left(40 \cos \dfrac{\pi}{6}\right)t, \ y = \left(40 \sin \dfrac{\pi}{6}\right)t - \dfrac{1}{2}gt^2$

22. $x = \left(20 \cos \dfrac{\pi}{3}\right)t, \; y = \left(20 \sin \dfrac{\pi}{3}\right)t - \dfrac{1}{2}gt^2$

23. $x = \left(10 \cos \dfrac{\pi}{4}\right)t, \; y = \left(10 \sin \dfrac{\pi}{4}\right)t - \dfrac{1}{2}gt^2$

24. $x = \left(24 \cos \dfrac{\pi}{6}\right)t, \; y = \left(24 \sin \dfrac{\pi}{6}\right)t - \dfrac{1}{2}gt^2$

▶ Each of the following pairs of parametric equations represents a cycloid. Find the coordinates of the point on the graph corresponding to the given value of t.

EXAMPLE

$x = 2(t - \sin t), \; y = 2(1 - \cos t); \quad t = \dfrac{\pi}{2}$

Solution

$x = 2\left(\dfrac{\pi}{2} - 1\right), \; y = 2(1 - 0)$

$x = \pi - 2 \qquad y = 2$

25. $x = 3(t - \sin t), \; y = 3(1 - \cos t); \quad t = \pi$

26. $x = 4(t - \sin t), \; y = 4(1 - \cos t); \quad t = \dfrac{3\pi}{2}$

27. $x = 5(t - \sin t), \; y = 5(1 - \cos t); \quad t = 2\pi$

28. $x = t - \sin t, \; y = 1 - \cos t; \quad t = \dfrac{\pi}{4}$

29. $x = 6(t - \sin t), \; y = 6(1 - \cos t); \quad t = 0$

30. $x = 8(t - \sin t), \; y = 8(1 - \cos t); \quad t = \dfrac{\pi}{6}$

CHECK LIST FOR THE STUDENT

If you have learned the material in Section 3.11, you should be able to do the following:

1. Eliminate the parameter from each of some given pairs of parametric equations, including those of the following types, and identify the graph. (The letters a, b, c, and d represent specific real numbers.)

(1) $\begin{cases} x = at + c \\ y = bt + d \end{cases}$

(2) $\begin{cases} x = a \cos t + b \\ y = a \sin t + c \end{cases}$ $\begin{cases} x = a \sin t + b \\ y = a \cos t + c \end{cases}$

(3) $\begin{cases} x = a \cos t + c \\ y = b \sin t + d \end{cases}$ $\begin{cases} x = a \sin t + c \\ y = b \cos t + d \end{cases}$

(4) $\begin{cases} x = a \sec t + c \\ y = b \tan t + d \end{cases}$ $\begin{cases} x = a \tan t + c \\ y = b \sec t + d \end{cases}$

(5) $\begin{cases} x = a \csc t + c \\ y = b \cot t + d \end{cases}$ $\begin{cases} x = a \cot t + c \\ y = b \csc t + d \end{cases}$

(6) $\begin{cases} x = at \\ y = bt^2 + ct + d \end{cases}$

2. Show by example that the same graph may be represented by more than one pair of parametric equations.

3. Show by example that a pair of parametric equations may represent only part of the graph of the equation obtained by eliminating the parameter.

4. Recognize that a pair of equations of the form $x = (v_0 \cos A)t$, $y = (v_0 \sin A)t - \frac{1}{2}gt^2$ represents the path of a projectile, eliminate the parameter, and identify the graph.

5. Tell how to trace out the points on a cycloid.

6. Take a pair of equations of the form $x = a(t - \sin t)$, $y = a(1 - \cos t)$, and find the coordinates of the point on its graph corresponding to a given value of t.

SECTION QUIZ 3.11

In each of the following, eliminate the parameter and identify the graph.

1. $x = 3t - 2$, $y = t + 4$

2. $x = 7 \cos t + 2$, $y = 7 \sin t - 1$

3. $x = 2 \cos t$, $y = 5 \sin t$

4. $x = 3 \csc t$, $y = 2 \cot t$

5. $x = 2 \tan t - 3$, $y = 4 \sec t + 1$

6. $x = 2t$, $y = 4t + t^2$

7. Show that the pair of equations $x = t + 1$, $y = 2t - 1$ represents the same graph as the pair $x = 4t$, $y = 8t - 3$.

8. Explain why the pair of equations $x = t^2$, $y = 3t^2 + 1$ represents only part of the line $y = 3x + 1$, and sketch the graph represented by $x = t^2$, $y = 3t^2 + 1$.

9. The pair of equations

$$x = \left(60 \cos \frac{\pi}{3}\right) t, \quad y = \left(60 \sin \frac{\pi}{3}\right) t - \frac{1}{2} g t^2$$

represents the path of a projectile. Eliminate the parameter, identify the graph, and make a very rough sketch of the graph.

10. The pair of equations $x = 2(t - \sin t)$, $y = 2(1 - \cos t)$ represents a cycloid. Find the coordinates of the point on its graph corresponding to $t = \pi$, and make a very rough sketch of the graph.

Part 3 Final Test

1. The radian measure of an angle in standard position is $9\pi/8$. Find the radian measure and then the degree measure of the reference angle.

2. (a) Find the length of arc subtended by a central angle of measure $80°$ in a circle with radius 9 cm.
   (b) Find the degree measure of the central angle subtending an arc of length 4 cm on the unit circle.

3. Use the association process of the text in each of the following.
   (a) Find the real number and the coordinates of the unit circle point associated with the standard-position angle whose measure is $900°$.
   (b) Find the degree measure of the standard-position angle and find the coordinates of the unit circle point associated with the real number $5\pi/2$.

4. Let $(1/\sqrt{10},\ -3/\sqrt{10})$ be the unit circle point associated with real number α. Find $\cos \alpha$, $\sin \alpha$, $\tan \alpha$, $\cot \alpha$, $\sec \alpha$, and $\csc \alpha$.

5. In right triangle ABC the right angle is at C and the lengths of the sides opposite A and B are $a = 2$ meters and $b = 3$ meters. Find $\sin A$, $\cos A$, $\tan A$, $\cot A$, $\sec A$, and $\csc A$. Find $\sin B$ and $\tan B$.

6. Give the value of each of the following.

   (a) $\sin\left(\dfrac{-7\pi}{6}\right)$ (b) $\sec \dfrac{11\pi}{4}$ (c) $\cot 420°$

   (d) $\tan 16\pi$ (e) $\cos \dfrac{7\pi}{2}$ (f) $\csc (-45°)$

7. Three streets make a right triangle ABC with right angle at C. If the lengths of the hypotenuse and the side opposite A are 200 meters and 100 meters, respectively, what is the angle at A?

8. Give the domain, range, and fundamental period of the function defined by each of the following.
   (a) $y = 3 \sin x$ (b) $y = 3 \cot x$ (c) $y = 3 \csc x$

Find the amplitude and fundamental period and sketch the graph of each of the following. Label any intercept(s) shown on your graph.

9. $y = \frac{1}{2} \sin 4x$

10. $y = 3 \cos\left(\dfrac{1}{2}x + \dfrac{\pi}{2}\right)$

Sketch the graph and give the range of the function defined by each of the following.

11. $y = \tan x$; $\dfrac{-3\pi}{2} < x < \dfrac{3\pi}{2}$; $x \neq \dfrac{-\pi}{2}$, $x \neq \dfrac{\pi}{2}$

12. $y = \csc x$; $0 < x < 3\pi$; $x \neq \pi$, $x \neq 2\pi$

13. If $\tan \theta = 12/5$ and $\cos \theta < 0$, what are the values of $\sin \theta$, $\cos \theta$, $\cot \theta$, $\sec \theta$, and $\csc \theta$?

14. (a) Use an addition formula to find the exact value of $\sin ((\pi/4) - (\pi/6))$.

(b) Use the fact that $\tan \theta = \frac{-3}{4}$ and $3\pi/2 < \theta < 2\pi$ to find $\sin 2\theta$.

15. Use the fact that $\cos 2\theta = \frac{-1}{4}$ and $3\pi < 2\theta < 4\pi$ to find $\sin \theta$.

16. Prove that $\cos^2\theta - \sin^2\theta \tan^2\theta = 1 - \tan^2\theta$ is an identity.

Let triangle ABC have angles of measure α, β, and γ at A, B, and C, respectively, and let a, b, and c represent the lengths of the sides opposite A, B, and C.

17. (a) If $\alpha = \pi/6$, $\gamma = \pi/4$, and $c = 100$ m, find a.
(b) If $\gamma = 45°$, $c = 5\sqrt{2}$ m, and $a = 5$, find α.

18. If $c = 5$ cm, $b = 8$ cm, and $\alpha = \pi/3$, find a.

19. Sketch the graph of $y = \text{Arcsin } x$, for $\frac{-1}{2} \le x \le 1$.

20. Find the value of each of the following.
(a) $\text{Arctan } \sqrt{3}$ (b) $\text{Arcsec } 2/\sqrt{3}$
(c) $\text{Arccsc } (-2/\sqrt{3})$ (d) $\cot (\text{Cot}^{-1} 2)$
(e) $\tan (\text{Sin}^{-1} \frac{1}{5})$

21. (a) Solve the equation $\sin \theta = -\sqrt{2}/2$ for $0 \le \theta \le \pi$.
(b) Solve the equation $\sin \theta = -\sqrt{2}/2$ for $0 \le \theta \le 2\pi$. Write an expression for *all* solutions of $\sin \theta = -\sqrt{2}/2$.

22. Solve the equation $3 \sec^2\theta - 2 \tan^2\theta = 6$.

23. (a) Find three other pairs of polar coordinates (r, θ) of the point $(7, -50°)$ such that $-360° \le \theta \le 360°$.
(b) Find the Cartesian coordinates of the point $(-3, 60°)$.
(c) Find a pair of polar coordinates (r, θ) of the point $(-5/2, 5\sqrt{3}/2)$, with (r, θ) such that $0° \le \theta \le 360°$ and $r \ge 0$.

24. Sketch the graph of each of the following.
(a) $r = 3 \sin \theta$ (b) $r = 3 + \cos \theta$

25. Eliminate the parameter and identify the graph of

$$\begin{cases} x = 5 \cos t \\ y = 2 \sin t. \end{cases}$$

Part 4 Precalculus College Algebra

Calculus students sometimes have to do more with algebra than just perform straightforward manipulations. For example, they might have to find the roots of a polynomial equation so that they can locate the x-intercepts of a graph, they might have to express a rational expression as a sum of partial fractions so that they can perform an integration, or they might be expected to apply the principle of mathematical induction so that they can derive a necessary formula or prove a basic theorem. Part 4 of this text presents topics from college algebra that are useful in calculus.

4.1 Polynomial Equations

The purpose of this section is to present some useful theorems about polynomial equations, including theorems that will enable us to determine the number of roots of a polynomial equation with real coefficients, to find the rational roots of a polynomial equation with integral coefficients, and to locate real roots of polynomial equations with real coefficients.

A **polynomial equation in the one variable** x is an equation that can be written in the form

$$P(x) = 0$$

where $P(x)$ is a polynomial. If $P(x)$ is of degree n, then the equation $P(x) = 0$ is an nth-degree equation. Linear equations such as

$$ax + b = 0$$

and quadratic equations such as

$$ax^2 + bx + c = 0$$

are polynomial equations of degree 1 and 2, respectively. The equation

$$x^4 + 4x^3 - 7x^2 - 22x + 24 = 0$$

is a polynomial equation of fourth degree. As we agreed in Section 1.4, unless otherwise specified, by a *polynomial* we mean a polynomial whose coefficients are all real numbers.

Exercise 4.1, 1–5

The number r is a *solution,* or *root,* of a polynomial equation $P(x) = 0$ provided that $P(r) = 0$. The terms *solution* and *root* are synonyms. If $P(r) = 0$, the number r is also called a *zero* of the polynomial $P(x)$.

We have already seen how to find the solutions or roots of linear and quadratic equations. In fact, in Section 1.7 we even produced formulas for the solutions of $ax + b = 0$ and $ax^2 + bx + c = 0$ where the coefficients a, b, and c are all real numbers and $a \neq 0$. The formulas for these solutions are:

$$ax + b = 0 \qquad\qquad ax^2 + bx + c = 0$$

$$x = \frac{-b}{a} \qquad\qquad x = \frac{-b \pm \sqrt{b^2 - 4ac}}{2a}.$$

It is also possible to find the solutions of any third- or fourth-degree polynomial equation by using only multiplication, division, addition, subtraction, and extraction of roots, but the methods are extremely long and complicated. Moreover, it has been shown that there is no such straightforward method for finding the solutions of polynomial equations of degree greater than 4. So we shall not be developing additional formulas for the solutions of polynomial equations. Instead, we shall be considering theorems that will enable us to locate or approximate the solutions.

The *remainder theorem,* which was stated and proved in Section 1.4, can be used to determine whether or not a particular number is actually a root of a given polynomial equation. According to the remainder theorem,

if the polynomial $P(x)$ is divided by $x - r$, the remainder equals $P(r)$.

It follows that one way to determine whether or not r is a root of $P(x) = 0$ is to divide $P(x)$ by $(x - r)$ and examine the remainder.

If the remainder $P(r)$ is zero, r must be a root of $P(x) = 0$. If the remainder $P(r)$ is not zero, r is not a root.

We saw in Section 1.4 that synthetic division may be used to advantage in computing the remainder $P(r)$. The remainder theorem and synthetic division are reviewed in the next example.

EXAMPLE 1

Use the remainder theorem and synthetic division to show that -3 is a root and 4 is not a root of $x^4 + 4x^3 - 7x^2 - 22x + 24 = 0$.

Solution

Let $P(x) = x^4 + 4x^3 - 7x^2 - 22x + 24$. To show that -3 is a root, we must show that $P(-3)$ is zero. By the remainder theorem we know that $P(-3)$ is the remainder after $P(x)$ is divided by $x - (-3)$, or $x + 3$. Using synthetic division to divide $P(x)$ by $x + 3$, we have

$$
\begin{array}{r|rrrrr}
-3 & 1 & 4 & -7 & -22 & 24 \\
   &   & -3 & -3 & 30 & -24 \\
\hline
   & 1 & 1 & -10 & 8 & 0
\end{array}
$$

The remainder $P(-3)$ is zero. The number -3 is a root.

To show that 4 is not a root, we compute $P(4)$ by dividing $P(x)$ by $x - 4$.

$$
\begin{array}{r|rrrrr}
4 & 1 & 4 & -7 & -22 & 24 \\
  &   & 4 & 32 & 100 & 312 \\
\hline
  & 1 & 8 & 25 & 78 & 336
\end{array}
$$

Exercise 4.1, 6–10

The remainder $P(4)$ is 336, not zero. The number 4 is not a root.

A very useful theorem called the *factor theorem* follows directly from the remainder theorem.

Factor Theorem   If r is a root of the polynomial equation $P(x) = 0$, then $(x - r)$ is a factor of $P(x)$.

Proof   If $P(x)$ is divided by $x - r$ and $Q(x)$ is the quotient, then

$$P(x) = Q(x)(x - r) + P(r)$$

by the remainder theorem. But if r is a root of $P(x) = 0$, $P(r)$ must be zero, so

$$P(x) = Q(x)(x - r).$$

Exercise 4.1, 11–15   In other words, if r is a root of $P(x) = 0$, then $(x - r)$ is a factor of $P(x)$.

The *converse of the factor theorem* follows directly from the zero factor law, and we have already used it in solving quadratic equations by factoring.

Converse of the   If $(x - r)$ is a factor of polynomial $P(x)$, then r is a root of the equation
Factor Theorem $P(x) = 0$.

Proof   If $(x - r)$ is a factor of $P(x)$, then

$$P(x) = Q(x)(x - r)$$

for some polynomial $Q(x)$. Therefore

$$P(r) = Q(r)(r - r)$$
$$= 0.$$

That is, r is a root of $P(x) = 0$.

EXAMPLE 2   Use the converse of the factor theorem to find the roots of the polynomial equation $(x - 3)(x + 5)x = 0$.

Solution   Let $P(x) = 0$ be the given equation.

Since $x - 3$ is a factor of $P(x)$, 3 is a root of $P(x) = 0$.

Since $x + 5$ is a factor of $P(x)$, -5 is a root of $P(x) = 0$.

Since x is a factor of $P(x)$, 0 is a root of $P(x) = 0$.

Exercise 4.1, 16–20   The roots are 3, -5, and 0.

It may be that a polynomial $P(x)$ with real coefficients has $(x - r)$ as a factor more than once. For example, the polynomial $x^2 - 6x + 9$ has $(x - 3)$ as a factor *twice,* in the sense that

$$x^2 - 6x + 9 = (x - 3)^2.$$

In this case we say that the factor $(x - 3)$ has multiplicity 2, or that the root 3 of the equation $x^2 - 6x + 9 = 0$ has multiplicity 2. In general, if

k is a positive integer, and if polynomial $P(x)$ has $(x - r)^k$ as a factor but has no higher power of $(x - r)$ as a factor, we say that the factor $(x - r)$ has multiplicity k and the solution or root r of the equation $P(x) = 0$ has multiplicity k.

EXAMPLE 3

Make up an equation whose roots are 2, -3, and 0 with multiplicity 1, 2, and 4, respectively.

Solution

Exercise 4.1, 21–30

$(x - 2)(x + 3)^2 x^4 = 0$

Now let us consider the question of how many solutions any particular polynomial equation may have. We have seen that any quadratic equation $ax^2 + bx + c = 0$ with real coefficients has either

1. Exactly one real-number solution,

2. Exactly two distinct real-number solutions, or

3. Exactly two distinct imaginary-number solutions.

In the case of exactly one real-number solution, say r, the quadratic is a perfect square, and the equation can be written in the form

$$(x - r)^2 = 0$$

so that the solution r is of multiplicity 2. Thus if we count each solution as many times as its multiplicity, and if we recall that real numbers and imaginary numbers are all complex numbers, we can say that any quadratic equation with real coefficients has exactly 2 complex-number solutions. This is a special case of the *number-of-roots theorem*.

Number-of-Roots Theorem

If each root is counted as many times as its multiplicity, any nth-degree polynomial equation $P(x) = 0$ with real coefficients has exactly n complex-number roots. It follows that any nth-degree polynomial equation $P(x) = 0$ with real coefficients has *at most n distinct* complex-number roots.

EXAMPLE 4

How many distinct roots has the equation $(x - 2)^5 (x + 4)^3 (x - 7) = 0$? What are these roots? If each root is counted as many times as its multiplicity, how many roots does the given equation have?

Solution

The given equation has three distinct roots. They are 2, -4, and 7. If each root is counted as many times as its multiplicity, there are $5 + 3 + 1$ or 9 roots. (Note that the equation is a ninth-degree polynomial equation.)

Exercise 4.1, 31–40

It can be shown that if one of the roots of polynomial equation $P(x) = 0$ is an imaginary number, say $a + bi$, then its *conjugate*, $a - bi$, is also a root of the equation. Imaginary roots occur in pairs.

Conjugate Root Theorem

If imaginary number $a + bi$ is a root of $P(x) = 0$, then $a - bi$ is also a root. Conversely, if $a - bi$ is a root of $P(x) = 0$, then $a + bi$ is also a root.

EXAMPLE 5

The imaginary number $2 - 3i$ is a root of the equation $x^2 - 4x + 13 = 0$. Show that its conjugate $2 + 3i$ is also a root.

Solution

Substituting $2 + 3i$ for x in the given equation, we see that $2 + 3i$ is a root.

$$(2 + 3i)^2 - 4(2 + 3i) + 13 = 0$$
$$4 + 12i + 9i^2 - 8 - 12i + 13 = 0$$
$$4 - 9 - 8 + 13 = 0 \quad \text{True}$$

EXAMPLE 6

If $3 + 7i$ is a root of a polynomial equation $P(x) = 0$, what are two factors of $P(x)$?

Solution

If $3 + 7i$ is a root, then $3 - 7i$ is also a root. Two factors are $(x - (3 + 7i))$ and $(x - (3 - 7i))$.

In Example 6, we noted that $3 + 7i$ and $3 - 7i$ are roots of $P(x) = 0$ so that two factors of $P(x)$ are $(x - (3 + 7i))$ and $(x - (3 - 7i))$. If we multiply these two factors together, we obtain the quadratic

$$x^2 - 6x + 58.$$

(Students should check this multiplication themselves.) This quadratic $x^2 - 6x + 58$ is a factor of $P(x)$ which has real coefficients, but it cannot be factored into linear factors with *real* coefficients. Its factors $(x - 3 - 7i)$ and $(x - 3 + 7i)$ have *imaginary* coefficients. Any quadratic with real coefficients that cannot be factored into linear factors with real coefficients is called an **irreducible** quadratic. Each pair of conjugate imaginary roots of the polynomial equation $P(x) = 0$ is associated with an irreducible quadratic factor of $P(x)$, just as each real root is associated with a linear factor with real coefficients.

Exercise 4.1, 41–50

Factorization Theorem

Any polynomial $P(x)$ with real coefficients can be factored into a product of linear factors with real coefficients and irreducible quadratic factors.

EXAMPLE 7

Solve $x^3 - 3x^2 + 4x - 12 = 0$ by factoring. (Hint: 3 is a root.)

Solution

If 3 is a root, $(x - 3)$ is a factor. We can factor out $(x - 3)$ by dividing $x^3 - 3x^2 + 4x - 12$ by $x - 3$. Using synthetic division, we have

$$
\begin{array}{r|rrrr}
3 & 1 & -3 & 4 & -12 \\
  &   & 3 & 0 & 12 \\
\hline
  & 1 & 0 & 4 & 0
\end{array}
$$

$$x^3 - 3x^2 + 4x - 12 = (x - 3)(x^2 + 0x + 4)$$

The given equation may be written

$$(x - 3)(x^2 + 4) = 0.$$

(Note that the quadratic $x^2 + 4$ is irreducible.) By the zero factor law, we have

$$x - 3 = 0 \quad \text{or} \quad x^2 + 4 = 0$$
$$x = 3 \qquad\qquad x^2 = -4$$
$$x = \pm 2i.$$

Exercise 4.1, 51–55 The roots are 3, $2i$, and $-2i$. (Note that $2i$ and $-2i$ are conjugates.)

If a given polynomial equation has integral coefficients, we can find any rational roots that it may have by employing the next theorem. (Recall that a rational number is a number that can be written in the form p/q where p and nonzero q are integers.)

Rational Root Theorem   If the nth-degree polynomial equation $P(x) = 0$ has integral coefficients, and if $P(x) = 0$ has a rational root p/q where p and q have no common factors other than $+1$ or -1, then the numerator p must be a factor of the constant term of $P(x)$, and the denominator q must be a factor of the coefficient of x^n in $P(x)$.

EXAMPLE 8   List all the possible rational roots of $2x^3 - x^2 - 15x + 18 = 0$.

Solution   Possible numerators are factors of 18.

Possible numerators are ± 1, ± 2, ± 3, ± 6, ± 9, ± 18.

Possible denominators are factors of 2.

Possible denominators are ± 1, ± 2.

Possible rational roots are

$$\pm \tfrac{1}{1}, \ \pm \tfrac{1}{2}, \ \pm \tfrac{2}{1}, \ \pm \tfrac{2}{2}, \ \pm \tfrac{3}{1}, \ \pm \tfrac{3}{2}, \ \pm \tfrac{6}{1}, \ \pm \tfrac{6}{2}, \ \pm \tfrac{9}{1}, \ \pm \tfrac{9}{2}, \ \pm \tfrac{18}{1}, \ \pm \tfrac{18}{2}.$$

If duplicates are omitted, the list of possible rational roots may be written more compactly as

$$\pm 1, \ \pm \tfrac{1}{2}, \ \pm 2, \ \pm 3, \ \pm \tfrac{3}{2}, \ \pm 6, \ \pm 9, \ \pm \tfrac{9}{2}, \ \pm 18.$$

To find any rational roots that a polynomial equation with integral coefficients may have, we first list all the candidates provided by the rational root theorem, and then we may use the remainder theorem and synthetic division to check out each of these candidates. Once one rational root has been found, the task of checking for additional roots is somewhat easier, because we can work with a polynomial equation of lower degree than the given equation. Indeed, suppose that r is found to be a root of

$$P(x) = 0.$$

Then $(x - r)$ is a factor of $P(x)$ so that

$$P(x) = Q(x)(x - r)$$

and the given equation can be written

$$Q(x)(x - r) = 0.$$

Any additional roots of $P(x) = 0$ will be roots of

$$Q(x) = 0,$$

and the root r should be checked in $Q(x) = 0$ to see if it is a multiple root of $P(x) = 0$. The equation $Q(x) = 0$ is called the *depressed equation*. Since $Q(x)$ is the quotient when $P(x)$ is divided by $(x - r)$, the coefficients of $Q(x)$ are the numbers in the third line of the synthetic division of $P(x)$ by $(x - r)$. This is demonstrated in the next example.

EXAMPLE 9

Solution

Find the rational roots of $2x^3 - x^2 - 15x + 18 = 0$.

By the rational root theorem, if the given equation has any rational roots, they must be among the numbers

$$\pm 1, \ \pm \tfrac{1}{2}, \ \pm 2, \ \pm 3, \ \pm \tfrac{3}{2}, \ \pm 6, \ \pm 9, \ \pm \tfrac{9}{2}, \ \pm 18.$$

We see that we have 18 possibilities to check. Using synthetic division to check ± 1, we find that neither of these are roots because the remainder is not zero for either of them. (It is also easy to show by direct substitution in the given equation that $+1$ and -1 are not roots.)

```
1)2  -1  -15   18        -1)2   -1   -15   18
      2    1  -14              -2    3    12
   ─────────────────        ────────────────────
   2    1  -14    4          2   -3   -12   30
```

Since it is usually easier to check integral roots, we next use synthetic division to check the candidate 2, and we find that 2 is a root.

```
2)2  -1  -15   18
      4    6  -18
   ─────────────────
   2    3   -9    0
```

The coefficients in the quotient are 2, 3, -9 and the quotient is $2x^2 + 3x - 9$, so the given equation can be written

$$(2x^2 + 3x - 9)(x - 2) = 0.$$

Any additional roots are roots of the depressed equation

$$2x^2 + 3x - 9 = 0.$$

This happens to be a quadratic that we can solve by factoring.

$$(2x - 3)(x + 3) = 0$$

$$2x - 3 = 0 \quad \text{or} \quad x + 3 = 0$$

$$x = \tfrac{3}{2} \qquad\qquad x = -3.$$

The rational roots of the given equation are $2, \frac{3}{2}$, and -3. Since the given equation is of third degree, we know by the number-of-roots theorem that it can have no more than 3 different roots. Consequently, we have found *all* the roots of the given equation.

Exercise 4.1, 56–75

Using the rational root theorem in Example 9, we had eighteen candidates to check. We were fortunate to find a rational root on only the third try. The process could be even longer and more tedious for some other given equation. There are several theorems that may sometimes shorten such calculations by eliminating some of the candidates provided by the rational root theorem. One of these is called the *location-of-roots theorem.*

Before formally stating the location-of-roots theorem, let us consider the matter geometrically. Note that the real roots of polynomial equation $P(x) = 0$ are the x-intercepts of the graph of the function defined by the equation $y = P(x)$. That is, they are the x-coordinates of the points of intersection of the x-axis and the graph of $y = P(x)$. For example, to find the x-intercepts of the graph of $y = 4x^3 + 3x^2 - 25x + 6$, we let $y = 0$ and then solve for x. The roots of $4x^3 + 3x^2 - 25x + 6 = 0$ are $-3, \frac{1}{4}$, and 2, so the x-intercepts of the graph of $y = 4x^3 + 3x^2 - 25x + 6$ are $-3, \frac{1}{4}$, and 2. The graph and its intercepts are shown in Figure 4.1.

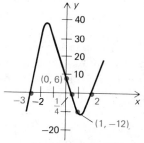

FIGURE 4.1   Graph of $y = 4x^3 + 3x^2 - 25x + 6$

We see in Figure 4.1 that the graph of $y = 4x^3 + 3x^2 - 25x + 6$ is *continuous;* the graph can be drawn without lifting the pencil off the paper. It can be shown that the graph of any *polynomial function*—that is, any function defined by an equation $y = P(x)$ where $P(x)$ is a polynomial with real coefficients—is continuous. Thus if a and b are real numbers such that $P(a)$ and $P(b)$ have opposite signs, the points $(a, P(a))$ and $(b, P(b))$ on the graph must be on opposite sides of the x-axis, and the graph must cross the x-axis at least once, or have at least one x-intercept, between a and b. That is, the polynomial equation $P(x) = 0$ must have at least one real root between a and b. This is the conclusion of the *location-of-roots theorem.*

If $P(x)$ is a polynomial with real coefficients, if a and b are real numbers, and if $P(a)$ and $P(b)$ have opposite signs, then at least one real root of $P(x) = 0$ is located between a and b.

EXAMPLE 10

Use the location-of-roots theorem to show that there is at least one root of $4x^3 + 3x^2 - 25x + 6 = 0$ between 0 and 1.

Solution

Let $P(x) = 4x^3 + 3x^2 - 25x + 6$.

$$P(0) = 6$$
$$P(1) = -12$$

$P(0)$ and $P(1)$ are of opposite signs.

There is at least one root of $4x^3 + 3x^2 - 25x + 6 = 0$ between 0 and 1. (Note in Figure 4.1 that the graph goes through $(0, 6)$ and $(1, -12)$, and since the graph is continuous it must cross the x-axis at least once between these two points. Actually, this graph crosses the x-axis *exactly* once between $(0, 6)$ and $(1, -12)$.)

Exercise 4.1, 76–85

If a root located by the location-of-roots theorem happens to be a rational root, it can be found by using the rational root theorem. If a root of $P(x) = 0$ is located between real numbers a and b and is *not* rational, it can be approximated by using any one of several straightforward methods. One approximating technique called *Newton's method* requires the use of a bit of calculus and is often taken up in beginning calculus courses.

There are several other theorems that may help locate the roots of a polynomial equation. One theorem says in part that if B is a positive real number, and if the numbers in the third line of the synthetic division of $P(x)$ by $(x - B)$ are all nonnegative, then all the real roots of $P(x) = 0$ are less than or equal to B. Another theorem, called *Descartes' rule of signs*, specifies how to determine the number of positive roots that $P(x) = 0$ may have by considering the signs of the coefficients of $P(x)$. Students who have to do a lot of problems requiring them to locate the real roots of a polynomial equation should look up these theorems in a text on the theory of equations or in a college algebra text that emphasizes the theory of equations.

EXERCISE 4.1 POLYNOMIAL EQUATIONS

▶ Tell the degree of each of the following equations.

EXAMPLE

$3x^4 - 5x^2 + 6x^9 - 2x + 5 = 0$

Solution

Degree 9

1. $2x^2 - 5x^7 - 3x^4 + 2x^4 - 7 = 0$

2. $x + 7 = 0$

3. $2x^2 - 5x + 7 = 0$

4. $x^3 - 2x^2 + 7x - 5 = 0$

5. $x - 7x^2 + 6x^4 - 2x^3 = 0$

▶ In each of the following, use the remainder theorem and synthetic division to show that the first number given is a root and the second number is not a root.

EXAMPLE

$2x^3 + 11x^2 + 2x - 15 = 0; \quad -5, 2$

Solution

$$
\begin{array}{r|rrrr}
-5) & 2 & 11 & 2 & -15 \\
 & & -10 & -5 & 15 \\
\hline
 & 2 & 1 & -3 & 0
\end{array}
\qquad
\begin{array}{r|rrrr}
2) & 2 & 11 & 2 & -15 \\
 & & 4 & 30 & 64 \\
\hline
 & 2 & 15 & 32 & 49
\end{array}
$$

Remainder = 0; -5 is a root. Remainder $\neq$ 0; 2 is not a root.

6. $3x^4 - 22x^3 + 19x^2 - 88x + 28 = 0; \quad 7, -3$

7. $x^4 + 6x^3 - 25x^2 - 54x + 144 = 0; \quad -8, -2$

8. $x^3 + 2x^2 + 15x + 14 = 0; \quad -1, 4$

9. $2x^3 + 3x^2 - 17x + 12 = 0; \quad \frac{3}{2}, 3$

10. $x^4 - 72x^2 - 729 = 0; \quad -9, 2$

▶ In each of the following, use the factor theorem to construct an equation with the lowest possible degree having the given roots.

EXAMPLE

$2, -7, 0$

Solution

$(x - 2)(x + 7)x = 0$

11. $3, -5, 4$

12. $0, -2, 1$

13. $3, -3, 4, -4$

14. $2 + 3i, 2 - 3i, 5$

15. $2i, -2i, 3$

▶ Use the converse of the factor theorem to find the roots of each of the following equations.

EXAMPLE

$x(x + 7)(x - 2)(x + 3) = 0$

Solution

$0, -7, 2, -3$

16. $(x - 3)(x + 2)(x + 5) = 0$

17. $x(x - 2)(x + 4) = 0$

18. $x(x^2 - 4) = 0$

19. $x(x^2 + x - 6) = 0$

20. $(x^2 - 9)(x^2 - 4x + 4) = 0$

EXAMPLE $x^3(x - 5)^2(x + 3) = 0$

Solution   0 is of multiplicity 3

5 is of multiplicity 2

−3 is of multiplicity 1

21. $x^2(x + 1)^3(x - 4)^5 = 0$ **22.** $x(x - 1)^2(x + 5)^7 = 0$

23. $(x - 3)^2(x^2 + 1)^3 = 0$ **24.** $x^3(x^2 - 6x + 9) = 0$

25. $x^4(x - 2)^3(x + 9) = 0$

▶ In each of the following, make up an equation having the given roots with the given multiplicities and having no other roots.

EXAMPLE   0, 3, −2 with multiplicity 4, 1, and 5, respectively

Solution $x^4(x - 3)(x + 2)^5 = 0$

26. 0, −5, 6 with multiplicity 1, 2, and 3, respectively

27. −5, −4, −1 with multiplicity 2, 3, and 4, respectively

28. 1, 2, −3 with multiplicity 1, 3, and 2, respectively

29. 4, 3i, −3i with multiplicity 3, 5, and 5, respectively

30. $\frac{2}{3}$, $\frac{4}{5}$, 6 with multiplicity 1, 2, and 3, respectively

▶ What is the greatest number of distinct complex-number roots that each of the following equations may have? If each root is counted as many times as its multiplicity, how many roots does the given equation have?

EXAMPLE $x^5 - 3x^2 + 2x - 7 = 0$

Solution   5, 5

31. $2x^3 - 5x^4 - 3x + 7 = 0$ **32.** $x^2 - 5x + 8 = 0$

33. $3x^7 - 2x^5 - 2x + 9x^2 + 8 = 0$

34. $x^3 - 3x^2 + 5x - 8 = 0$ **35.** $2x - 3x^2 + 5x^5 - 3x^4 = 0$

▶ How many distinct roots does each of the following equations have? What are these roots? If each root is counted as many times as its multiplicity, how many roots does the given equation have?

EXAMPLE $x(x - 2)^3(x + 1)^5(x - 3) = 0$

Solution   Four distinct roots: 0, 2, -1, 3; $1 + 3 + 5 + 1 = 10$ roots

36. $x^2(x - 1)^3(x + 2) = 0$ **37.** $(x + 1)^4(x - 3)^5(x + 4) = 0$

38. $x^3(x - 1)(x + 3)^2 = 0$ **39.** $(x + 5)^2(x - 2)^3(x^2 + 1) = 0$

40. $x^5(x - 1)^4(x + 2)^3 = 0$

▶ In each of the following, show that the conjugate of the given root is also a root.

EXAMPLE $x^2 + 9 = 0$, $3i$

Solution   The conjugate $-3i$ is also a root because $(-3i)^2 + 9 = 0$.

41. $x^2 + 25 = 0$, $5i$ **42.** $x^2 - 6x + 34 = 0$, $3 - 5i$

43. $x^2 - 4x + 20 = 0$, $2 + 4i$ **44.** $x^3 + 36x = 0$, $-6i$

45. $x^2 - 2x + 5 = 0$, $1 - 2i$

▶ In each of the following, the given number is a root of a polynomial equation $P(x) = 0$. Find an irreducible quadratic factor of $P(x)$.

EXAMPLE $2 - 3i$

Solution $2 + 3i$ is also a root, $(x - 2 + 3i)$ and $(x - 2 - 3i)$ are factors.

$$(x - 2 + 3i)(x - 2 - 3i) = (x - 2)^2 - 9i^2$$
$$= x^2 - 4x + 4 + 9$$

$x^2 - 4x + 13$ is an irreducible quadratic factor.

46. $3 + 4i$ **47.** $2 - 5i$ **48.** $3 + 7i$ **49.** $2 - 6i$ **50.** $9i$

▶ Knowing that the given number is a root of the given equation, solve each of the following by factoring.

EXAMPLE 2, $x^3 - x - 6 = 0$

Solution

$$\begin{array}{r|rrr} 2 & 1 & 0 & -1 & -6 \\ & & 2 & 4 & 6 \\ \hline & 1 & 2 & 3 & 0 \end{array}$$

$x^3 - x - 6 = (x - 2)(x^2 + 2x + 3)$

$x^2 + 2x + 3 = 0$

$$x = \frac{-2 \pm \sqrt{4 - 12}}{2}$$

$$= \frac{-2 \pm \sqrt{8}i}{2}$$

$$= \frac{-2 \pm 2\sqrt{2}i}{2}$$

The roots are 2 and $-1 \pm \sqrt{2}i$.

51. 1, $x^3 + 2x^2 + 2x - 5 = 0$ **52.** 3, $x^3 - 2x^2 - 2x - 3 = 0$

53. -2, $x^3 + 3x^2 + 4x + 4 = 0$

54. -1, $x^3 + x^2 + 16x + 16 = 0$

55. 4, $x^3 - 2x^2 - 3x - 20 = 0$

▶ List all the possible rational roots of each of the following.

EXAMPLE $2x^3 - 3x^2 + 5x - 6 = 0$

Solution   Possible numerators: ± 1, ± 2, ± 3, ± 6

Possible denominators: ± 1, ± 2

Possible rational roots: ± 1, $\pm\frac{1}{2}$, ± 2, ± 3, $\pm\frac{3}{2}$, ± 6

56. $3x^3 - 2x^2 + 5x - 10 = 0$ **57.** $x^4 - 3x^3 + 2x^2 - 8 = 0$

58. $5x^2 - 2x + 7 = 0$ **59.** $7x^3 - 3x^2 + 2x + 15 = 0$

60. $3x^3 - 2x^2 + 7x - 1 = 0$

▶ Find the rational roots of each of the following.

EXAMPLE $x^3 - 3x - 2 = 0$

Solution   Possible rational roots are ± 2 and ± 1.

$$
\begin{array}{r|rrrr}
2) & 1 & 0 & -3 & -2 \\
   &   & 2 & 4 & 2 \\
\hline
   & 1 & 2 & 1 & 0 \\
\end{array}
$$

$(x - 2)(x^2 + 2x + 1) = 0$

$(x - 2)(x + 1)^2 = 0$

2 is a root.   The roots are 2 and -1.

61. $2x^3 - x^2 - 7x + 6 = 0$

62. $12x^3 + 29x^2 - 31x - 30 = 0$

63. $10x^3 + 63x^2 + 87x - 20 = 0$

64. $2x^4 - 15x^3 - 11x^2 + 135x - 63 = 0$

65. $x^4 - 10x^3 + 37x^2 - 60x + 36 = 0$

66. $3x^5 - 142x^3 + 7x^4 - 342x^2 - 245x - 49 = 0$

67. $4x^4 - 5x^3 - 83x^2 + 201x - 45 = 0$

68. $x^5 + 7x^4 + 19x^3 + 25x^2 + 16x + 4 = 0$

69. $x^7 + 2x^6 - 39x^5 - 40x^4 + 400x^3 = 0$

70. $3x^3 - 34x^2 + 115x - 100 = 0$

► Find all the roots of each of the following.

EXAMPLE

Solution

$5x^3 + 7x^2 + 19x - 15 = 0$

Possible rational roots: $\pm 1,\ \pm \frac{1}{5},\ \pm 3,\ \pm \frac{3}{5},\ \pm 5,\ \pm 15$

$$\frac{3}{5}\overline{)\begin{array}{rrrr} 5 & 7 & 19 & -15 \\ & 3 & 6 & 15 \\ \hline 5 & 10 & 25 & 0 \end{array}}$$

$(x - \frac{3}{5})(5x^2 + 10x + 25) = 0$

$x^2 + 2x + 5 = 0$

$$x = \frac{-2 \pm \sqrt{4 - 20}}{2}$$

$$= -1 \pm 2i$$

The roots are $\frac{3}{5}$ and $-1 \pm 2i$.

71. $11x^3 + 9x^2 + 31x - 6 = 0$ **72.** $9x^3 + 22x^2 + 35x + 12 = 0$

73. $x^3 + 11x^2 + 31x + 6 = 0$ **74.** $14x^3 - 19x^2 + 11x + 2 = 0$

75. $24x^3 - 25x^2 + 14x - 3 = 0$

► In each of the following, show that there is at least one real root between the given numbers.

EXAMPLE

Solution

$4x^3 - 5x^2 - 47x + 12 = 0$; 0 and 1

Let $P(x) = 4x^3 - 5x^2 - 47x + 12$

$P(0) = 12$ positive

$P(1) = -36$ negative

By the location-of-roots theorem, there is at least one real root between 0 and 1.

76. $4x^3 - 5x^2 - 47x + 12 = 0$; 3 and 5

77. $4x^3 - 5x^2 - 47x + 12 = 0$; 0 and 1

78. $4x^3 - 9x^2 - 4x + 9 = 0$; 2 and 3

79. $4x^3 - 9x^2 - 4x + 9 = 0$; 0 and 2

80. $4x^4 - 47x^3 + 95x^2 + 200x = 0$; -1 and -2

81. $2x^3 - 6x^2 - 14x + 15 = 0$; 0 and 2

82. $2x^4 - 4x^3 - 20x^2 + x + 15 = 0$; -2 and -3

83. $3x^3 + 7x^2 - 16x - 20 = 0$; -3 and -4

84. $3x^4 + 7x^3 - 16x^2 - 20x = 0$; 1 and 3

85. $3x^4 - 2x^3 - 37x^2 + 28x + 60 = 0$; 0 and -2

CHECK LIST FOR THE STUDENT

If you have learned the material in Section 4.1, you should be able to do the following:

1. Give an example of a polynomial equation of a specified degree.
2. Tell the degree of a given polynomial equation.
3. State and use the remainder theorem.
4. State and use the factor theorem.
5. State and use the converse of the factor theorem.
6. Tell what is meant by a root of multiplicity k where k is a positive integer.
7. Make up an equation with given roots having given multiplicities.
8. State and use the number-of-roots theorem.
9. State and use the conjugate root theorem.
10. Define an irreducible quadratic.
11. State and use the factorization theorem.
12. State the rational root theorem.
13. Tell what is meant by the *depressed equation*, if $P(x)$ is divided by $(x - r)$.
14. Find the rational roots of a polynomial equation with integral coefficients.
15. State and use the location-of-roots theorem.
16. Give a geometric interpretation of the location-of-roots theorem.

SECTION QUIZ 4.1

1. What is the degree of the equation $3x^5 - 2x^6 + x^4 - 15x^3 - 8x + 9 = 0$?
2. Give the formula for the solutions of the equation $ax^2 + bx + c = 0$ where a, b, and c are real numbers and $a \neq 0$.
3. Use the remainder theorem to determine whether or not 2 is a root of the equation $4x^3 - 3x^2 + 5x - 30 = 0$.
4. Use the factor theorem to construct an equation with the lowest possible degree having the roots 3, −4, 0, and 1.
5. What are the roots and what is the multiplicity of each root of the equation $x^2(x - 3)^4(x + 1) = 0$?
6. Make up an equation having the roots 0, −4, and 6 with multiplicity 3, 2, and 1, respectively, and having no other roots.
7. What is the greatest number of distinct complex-number roots that the equation $3x^4 - 2x^3 + 5x^2 - 7x + 8 = 0$ may have?
8. If each root is counted as many times as its multiplicity, how many roots does the equation $2x^5 - 3x^4 + 5x^3 - 7 = 0$ have?

9. Show that $2 - i$ and its conjugate are both roots of $x^2 - 4x + 5 = 0$.

10. If $3 - 4i$ is a root of a polynomial equation $P(x) = 0$, what is an irreducible quadratic factor of $P(x)$?

11. Verify that 1 is a root of the equation $x^3 - 5x^2 + 11x - 7 = 0$, and find the other roots.

12. List all the possible rational roots of the equation $7x^5 - 2x^3 + 6x - 15 = 0$.

13. Find the rational roots of the equation $2x^3 - 3x^2 - 11x + 6 = 0$.

14. Find all the roots of the equation $x^3 - 2x^2 + 9x - 18 = 0$.

15. Use the location-of-roots theorem to show that there is at least one root of $5x^3 + 2x^2 - 16x + 4 = 0$ between 0 and 1.

4.2 Matrices and Simultaneous Linear Equations

The purpose of this section is to introduce matrices and to show how matrices can be used to solve simultaneous linear equations.

A **matrix** is a rectangular array of symbols, usually symbols for numbers. It is conventional to enclose the array in large parentheses or brackets. For example, the coefficients in the system of equations

$$3x - 2y + 6z = -6$$
$$-3x + 10y + 11z = 13$$
$$x - 2y - z = -3$$

form the matrix

$$\begin{pmatrix} 3 & -2 & 6 \\ -3 & 10 & 11 \\ 1 & -2 & -1 \end{pmatrix}.$$

This matrix is called the *coefficient matrix of the system.* The matrix obtained by including the column of constant terms is called the *augmented matrix* or simply *the* matrix of the system. The matrix

$$\begin{pmatrix} 3 & -2 & 6 & -6 \\ -3 & 10 & 11 & 13 \\ 1 & -2 & -1 & -3 \end{pmatrix}$$

is the augmented matrix of the given system.

The numbers making up a matrix are called the **elements** of the matrix. The elements are said to be arranged in *rows* and *columns.* In the matrix

$$\begin{pmatrix} 3 & -2 & 6 & -6 \\ -3 & 10 & 11 & 13 \\ 1 & -2 & -1 & -3 \end{pmatrix}$$

there are three rows and four columns. The first row is   3 -2 6 -6, the second row is -3 10   11   13, etc. The first column is

$$3$$
$$-3$$
$$1,$$

etc. A matrix with three rows and four columns is known as a three-by-four matrix. A matrix with m *rows* and n *columns* is an m-*by*-n matrix. The expression m by n is sometimes written $m \times n$. If $m = n$ so that the matrix has the same number of rows as columns, it is said to be a *square* matrix.

The rows and columns of a matrix are numbered from top to bottom and from left to right. The element 8 in the matrix

$$\begin{pmatrix} 2 & 1 & 3 & -5 \\ 6 & 4 & 8 & 2 \\ 5 & -3 & 9 & 7 \end{pmatrix}$$

is in the second row and third column; the element 7 is in the third row and fourth column. The elements of a square matrix located in the first row and first column, the second row and second column, the third row and third column, etc. are said to be in the *main diagonal.* The elements 2 4 9 are in the main diagonal of the matrix

$$\begin{pmatrix} 2 & 1 & 3 \\ 6 & 4 & 8 \\ 5 & -3 & 9 \end{pmatrix}.$$

For convenience, we sometimes refer to the first row of a matrix as R_1, the second row as R_2, the third row as R_3, etc. Likewise, the symbols C_1, C_2, C_3, etc. are used for Column 1, Column 2, Column 3, etc.

Matrices may be used to solve systems of n linear equations in n unknowns. One way to solve such a system is to work with the augmented matrix so as to replace it with the matrix of an equivalent system whose solution is obvious. (Equivalent systems of equations have the same solution.) We work with the rows of the matrix just as we would work with the equations of the system. The following *row operations* replace an augmented matrix with the matrix of an equivalent system.

Row Operations Producing the Matrix of an Equivalent System

1. Interchanging any two rows
2. Multiplying each element of any row by a nonzero real number k
3. Adding to each element of a row the product of a real number k and the corresponding element of some other row

For simplicity, we describe the second of these operations as multiplying the row by k, and we describe the third as replacing the row by the sum of it and k times some other row. We say that these three row operations produce *row-equivalent* matrices.

If we use row operations to replace the augmented matrix of a system by the matrix of an equivalent system whose coefficient matrix has all ones in the main diagonal and zeros elsewhere, we will be able to read off the solution of the system. For example, suppose we used row operations to replace the augmented matrix of a given system of three equations in three unknowns by the matrix

$$\begin{pmatrix} 1 & 0 & 0 & 5 \\ 0 & 1 & 0 & -7 \\ 0 & 0 & 1 & 4 \end{pmatrix}.$$

The given system would then be equivalent to the system

$$1 \cdot x + 0 \cdot y + 0 \cdot z = 5$$
$$0 \cdot x + 1 \cdot y + 0 \cdot z = -7$$
$$0 \cdot x + 0 \cdot y + 1 \cdot z = 4.$$

In other words, the given system would be equivalent to the system

$$x \qquad\qquad = 5$$
$$y \qquad = -7$$
$$z = 4,$$

whose solution is obviously $x = 5$, $y = -7$, $z = 4$. This method is demonstrated in the next example. It is called the **Gaussian elimination method.** In some cases this method takes longer to do by hand than the elimination method presented in Section 1.10, but it is more easily programmed for a high-speed computer.

EXAMPLE 1

Use the Gaussian elimination method to solve the following system of equations.

$$3x - 2y + 6z = -6$$
$$-3x + 10y + 11z = 13$$
$$x - 2y - z = -3$$

(This system of equations was solved in Example 1 of Section 1.10 by using the elimination method presented in that section.)

Solution

The augmented matrix is

$$\begin{pmatrix} 3 & -2 & 6 & -6 \\ -3 & 10 & 11 & 13 \\ 1 & -2 & -1 & -3 \end{pmatrix}.$$

We shall use row operations to produce row-equivalent matrices until we produce the matrix of a system whose coefficient matrix has all ones in the main diagonal and zeros elsewhere. The row operations used will be indicated to the right. For convenience, we shall use the symbols R_1, R_2, and R_3 for Row 1, Row 2, and Row 3, respectively.

$$\begin{pmatrix} 3 & -2 & 6 & -6 \\ -3 & 10 & 11 & 13 \\ 1 & -2 & -1 & -3 \end{pmatrix} \rightarrow \begin{pmatrix} 1 & -2 & -1 & -3 \\ -3 & 10 & 11 & 13 \\ 3 & -2 & 6 & -6 \end{pmatrix}$$ Interchanging R_1 and R_3

$$\rightarrow \begin{pmatrix} 1 & -2 & -1 & -3 \\ 0 & 4 & 8 & 4 \\ 0 & 4 & 9 & 3 \end{pmatrix} \quad \begin{array}{l} 3R_1 + R_2 \\ -3R_1 + R_3 \end{array}$$

$$\rightarrow \begin{pmatrix} 1 & -2 & -1 & -3 \\ 0 & 1 & 2 & 1 \\ 0 & 4 & 9 & 3 \end{pmatrix} \quad \left(\frac{1}{4}\right) R_2$$

$$\rightarrow \begin{pmatrix} 1 & 0 & 3 & -1 \\ 0 & 1 & 2 & 1 \\ 0 & 0 & 1 & -1 \end{pmatrix} \quad \begin{array}{l} 2R_2 + R_1 \\ \\ -4R_2 + R_3 \end{array}$$

$$\rightarrow \begin{pmatrix} 1 & 0 & 0 & 2 \\ 0 & 1 & 0 & 3 \\ 0 & 0 & 1 & -1 \end{pmatrix} \quad \begin{array}{l} -3R_3 + R_1 \\ -2R_3 + R_2 \end{array}$$

The given system is equivalent to the system

$$\begin{aligned} x & = 2 \\ y & = 3 \\ z & = -1, \end{aligned}$$

whose solution is obviously $x = 2$, $y = 3$, $z = -1$.

Notice that in performing the various row operations to find the solution, we first got a 1 in the first row and first column, and then we successively used row operations to get zeros in the rest of the first column. Next we got a 1 in the second row and second column, and then we successively used row operations to get zeros in the rest of the second column. Finally we got a 1 in the third row and third column and used row operations to get zeros in the rest of the third column.

The system of equations in Example 1 has exactly one solution, namely $x = 2$, $y = 3$, $z = -1$. As we said in Section 1.10, a system of n simultaneous linear equations in n unknowns has either

1. Exactly one solution,

2. No solution, or

3. An unlimited number of solutions.

If we attempt to use the Gaussian elimination method on a system of n linear equations in n unknowns having no solution or having an unlimited number of solutions, we will not be able to produce the matrix of

Exercise 4.2, 1–20

an equivalent system having all ones in the main diagonal and zeros elsewhere in its coefficient matrix. Instead, we can produce the matrix of an equivalent system having a row of all zeros in its coefficient matrix. If we attempt to use the Gaussian elimination method and manage to produce such a row-equivalent matrix, we will know that there is not exactly one solution.

EXAMPLE 2   Try to solve the following system with the Gaussian elimination method.

$$x - 5y = 4$$
$$2x - 10y = 8$$

Solution   The augmented matrix is

$$\begin{pmatrix} 1 & -5 & 4 \\ 2 & -10 & 8 \end{pmatrix}.$$

Multiplying Row 1 by -2 and adding the product to Row 2 produces

$$\begin{pmatrix} 1 & -5 & 4 \\ 0 & 0 & 0 \end{pmatrix}.$$

The row of all zeros in the coefficient matrix indicates that there is not exactly one solution. In fact, there is an unlimited number of solutions. The two given equations are equivalent. (The second may be obtained from the first by multiplying through by 2.) An unlimited number of solutions may be found by arbitrarily assigning values to one of the variables and then solving for the other.

EXAMPLE 3   Try to solve the following system with the Gaussian elimination method.

$$x - 5y = 4$$
$$2x - 10y = 7$$

Solution   The augmented matrix is

$$\begin{pmatrix} 1 & -5 & 4 \\ 2 & -10 & 7 \end{pmatrix}.$$

Multiplying Row 1 by -2 and adding the product to Row 2 produces

$$\begin{pmatrix} 1 & -5 & 4 \\ 0 & 0 & -1 \end{pmatrix}.$$

The row of all zeros in the coefficient matrix indicates that there is not exactly one solution. In this example there is no solution. The given equations are *inconsistent*. The first one tells us that

$$x - 5y = 4$$

or that $2x - 10y$ is 8,

whereas according to the second one,

$$2x - 10y \text{ is } 7.$$

EXERCISE 4.2 MATRICES AND SIMULTANEOUS LINEAR EQUATIONS

▶ Use the Gaussian elimination method to solve each of the following systems of equations.

EXAMPLE

$$x - 5y = 4$$
$$3x - 10y = 2$$

Solution

$$\begin{pmatrix} 1 & -5 & 4 \\ 3 & -10 & 2 \end{pmatrix} \rightarrow \begin{pmatrix} 1 & -5 & 4 \\ 0 & 5 & -10 \end{pmatrix} \quad -3R_1 + R_2$$

$$\rightarrow \begin{pmatrix} 1 & -5 & 4 \\ 0 & 1 & -2 \end{pmatrix} \quad \left(\frac{1}{5}\right)R_2$$

$$\rightarrow \begin{pmatrix} 1 & 0 & -6 \\ 0 & 1 & -2 \end{pmatrix} \quad 5R_2 + R_1$$

$$x = -6$$
$$y = -2$$

1. $2x - 3y = 9$
  $x + 7y = -4$

2. $5x + 2y = 13$
  $3x - 6y = -21$

3. $4x - 3y = -17$
  $x + y = 1$

4. $6x - 5y = 10$
  $8x - 7y = 14$

5. $5x - 4y = 11$
  $3x + 6y = -27$

6. $x - y = -7$
  $3x + 5y = 11$

7. $11x - 9y = 20$
  $6x + 5y = 1$

8. $8x - 6y = 8$
  $3x + 7y = 3$

9. $-x + 5y + 5 = 0$
  $2x + 3y + 16 = 0$

10. $4x - 8y - 4 = 0$
  $4x - 6y - 8 = 0$

11. $x - 6y - 4z = -36$
  $x - 3y - z = -30$
  $-x + 9y + 8z = 41$

12. $x + 2y + 3z = -1$
  $x + 4y + 5z = -3$
  $x + 4y + 6z = 0$

13. $4x + 5y + 23z = -83$
  $x + y + 5z = -18$
  $-3x - 2y - 11z = 40$

14. $5x + 12y - 2z = 6$
  $6x + 18y - 5z = -11$
  $x + 2y = 4$

15. $A + C = 4$

$3A + B + 2C + D = 6$

$3A + 3B + C + D + E = 3$

$A + 3B = -5$

$B = -2$

(This system was solved in Example 2 of Section 1.10 by another method.)

16. $-A + 4B + 9C = 4$

$A - B - 2C = 6$

$-2A + 4B + 11C = 1$

17. $-x - 2y - 5z = 3$

$x + 3y + 10z = -8$

$-2x - 4y - 5z = 6$

18. $x - 2y - 8z = -32$

$x - y - z = -5$

$2x - 3y - 8z = -33$

19. $A - 3B + C = -3$

$A + C = 6$

$3A + B = 9$

20. $A + D = -1$

$3A + B - 2C + D = -4$

$3B + C = -1$

$2A - B + D = 1$

▶ If possible, solve each of the following systems with the Gaussian elimination method. If there is not exactly one solution, say so.

21. $2x + 4y = 8$

$6x + 12y = 24$

22. $2x + 4y = 8$

$6x + 12y = 20$

23. $2x + 4y = 8$

$6x + 10y = 22$

24. $3x - 5y = 1$

$6x - 10y = 2$

25. $3x - 5y = 1$

$6x - 10y = 5$

26. $2x + 3y + z = 1$

$-3x + 5y + 3z = -2$

$4x + 6y + 2z = 2$

27. $2x + 3y + z = 1$

$-3x + 5y + 3z = -2$

$4x + 6y + 2z = 4$

28. $2x + 3y + z = 1$

$-3x + 5y + 3z = -2$

$4x + 6y + 4z = 6$

29. $-A - D = 1$

$B + 2C = -1$

$-2A - 2D = 3$

$2C + 3D = 0$

30. $A + D = 2$

$3B + C = -1$

$2A + 2D = 4$

$4C + 3D = 0$

CHECK LIST FOR THE STUDENT

If you have learned the material in Section 4.2, you should be able to do the following:

1. Tell how many rows and how many columns there are in a given matrix.

2. Tell which element is in a specified row and a specified column of a given matrix.

3. List the elements in the main diagonal of a given square matrix.

4. Describe three row operations that produce the matrix of an equivalent system.

5. Tell what is meant by row-equivalent matrices.

6. Use the Gaussian elimination method to solve a system of n linear equations in n unknowns.

7. Tell what is indicated about the solutions when row operations on the matrix of a system produce the matrix of an equivalent system having a row of all zeros in its coefficient matrix.

SECTION QUIZ 4.2

1. How many rows and how many columns are there in the following matrix?

$$\begin{pmatrix} 2 & 7 & 8 & -3 \\ 1 & 0 & 4 & 5 \\ 0 & 1 & -2 & 6 \end{pmatrix}$$

2. What is the element in the second row and third column of the following matrix?

$$\begin{pmatrix} 3 & -4 & 5 \\ 2 & 6 & 9 \\ 1 & 0 & 8 \end{pmatrix}$$

3. What are the elements in the main diagonal of the following matrix?

$$\begin{pmatrix} 2 & -1 & 6 \\ 4 & 7 & 3 \\ 0 & 1 & 5 \end{pmatrix}$$

4. Replace the following matrix with a row-equivalent matrix by interchanging the second and third rows.

$$\begin{pmatrix} 1 & 0 & 6 \\ 0 & 0 & 5 \\ 0 & 1 & 4 \end{pmatrix}$$

5. Replace the following matrix with a row-equivalent matrix by multiplying the second row by -2.

$$\begin{pmatrix} 4 & -3 & 7 \\ 2 & -2 & 3 \\ 5 & 9 & -8 \end{pmatrix}$$

6. Replace the following matrix with a row-equivalent matrix by replacing the third row by the sum of it and two times the first row.

$$\begin{pmatrix} 1 & -1 & 2 \\ 0 & 7 & 9 \\ -2 & 3 & 4 \end{pmatrix}$$

Use the Gaussian elimination method to solve each of the following systems of equations.

7. $3x - 2y = 12$
   $x + 4y = -10$

8. $2x - 3y + 4z = -4$
   $x + y - 2z = 5$
   $4x - y + z = 3$

If possible, solve each of the following systems with the Gaussian elimination method. If there is not exactly one solution, say so.

9. $3x - 4y = 2$
   $6x - 8y = 1$

10. $x - 2y + z = 4$
    $2x - 4y + 2z = 8$
    $x - 3y + 5z = 7$

4.3 Determinants and Simultaneous Linear Equations

The purpose of this section is to introduce determinants, to show how to find the value of a determinant, and to show how determinants can be used to solve systems of simultaneous linear equations.

Associated with each square matrix whose elements are numbers there is a number called the *determinant* of the matrix. The determinant of a matrix is designated by using vertical bars to enclose the matrix rather than large parentheses or brackets. For example, the determinant of the matrix

$$\begin{pmatrix} 3 & -1 & 2 \\ 4 & 5 & 7 \\ 1 & 6 & -8 \end{pmatrix}$$

is written

$$\begin{vmatrix} 3 & -1 & 2 \\ 4 & 5 & 7 \\ 1 & 6 & -8 \end{vmatrix}$$

If a square matrix is represented by a capital letter, some authors represent its determinant by writing "det" in front of the letter; det A represents the determinant of matrix A. The determinant of an $n \times n$ matrix is said to be of **order** n. A determinant is really a number; sometimes that number is referred to as the *value* of the collection of symbols representing the determinant.

The value of any 2×2 determinant

$$\begin{vmatrix} a & b \\ c & d \end{vmatrix}$$

where a, b, c, and d are numbers is defined as follows.

$$\begin{vmatrix} a & b \\ c & d \end{vmatrix} = ad - bc$$

For example,

$$\begin{vmatrix} 3 & -2 \\ 4 & 5 \end{vmatrix} = (3)(5) - (-2)(4)$$
$$= 15 + 8$$
$$= 23.$$

The value of any $n \times n$ determinant for integral n greater than 2 can be defined concisely after first introducing a subscript notation for the elements in any determinant. According to this notation, the symbol a_{ij}

represents the element in the *ith row* and *jth column* of a matrix. For example, a_{23} represents the element in the second row and third column. Using this notation, any 2×2 determinant is written

$$\begin{vmatrix} a_{11} & a_{12} \\ a_{21} & a_{22} \end{vmatrix},$$

and any 3×3 determinant is written

$$\begin{vmatrix} a_{11} & a_{12} & a_{13} \\ a_{21} & a_{22} & a_{23} \\ a_{31} & a_{32} & a_{33} \end{vmatrix}.$$

Using this notation, the value of any 2×2 determinant is

$$\begin{vmatrix} a_{11} & a_{12} \\ a_{21} & a_{22} \end{vmatrix} = a_{11}a_{22} - a_{12}a_{21}.$$

Any element a_{ij} is said to be in *plus* position if the sum $i + j$ is even, and a_{ij} is said to be in *minus* position if the sum $i + j$ is odd. We see that the positions alternate as shown in the following charts for 2×2, 3×3, and 4×4 determinants.

$$\begin{vmatrix} + & - \\ - & + \end{vmatrix} \quad \begin{vmatrix} + & - & + \\ - & + & - \\ + & - & + \end{vmatrix} \quad \begin{vmatrix} + & - & + & - \\ - & + & - & + \\ + & - & + & - \\ - & + & - & + \end{vmatrix}$$

The **minor** of any element a_{ij} in a given determinant is the determinant that is left after the *i*th row and *j*th column are deleted. For example, to find the minor of a_{12} in a 3×3 determinant, simply delete the first row and second column as follows:

$$\begin{vmatrix} a_{11} & a_{12} & a_{13} \\ a_{21} & a_{22} & a_{23} \\ a_{31} & a_{32} & a_{33} \end{vmatrix}$$

The minor of a_{12} in a 3×3 determinant is

Exercise 4.3, 1–15

$$\begin{vmatrix} a_{21} & a_{23} \\ a_{31} & a_{33} \end{vmatrix}.$$

The value of any $n \times n$ determinant can be defined in terms of minors. To find the value of any $n \times n$ determinant, we can do the following:

1. Choose any row or column.

2. Multiply each element in the chosen row or column by $+1$ (if it is in plus position) or by -1 (if it is in minus position), and then multiply it by its minor; add all these products.

This method of finding the value of a determinant is called *expansion by minors* of the elements in the chosen row or column.

EXAMPLE 1

Write down an expression for the value of the following determinant using expansion by minors of the elements in the first row.

$$\begin{vmatrix} a_{11} & a_{12} & a_{13} \\ a_{21} & a_{22} & a_{23} \\ a_{31} & a_{32} & a_{33} \end{vmatrix}$$

Solution

$$\begin{vmatrix} a_{11} & a_{12} & a_{13} \\ a_{21} & a_{22} & a_{23} \\ a_{31} & a_{32} & a_{33} \end{vmatrix} = a_{11}\begin{vmatrix} a_{22} & a_{23} \\ a_{32} & a_{33} \end{vmatrix} - a_{12}\begin{vmatrix} a_{21} & a_{23} \\ a_{31} & a_{33} \end{vmatrix} + a_{13}\begin{vmatrix} a_{21} & a_{22} \\ a_{31} & a_{32} \end{vmatrix}$$

$$= a_{11}(a_{22}a_{33} - a_{23}a_{32}) - a_{12}(a_{21}a_{33} - a_{23}a_{31})$$
$$+ a_{13}(a_{21}a_{32} - a_{22}a_{31})$$

EXAMPLE 2

Find the value of the following determinant using expansion by minors of the elements in the first column.

$$\begin{vmatrix} 3 & 4 & 5 \\ -2 & 1 & -3 \\ -7 & 2 & 6 \end{vmatrix}$$

Solution

$$\begin{vmatrix} 3 & 4 & 5 \\ -2 & 1 & -3 \\ -7 & 2 & 6 \end{vmatrix} = 3\begin{vmatrix} 1 & -3 \\ 2 & 6 \end{vmatrix} - (-2)\begin{vmatrix} 4 & 5 \\ 2 & 6 \end{vmatrix} + (-7)\begin{vmatrix} 4 & 5 \\ 1 & -3 \end{vmatrix}$$

$$= 3(6 + 6) \quad + 2(24 - 10) \quad - 7(-12 - 5)$$
$$= 3(12) \qquad + 2(14) \qquad - 7(-17)$$
$$= 36 \qquad\quad + 28 \qquad\quad + 119$$
$$= 183$$

EXAMPLE 3

Find the value of the following determinant.

$$\begin{vmatrix} 1 & 2 & -1 & 0 \\ 4 & 0 & 5 & 1 \\ 0 & -3 & 1 & 2 \\ 3 & 0 & 6 & 7 \end{vmatrix}$$

Solution

Since we are free to choose any row or column for expanding by minors, here we should choose the second column. It has the most zeros in it, so the computations will be easiest with this choice. Expanding by minors of the elements in the second column, we have

$$\begin{vmatrix} 1 & 2 & -1 & 0 \\ 4 & 0 & 5 & 1 \\ 0 & -3 & 1 & 2 \\ 3 & 0 & 6 & 7 \end{vmatrix} = -(2)\begin{vmatrix} 4 & 5 & 1 \\ 0 & 1 & 2 \\ 3 & 6 & 7 \end{vmatrix} + 0 - (-3)\begin{vmatrix} 1 & -1 & 0 \\ 4 & 5 & 1 \\ 3 & 6 & 7 \end{vmatrix} + 0$$

This reduces our problem to that of computing the values of two 3×3 determinants. Expanding the first of these by minors of the elements in the first column and the second by minors of the elements in the third column, we have

$$\begin{vmatrix} 1 & 2 & -1 & 0 \\ 4 & 0 & 5 & 1 \\ 0 & -3 & 1 & 2 \\ 3 & 0 & 6 & 7 \end{vmatrix} = (-2)\left[4\begin{vmatrix} 1 & 2 \\ 6 & 7 \end{vmatrix} + 3\begin{vmatrix} 5 & 1 \\ 1 & 2 \end{vmatrix} \right] + 3\left[-(1)\begin{vmatrix} 1 & -1 \\ 3 & 6 \end{vmatrix} + 7\begin{vmatrix} 1 & -1 \\ 4 & 5 \end{vmatrix} \right]$$

$$= (-2)[4(7 - 12) + 3(10 - 1)] + 3[-(6 + 3) + 7(5 + 4)]$$
$$= (-2)[4(-5) + 3(9)] + 3[-9 + 7(9)]$$
$$= (-2)[-20 + 27] + 3[-9 + 63]$$
$$= (-2)[7] + 3[54]$$
$$= -14 + 162$$

Exercise 4.3, 16–30

$$= 148.$$

The following properties of determinants will not be proved here, but they are listed because they can sometimes be used to simplify computations.

PROPERTIES OF DETERMINANTS

1. If two rows or two columns are identical, the value of the determinant is zero.

Example:
$$\begin{vmatrix} 1 & 3 & 7 \\ -2 & 4 & 9 \\ 1 & 3 & 7 \end{vmatrix} = 0 \qquad \begin{vmatrix} 3 & 3 & 2 \\ 8 & 8 & 7 \\ 4 & 4 & 5 \end{vmatrix} = 0$$

2. If every element in some row or column is zero, the value of the determinant is zero.

Example:
$$\begin{vmatrix} 1 & 3 & 7 \\ 0 & 0 & 0 \\ 5 & 8 & -4 \end{vmatrix} = 0 \qquad \begin{vmatrix} 3 & 2 & 0 \\ 8 & 6 & 0 \\ 4 & -5 & 0 \end{vmatrix} = 0$$

3. If each element of any row or column is multiplied by a nonzero real number k, the resulting determinant is k times the original one.

Example:
$$\begin{vmatrix} 6 & 1 & 5 \\ -3 & 7 & 8 \\ 9 & 4 & -2 \end{vmatrix} = 3 \begin{vmatrix} 2 & 1 & 5 \\ -1 & 7 & 8 \\ 3 & 4 & -2 \end{vmatrix}$$

$$\begin{vmatrix} 6 & 3 & 15 \\ -1 & 7 & 8 \\ 3 & 4 & -2 \end{vmatrix} = 3 \begin{vmatrix} 2 & 1 & 5 \\ -1 & 7 & 8 \\ 3 & 4 & -2 \end{vmatrix}$$

4. If all the corresponding rows and columns are interchanged, the resulting determinant is equal to the original one.

Example:
$$\begin{vmatrix} 2 & 1 & 5 \\ -1 & 7 & 8 \\ 3 & 4 & -2 \end{vmatrix} = \begin{vmatrix} 2 & -1 & 3 \\ 1 & 7 & 4 \\ 5 & 8 & -2 \end{vmatrix}$$

5. If exactly two rows or exactly two columns are interchanged, the resulting determinant is the negative of the original one.

Example:
$$\begin{vmatrix} 2 & 1 & 5 \\ -1 & 7 & 8 \\ 3 & 4 & -2 \end{vmatrix} = - \begin{vmatrix} 3 & 4 & -2 \\ -1 & 7 & 8 \\ 2 & 1 & 5 \end{vmatrix}$$

$$\begin{vmatrix} 2 & 1 & 5 \\ -1 & 7 & 8 \\ 3 & 4 & -2 \end{vmatrix} = - \begin{vmatrix} 2 & 5 & 1 \\ -1 & 8 & 7 \\ 3 & -2 & 4 \end{vmatrix}$$

6. If the product of a nonzero real number k and each element of some row or column is added to the corresponding element in some other row or column, the resulting determinant is equal to the original one.

Example:
$$\begin{vmatrix} 2 & 1 & 5 \\ -1 & 7 & 8 \\ 3 & 4 & -2 \end{vmatrix} = \begin{vmatrix} 2 & 1 & 5 \\ -1 & 7 & 8 \\ -5 & 0 & -22 \end{vmatrix}$$
$(-4)\,R_1 + R_3$ (The product of (-4) and each element of R_1 was added to the corresponding element of R_3.)

$$\begin{vmatrix} 2 & 1 & 5 \\ -1 & 7 & 8 \\ 3 & 4 & -2 \end{vmatrix} = \begin{vmatrix} 2 & 15 & 5 \\ -1 & 0 & 8 \\ 3 & 25 & -2 \end{vmatrix}$$
$7\,C_1 + C_2$ (The product of 7 and each element of C_1 was added to the corresponding element of C_2.)

EXAMPLE 4

Find the value of the following determinant. Use properties of determinants to simplify computation by first reducing the problem to that of finding the value of a 2×2 determinant.

$$\begin{vmatrix} 3 & 1 & 7 \\ 2 & -4 & 5 \\ -6 & 9 & 8 \end{vmatrix}$$

Solution

There is an unlimited number of ways to use properties of determinants in finding the value of the given determinant. Two ways are demonstrated here. The various operations on rows and columns are indicated to the right. Here is one way to find the value.

$$\begin{vmatrix} 3 & 1 & 7 \\ 2 & -4 & 5 \\ -6 & 9 & 8 \end{vmatrix} = \begin{vmatrix} 1 & 5 & 2 \\ 2 & -4 & 5 \\ -6 & 9 & 8 \end{vmatrix} \quad (=1)R_2 + R_1$$

$$= \begin{vmatrix} 1 & 5 & 2 \\ 0 & -14 & 1 \\ 0 & 39 & 20 \end{vmatrix} \quad \begin{matrix} (=2)R_1 + R_2 \\ 6R_1 + R_3 \end{matrix}$$

$$= \begin{vmatrix} -14 & 1 \\ 39 & 20 \end{vmatrix} \quad \begin{matrix} \text{Expanding by} \\ \text{minors of } C_1 \end{matrix}$$

$$= (-14)(20) - (39)$$
$$= -280 - 39$$
$$= -319$$

Here is another way to find the value required.

$$\begin{vmatrix} 3 & 1 & 7 \\ 2 & -4 & 5 \\ -6 & 9 & 8 \end{vmatrix} = - \begin{vmatrix} 1 & 3 & 7 \\ -4 & 2 & 5 \\ 9 & -6 & 8 \end{vmatrix} \quad \text{Interchanging } C_1 \text{ and } C_2$$

$$= - \begin{vmatrix} 1 & 0 & 0 \\ -4 & 14 & 33 \\ 9 & -33 & -55 \end{vmatrix} \quad \begin{matrix} (=3)C_1 + C_2 \\ (=7)C_1 + C_3 \end{matrix}$$

$$= - \begin{vmatrix} 14 & 33 \\ -33 & -55 \end{vmatrix} \quad \begin{matrix} \text{Expanding by} \\ \text{minors of } R_1 \end{matrix}$$

$$= -[(14)(-55) - (33)(-33)]$$
$$= -[-770 + 1089]$$
$$= -319$$

Exercise 4.3, 31–35

Determinants may be used to solve systems of n simultaneous linear equations in n unknowns. *Cramer's rule* tells how to do this when the determinant of the coefficient matrix is not zero.

Cramer's Rule

If the determinant of the coefficient matrix of a system of n simultaneous linear equations in n unknowns is not zero, then the system has exactly one solution, and the value of each unknown is equal to a quotient of two determinants. The determinant in the denominator is the determinant of the coefficient matrix. The determinant in the numerator is derived from the determinant of the coefficient matrix by replacing the column of coefficients of the unknown sought by the column of constant terms as they appear isolated on the right sides of the equals signs in the equations of the system.

Let us express Cramer's rule symbolically for the case of *two* linear equations in *two* unknowns.

Cramer's Rule for 2 Equations in 2 Unknowns

If $\begin{vmatrix} a_{11} & a_{12} \\ a_{21} & a_{22} \end{vmatrix} \neq 0$, the system of equations

$$a_{11}x + a_{12}y = k_1$$
$$a_{21}x + a_{22}y = k_2$$

has exactly one solution. That solution is

$$x = \frac{\begin{vmatrix} k_1 & a_{12} \\ k_2 & a_{22} \end{vmatrix}}{\begin{vmatrix} a_{11} & a_{12} \\ a_{21} & a_{22} \end{vmatrix}}, \qquad y = \frac{\begin{vmatrix} a_{11} & k_1 \\ a_{21} & k_2 \end{vmatrix}}{\begin{vmatrix} a_{11} & a_{12} \\ a_{21} & a_{22} \end{vmatrix}}.$$

EXAMPLE 5

Use Cramer's rule to find the solution of the following system.

$$x - 5y - 4 = 0$$
$$3x - 10y - 2 = 0$$

Solution

In order to apply Cramer's rule, we first rewrite the equations with the constant terms isolated on the right sides of the equations.

$$x - 5y = 4$$
$$3x - 10y = 2$$

This same system was solved with the Gaussian elimination method in the first example presented in Exercise 4.2. Solving it with Cramer's rule, we have

$$x = \frac{\begin{vmatrix} 4 & -5 \\ 2 & -10 \end{vmatrix}}{\begin{vmatrix} 1 & -5 \\ 3 & -10 \end{vmatrix}} = \frac{-30}{5} = -6,$$

$$y = \frac{\begin{vmatrix} 1 & 4 \\ 3 & 2 \end{vmatrix}}{\begin{vmatrix} 1 & -5 \\ 3 & -10 \end{vmatrix}} = \frac{-10}{5} = -2.$$

This system could also be solved readily with the methods presented in Section 1.10.

In view of the fact that we can solve systems of n simultaneous equations in n unknowns with the methods of Section 1.10, it may seem that it was unnecessary to introduce additional methods for solving systems of linear equations. Cramer's rule and matrix methods are extremely important, however, because they arise in other contexts in mathematics and they are readily programmable for computation on a high-speed computer.

Note that Cramer's rule applies only to the case in which the determinant of the coefficient matrix is not zero. If the determinant of the coefficient matrix is zero, the system fails to have exactly one solution.

Cramer's rule is used to solve *three* linear equations in *three* unknowns in the next example.

EXAMPLE 6

Use Cramer's rule to find the solution of the following system.

$$3x - 2y + 6z = -6$$
$$-3x + 10y + 11z = 13$$
$$x - 2y - z = -3$$

Solution

(Note that this system of equations was solved by other methods in Example 1 of Section 1.10 and Example 1 of Section 4.2.) The determinant of the coefficient matrix is

$$\begin{vmatrix} 3 & -2 & 6 \\ -3 & 10 & 11 \\ 1 & -2 & -1 \end{vmatrix}.$$

Evaluating this determinant, we have

$$\begin{vmatrix} 3 & -2 & 6 \\ -3 & 10 & 11 \\ 1 & -2 & -1 \end{vmatrix} = \begin{vmatrix} 0 & 4 & 9 \\ 0 & 4 & 8 \\ 1 & -2 & -1 \end{vmatrix} \qquad \begin{matrix} (-3)R_3 + R_1 \\ 3R_3 + R_2 \end{matrix}$$

$$= \begin{vmatrix} 4 & 9 \\ 4 & 8 \end{vmatrix}$$

$$= -4.$$

Since the value of the coefficient matrix is not zero, the given system has exactly one solution. That solution is

$$x = \frac{\begin{vmatrix} -6 & -2 & 6 \\ 13 & 10 & 11 \\ -3 & -2 & -1 \end{vmatrix}}{-4} = \frac{-8}{-4} = 2,$$

$$y = \frac{\begin{vmatrix} 3 & -6 & 6 \\ -3 & 13 & 11 \\ 1 & -3 & -1 \end{vmatrix}}{-4} = \frac{-12}{-4} = 3,$$

$$z = \frac{\begin{vmatrix} 3 & -2 & -6 \\ -3 & 10 & 13 \\ 1 & -2 & -3 \end{vmatrix}}{-4} = \frac{4}{-4} = -1.$$

Exercise 4.3, 36–55 Students should verify that the values of the determinants in the numerators are actually as we have just stated.

EXERCISE 4.3 DETERMINANTS AND SIMULTANEOUS LINEAR EQUATIONS

▶ In each of the following, name the element in the indicated row and column of the determinant.

$$\begin{vmatrix} 2 & -1 & 5 \\ 3 & 0 & 4 \\ 6 & -7 & 8 \end{vmatrix}.$$

EXAMPLE

First row, third column

Solution

5

1. First row, second column

2. Second row, first column

3. Second row, third column

4. Third row, second column

5. Third row, third column

▶ Using the subscript notation presented in the text, state the row and column in which each of the following elements lies, and state whether the element is in plus position or minus position.

EXAMPLE

a_{34}

Solution

Third row, fourth column

$3 + 4 = 7$, minus position

6. a_{43} **7.** a_{23} **8.** a_{14} **9.** a_{41} **10.** a_{13}

▶ Find the minor of each of the following elements in the determinant

$$\begin{vmatrix} 2 & -1 & 5 \\ 3 & 0 & 4 \\ 6 & -7 & 8 \end{vmatrix}.$$

EXAMPLE   4

Solution $\begin{vmatrix} 2 & -1 \\ 6 & -7 \end{vmatrix}$

11. 5   **12.** −7   **13.** 6   **14.** −1   **15.** 3

▶ Find the value of each of the following.

EXAMPLES   (a) $\begin{vmatrix} -3 & 8 \\ -2 & 5 \end{vmatrix}$

(b) $\begin{vmatrix} 2 & 4 & 1 \\ 1 & -3 & -2 \\ 3 & 5 & -1 \end{vmatrix}$

Solutions   (a) $(-15) - (-16) = 1$

(b) $\begin{vmatrix} 2 & 4 & 1 \\ 1 & -3 & -2 \\ 3 & 5 & -1 \end{vmatrix} = 2\begin{vmatrix} -3 & -2 \\ 5 & -1 \end{vmatrix} - (1)\begin{vmatrix} 4 & 1 \\ 5 & -1 \end{vmatrix} + 3\begin{vmatrix} 4 & 1 \\ -3 & -2 \end{vmatrix}$

$$= 2(13) - (-9) + 3(-5)$$
$$= 26 + 9 - 15$$
$$= 20$$

16. $\begin{vmatrix} 3 & -5 \\ -4 & 6 \end{vmatrix}$ **17.** $\begin{vmatrix} -1 & 7 \\ 6 & 0 \end{vmatrix}$

18. $\begin{vmatrix} -8 & 9 \\ 6 & 3 \end{vmatrix}$ **19.** $\begin{vmatrix} 1 & -8 \\ 0 & 9 \end{vmatrix}$

20. $\begin{vmatrix} 3 & -6 \\ 5 & -2 \end{vmatrix}$ **21.** $\begin{vmatrix} 4 & 1 & 2 \\ 1 & -5 & 6 \\ -4 & 3 & -5 \end{vmatrix}$

22. $\begin{vmatrix} 1 & 0 & 3 \\ 2 & -5 & 4 \\ 6 & 0 & 8 \end{vmatrix}$ **23.** $\begin{vmatrix} 5 & 7 & 1 \\ -5 & 6 & 2 \\ 1 & 7 & -3 \end{vmatrix}$

24. $\begin{vmatrix} 3 & 7 & -1 \\ 2 & -6 & 5 \\ 0 & 4 & -3 \end{vmatrix}$ **25.** $\begin{vmatrix} 2 & -2 & 1 \\ 4 & -3 & 7 \\ -1 & 3 & -4 \end{vmatrix}$

26.
$$\begin{vmatrix} 3 & 1 & 0 & 2 \\ 4 & -5 & 6 & 0 \\ 2 & -3 & 0 & 1 \\ -3 & -1 & 0 & 7 \end{vmatrix}$$

27.
$$\begin{vmatrix} -3 & 2 & -1 & 5 \\ 0 & 3 & 5 & 6 \\ 1 & 4 & -2 & -4 \\ 0 & 6 & 0 & 8 \end{vmatrix}$$

28.
$$\begin{vmatrix} 0 & 1 & 0 & 1 \\ 2 & 0 & 3 & 0 \\ -1 & 0 & 0 & 1 \\ 4 & -2 & 6 & 0 \end{vmatrix}$$

29.
$$\begin{vmatrix} 1 & -3 & 4 & 5 \\ -2 & 3 & 0 & 7 \\ 6 & -7 & 8 & 2 \\ -1 & -4 & 6 & 1 \end{vmatrix}$$

30.
$$\begin{vmatrix} 1 & 7 & 6 & -3 \\ 2 & 1 & 4 & 3 \\ -2 & 3 & 1 & -4 \\ 6 & 5 & 7 & 1 \end{vmatrix}$$

▶ Using properties of determinants, fill in the blank parentheses in each of the following.

EXAMPLES

(a) $\begin{vmatrix} 3 & 5 & 8 \\ 1 & -3 & 2 \\ 3 & 5 & 8 \end{vmatrix} = (\quad)$

(b) $\begin{vmatrix} 3 & 5 & 8 \\ 1 & -3 & 2 \\ 3 & 5 & 7 \end{vmatrix} = (\quad) \begin{vmatrix} 3 & 1 & 3 \\ 5 & -3 & 5 \\ 8 & 2 & 7 \end{vmatrix}$

Solutions

(a) 0   (Two rows are identical.)

(b) +

31. $\begin{vmatrix} 2 & 5 & 6 \\ 3 & 4 & 1 \\ 4 & -1 & 7 \end{vmatrix} = (\quad) \begin{vmatrix} 5 & 2 & 6 \\ 4 & 3 & 1 \\ -1 & 4 & 7 \end{vmatrix}$

32. $\begin{vmatrix} 2 & 5 & 6 \\ 3 & 4 & 1 \\ 4 & -1 & 7 \end{vmatrix} = (\quad) \begin{vmatrix} 3 & 4 & 1 \\ 2 & 5 & 6 \\ 4 & -1 & 7 \end{vmatrix}$

33. $\begin{vmatrix} 2 & 5 & 6 \\ 3 & 4 & 2 \\ 4 & -1 & 4 \end{vmatrix} = (\quad) \begin{vmatrix} 2 & 5 & 3 \\ 3 & 4 & 1 \\ 4 & -1 & 2 \end{vmatrix}$

34. $\begin{vmatrix} 2 & 5 & 6 \\ 3 & 4 & 1 \\ 0 & 0 & 0 \end{vmatrix} = (\quad)$

35. $\begin{vmatrix} 3 & 7 & 3 \\ 5 & 6 & 5 \\ 1 & 4 & 1 \end{vmatrix} = (\quad)$

36–55. Use Cramer's rule to solve each of the systems in items 1–20 of Exercise 4.2.

CHECK LIST FOR THE STUDENT

If you have learned the material in Section 4.3, you should be able to do the following:

1. Show how to denote the determinant of a given square matrix.

2. Give the order of the determinant of an $n \times n$ matrix.

3. Find the value of any 2×2 determinant.

4. State the row and column in which a specified element lies in a given determinant.

5. Use the subscript notation presented in the text, and specify the row and column in which the element a_{ij} lies.

6. Find the minor of a specified element in a given determinant.

7. State whether a specified element in a given determinant is in plus position or in minus position.

8. Use expansion by minors to find the value of a given determinant.

9. Use the properties of determinants given in the text.

10. Use Cramer's rule to solve a given system of n simultaneous linear equations in n unknowns.

SECTION QUIZ 4.3

1. In what row and column does 7 lie in the following determinant?

$$\begin{vmatrix} 2 & -1 & 3 \\ 4 & 8 & 5 \\ -2 & 7 & 9 \end{vmatrix}$$

2. Using the subscript notation presented in the text, in what row and column does the element a_{23} lie, and is a_{23} in plus position or minus position?

3. What is the minor of 9 in the following determinant?

$$\begin{vmatrix} 3 & -1 & 4 \\ 0 & 2 & 8 \\ 4 & 9 & 1 \end{vmatrix}$$

4. Find the value of the following determinant.

$$\begin{vmatrix} 3 & 7 \\ -1 & 8 \end{vmatrix}$$

5. Using expansion by minors of the elements in the first column, find the value of the following determinant.

$$\begin{vmatrix} 3 & 4 & 2 \\ 1 & 7 & 6 \\ -2 & 5 & -1 \end{vmatrix}$$

Using properties of determinants, fill in the blank parentheses in each of the following.

6. (a) $\begin{vmatrix} 2 & 0 & 4 \\ 3 & 0 & 5 \\ 1 & 0 & 7 \end{vmatrix} = (\quad)$ (b) $\begin{vmatrix} 2 & 4 & 4 \\ 3 & 5 & 5 \\ 1 & 7 & 7 \end{vmatrix} = (\quad)$

7. (a) $\begin{vmatrix} 3 & 5 & 7 \\ -2 & 1 & 4 \\ 6 & 8 & -3 \end{vmatrix} = (\quad) \begin{vmatrix} 3 & -2 & 6 \\ 5 & 1 & 8 \\ 7 & 4 & -3 \end{vmatrix}$

(b) $\begin{vmatrix} 3 & 5 & 7 \\ -4 & 2 & 8 \\ 6 & 8 & -3 \end{vmatrix} = (\quad) \begin{vmatrix} 3 & 5 & 7 \\ -2 & 1 & 4 \\ 6 & 8 & -3 \end{vmatrix}$

8. (a) $\begin{vmatrix} 3 & 5 & 9 \\ -2 & 8 & 7 \\ 5 & 1 & 4 \end{vmatrix} = (\quad) \begin{vmatrix} 5 & 1 & 4 \\ -2 & 8 & 7 \\ 3 & 5 & 9 \end{vmatrix}$

(b) $\begin{vmatrix} 3 & 5 & 9 \\ -2 & 8 & 7 \\ 5 & 1 & 4 \end{vmatrix} = (\quad) \begin{vmatrix} 5 & -3 & 2 \\ -2 & 8 & 7 \\ 5 & 1 & 4 \end{vmatrix}$

Use Cramer's rule to find the solution of each of the following systems.

9. $3x - 2y = 2$

$5x - y = -6$

10. $2x - 3y + 5z = -3$

$x - 2y + 4z = -2$

$3x + y - 2z = 5$

4.4 Partial Fractions

The purpose of this section is to present the partial fraction theorem and to demonstrate an application of methods of solving systems of simultaneous linear equations.

In basic algebra we learn how to express a sum of fractions as a single fraction. As we observed in Sections 1.6 and 1.10, in calculus it is sometimes necessary to work in reverse and express a fraction whose numerator and denominator are polynomials as a sum of "partial" fractions whose denominators are of lower degree than the denominator of the given fraction. This decomposition can be effected by applying the *partial fraction theorem.*

Before stating the partial fraction theorem, let us review some facts about polynomials.

1. Where a and b are real numbers, and $a \neq 0$, the expression $ax + b$ is a **linear** or *first-degree* polynomial.

Example: $3x + 7$

2. Where a, b, and c are real numbers, and $a \neq 0$, $ax^2 + bx + c$ is a **quadratic** or *second-degree* polynomial.

Example: $3x^2 + 5x - 2$

3. Any quadratic with real coefficients that cannot be factored into linear factors with real coefficients is called an **irreducible quadratic.**

Example: $3x^2 + 2x + 1$

4. If the quadratic equation $ax^2 + bx + c = 0$ has only imaginary roots, then $ax^2 + bx + c$ is an irreducible quadratic.

Example: $3x^2 + 2x + 1 = 0$

5. Any polynomial with real coefficients can be factored into a product of linear factors with real coefficients and irreducible quadratic factors. (See the factorization theorem, Section 4.1.)

Example: $x^4 + 3x^3 + 5x^2 + x - 10 = (x + 2)(x - 1)(x^2 + 2x + 5)$

6. In the factorization of a polynomial $P(x)$ into a product of linear factors and irreducible quadratic factors, one or more of the factors may be repeated. If k is a positive integer and $P(x)$ has $(ax + b)^k$ or $(ax^2 + bx + c)^k$ as a factor but has no higher power of $ax + b$ or of $ax^2 + bx + c$ as a factor, then the factor $(ax + b)$ or $(ax^2 + bx + c)$ is said to be of **multiplicity k**.

Example:

$$P(x) = (x + 2)^3(x - 1)(x^2 + 2x + 5)^2$$

$(x + 2)$ of multiplicity 3

$(x^2 + 2x + 5)$ of multiplicity 2

Now let us state the partial fraction theorem. (The proof will not be included in this text.)

Partial Fraction Theorem   If $N(x)$ and nonzero $D(x)$ are polynomials in x with real coefficients and with $N(x)$ of lower degree than $D(x)$, then $N(x)/D(x)$ can be expressed as a sum of partial fractions as follows:

1. For each unrepeated linear factor $(ax + b)$ of $D(x)$, there is a partial fraction of the form

$$\frac{A}{ax + b}$$

where A is a specific real number.

Example: For unrepeated factor $(2x + 7)$, there is a partial fraction $A/(2x + 7)$.

2. For each repeated linear factor $(ax + b)$ of multiplicity k, there is a sum of partial fractions of the form

$$\frac{A_1}{ax + b} + \frac{A_2}{(ax + b)^2} + \frac{A_3}{(ax + b)^3} + \cdots + \frac{A_k}{(ax + b)^k}$$

where $A_1, A_2, A_3, \ldots, A_k$ are specific real numbers.

Example: For factor $(2x + 7)$ of multiplicity 3, there is a sum of 3 partial fractions of the form

$$\frac{A_1}{2x + 7} + \frac{A_2}{(2x + 7)^2} + \frac{A_3}{(2x + 7)^3}.$$

3. For each unrepeated irreducible quadratic factor $(ax^2 + bx + c)$, there is a partial fraction of the form

$$\frac{Ax + B}{ax^2 + bx + c}$$

where A and B are specific real numbers.

Example: For unrepeated factor $(3x^2 + 2x + 1)$, there is a partial fraction

$$\frac{Ax + B}{3x^2 + 2x + 1}.$$

4. For each repeated irreducible quadratic factor $(ax^2 + bx + c)$ of multiplicity k, there is a sum of k partial fractions of the form

$$\frac{A_1 x + B_1}{ax^2 + bx + c} + \frac{A_2 x + B_2}{(ax^2 + bx + c)^2} + \cdots + \frac{A_k x + B_k}{(ax^2 + bx + c)^k}$$

where $A_1, B_1, A_2, B_2, \ldots, A_k, B_k$ are specific real numbers.

Example: For factor $(x^2 + 2x + 5)$ of multiplicity 3, there is a sum of 3 partial fractions of the form

$$\frac{A_1 x + B_1}{x^2 + 2x + 5} + \frac{A_2 x + B_2}{(x^2 + 2x + 5)^2} + \frac{A_3 x + B_3}{(x^2 + 2x + 5)^3}.$$

Note that the partial fraction theorem is stated only for the case in which the numerator is of lower degree than the denominator. If the numerator $N(x)$ is of the same or of higher degree than the denominator $D(x)$, we can divide $N(x)$ by $D(x)$ until the remainder $R(x)$ is of lower degree than $D(x)$; we can then apply the partial fraction theorem to the fraction $R(x)/D(x)$.

EXAMPLE 1

Write out the form of the partial-fraction decomposition of

$$\frac{2x^3 + 3x^2 - 24x - 23}{2x^2 - 5x - 7}.$$

Solution

Since the numerator is not of lower degree than the denominator, we must first divide.

$$
\begin{array}{r}
x + 4 \\
2x^2 - 5x - 7 \overline{\smash{\big)}\ 2x^3 + 3x^2 - 24x - 23} \\
\underline{2x^3 - 5x^2 - 7x} \\
8x^2 - 17x - 23 \\
\underline{8x^2 - 20x - 28} \\
3x + 5
\end{array}
$$

$$\frac{2x^3 + 3x^2 - 24x - 23}{2x^2 - 5x - 7} = x + 4 + \frac{3x + 5}{2x^2 - 5x - 7}$$

$$= x + 4 + \frac{3x + 5}{(2x - 7)(x + 1)}$$

We may now express

$$\frac{3x + 5}{2x^2 - 5x - 7}$$

as a sum of partial fractions. The factors of the denominator are $(2x - 7)$ and $(x + 1)$.

For factor $(2x - 7)$ there is a partial fraction $A/(2x - 7)$.

For factor $(x + 1)$ there is a partial fraction $B/(x + 1)$.

The decomposition of the given fraction must be of the form

$$\frac{2x^3 + 3x^2 - 24x - 23}{2x^2 - 5x - 7} = x + 4 + \frac{A}{2x - 7} + \frac{B}{x + 1}.$$

EXAMPLE 2   Write out the form of the partial-fraction decomposition of

$$\frac{x^2 - 4x - 5}{(x - 3)(x - 1)^2(x^2 + 2x + 5)}.$$

Solution   The numerator is of lower degree than the denominator; the partial fraction theorem applies. The factors of the denominator are $(x - 3)$, $(x - 1)^2$, and $(x^2 + 2x + 5)$.

For factor $(x - 3)$ there is a partial fraction $A/(x - 3)$.

For factor $(x - 1)^2$ there is a sum of two partial fractions,

$$\frac{B}{x - 1} + \frac{C}{(x - 1)^2}.$$

For factor $(x^2 + 2x + 5)$ there is a partial fraction

$$\frac{Dx + E}{x^2 + 2x + 5}.$$

$$\frac{x^2 - 4x - 5}{(x - 3)(x - 1)^2(x^2 + 2x + 5)} = \frac{A}{x - 3} + \frac{B}{x - 1} + \frac{C}{(x - 1)^2}$$

$$+ \frac{Dx + E}{x^2 + 2x + 5}$$

EXAMPLE 3   Write out the form of the partial-fraction decomposition of

$$\frac{2x^2 - 5x - 3}{(x + 5)(x^2 + 9)^2}.$$

Solution   The numerator is of lower degree than the denominator; the partial fraction theorem applies. The factors of the denominator are $(x + 5)$ and $(x^2 + 9)^2$.

For factor $(x + 5)$ there is a partial fraction $A/(x + 5)$.

For factor $(x^2 + 9)^2$ there is a sum of two partial fractions,

$$\frac{Bx + C}{x^2 + 9} + \frac{Dx + E}{(x^2 + 9)^2}.$$

$$\frac{2x^2 - 5x - 3}{(x + 5)(x^2 + 9)^2} = \frac{A}{x + 5} + \frac{Bx + C}{x^2 + 9} + \frac{Dx + E}{(x^2 + 9)^2}.$$

Exercise 4.4, 1–5

The next example illustrates an application of methods of solving systems of simultaneous linear equations. We have had several of these methods. Elementary methods were presented in Section 1.10, and methods using matrices and determinants were presented in Sections 4.2 and 4.3.

EXAMPLE 4

Express the following as a sum of partial fractions.

$$\frac{6x^2 + 7x - 4}{x(x - 1)(x + 2)}$$

Solution

The numerator is of lower degree than the denominator; the partial fraction theorem applies. The factors of the denominator are the un-repeated linear factors x, $x - 1$, and $x + 2$. The decomposition required is of the form

$$\frac{6x^2 + 7x - 4}{x(x - 1)(x + 2)} = \frac{A}{x} + \frac{B}{x - 1} + \frac{C}{x + 2}.$$

In Example 3 of Section 1.10, we assumed that this was the form of the decomposition of the given fraction, and we found A, B, and C by two methods. The first method, *equating coefficients of like powers,* led to a system of three equations in the three "unknowns" A, B, and C. We obtained that system by first combining the three partial fractions on the right and then equating coefficients of like terms in the numerators on the left and right of the equation. In particular, we obtained the system

$$\begin{aligned} A + B + C &= 6 \\ A + 2B - C &= 7 \\ -2A &= -4. \end{aligned}$$

We solved this system for A, B, and C by the methods of Section 1.10. We may now solve this system by using Gaussian elimination or by using Cramer's rule.

The second method presented in Section 1.10 for finding A, B, and C, *substituting convenient values,* is easier in this case. If repeated factors appear in the denominator, however, it might be easier to use the first method or a combination of the two methods.

Exercise 4.4, 6–15

EXERCISE 4.4 PARTIAL FRACTIONS

▶ Write out the form of the partial-fraction decomposition of each of the following.

EXAMPLES

(a) $\dfrac{3x^2 - 5x + 9}{(x - 3)^2(x^2 + 2x + 3)} = \dfrac{A}{x - 3} + \dfrac{B}{(x - 3)^2} + \dfrac{Cx + D}{x^2 + 2x + 3}$

(b) $\dfrac{2x^3 + 1}{(x - 5)(x^2 + 2x + 3)^2} = \dfrac{A}{x - 5} + \dfrac{Bx + C}{x^2 + 2x + 3} + \dfrac{Dx + E}{(x^2 + 2x + 3)^2}$

1. $\dfrac{x^2 - 7x + 5}{x(x - 3)(x + 7)}$

2. $\dfrac{3x^2 - 5x + 1}{(x - 2)^3(x^2 + 1)}$

3. $\dfrac{2x + 1}{x^2 - 6x + 9}$

4. $\dfrac{x^3 - 5x}{(x^2 + 1)^3(x - 1)}$

5. $\dfrac{3x^2 - 2}{(x - 1)^2(3x^2 + x + 1)}$

▶ Express each of the following as a sum of partial fractions. Unless otherwise specified by your instructor, use any method you wish.

EXAMPLE

$\dfrac{2x + 1}{(x - 1)^2}$

Solution

$$\dfrac{2x + 1}{(x - 1)^2} = \dfrac{A}{x - 1} + \dfrac{B}{(x - 1)^2}$$

$$= \dfrac{A(x - 1) + B}{(x - 1)^2}$$

$$2x + 1 = A(x - 1) + B$$

Let $x = 1$. $\qquad\qquad 3 = B$

Equate coefficients of x. $\quad 2 = A$

$$\dfrac{2x + 1}{(x - 1)^2} = \dfrac{2}{x - 1} + \dfrac{3}{(x - 1)^2}$$

6. $\dfrac{4x - 5}{(x - 2)^2}$

7. $\dfrac{2x^2 - 11x + 3}{x(x - 1)(x - 3)}$

8. $\dfrac{-x^2 + 4x + 2}{(x - 1)(x^2 + x + 1)}$

9. $\dfrac{-5x^2 - 11x - 7}{(x - 2)(x^2 + x + 1)^2}$

10. $\dfrac{x^3 + 5x^2 + 13x + 7}{(x^2 + 3x + 4)^2}$

11. $\dfrac{x^3 + 7x^2 + 6x - 42}{x^2 + 2x - 8}$

12. $\dfrac{2x^3 - x^2 + 4x - 7}{x^2 + x + 2}$

13. $\dfrac{x^2 - 22x - 30}{x(x^2 + 3x - 10)}$

14. $\dfrac{x^2 - 2x + 2}{(x - 3)^3}$

15. $\dfrac{x^4 - 3x^3 + x^2 - 19x - 22}{(x - 2)(x^2 + 4)^2}$

CHECK LIST FOR THE STUDENT

If you have learned the material in Section 4.4, you should be able to do the following:

1. Give the form of the partial fraction(s) associated with each of the following types of factors of $D(x)$, where $N(x)$ and nonzero $D(x)$ are polynomials with real coefficients, and where the numerator of $N(x)/D(x)$ is of lower degree than the denominator.

(a) Unrepeated linear: $ax + b$
(b) Repeated linear of multiplicity k: $(ax + b)^k$
(c) Unrepeated irreducible quadratic: $ax^2 + bx + c$
(d) Repeated irreducible quadratic of multiplicity k: $(ax^2 + bx + c)^k$

2. Write out the form of the partial-fraction decomposition of a given $N(x)/D(x)$ where $N(x)$ and nonzero $D(x)$ are polynomials with real coefficients, where the degree of $N(x)$ is not necessarily lower than the degree of $D(x)$, and where $D(x)$ has been factored into a product of linear and irreducible quadratic factors.

3. Express a given fraction whose numerator and nonzero denominator are polynomials with real coefficients as a sum of partial fractions, and show how this could involve the solution of a system of simultaneous linear equations.

SECTION QUIZ 4.4

Write out the form of the partial-fraction decomposition of each of the following.

1. $\dfrac{2x^2 - 4x + 7}{(x - 4)(x + 5)^2}$

2. $\dfrac{5x^2 - 3x + 8}{(x - 3)(x^2 - 4x + 29)}$

3. $\dfrac{2x^4 + 1}{(x - 3)(x^2 - 4x + 29)^2}$

Express each of the following as a sum of partial fractions.

4. $\dfrac{2x - 3}{(x - 3)^2}$

5. $\dfrac{21x + 15}{x(x + 1)(x - 5)}$

4.5 Inequalities in Two Variables; Systems of Inequalities

The purpose of this section is to introduce inequalities in two variables and their graphs, to show how to solve some systems of simultaneous inequalities in two variables, and to suggest some applications.

Properties of inequalities and linear inequalities in one variable were considered in Section 1.9. In this section we shall consider inequalities in two variables. We shall consider inequalities such as

$$2x - y + 1 \geq 0, \quad y > x^2, \quad \text{and} \quad x^2 + y^2 \leq 9.$$

A **solution of an inequality in exactly two variables** with real numbers for coefficients is an ordered pair of real numbers that *satisfy* the inequality in the sense that the inequality becomes a true statement when the members of the ordered pair replace the variables. The *solution set* or simply *the solution* is the set of all ordered pairs of real numbers that satisfy the inequality. The pairs $(1, -1)$, $(3, 5)$, and $(4, 9)$ are some of the solutions (x, y) of

$$y \leq 2x + 1$$

because the members of each of these ordered pairs satisfy the inequality. The statements

$$-1 \leq 2(1) + 1, \quad 5 \leq 2(3) + 1, \quad \text{and} \quad 9 \leq 2(4) + 1$$

are all true. Similarly, $(2, 6)$ and $(3, 10)$ are solutions (x, y) of

$$y > x^2.$$

The statements

$$6 > 2^2 \quad \text{and} \quad 10 > 3^2$$

are both true.

The graph of an inequality in exactly two variables with real numbers for coefficients is the set of all points whose Cartesian coordinates satisfy the inequality. We shall see that the graph of $y \leq 2x + 1$ consists of the line $y = 2x + 1$ and the half-plane shaded in Figure 4.2; the graph of $y > x^2$ is the shaded region in Figure 4.3.

A **linear inequality in two variables**, x and y, is an inequality that can be written in one of the forms

$$ax + by + c < 0 \qquad ax + by + c > 0$$
$$ax + by + c \leq 0 \qquad ax + by + c \geq 0$$

where $a \neq 0$, $b \neq 0$, and a, b, and c are specific real numbers. For example, $y \leq 2x + 1$ is a linear inequality in the two variables x and y because it can be written in one of the prescribed forms.

The **graph** of the linear inequality

$$ax + by + c \leq 0 \quad \text{or of} \quad ax + by + c \geq 0$$

consists of all points on the line $ax + by + c = 0$, together with all points in one of the half-planes determined by the line. Likewise, the graph of the linear inequality

$$ax + by + c < 0 \quad \text{or of} \quad ax + by + c > 0$$

consists only of all points in one of the half-planes determined by the line $ax + by + c = 0$. To find out which half-plane, simply choose a convenient point in one half-plane and substitute its coordinates in the given inequality to determine whether or not they satisfy the inequality.

EXAMPLE 1

Sketch the graph of $y \leq 2x + 1$.

Solution

Every point on the line $y = 2x + 1$ has coordinates that satisfy the inequality, so the line $y = 2x + 1$ is part of the graph. It is shown in Figure 4.2. The rest of the graph consists of all the points on one "side" of the line $y = 2x + 1$. To determine which side, choose a convenient point and substitute its coordinates in the inequality. Let us choose the point $(0, 0)$. Since the inequality

$$0 \leq 2(0) + 1$$

is true, the point $(0, 0)$ is on the graph and so also are all the other points on that side of the line. This is shown by the shading in Figure 4.2.

Notice that the line $y = 2x + 1$ is very definitely included in the graph of $y \leq 2x + 1$. If the inequality had been $y < 2x + 1$, the line would not have been part of the graph. To show this, we would have drawn it as a dashed line instead of a solid line.

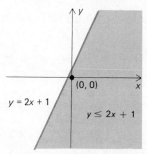

Exercise 4.5, 1–10

FIGURE 4.2   Graph of $y \leq 2x + 1$

A **polynomial inequality in two variables**, x and y, is an inequality that can be written in one of the forms

$$y < P(x) \qquad y > P(x)$$
$$y \leq P(x) \qquad y \geq P(x)$$

PART FOUR | PRECALCULUS COLLEGE ALGEBRA

where $P(x)$ is a polynomial with real coefficients. If the polynomial is of second degree, the inequality is called a *quadratic inequality in two variables*. For example,

$$y > x^2$$

is a quadratic inequality in two variables.

The graph of polynomial inequality $y \leq P(x)$ or $y \geq P(x)$ consists of all points on the graph of $y = P(x)$ together with all the points on one "side" of $y = P(x)$. Likewise, the graph of $y < P(x)$ or $y > P(x)$ consists only of all points on one side of $y = P(x)$. As with linear inequalities in two variables, to determine which side, simply choose a convenient point and substitute its coordinates in the inequality to see whether or not they satisfy it.

EXAMPLE 2   Sketch the graph of $y > x^2$.

Solution   As a reference, consider the graph of the parabola $y = x^2$ shown in Figure 4.3. The graph of the inequality $y > x^2$ does not include the points *on* the parabola, so the parabola has been drawn dashed instead of solid. The points on the graph of $y > x^2$ are those points whose y-coordinates are *greater than* x^2. We suspect that they must be the points *above* $y = x^2$. To verify this, choose a convenient point above the parabola and substitute its coordinates in the inequality. Choose $(0, 1)$. The statement

$$1 > 0^2$$

is true. The point $(0, 1)$ is on the graph. The graph is shaded in Figure 4.3.

The next two examples provide further illustrations of graphs of inequalities in two variables.

EXAMPLE 3   Sketch the graph of $x^2 + y^2 \leq 9$.

Solution   The graph is the set of all points within or on the circle $x^2 + y^2 = 9$. It is shown in Figure 4.4.

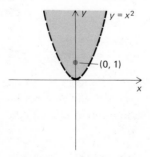

FIGURE 4.3   Graph of $y > x^2$

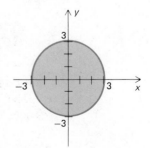

FIGURE 4.4   Graph of $x^2 + y^2 \leq 9$

EXAMPLE 4   Sketch the graph of $y \le 2^x$.

Solution   Every point on the graph of $y = 2^x$ has coordinates that satisfy the inequality, so $y = 2^x$ is part of the graph. The rest of the graph consists of all the points *below* $y = 2^x$. As a check, notice that the point $(0, 0)$ is below $y = 2^x$ and its coordinates satisfy the inequality. Indeed, since $2^0 = 1$, the statement

$$0 \le 2^0$$

is true. The graph is shown in Figure 4.5.

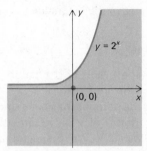

Exercise 4.5, 11–30   **FIGURE 4.5** Graph of $y \le 2^x$

Sometimes it is necessary to consider a system of inequalities in two variables. The *solution set* or simply *the solution* of a system of inequalities in two variables with real-number coefficients is the set of all ordered pairs whose coordinates satisfy each and every one of the inequalities of the system. It is the *intersection* of the solution sets of all the individual inequalities in the system. One way to find the solution of a system graphically is to sketch the graphs of all the inequalities in the system on the same coordinate system, using a different color or different shading for each graph, and then note the intersection of all these graphs. This is illustrated in the next two examples.

EXAMPLE 5   Sketch the graph of the solution of the system

$$y \le x - 3$$
$$y < -x + 5$$
$$y \ge 0.$$

Solution   We first sketch the graphs of the lines $y = x - 3$, $y = -x + 5$, and $y = 0$. They are shown in Figure 4.6. We drew the lines $y = x - 3$ and $y = 0$ solid because points on them are included on the graph; we drew the line $y = -x + 5$ dashed because no points on it are included on the graph.

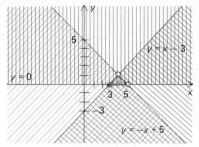

FIGURE 4.6 Graph of
$$\begin{cases} y \le x - 3 \\ y < -x + 5 \\ y \ge 0 \end{cases}$$

The graph of $y \le x - 3$ consists of all points on or below $y = x - 3$; the graph of $y < -x + 5$ consists of all points below $y = -x + 5$; the graph of $y \ge 0$ consists of all points on or above the x-axis. The set of all points common to all three of these graphs is the graph of the solution of the system. It is shown shaded in Figure 4.6. It consists of the interior, two sides, and one vertex of a triangle.

To show that the points of intersection of the excluded line $y = -x + 5$ and the lines $y = x - 3$ and $y = 0$ are not included, we put open dots around them. To show that the point of intersection of $y = x - 3$ and $y = 0$ *is* included, we put a solid dot there.

Exercise 4.5, 16–20

EXAMPLE 6

Sketch the graph of the solution of the system

$$y \le x + 2$$
$$y \ge x^2.$$

Label the points of intersection of $y = x + 2$ and $y = x^2$.

Solution

We first sketch the line $y = x + 2$ and the parabola $y = x^2$. Both are drawn solid in Figure 4.7 because points on both are on the graph of the system.

The graph of $y \le x + 2$ consists of all points on or to the right of $y = x + 2$. The graph of $y \ge x^2$ consists of all points on or above $y = x^2$. The intersection of these two sets of points is the graph of the system. It is shown in Figure 4.7.

We can find the points of intersection of the line and the parabola by substituting $y = x + 2$ for y in the equation $y = x^2$ and then solving for x. We have

$$x + 2 = x^2$$
$$x^2 - x - 2 = 0$$
$$(x - 2)(x + 1) = 0$$

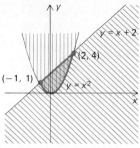

FIGURE 4.7 Graph of
$$\begin{cases} y \le x + 2 \\ y \ge x^2 \end{cases}$$

Exercise 4.5, 31–40

It follows that $x = 2$ or $x = -1$. We can find the corresponding values of y by substituting these values of x in either $y = x + 2$ or $y = x^2$. This would yield the points of intersection $x = 2$, $y = 4$ and $x = -1$, $y = 1$.

The system given in Example 6 is typical of those encountered in application problems in calculus. For example, one might need to use calculus to find the number of units in the area shaded in Figure 4.7, one might be required to find the volume of the solid generated by revolving that area about the x-axis, or one might be told a thin plate has that area and might be required to find the center of gravity of the plate.

Systems of inequalities in two variables arise in many other contexts than in calculus. The next two examples suggest how they might arise in such diverse areas as manufacturing and archeological research. In each of these examples, two variables are subject to certain constraints that form a system of linear inequalities, and the problem is to find values of the variables that will make a certain linear expression in these variables take on its maximum value subject to the given constraints. In the mathematical subject called *linear programming*, it is shown that the maximum (or minimum) value of a linear expression that is defined subject to such constraints is taken on at the "corners" or vertices of the graph of the system of inequalities specifying the constraints. Accordingly, to solve problems such as those illustrated in the next two examples, we need only set up the system of inequalities specifying the constraints, sketch the graph of the system, locate the "corners," and then systematically compute the expression that is to be made a maximum (or minimum) at these corners.

EXAMPLE 7

A company makes flutes and piccolos. They can make no more than 500 instruments a year. They must make at least 200 flutes and at least 100 piccolos to satisfy their regular annual orders. They make a profit of

$250 on each flute and $200 on each piccolo. If they sell all their instruments, how many of each kind must they make in a given year in order to realize the maximum profit?

Solution

Let x be the number of flutes.

Let y be the number of piccolos.

The annual profit in dollars is

$$P = 250x + 200y.$$

The constraints are

$$x + y \leq 500$$
$$x \geq 200$$
$$y \geq 100.$$

According to the theorem from linear programming that we referred to earlier, the maximum value of P will be taken on at one of the "corners" or vertices of the graph of this system of inequalities.

The graph of the system is shown in Figure 4.8.

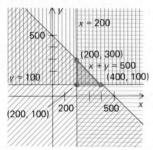

FIGURE 4.8 Graph of
$$\begin{cases} x + y \leq 500 \\ x \geq 200 \\ y \geq 100 \end{cases}$$

The vertices of the graph are

(200, 100),   (200, 300),   and   (400, 100).

The values of the annual profit P at these vertices are given in the following table.

Vertices	$P = 250x + 200y$
(200, 100)	$50,000 + 20,000 = 70,000$
(200, 300)	$50,000 + 60,000 = 110,000$
(400, 100)	$100,000 + 20,000 = 120,000$

The maximum value of P is taken on at the vertex (400, 100). To realize the maximum profit, the company should make 400 flutes and 100 piccolos.

(The solution to this problem might appear obvious, but the problem presents an application and serves as an illustration of a technique that can be used in more complicated problems whose solutions are not obvious.)

EXAMPLE 8

Professor Carberry, the psychoceramist, was granted $500 and 100 sections of a desert to search for archaic ceramics. He planned to use two research techniques: the new Smiley method, and the traditional shovel method. He estimated that the Smiley method would locate 1 pot per 5 sections at a cost of $50 per pot, and the value of each pot located this way would be $10. (These pots would probably not be cracked.) The traditional method locates 1 pot per section at a cost of only $2 per pot, but the pots located this way are in poorer condition and are worth only $1 each. On how many sections should Carberry use each of these methods in order for the collection of pots located to have the maximum value?

Solution

Let x be the number of sections researched by the Smiley method.

Let y be the number of sections researched by the traditional shovel method.

The number of pots located by each of these methods is $x/5$ and $y/1$, respectively. The value of the located pots in dollars is

$$V = 10\left(\frac{x}{5}\right) + \left(\frac{y}{1}\right)$$

$$= 2x + y.$$

The cost of the project is

$$50\left(\frac{x}{5}\right) + 2\left(\frac{y}{1}\right)$$

or
$$10x + 2y.$$

The constraints of the problem are

$$x + y \leq 100$$
$$10x + 2y \leq 500$$
$$x \geq 0 \qquad \text{(Number of sections is nonnegative.)}$$
$$y \geq 0.$$

The linear programming theorem we have been using assures us that the maximum value of V will occur at one of the "corners" or vertices of the graph of this system of inequalities. The graph of the system is shown in

Figure 4.9. Three of the vertices are at (0, 0), (0, 100), and (50, 0). The fourth vertex is the point of intersection of the lines $x + y = 100$ and $10x + 2y = 500$ or $5x + y = 250$. Solving the equations of these lines simultaneously, we have

$$
\begin{array}{rl}
x + y = & 100 \\
5x + y = & 250 \\
\hline
-4x \qquad = & -150
\end{array}
$$

$$x = \frac{150}{4} = \frac{75}{2}$$

$$y = 100 - \frac{75}{2} = \frac{125}{2}.$$

The fourth vertex is at $(\frac{75}{2}, \frac{125}{2})$.

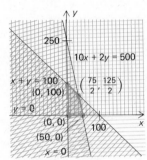

FIGURE 4.9   Graph of
$$
\begin{array}{rl}
x + y & \leq 100 \\
10x + 2y & \leq 500 \\
x & \geq 0 \\
y & \geq 0
\end{array}
$$

The values of $V = 2x + y$ at the four vertices are listed in the following table.

Vertices	$V = 2x + y$
(0, 0)	0
(0, 100)	$100
(50, 0)	$100
$(\frac{75}{2}, \frac{125}{2})$	$137.50

The maximum value of V is at $(\frac{75}{2}, \frac{125}{2})$, that is, at $(37\frac{1}{2}, 62\frac{1}{2})$. In order to obtain the most valuable collection of pots within the given constraints, Carberry should use the Smiley method on $37\frac{1}{2}$ sections and the traditional shovel method on $62\frac{1}{2}$ sections.

Exercise 4.5, 41–45

EXERCISE 4.5 INEQUALITIES IN TWO VARIABLES; SYSTEMS OF INEQUALITIES

▶ Sketch the graph of each of the following.

EXAMPLES

(a) $3x + 2y - 12 > 0$ (b) $y \geq x^3$

Solutions

(a) The line $3x + 2y - 12 = 0$ is not part of the graph so it has been drawn dashed. Choose the point $(0, 0)$ on one side. The statement $0 + 0 - 12 > 0$ is not true, so $(0, 0)$ is not on the graph. The graph must consist of all points on the other side.

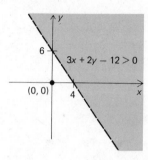

(b) The points on $y = x^3$ are part of the graph. Choose the point $(0, 2)$ on one side of $y = x^3$. The coordinates of $(0, 2)$ satisfy $y \geq x^3$, so all the points on that side are part of the graph.

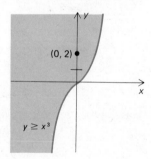

1. $3x + 2y - 12 < 0$ 2. $3x + 2y - 12 \leq 0$

3. $3x + 2y - 12 \geq 0$ 4. $x \leq 5$

5. $y > 3$ 6. $y \geq 2x + 5$

7. $x > 0$ 8. $y < 3x - 2$

9. $y \geq -1$ 10. $x + y \leq 4$

11. $y \geq x^2$ 12. $y < x^2$

13. $y \leq -x^2$ 14. $y > x^3$

15. $y \leq x^3$ **16.** $y < x^3$

17. $y \leq x^2 - 6x + 10$ **18.** $y > x^2 - 6x + 10$

19. $y \geq x^2 - 2x + 4$ **20.** $y < x^2 - 2x + 4$

21. $y > 2^x$ **22.** $y \leq 3^x$

23. $y \geq 3^x$ **24.** $y \leq \log_{10} x$

25. $y > \log_{10} x$ **26.** $x^2 + y^2 \geq 9$

27. $x^2 + y^2 < 4$ **28.** $x^2 + y^2 \leq 4$

29. $y^2 \leq x$ **30.** $y^2 \geq x$

▶ Sketch the graph of the solution set of each of the following systems.

EXAMPLE

$$x + y \leq 4$$
$$y \geq 0$$
$$y \leq x + 2$$

Solution

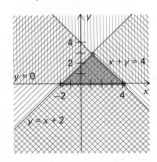

31. $x + y \leq 4$ **32.** $x + y < 4$ **33.** $x + 2y \leq 4$
  $\quad y \geq 0$ $\quad y > 2$ $\quad x \geq 0$
  $\quad x \geq 0$ $\quad y < x + 2$ $\quad y > 1$
  $\quad y \leq x + 2$

34. $3x + 2y \leq 12$ **35.** $4x - 3y \geq 12$ **36.** $y \leq 3$
  $\quad y < x + 1$ $\quad x + y \leq 3$ $\quad y \geq x^2$
  $\quad y \geq 0$

37. $y \leq -x^2$ **38.** $x^2 + y^2 \leq 16$ **39.** $y \leq x^3$
  $\quad y \geq -3$ $\quad x \leq 2$ $\quad x \leq 2$

40. $y \leq e^x$
  $\quad x \geq 0$

41. A farmer has no more than 500 acres on which to plant corn and rye. He must use at least 100 acres for corn and at least 200 acres for rye. He can make a profit of $\$80$ per acre on corn and a profit of $\$75$ per acre on rye. If he sells all of his corn and rye, how many acres of each should he raise to make the greatest profit?

42. A 4-H club was granted $500 with which to experiment with two kinds of fertilizer on 100 acres of land. If they use Type X, they will need 2 acres of land for each carload of produce at a cost of $50 per carload, and the profit on this produce will be $10 per carload more than the average profit. If they use Type Y, they will need only 1 acre of land for each carload of produce at a cost of only $5 per carload, but the profit on this produce would be only $2 per carload more than the average profit. On how many acres should they use each type of fertilizer in order to obtain the maximum profit more than the average profit?

43. A warehouse fire forced a manufacturer to sell at a loss at least enough boxes of product A and product B to make up 40,000 cubic meters of space. The manufacturer lost $1 on each box sold of product A and $2 on each box sold of product B. It was necessary to sell at least 1000 boxes of product A. If each box of product A takes 8 cubic meters and each box of product B takes 4 cubic meters, how many boxes of each product should he sell in order to minimize his loss?

44. A publisher has 4 square meters at a convention to display textbooks. She wants to display two kinds, school books and college books. She has no more than 8 square meters of school books to display and no more than 6 square meters of college books to display. She estimates that each square meter of school books on display attracts 4 requests per hour for sample copies, and each square meter of college books attracts 3 requests per hour for sample copies. How many square meters of each should she display in order to attract the maximum number of requests per hour for sample copies?

45. Dr. Trebbin can see no more than 24 of his patients per day. He must schedule at least 10 regular check-ups per day. He always schedules at least 3 emergencies per day. Suppose that he spends $\frac{1}{3}$ hour with each regular check-up and $\frac{1}{2}$ hour with each emergency. If the number of regular check-ups per day is at least 3 times the number of emergencies per day, how many of each kind of patient should he schedule a day in order for him to spend the maximum amount of time per day treating his patients? In that case, what would be the number of hours per day he would be working with patients?

CHECK LIST FOR THE STUDENT

If you have learned the material in Section 4.5, you should be able to do the following:

1. Sketch the graph of a given linear or polynomial inequality in two variables.
2. Sketch the graphs of other given inequalities in two variables, such as

$$y \leq 2^x, \quad y > \log_{10}x, \quad \text{and} \quad x^2 + y^2 \leq 25.$$

3. Sketch the graph of a given system of inequalities in two variables.
4. Use a basic theorem from linear programming to find the maximum or minimum value of a given linear expression in two variables, subject to some given constraints expressed as a system of linear inequalities in the two variables.

SECTION QUIZ 4.5

Sketch the graph of each of the following.
1. $2x - 3y - 6 \leq 0$

2. $2x - 3y - 6 < 0$

3. $y < 2$

4. $x \geq -1$

5. $y \leq x^2$

6. $y > 3^x$

7. $x^2 + y^2 < 4$

Sketch the graph of the solution set of each of the following systems.
8. $y \leq x + 3$

$\quad y \geq 0$

$\quad x \leq 1$

9. $x^2 + y^2 \leq 9$

$\quad\quad y < x$

10. A family raises and sells lettuce and beans. They have room for no more than 20 rows of vegetables. In one summer they must raise at least 1 row of beans and 2 rows of lettuce to satisfy their regular customers. They make a profit of $22 on each row of lettuce and $25 on each row of beans.
   (a) Let x be the number of rows of lettuce, let y be the number of rows of beans, and set up a system of inequalities expressing the given constraints.
   (b) Graph the system of inequalities you have set up.
   (c) Assume that the family sells all their vegetables and compute their profit at each "corner" of the graph you have obtained.
   (d) How many rows of each vegetable must the family raise in a summer in order to realize the maximum profit?

4.6 Sequences and Series

The purpose of this section is to introduce sequences and series, to introduce the summation notation, to consider geometric and arithmetic sequences and series, and to suggest some applications.

INTRODUCTION TO SEQUENCES

The expression

$$\tfrac{1}{2}, \tfrac{1}{4}, \tfrac{1}{8}, \tfrac{1}{16}, \ \cdots \ , (\tfrac{1}{2})^n, \ \cdots$$

represents an **infinite sequence**. The letter n represents any positive integer. The numbers separated by the commas are *terms* of the sequence. The three dots at the end indicate that there are an unlimited number of terms formed according to the rule $(\tfrac{1}{2})^n$ by successively letting n take on the values 1, 2, 3, 4, etc. of all the positive integers.

It is convenient to represent an entire sequence simply by writing the rule of formation in braces. For instance,

$$\{(\tfrac{1}{2})^n\}$$

represents the sequence

$$\tfrac{1}{2}, \tfrac{1}{4}, \tfrac{1}{8}, \tfrac{1}{16}, \ \cdots \ , (\tfrac{1}{2})^n, \ \cdots$$

Here are some more examples of expressions that represent infinite sequences.

$$2, 2^2, 2^3, 2^4, \ \ldots \ , 2^n, \ \ldots \qquad \{2^n\}$$
$$0.8, (0.8)^2, (0.8)^3, (0.8)^4, \ \ldots \ , (0.8)^n, \ \ldots \qquad \{(0.8)^n\}$$
$$3 - \tfrac{1}{2}, 3 - \tfrac{1}{4}, 3 - \tfrac{1}{8}, \ \ldots \ , 3 - (\tfrac{1}{2})^n, \ \ldots \qquad \{3 - (\tfrac{1}{2})^n\}$$
$$2, 5, 8, 11, \ \ldots \ , 2 + 3(n - 1), \ \ldots \qquad \{2 + 3(n - 1)\}$$
$$1, -1, 1, -1, \ \ldots \ , (-1)^{n+1}, \ \ldots \qquad \{(-1)^{n+1}\}$$

In each case, there is given a rule of formation expressed in terms of n with the understanding that the terms are formed by letting n successively take on the values 1, 2, 3, 4, etc. of all the positive integers.

Technically speaking, an infinite sequence is a function whose domain is the set of all positive integers and whose rule of assignment is the rule of formation of the terms. For example, the rule of assignment of the sequence

$$\tfrac{1}{2}, \tfrac{1}{4}, \tfrac{1}{8}, \tfrac{1}{16}, \ \cdots \ , (\tfrac{1}{2})^n, \ \cdots$$

assigns the number $(\tfrac{1}{2})^n$ to positive integer n. If we called this sequence the function f and used the usual function notation, we would write

$$f(n) = (\tfrac{1}{2})^n.$$

In working with infinite sequences, however, it is customary to let n represent any element in the domain (the positive integers) and to use the notation a_n for the rule of assignment rather than the notation $f(n)$. For our example, it is customary to write

$$a_n = (\tfrac{1}{2})^n.$$

The general expression

$$a_1, a_2, a_3, a_4, \ldots, a_n, \ldots$$

represents an infinite sequence whose rule assigns a_n to positive integer n. The a_1, a_2, a_3, a_4, etc. are called the *terms* of the sequence. The rule a_n is sometimes called the *general term* or the *nth* term of the sequence. The notation

$$\{a_n\}$$

represents the entire sequence.

In specifying an infinite sequence, it is not sufficient to give merely the first few terms of the sequence. In order to specify a sequence, *the rule of assignment* (the general term) *must be given*. For example, just knowing that

$$2, 4, 8$$

are the first three terms of a sequence is not enough to tell us the next term, much less the entire sequence. There is an unlimited number of infinite sequences starting out with these three terms. Some of them are

$$2, 4, 8, 16, \quad 32, \ldots, 2^n, \ldots$$
$$2, 4, 8, 22, \quad 56, \ldots, 2^n + (n-1)(n-2)(n-3), \ldots$$
$$2, 4, 8, 34, \quad 128, \ldots, 2^n + (n-1)^2(n-2)(n-3), \ldots.$$

We observed in Part 2 of this text that there are *two* essential ingredients in the definition of any function, the domain and the rule of assignment. Both of these must be given in order for the function to be defined. An infinite sequence is a function whose domain is the set of all positive integers, and its rule of assignment (its general term) *must* be given or the sequence itself has not been given.

The expression

$$a_1, a_2, a_3, \ldots, a_n, \ldots a_k$$

represents a sequence with exactly k terms. Its domain is not the set of *all* positive integers but rather its domain is the set of all positive integers 1, 2, 3, etc. up to and including the positive integer k. Its terms are formed according to its rule of assignment, its general term a_n, by letting

n successively take on the values of the integers 1 through k. The expression

$$2, 2^2, 2^3, \ldots, 2^n, \ldots 2^8$$

represents the sequence

$$2, 2^2, 2^3, 2^4, 2^5, 2^6, 2^7, 2^8$$

whose domain is the set of positive integers 1, 2, 3, 4, 5, 6, 7, 8.

Sequences arise quite naturally in everyday life as well as in science and pure mathematics. The next few examples illustrate this.

EXAMPLE 1

A chemical evaporates in such a way that it loses half of its volume each day. If it initially has 1 unit of volume, during the first day it loses $\frac{1}{2}$ of that, leaving $\frac{1}{2}$ of a unit; during the second day it loses $\frac{1}{2}$ of the remains, leaving $\frac{1}{4}$ of a unit; the third day it loses $\frac{1}{2} \times \frac{1}{4}$, leaving $\frac{1}{8}$ of a unit, etc. The volume remaining at the end of the nth day is $(\frac{1}{2})^n$ of a unit. The volumes remaining at the ends of the successive days are the terms in the infinite sequence

$$\tfrac{1}{2}, \tfrac{1}{4}, \tfrac{1}{8}, \tfrac{1}{16}, \ldots, (\tfrac{1}{2})^n, \ldots.$$

In this example, for that matter, the volumes *lost* each successive day are the terms in the same infinite sequence.

EXAMPLE 2

A concerned ecologist in Northern Alaska noticed that each year his community has twice as much nonrecyclable garbage as it had the previous year. He wondered how long the town dump could accommodate this. He started to make a study of the situation. The first year of the study, the community deposited 2 tons of the garbage, the second year it deposited $2 \times 2 = 2^2$ tons, the third year it deposited $2 \times 2^2 = 2^3$ tons, etc. At this rate, during the nth year 2^n tons would be deposited. If the dumping continues in this manner indefinitely, the number of tons deposited in succeeding years form the infinite sequence

$$2, 2^2, 2^3, 2^4, \ldots, 2^n, \ldots.$$

EXAMPLE 3

A congressional committee proposed to give a rebate on federal income taxes in order to bolster the economy. An economist estimated that of each 1 million dollars in rebate, 80% would be spent in the United States (fed back into the local economy), and 20% would be either saved or spent abroad. Of each sum spent in the United States, 80% would again be spent in the United States and 20% would either be saved or spent abroad, etc. In other words, the economist estimated that for each million-dollar rebate, there would be a sequence of spending "sprees," and that in millions of dollars the amounts fed back into the local economy would be

$$0.8 \quad \text{on the first spree,}$$
$$0.8(0.8) = (0.8)^2 \quad \text{on the second spree,}$$
$$0.8(0.8)^2 = (0.8)^3 \quad \text{on the third spree,}$$

$$\vdots \qquad \qquad \vdots$$

$$(0.8)^n \quad \text{on the } n\text{th spree.}$$

In millions of dollars, the successive amounts fed back into the local economy for each million-dollar rebate form the infinite sequence

$$0.8, \ (0.8)^2, \ (0.8)^3, \ \ldots, \ (0.8)^n, \ \ldots.$$

EXAMPLE 4

If a person deposited $100 in an account paying 6% interest compounded annually, at the end of the first year the number of dollars in the account would be the original 100 plus the interest.

$$100 + 100(0.06) = 100(1.06)$$

At the end of the second year, interest would be computed on the amount 100(1.06), and the number of dollars in the account would then be

$$100(1.06) + 100(1.06)(.06) = 100(1.06)(1.06)$$
$$= 100(1.06)^2.$$

Similarly, at the end of the third year, the number of dollars in the account would be

$$100(1.06)^3.$$

At the end of the nth year, the number of dollars in the account would be

$$100(1.06)^n.$$

The numbers of dollars in the account at the end of each successive year are the terms in the infinite sequence

$$100(1.06), \ 100(1.06)^2, \ 100(1.06)^3, \ \ldots, \ 100(1.06)^n, \ \ldots.$$

EXAMPLE 5

An organization established an annual prize with the agreement that the prize would be $2 the first year with an annual increase of $3 each year thereafter. In dollars, the prize the second year would be $2 + 3 = 5$, the prize the third year would be $2 + (2)(3) = 8$, the prize the fourth year would be $2 + (3)(3) = 11$, etc. The nth year the prize would be

$$2 + (n - 1)(3) = 2 + 3(n - 1).$$

The dollar values of the prize in succeeding years are the terms in the infinite sequence

$$2, \ 5, \ 8, \ 11, \ \ldots, \ 2 + 3(n - 1), \ \ldots.$$

EXAMPLE 6

As x takes on the values

$$\frac{\pi}{2}, \frac{3\pi}{2}, \frac{5\pi}{2}, \frac{7\pi}{2}, \ldots, (2n-1)\frac{\pi}{2}, \ldots,$$

$\sin x$ takes on the values

Exercise 4.6, 1–15

$$1, -1, 1, -1, \ldots, (-1)^{n+1}, \ldots.$$

CONVERGENCE OF SEQUENCES

As we go further and further out in some sequences (as n takes on larger and larger values), the terms get closer and closer in value to a particular number. For example, as we go further and further out in the sequence

$$\tfrac{1}{2}, \tfrac{1}{4}, \tfrac{1}{8}, \tfrac{1}{16}, \tfrac{1}{32}, \ldots, (\tfrac{1}{2})^n, \ldots,$$

the terms get closer and closer in value to the number 0. We might say that as n increases, the terms *approach* 0. Similarly, as we go further and further out in the sequence

$$3 - \tfrac{1}{2}, 3 - \tfrac{1}{4}, 3 - \tfrac{1}{8}, 3 - \tfrac{1}{16}, 3 - \tfrac{1}{32}, \ldots, 3 - (\tfrac{1}{2})^n, \ldots,$$

the terms get closer and closer in value to the number 3. The terms of this sequence *approach* 3 as n increases. We say that the sequence $\{(\tfrac{1}{2})^n\}$ *converges* to 0; we say that the sequence $\{3 - (\tfrac{1}{2})^n\}$ *converges* to 3. We say that 0 is the *limit* of the sequence $\{(\tfrac{1}{2})^n\}$; we say that 3 is the *limit* of the sequence $\{3 - (\tfrac{1}{2})^n\}$. We express this symbolically by writing

$$\lim_{n \to \infty} (\tfrac{1}{2})^n = 0$$

and

$$\lim_{n \to \infty} 3 - (\tfrac{1}{2})^n = 3.$$

For any general term a_n, the symbols

$$\lim_{n \to \infty} a_n$$

are read

"the limit as n approaches infinity of a_n"

or "the limit as n increases indefinitely of a_n."

In general, the sequence $\{a_n\}$ is said to **converge** to L provided that as we go further and further out in the sequence, the terms get closer and closer in value to the number L. We say that L is the **limit** of the sequence. We may write

$\lim_{n \to \infty} a_n = L$ **provided that** we can find an n large enough so that each of the terms beyond a_n will be as close to L in value as we please.

More technically,

$$\lim_{n \to \infty} a_n = L \quad \text{provided that}$$ there exists a number N such that for all $n > N$, $|a_n - L|$ is less than any pre-assigned and arbitrarily small number.

If the terms of an infinite sequence $\{a_n\}$ do not approach a number L as n increases indefinitely (that is, if the sequence does not converge to a limit), then the sequence is said to **diverge**. The sequence

$$2, 2^2, 2^3, 2^4, \ldots, 2^n, \ldots$$

diverges. There is no number that the terms approach in value as n increases. The sequence

$$1, -1, 1, -1, \ldots, (-1)^{n-1}, \ldots$$

also diverges. As n increases, the terms just fluctuate in value from 1 to -1; they do not all approach a particular number L.

EXAMPLE 7

Write out the first four terms and the general term of $\{(\frac{1}{3})^n\}$ and of $\{7 - (\frac{1}{3})^n\}$. What is $\lim_{n \to \infty} (\frac{1}{3})^n$, and what is $\lim_{n \to \infty} 7 - (\frac{1}{3})^n$?

Solution

$$\{(\tfrac{1}{3})^n\} \qquad \tfrac{1}{3}, \tfrac{1}{9}, \tfrac{1}{27}, \tfrac{1}{81}, \ldots, (\tfrac{1}{3})^n, \ldots$$

$$\{7 - (\tfrac{1}{3})^n\} \qquad 7 - \tfrac{1}{3}, 7 - \tfrac{1}{9}, 7 - \tfrac{1}{27}, 7 - \tfrac{1}{81}, \ldots, 7 - (\tfrac{1}{3})^n, \ldots$$

As n increases indefinitely, the terms of $\{(\frac{1}{3})^n\}$ get successively closer in value to 0, so

$$\lim_{n \to \infty} (\tfrac{1}{3})^n = 0.$$

As n increases indefinitely, the terms of $\{7 - (\frac{1}{3})^n\}$ get successively closer in value to 7, so

Exercise 4.6, 16–20

$$\lim_{n \to \infty} 7 - (\tfrac{1}{3})^n = 7.$$

GEOMETRIC SEQUENCES

Several of the sequences that we have used as examples are *geometric sequences*. In particular,

$$\tfrac{1}{2}, \tfrac{1}{4}, \tfrac{1}{8}, \tfrac{1}{16}, \ldots, (\tfrac{1}{2})^n, \ldots,$$

$$2, 2^2, 2^3, 2^4, \ldots, 2^n, \ldots,$$

$$\tfrac{1}{8}, \tfrac{1}{16}, \tfrac{1}{32}, \tfrac{1}{64}, \ldots, (\tfrac{1}{2})^{n+2}, \ldots,$$

and

$$1, -1, 1, -1, \ldots, (-1)^{n+1}, \ldots$$

are geometric sequences. After the first term, each of the terms of a geometric sequence can be formed by multiplying the preceding term by a nonzero constant called the *common ratio*. In other words, after the first term, the ratio of any term to the preceding term is a constant called the **common ratio**. The sequence $\{ar^{n-1}\}$,

$$a, \ ar, \ ar^2, \ ar^3, \ \ldots, \ ar^{n-1}, \ \ldots,$$

is a **geometric sequence** with first term $a \neq 0$ and common ratio r.

The first term of $\{(\frac{1}{2})^n\}$ is $\frac{1}{2}$, and the common ratio is $\frac{1}{2}$.

The first term of $\{2^n\}$ is 2, and the common ratio is 2.

The first term of $\{(\frac{1}{2})^{n+2}\}$ is $\frac{1}{8}$, and the common ratio is $\frac{1}{2}$.

The first term of $\{(-1)^{n+1}\}$ is 1, and the common ratio is -1.

Geometric sequences are sometimes called *geometric progressions*.

We saw that the geometric sequence $\{(\frac{1}{2})^n\}$ with common ratio $\frac{1}{2}$ converges to 0, whereas the geometric sequence $\{2^n\}$ with common ratio 2 does not converge, and the geometric sequence $\{(-1)^{n+1}\}$ with common ratio -1 does not converge. It is true that for any geometric sequence with common ratio r,

if $|r| < 1$, the sequence converges to 0, and

if $|r| > 1$, the sequence does not converge (diverges).

This is because

if $|r| < 1$, $\lim_{n \to \infty} r^n = 0$, and

if $|r| > 1$, $\lim_{n \to \infty} r^n$ fails to exist.

If $r = 1$ and the first term is a, the sequence is simply $a, a, a, a, \ldots,$ $a, \ldots,$ which converges to a. If $r = -1$ and the first term is a, the sequence is $a, -a, a, -a, \ldots, (-1)^{n+1}a, \ldots,$ which does not converge.

EXAMPLE 8   State whether or not each of the following sequences converges. If it converges, to what does it converge?
(a) $\{(\frac{-1}{3})^{n-1}\}$ (b) $\{3(-1)^n\}$ (c) $\{(-3)^{n-1}\}$

Solution   (a) $\{(\frac{-1}{3})^{n-1}\}$ is a geometric sequence with common ratio $r = \frac{-1}{3}$. Since $|r| < 1$, the sequence converges to 0.
(b) $\{3(-1)^n\}$ is a geometric sequence with common ratio $r = -1$. Since $r = -1$, the sequence does not converge.
(c) $\{(-3)^{n-1}\}$ is a geometric sequence with common ratio $r = -3$. Since $|r| > 1$, the sequence does not converge.

We can derive a very useful formula for the sum of the first n terms of a geometric sequence with common ratio other than one. In fact, if S_n is the sum of the first n terms of the geometric sequence $\{ar^{n-1}\}$ with $r \neq 1$, that is, if

$$S_n = a + ar + ar^2 + ar^3 + \cdots + ar^{n-1},$$

and if $r \neq 1$, we can show that

$$S_n = \frac{a - ar^n}{1 - r}.$$

To show this, we first write out S_n, putting in the next to the last term of it for future reference. Next we multiply both sides of the equation defining S_n by the common ratio r. (In so doing, we use the fact that $ar^{n-2} \times r = ar^{n-1}$ and $ar^{n-1} \times r = ar^n$.)

$$S_n = a + ar + ar^2 + ar^3 + \cdots + ar^{n-2} + ar^{n-1}$$
$$S_n r = ar + ar^2 + ar^3 + ar^4 + \cdots + ar^{n-1} + ar^n$$

Now we subtract the second of these equations from the first. When we do this subtraction, all the terms on the right of the equals signs drop out except for the first term of S_n and the last term of $S_n r$. We get

$$S_n - S_n r = a - ar^n.$$

Finally we solve for S_n, thereby obtaining the formula we seek.

$$S_n(1 - r) = a - ar^n$$

$$S_n = \frac{a - ar^n}{1 - r} \qquad \text{if } r \neq 1$$

If $r = 1$, then $S_n = a + a \cdot 1 + a \cdot 1^2 + a \cdot 1^3 + \cdots + a \cdot 1^{n-1}$, so S_n is just the sum of n terms each of which is a. That is,

$$S_n = na, \qquad \text{if } r = 1.$$

EXAMPLE 9

(a) Write out a formula for the sum of the first n terms of $\{(\frac{1}{2})^n\}$. Use this formula to find the sum of the first 5 terms.

(b) Write a formula for the sum of the first n terms of $\{3 \times (1)^n\}$, and find the sum of the first 5 terms.

Solution

(a) In the sequence $\{(\frac{1}{2})^n\}$, $a = \frac{1}{2}$ and $r = \frac{1}{2}$.

$$S_n = \frac{a - ar^n}{1 - r}$$

$$= \frac{\frac{1}{2} - \frac{1}{2}(\frac{1}{2})^n}{1 - \frac{1}{2}}$$

$$= \frac{\frac{1}{2} - \frac{1}{2}(\frac{1}{2})^n}{\frac{1}{2}}$$

$$= 1 - (\tfrac{1}{2})^n$$

$$S_5 = 1 - (\tfrac{1}{2})^5$$

$$= 1 - \tfrac{1}{32}$$

$$= \tfrac{31}{32}$$

(b) In the sequence $\{3 \times (1)^n\}$, $a = 3$ and $r = 1$, so every term is 3.

$$S_n = 3n$$
$$S_5 = 15$$

Exercise 4.6, 21–40

ARITHMETIC SEQUENCES

The sequence

$$2, 5, 8, 11, \ldots , 2 + 3(n - 1), \ldots$$

is an *arithmetic sequence.* After the first term, each term of an arithmetic sequence can be formed by adding a nonzero constant called the *common difference* to the preceding term. The sequence

$$a, a + d, a + 2d, \ldots , a + (n - 1)d, \ldots$$

is an **arithmetic sequence** with first term a and common difference d. The first term of

$$2, 5, 8, 11, \ldots , 2 + 3(n - 1), \ldots$$

is 2, and the common difference is 3. Arithmetic sequences are sometimes called *arithmetic progressions.*

Every infinite arithmetic sequence diverges. There is no number that the terms approach in value as n increases indefinitely.

There is a very useful formula for the sum of the first n terms of an arithmetic sequence. To see this, let S_n be the sum of the first n terms of

$$a, a + d, a + 2d, \ldots , a + (n - 1)d, \ldots$$

so that

$$S_n = a + (a + d) + (a + 2d) + \cdots + a + (n - 1)d.$$

If we call the last term l, we have

$$l = a + (n - 1)d$$

and

$$S_n = a + (a + d) + (a + 2d) + \cdots + l$$

It can be shown that

$$S_n = \frac{n(a + l)}{2}.$$

The proof is outlined in Exercise 4.6.

EXAMPLE 10

Write out a formula for the sum of the first n terms of the following arithmetic sequence.

$$2, 5, 8, \ldots , 2 + 3(n - 1), \ldots.$$

Use this formula to find the sum of the first five terms.

Solution

Let S_n be the sum of the first n terms. The first term $a = 2$; the common difference $d = 3$; the last term (the nth term) of S_n is

$$l = a + (n - 1)d$$
$$= 2 + 3(n - 1).$$

$$S_n = \frac{n(a + l)}{2}$$

$$= \frac{n(2 + 2 + 3(n - 1))}{2}$$

$$= \frac{4n + 3n(n - 1)}{2}$$

$$S_5 = \frac{20 + 15(4)}{2}$$

$$= \frac{80}{2}$$

Exercise 4.6, 41–50

$$= 40$$

INTRODUCTION TO SERIES AND SUMMATION NOTATION

Sometimes we are more interested in the *sum* of the terms of a sequence than we are in the sequence itself. For instance, the economist in Example 3 might be more interested in knowing the total effect each million-dollar rebate would have on the local economy than in knowing the effect in any one particular year. That is, the economist might want to know the total amount

$$0.8 + (0.8)^2 + (0.8)^3 + \cdots + (0.8)^n + \cdots$$

fed back into the local economy if the spending sprees went on indefinitely as predicted. In order to provide this information, we must first know what is meant by the sum of an unlimited number of terms. We must consider *infinite series*.

If we replace the commas in the expression

$$a_1, a_2, a_3, \ldots, a_n \ldots$$

by plus signs and raise the dots to align with the plus signs, we have an expression for an **infinite series**.

$$a_1 + a_2 + a_3 + \cdots + a_n + \cdots \qquad \text{Infinite series}$$

Such a series can be written more compactly as

$$\sum_{n=1}^{\infty} a_n$$

where it is understood that this means the sum of terms formed by successively letting n take on the values 1, 2, 3, 4, etc. of all the positive integers. The general term a_n is displayed in the **summation notation** $\sum_{n=1}^{\infty} a_n$. When this sum is written out, the general term must be written in so that the rule of formation of the terms will be known.

$$\sum_{n=1}^{\infty} a_n = a_1 + a_2 + a_3 + \cdots + a_n + \cdots$$

EXAMPLE 11   Write out $\displaystyle\sum_{n=1}^{\infty} 2^n$.

Solution

$$\sum_{n=1}^{\infty} 2^n = 2 + 2^2 + 2^3 + 2^4 + \cdots + 2^n + \cdots$$

We may use summation notation to indicate the sum of the first k terms of a series, where k is a particular positive integer. For example, the sum of the first five terms of $\displaystyle\sum_{n=1}^{\infty} a_n$ may be written $\displaystyle\sum_{n=1}^{5} a_n$.

$$\sum_{n=1}^{5} a_n = a_1 + a_2 + a_3 + a_4 + a_5$$

We may also use summation notation to indicate a sum of k successive terms that does not begin with the first term. For example, the sum of the five terms beginning with a_3 may be written $\displaystyle\sum_{n=3}^{7} a_n$.

$$\sum_{n=3}^{7} a_n = a_3 + a_4 + a_5 + a_6 + a_7$$

EXAMPLE 12   Write out $\displaystyle\sum_{n=1}^{5} 2^n$. Write out $\displaystyle\sum_{n=4}^{6} 2^n$.

Solution

$$\sum_{n=1}^{5} 2^n = 2 + 2^2 + 2^3 + 2^4 + 2^5$$

$$\sum_{n=4}^{6} 2^n = 2^4 + 2^5 + 2^6$$

In summation notation, there is nothing sacred about the letter n. We may use any letter instead of n. Many authors use i and write $\sum_{i=1}^{\infty} a_i$ instead of $\sum_{n=1}^{\infty} a_n$. Other authors use k and write $\sum_{k=1}^{\infty} a_k$. They all represent the same series.

Exercise 4.6, 51–65

PARTIAL SUMS, CONVERGENCE, AND SUMS OF INFINITE SERIES

To assign a meaning to the sum of an unlimited number of terms, we first set up a sequence of so-called *partial sums* of the terms. The first partial sum S_1 of

$$\sum_{n=1}^{\infty} a_n = a_1 + a_2 + a_3 + \cdots + a_n + \cdots$$

is just the first term.

$$S_1 = a_1$$

The second partial sum S_2 is the sum of the first two terms.

$$S_2 = a_1 + a_2$$

The third partial sum S_3 is the sum of the first three terms.

$$S_3 = a_1 + a_2 + a_3$$

The nth partial sum S_n is the sum of the first n terms.

$$S_n = a_1 + a_2 + a_3 + \cdots + a_n$$

If the sequence $\{S_n\}$ of partial sums converges to some number L, the infinite series $\sum_{n=1}^{\infty} a_n$ is said to converge to L; the number L is defined to be the **sum** of the infinite series. If the sequence $\{S_n\}$ of partial sums diverges, then the series $\sum_{n=1}^{\infty} a_n$ is said to diverge.

If $\{S_n\}$ converges to L, then $\sum_{n=1}^{\infty} a_n$ converges to L.

If $\{S_n\}$ diverges, then $\sum_{n=1}^{\infty} a_n$ diverges.

EXAMPLE 13

Write out the first three terms and write the nth term of the sequence of partial sums of $\sum_{n=1}^{\infty} \left(\frac{1}{2}\right)^n$. Does the series $\sum_{n=1}^{\infty} \left(\frac{1}{2}\right)^n$ converge? If so, what is its sum?

Solution

$S_1 = \frac{1}{2}$

$S_2 = \frac{1}{2} + \left(\frac{1}{2}\right)^2$

$S_3 = \frac{1}{2} + \left(\frac{1}{2}\right)^2 + \left(\frac{1}{2}\right)^3$

$\vdots$

$S_n = \frac{1}{2} + \left(\frac{1}{2}\right)^2 + \left(\frac{1}{2}\right)^3 + \cdots + \left(\frac{1}{2}\right)^n$

Here S_n is the sum of the first n terms of the geometric sequence $\{\left(\frac{1}{2}\right)^n\}$ with first term $a = \frac{1}{2}$ and common ratio $r = \frac{1}{2}$.

$$S_n = \frac{a - ar^n}{1 - r}$$

$$= \frac{\frac{1}{2} - \frac{1}{2}\left(\frac{1}{2}\right)^n}{1 - \frac{1}{2}}$$

$$= 1 - \left(\frac{1}{2}\right)^n$$

The series $\sum_{n=1}^{\infty} \left(\frac{1}{2}\right)^n$ converges to the number L provided that the sequence of partial sums $\{1 - \left(\frac{1}{2}\right)^n\}$ converges to L. The sequence of partial sums $\{1 - \left(\frac{1}{2}\right)^n\}$ converges to 1 because

$$\lim_{n \to \infty} 1 - \left(\frac{1}{2}\right)^n = 1.$$

Hence the series $\sum_{n=1}^{\infty} \left(\frac{1}{2}\right)^n$ converges to 1. We say that the sum

$$\frac{1}{2} + \left(\frac{1}{2}\right)^2 + \left(\frac{1}{2}\right)^3 + \cdots + \left(\frac{1}{2}\right)^n + \cdots$$

Exercise 4.6, 66–70   is 1.

GEOMETRIC AND ARITHMETIC SERIES

Any series such as $\sum_{n=1}^{\infty} \left(\frac{1}{2}\right)^n$ that is the sum of the terms of a geometric sequence is called a *geometric series*. The series

$$\sum_{n=1}^{\infty} ar^{n-1} = a + ar + ar^2 + ar^3 + \cdots + ar^{n-1} + \cdots$$

is the **geometric series** with first term a and common ratio r. The nth term S_n of its sequence of partial sums — that is, the sum S_n of its first n terms — is

$$S_n = \frac{a - ar^n}{1 - r}$$

$$= \frac{a}{1 - r} - \frac{ar^n}{1 - r}.$$

If $|r| < 1$, this nth term S_n of the sequence $\{S_n\}$ approaches $a/(1 - r)$ as n increases, because r^n approaches 0. That is, if $|r| < 1$, the sequence $\{S_n\}$ of partial sums converges to $a/(1 - r)$. By definition, this means that if $|r| < 1$, the geometric series itself converges to $a/(1 - r)$. It can be shown that the series diverges for $|r| \geq 1$.

If $|r| < 1$, the geometric series $\sum_{n=1}^{\infty} ar^{n-1}$ converges to $a/(1 - r)$.

If $|r| \geq 1$, the geometric series $\sum_{n=1}^{\infty} ar^{n-1}$ diverges.

EXAMPLE 14

State whether or not the series

$$\sum_{n=1}^{\infty} \left(\tfrac{-1}{2}\right)^{n-1} = 1 - \tfrac{1}{2} + \tfrac{1}{4} - \tfrac{1}{8} + \tfrac{1}{16} + \cdots + \left(\tfrac{-1}{2}\right)^{n-1} + \cdots$$

converges. If it converges, what is its sum?

Solution

This is a geometric series with first term $a = 1$ and common ratio $r = \tfrac{-1}{2}$. Since $|r| < 1$, the series converges. It converges to

$$\frac{a}{1 - r} = \frac{1}{1 - \left(\tfrac{-1}{2}\right)}$$

$$= \frac{1}{1 + \tfrac{1}{2}}$$

$$= \tfrac{2}{3}.$$

EXAMPLE 15

State whether or not the series $\sum_{n=1}^{\infty} 2^n$ converges.

Solution

The series $\sum_{n=1}^{\infty} 2^n$ is a geometric series with common ratio $r = 2$. Since $|r| \geq 1$, the series diverges (does not converge).

EXAMPLE 16

Find the total amount fed into the local economy for each million-dollar tax rebate as predicted by the economist in Example 3.

Solution

The total amount will be the initial 1 million plus the amount fed *back* into the economy. In millions of dollars, that will be 1 plus

$$\sum_{n=1}^{\infty} (0.8)^n = 0.8 + (0.8)^2 + (0.8)^3 + \cdots + (0.8)^n + \cdots .$$

The series $\sum_{n=1}^{\infty} (0.8)^n$ is a geometric series with first term $a = 0.8$ and common ratio $r = 0.8$. Since $|r| < 1$, the series converges. Its sum is

$$\frac{a}{1-r} = \frac{0.8}{1-0.8}$$

$$= \frac{0.8}{0.2}$$

$$= 4.$$

The total amount fed into the economy by the million-dollar rebate is therefore

$$1 + 4 = 5 \text{ million dollars.}$$

Any series such as

$$2 + 5 + 8 + 11 + \cdots + [2 + 3(n-1)] + \cdots$$

whose terms are the terms of an arithmetic sequence is called an *arithmetic series*. The nth term S_n of the sequence of partial sums of the **arithmetic series**

$$a + (a+d) + (a+2d) + \cdots + [a + (n-1)d] + \cdots$$

is

$$S_n = \frac{n(a + \ell)}{2}$$

where

$$\ell = a + (n-1)d.$$

Every arithmetic series diverges because its sequence of partial sums $\{S_n\}$ fails to converge. (As n increases indefinitely, S_n fails to approach a particular number.)

The series

$$2 + 2^2 + 2^3 + \cdots + 2^n + \cdots$$

and

$$3 + 5 + 8 + \cdots + [2 + 3(n-1)] + \cdots$$

are both series of positive terms that diverge. In each of these series, the general term increases indefinitely as n increases indefinitely. In particular, as n increases indefinitely, neither of these general terms approaches zero. In calculus, it is shown that in order for a series of positive terms to converge, the *general term* a_n *must approach zero* as n increases indefinitely. Many tests for convergence and divergence of various types of series are developed in calculus.

Exercise 4.6, 71–82

USES OF SEQUENCES AND SERIES IN CALCULUS

One of the most basic concepts of calculus, the *definite integral,* is defined in terms of a series. The next example utilizes notation commonly used in stating that definition and illustrates one situation in which that definition appears.

EXAMPLE 17

A calculus student approximated the area above the x-axis and under the graph of $y = x^2$ from $x = 0$ to $x = 3$ with the sum of the areas of the three rectangles shown in Figure 4.10. The base of each rectangle is of length 1 unit. The points at x_1, x_2, and x_3 were chosen at random, one point on each base. The height of each rectangle is the y-coordinate of the point on the graph of $y = x^2$ determined by the point chosen at random on its base. The area of the first rectangle is the length of its base times its height or

$$(1)(x_1^2).$$

The sum of the areas of the three rectangles is

$$(1)(x_1^2) + (1)(x_2^2) + (1)(x_3^2).$$

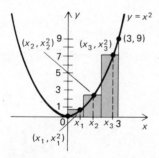

FIGURE 4.10   Area approximated with three rectangles

(a) Write this sum with summation notation.

(b) If n rectangles of equal base length were used instead of 3, the length of the base of each would be $3/n$ instead of 1. If points at x_1, x_2, x_3, ... , x_i, ... , x_n were then chosen at random, one point on each base as indicated in Figure 4.11, and if the height of each rectangle was taken to be the y-coordinate of the point on the graph of $y = x^2$ determined by the point chosen at random on its base, the area of the first rectangle would be

$$\left(\frac{3}{n}\right)(x_1^2).$$

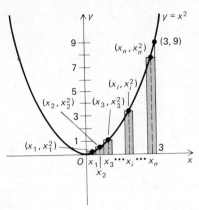

FIGURE 4.11   Area approximated with n rectangles

The area of the ith rectangle would be

$$\left(\frac{3}{n}\right)(x_i^2).$$

The sum of the areas of the n rectangles would be

$$\left(\frac{3}{n}\right)(x_1^2) + \left(\frac{3}{n}\right)(x_2^2) + \cdots + \left(\frac{3}{n}\right)(x_i^2) + \cdots + \left(\frac{3}{n}\right)(x_n^2)$$

$$= \frac{3}{n}[x_1^2 + x_2^2 + \cdots + x_i^2 + \cdots + x_n^2].$$

Write this sum with summation notation.

Solution

(a) $x_1^2 + x_2^2 + x_3^2 = \displaystyle\sum_{i=1}^{3} x_i^2$

(b) $\dfrac{3}{n}[x_1^2 + x_2^2 + \cdots + x_i^2 + \cdots + x_n^2] = \dfrac{3}{n}\displaystyle\sum_{i=1}^{n} x_i^2.$

(In calculus it is shown that as the number n of such rectangles increases indefinitely, the sum of their areas approaches a number. That number is defined as the area above the x-axis and under the graph of $y = x^2$ from $x = 0$ to $x = 3$. That number is also defined as a certain **definite integral**.)

In calculus, infinite series are sometimes used to represent functions. For example, the logarithmic function whose rule of assignment is $y = \ln(1 + x)$ can be represented by an infinite series for certain values of x. (Recall that $\ln(1 + x)$ means $\log_e(1 + x)$. If necessary, refer to Sections

2.8 and 2.9.) In fact, it is shown in calculus that

$$\ln(1+x) = x - \frac{x^2}{2} + \frac{x^3}{3} + \cdots + (-1)^{n+1}\frac{x^n}{n} + \cdots$$

$$\text{for } -1 < x \le 1$$

The next example shows how such series representations can be used to compute entries for logarithm tables.

EXAMPLE 18   Use the fact that

$$\ln(1+x) = x - \frac{x^2}{2} + \frac{x^3}{3} + \cdots + (-1)^{n+1}\frac{x^n}{n} + \cdots$$

for $-1 < x \le 1$, and write out an infinite series representing ln 2. How could you use this series to approximate the value of ln 2?

Solution   Let $x = 1$ in the given series representation for $\ln(1+x)$. Then

$$\ln 2 = 1 - \frac{1}{2} + \frac{1}{3} - \frac{1}{4} + \cdots + \frac{(-1)^{n+1}}{n} + \cdots.$$

We can approximate ln 2 by successively adding the terms in the series. The more terms we add, the better the approximation.

In calculus, it is shown that the exponential function whose rule of assignment is $y = e^x$ can also be represented by an infinite series. In fact, where $(n-1)!$, which is read $(n-1)$ *factorial,* means the product $1 \times 2 \times 3 \times \cdots \times (n-1)$, it is true that

$$e^x = 1 + x + \frac{x^2}{2} + \frac{x^3}{2 \times 3} + \frac{x^4}{2 \times 3 \times 4} + \cdots + \frac{x^{n-1}}{(n-1)!} + \cdots$$

for all real numbers x. This series representation enables us to find the value of e to as many decimal places as we please. Letting $x = 1$ in the series for e^x, we have

$$e = 1 + 1 + \frac{1}{2} + \frac{1}{2 \times 3} + \frac{1}{2 \times 3 \times 4} + \cdots + \frac{1}{(n-1)!} + \cdots.$$

Exercise 4.6, 83–85   We said in Section 1.1 that e is approximately 2.7182818284. This can be checked by successively adding terms of the series for e.

EXERCISE 4.6 SEQUENCES AND SERIES

INTRODUCTION TO SEQUENCES

▶ Using braces, rewrite each of the following sequences. Call each sequence f, and write its rule of assignment with the usual function notation.

EXAMPLE   3, 5, 7, 9, . . . , 2n + 1, . . .

Solution   {2n + 1}; f(n) = 2n + 1

> 1. 3, 6, 9, . . . , 3n, . . .   **2.** 4, 7, 10, . . . , 3n + 1, . . .
> **3.** 2, 5, 10, . . . , n² + 1, . . .   **4.** 1, 8, 27, . . . , n³, . . .
> **5.** 3, 9, 19, . . . , 2n² + 1, . . .

▶ For each of the following, write an expression that does not use braces and includes at least the first three terms.

EXAMPLE   {5n + 1}

Solution   6, 11, 16, . . . , 5n + 1, . . .

> **6.** {6n − 1}   **7.** {2n³ + 1}   **8.** {n² + 2n}
> **9.** {n(n + 1)}   **10.** {2n − 1}

11. If the chemical of Example 1 lost one quarter of its volume each day, what would be the sequence of volumes remaining at the end of successive days?

12. If the community in Example 2 deposited 3 tons of nonrecyclable garbage the first year and $3 \times 2^{n-1}$ tons the nth year, what is the sequence whose terms are the number of tons deposited in succeeding years?

13. Suppose that the economist in Example 3 is wrong, and of each 1 million dollars in rebate, only 75% would be spent in the States. Write an infinite sequence whose terms are the successive amounts, in millions of dollars, fed back into the local economy for each million-dollar rebate.

14. Suppose that the person in Example 4 deposited $300 in an account paying 5% interest compounded annually. Write out an infinite sequence whose terms are the numbers of dollars in the account at the end of each successive year.

15. Suppose that the prize in Example 5 was $5 the first year with an annual increase of $3 each year thereafter. Write out an infinite sequence whose terms are the dollar values of the prize in succeeding years.

CONVERGENCE OF SEQUENCES

▶ Write out the first four terms and the general term of each of the following, and find $\lim_{n\to\infty}$ of the general term.

EXAMPLES   (a) $\{(\tfrac{1}{4})^n\}$ (b) $\{8 + (\tfrac{1}{4})^n\}$

Solutions   (a) $\tfrac{1}{4}, \tfrac{1}{16}, \tfrac{2}{64}, \tfrac{1}{256}, \ldots, (\tfrac{1}{4})^n, \ldots$

$$\lim_{n\to\infty} (\tfrac{1}{4})^n = 0$$

(b) $8 + \frac{1}{4}$, $8 + \frac{1}{16}$, $8 + \frac{1}{64}$, $8 + \frac{1}{256}$, ... , $8 + (\frac{1}{4})^n$, ...

$$\lim_{n \to \infty} 8 + (\tfrac{1}{4})^n = 8$$

16. $\{(\frac{1}{5})^n\}$ **17.** $\{3 + (\frac{1}{5})^n\}$ **18.** $\{(\frac{1}{6})^n\}$

19. $\{1 + (\frac{1}{6})^n\}$ **20.** $\{(\frac{1}{2})^n + (\frac{1}{3})^n\}$

GEOMETRIC SEQUENCES

▶ Find the common ratio of each of the following geometric sequences and state whether or not the sequence converges. If it converges, to what does it converge?

EXAMPLES

(a) -3, 9, -27, 81, ... , $(-3)^n$, ...

(b) 1, $\frac{-1}{3}$, $\frac{1}{9}$, $\frac{-1}{27}$, ... , $(\frac{-1}{3})^{n-1}$, ...

Solutions

(a) $r = \frac{9}{-3} = -3$; $|r| > 1$, diverges

(b) $r = \frac{-1}{3}$; $|r| < 1$, converges to 0.

21. -2, 4, -8, 16, ... , $(-2)^n$, ...

22. 1, $\frac{1}{3}$, $\frac{1}{9}$, $\frac{1}{27}$, ... , $(\frac{1}{3})^{n-1}$, ...

23. 3, -3, 3, -3, ... , $3 \times (-1)^{n+1}$, ...

24. 3, 6, 12, 24, ... , $3 \times 2^{n-1}$, ...

25. 4, 2, 1, $\frac{1}{2}$, ... , $4 \times (\frac{1}{2})^{n-1}$, ...

26. -1, 1, -1, 1, ... , $(-1)(-1)^{n+1}$, ...

27. 5, -5, 5, -5, ... , $5 \times (-1)^{n+1}$, ...

28. $\frac{1}{4}$, $\frac{-1}{16}$, $\frac{1}{64}$, $\frac{-1}{256}$, ... , $\frac{1}{4} \times (\frac{-1}{4})^{n-1}$, ...

29. 1, 2, 4, 8, ... , 2^{n-1}, ...

30. 0.1, 0.01, 0.001, 0.0001, ... , $(0.1)^n$, ...

▶ Write out a formula for the sum of the first n terms of each of the following. Use this formula to find the sum of the first n terms for the given particular value of n.

EXAMPLE

$\{5^{n-1}\}$; $n = 4$

Solution

$$a = 1, r = 5, \qquad S_n = \frac{1 - 5^n}{1 - 5} = \frac{1 - 5^n}{-4}$$

$$S_4 = \frac{1 - 5^4}{-4} = \frac{1 - 625}{-4} = 156$$

31. $\{4^{n-1}\}$, $n = 5$ **32.** $\{(\frac{-1}{3})^{n-1}\}$; $n = 2$

33. $\{(\frac{1}{3})^n\}$; $n = 5$

34. $\{(\frac{-1}{3})^n\}$; $n = 5$

35. $\{(-2)^n\}$; $n = 5$

36. $\{3 \times (-1)^{n-1}\}$; $n = 4$

37. $\{(-1)^{n-1}\}$; $n = 6$

38. $\{5 \times (-1)^{n+1}\}$; $n = 4$

39. $\{4 \times (\frac{1}{2})^{n-1}\}$; $n = 3$

40. $\{(0.1)^n\}$; $n = 5$

ARITHMETIC SEQUENCES

▶ Find the common difference of each of the following arithmetic sequences. Write out a formula for the sum of the first n terms. Use this formula to find the sum of the first five terms.

EXAMPLE

$1, 3, 5, 7, \ldots, 1 + 2(n - 1), \ldots$

Solution

$d = 3 - 1 = 2, a = 1, \ell = a + (n - 1)d$
$$= 1 + 2(n - 1)$$

$$S_n = \frac{n(a + \ell)}{2}$$

$$= \frac{n[2 + 2(n - 1)]}{2}$$

$$= \frac{2n[1 + (n - 1)]}{2}$$

$$= n^2$$

$$S_5 = 5^2 = 25$$

41. $2, 4, 6, 8, \ldots, 2 + 2(n - 1), \ldots$

42. $3, 7, 11, 15, \ldots, 3 + 4(n - 1), \ldots$

43. $1, -1, -3, -5, \ldots, 1 + (-2)(n - 1), \ldots$

44. $5, 10, 15, 20, \ldots, 5 + 5(n - 1), \ldots$

45. $3, 0, -3, -6, \ldots, 3 + (-3)(n - 1), \ldots$

▶ Classify each of the following as either geometric or arithmetic. If it is geometric, give the common ratio. If it is arithmetic, give the common difference.

EXAMPLES

(a) $3, 6, 9, 12, \ldots, 3 + 3(n - 1), \ldots$
(b) $3, 6, 12, 24, \ldots, 3 \times 2^{n-1}, \ldots$

Solutions

(a) Arithmetic, $d = 3$
(b) Geometric, $r = 2$

46. $4, 8, 12, 16, \ldots, 4 + 4(n - 1), \ldots$

47. 4, 8, 16, 32, . . . , $4 \times 2^{n-1}$, . . .

48. $\frac{1}{2}$, 1, $\frac{3}{2}$, 2, . . . , $\frac{1}{2} + \frac{1}{2}(n - 1)$, . . .

49. $\frac{1}{2}$, 2, 8, 32, . . . , $\frac{1}{2} \times (4)^{n-1}$, . . .

50. 5, 10, 20, 40, . . . , $5 \times 2^{n-1}$, . . .

INTRODUCTION TO SERIES AND SUMMATION NOTATION

▶ Use summation notation to write each of the following.

EXAMPLE

$2 + 6 + 12 + 20 + \cdots + n(n + 1) + \cdots$

Solution

$$\sum_{n \to 1}^{\infty} n(n + 1)$$

51. $3 + 6 + 12 + 24 + \cdots + 3 \times 2^{n-1} + \cdots$

52. $4 + 8 + 12 + 16 + \cdots + [4 + 4(n - 1)] + \cdots$

53. $1 + 4 + 9 + 16 + \cdots + n^2 + \cdots$

54. $1 + 8 + 27 + 64 + \cdots + n^3 + \cdots$

55. $4 + 10 + 18 + 28 + \cdots + n(n + 3) + \cdots$

▶ Write out each of the following.

EXAMPLE

$$\sum_{n=1}^{\infty} 3^n$$

Solution

$3 + 3^2 + 3^3 + 3^4 + \cdots + 3^n + \cdots$

56. $\displaystyle\sum_{n=1}^{5} 3^n$

57. $\displaystyle\sum_{n=4}^{9} 3^n$

58. $\displaystyle\sum_{n=1}^{\infty} \frac{n}{n + 1}$

59. $\displaystyle\sum_{n=1}^{\infty} \frac{1}{2n}$

60. $\displaystyle\sum_{n=1}^{\infty} \frac{n}{n^3 + 2}$

61. $\displaystyle\sum_{n=3}^{7} \frac{n}{n^2 + 1}$

62. $\displaystyle\sum_{n=1}^{5} n^4$

63. $\displaystyle\sum_{n=1}^{\infty} \frac{n^3}{3^n}$

64. $\displaystyle\sum_{n=1}^{\infty} \frac{2^n}{n}$

65. $\displaystyle\sum_{n=1}^{9} \sqrt{n}$

PARTIAL SUMS, CONVERGENCE, AND SUMS OF INFINITE SERIES

▶ Write out the first three terms and write the nth term S_n of the sequence of partial sums of each of the following. Find the sum of the series by finding $\lim_{n\to\infty} S_n$.

EXAMPLE

$$\sum_{n=1}^{\infty} \left(\tfrac{1}{5}\right)^{n-1}$$

Solution

$$S_1 = 1, S_2 = 1 + \tfrac{1}{5}, S_3 = 1 + \tfrac{1}{5} + \left(\tfrac{1}{5}\right)^2$$

$$S_n = 1 + \tfrac{1}{5} + \left(\tfrac{1}{5}\right)^2 + \cdots + \left(\tfrac{1}{5}\right)^{n-1}$$

$$S_n = \frac{1 - \left(\tfrac{1}{5}\right)^n}{1 - \tfrac{1}{5}}$$

$$= \tfrac{5}{4}\left[1 - \left(\tfrac{1}{5}\right)^n\right]$$

$$\lim_{n\to\infty} S_n = \tfrac{5}{4}$$

66. $\displaystyle\sum_{n=1}^{\infty} \left(\tfrac{1}{4}\right)^{n-1}$ **67.** $\displaystyle\sum_{n=1}^{\infty} \left(\tfrac{-1}{3}\right)^{n-1}$ **68.** $\displaystyle\sum_{n=1}^{\infty} \left(\tfrac{1}{3}\right)^n$

69. $\displaystyle\sum_{n=1}^{\infty} \left(\tfrac{1}{6}\right)^{n-1}$ **70.** $\displaystyle\sum_{n=1}^{\infty} \left(\tfrac{-1}{4}\right)^{n-1}$

GEOMETRIC AND ARITHMETIC SERIES

▶ Tell whether or not each of the following converges. If it converges, find its sum.

EXAMPLES

(a) $\displaystyle\sum_{n=1}^{\infty} \left(\tfrac{1}{5}\right)^{n-1}$ (b) $\displaystyle\sum_{n=1}^{\infty} 5^{n-1}$

Solutions

(a) $\displaystyle\sum_{n=1}^{\infty} \left(\tfrac{1}{5}\right)^{n-1}$ is a geometric series with $r = \tfrac{1}{5}$.

$r < 1$, series converges to $\dfrac{a}{1 - r} = \dfrac{1}{\tfrac{4}{5}} = \tfrac{5}{4}$

(b) $\displaystyle\sum_{n=1}^{\infty} 5^{n-1}$ is a geometric series with $r = 5$.

$r > 1$, series diverges

71. $\displaystyle\sum_{n=1}^{\infty} \left(\tfrac{1}{4}\right)^{n-1}$ **72.** $\displaystyle\sum_{n=1}^{\infty} 4^{n-1}$ **73.** $\displaystyle\sum_{n=1}^{\infty} \left(\tfrac{-1}{3}\right)^{n-1}$

74. $\displaystyle\sum_{n=1}^{\infty} (-2)^n$ **75.** $\displaystyle\sum_{n=1}^{\infty} (-1)^{n-1}$

76. Assume that the evaporation of the chemical in Example 1 continues indefinitely, and find the sum of the volumes lost each day.

77. How many tons of nonrecyclable garbage would the community of Example 2 deposit in the first 10 years of the ecologist's study?

78. If the economist in Example 3 estimated that only 70% of money fed into the local economy would be spent locally, and if this situation continued indefinitely, what would be the total amount fed back into the local economy for each million-dollar tax rebate?

79. How much would the person of Example 4 have in the account at the end of 10 years?

80. Suppose that the dump in Example 2 can accommodate no more than 2044 tons of nonrecyclable garbage, and suppose that the people decided they would wait until the dump was half full before removing any garbage. How many years would they have to wait before the dump would be half full? How many years would they then have to remove the garbage before the dump was full? If they waited until the dump was half full, and if they can remove only 500 tons of garbage per year, within how many years from then would they have more garbage than the dump can accommodate?

81. A farmer noticed that 1 square meter of his 512-square meter pond was covered with weeds that were doubling the area they covered each day. He decided to wait until the weeds covered half the area of his pond before removing them. How many days would he have to wait for the weeds to cover half of the pond? How many days would he then have before they covered the entire pond?

82. Let S_n be the sum of the first n terms of the arithmetic sequence

$$a, a + d, a + 2d, \ldots , a + (n - 1)d, \ldots ,$$

and call the nth term l so that $S_n = a + (a + d) + (a + 2d) + \cdots + (l - 2d) + (l - d) + l$. (Note that the next to the last term is $l - d$, and the term before that is $l - 2d$.) Derive the formula

$$S_n = \frac{n(a + l)}{2}.$$

Hint   Rewrite the equation defining S_n by reversing the order of the terms on the right so that

$$S_n = l + (l - d) + (l - 2d) + \cdots + (a + 2d) + (a + d) + a.$$

Next add the two equations defining S_n; note that the terms involving d drop out, and note that

$$2S_n = n(a + l).$$

USES OF SEQUENCES AND SERIES IN CALCULUS

83. Suppose that the calculus student of Example 17 used six rectangles of equal base length instead of three. The length of the base of each rectangle would then be $\frac{1}{2}$ unit. Using six such rectangles, choose points $x_1, x_2, \ldots,$ x_6 at random, one point on each base. Take the height of each rectangle to be the y-coordinate of the point on the graph of $y = x^2$ determined by the point chosen at random on its base. Write out an expression for the sum of the areas of the six rectangles, and write this sum with summation notation.

84. Approximate each of the first 12 terms of the series for ln 2 in Example 18 with a four-place decimal, and use these 12 terms to find an approximation to ln 2. Use this approximation and use log $x = (0.4343)$ ln x to find an approximation to log 2. Compare your results with Tables 1 and 2 in Appendix B. How could you use the given series and obtain a better approximation to ln 2?

85. Use the first four terms of

$$e^x = 1 + x + \frac{x^2}{2} + \frac{x^3}{2 \times 3} + \cdots + \frac{x^{n-1}}{(n-1)^1} + \cdots ,$$

use a four-decimal-place approximation to $\frac{1}{6}$, and obtain a one-decimal-place approximation to e.

CHECK LIST FOR THE STUDENT

If you have learned the material in Section 4.6, you should be able to do the following:

1. Write one expression for an infinite sequence using braces and one without using braces.

2. Recognize that a sequence is a function, and give the domain and rule of assignment of a given sequence.

3. Tell what is meant by the terms and the general term of a sequence.

4. Tell why it is not sufficient to give merely the first few terms in specifying a sequence.

5. Give instances in which sequences arise in everyday life, in science, or in mathematics.

6. Write out the first four terms and the general term of a sequence written with braces.

7. Give an example of a convergent sequence, and tell what is meant by its limit.

8. Tell what is meant when a sequence is said to diverge.

9. Find $\lim_{n \to \infty}$ of the general term of some given sequence.

10. Tell how the second term and the succeeding terms of a geometric sequence can be formed.

11. Find the common ratio of a given geometric sequence.

12. Complete the statements

If $|r| < 1$, a geometric sequence with common ratio r _____.
If $|r| > 1$, a geometric sequence with common ratio r _____.
If $r = 1$, a geometric sequence with common ratio r _____.
If $r = -1$, a geometric sequence with common ratio r _____.

13. Complete the statements.

If $|r| < 1$, $\lim_{n \to \infty} r^n =$ _____.
If $|r| > 1$, $\lim_{n \to \infty} r^n =$ _____.

14. State whether or not a given geometric sequence converges, and tell to what it converges if it *does* converge.

15. Write and derive a formula for the sum S_n of the first n terms of a geometric sequence with first term a and common ratio $r \neq 1$; write a formula for S_n if $r = 1$.

16. Find the sum of the first n terms of a given geometric sequence for a particular given n.

17. Tell how the second term and the succeeding terms of an arithmetic sequence can be formed.

18. Find the common difference of a given arithmetic sequence.

19. Write a formula for the sum of the first n terms of an arithmetic sequence with first term a, with nth term l, and with common difference d.

20. Find the sum of the first n terms of a given arithmetic sequence for a particular n.

21. Write one expression for an infinite series using summation notation and one without using summation notation.

22. Write out an expression for a series written with summation notation.

23. Tell what is meant by the nth partial sum of an infinite series.

24. Write the first three terms and the nth term of the sequence of partial sums of a given infinite series.

25. Tell what is meant when an infinite series is said to converge and when an infinite series is said to diverge.

26. Tell what is meant by the sum of an infinite series.

27. Complete the statements, where $\{S_n\}$ is the sequence of partial sums of $\sum_{n=1}^{\infty} a_n$.

If $\{S_n\}$ converges to L, then $\sum_{n=1}^{\infty} a_n$ _____.

If $\{S_n\}$ diverges, then $\sum_{n=1}^{\infty} a_n$ _____.

28. Give the definition of a geometric series.

29. Find the common ratio of a given geometric series.

30. Complete the statements.
    If $|r| < 1$, the geometric series with first term a and common ratio r converges to _____.
    If $|r| \geq 1$, a geometric series with common ratio r _____.

31. State whether or not a given geometric series converges, and find its sum if it *does* converge.

32. Give the definition of an arithmetic series.

33. Tell whether or not an arithmetic series converges.

34. Give an instance of how series are used in calculus in approximating areas.

35. Give an instance of how series are used to represent functions.

SECTION QUIZ 4.6

1. (a) Using braces, rewrite the sequence 5, 7, 9, 11, . . . , $2n + 3$, ...
   (b) Without using braces, write an expression for $\{3n - 2\}$. Include at least the first three terms.

2. Write the first four terms and the general term a_n of $\{(\frac{1}{3})^n\}$, and give $\lim\limits_{n \to \infty} a_n$.

Classify each of the following as either a geometric sequence or an arithmetic sequence. If it is a geometric sequence, give its common ratio; if it is an arithmetic sequence, give its common difference.

3. $1, 5, 9, 13, \ldots, 1 + 4(n - 1), \ldots$

4. $1, 3, 9, 27, \ldots, 3^{n-1}, \ldots$

Write out a formula for the sum of the first n terms of each of the following.

5. $5, 5^2, 5^3, \ldots, 5^n, \ldots$

6. $3, 6, 9, \ldots, 3n, \ldots$

State whether or not each of the following converges. If it converges, find its sum.

7. $1 + \frac{1}{3} + \frac{1}{9} + \frac{1}{27} + \cdots + (\frac{1}{3})^{n-1} + \cdots$

8. $1 + 3 + 9 + 27 + \cdots + (3)^{n-1} + \cdots$

9. (a) Write out the first four terms, and write the nth term of

$$\sum_{n=1}^{\infty} \frac{n}{n + 2}.$$

(b) Write the first, second, and third partial sums of

$$\sum_{n=1}^{\infty} \frac{n}{n + 2}.$$

10. A pollution-free energy source is being studied. It is found that in one application a measure of the energy lost during the first hour is 0.1, during the second hour it is $(0.1)^2$, during the third hour it is $(0.1)^3$, etc. Assume that the energy lost during the nth hour is $(0.1)^n$.

(a) What is a formula for the sum of the energy lost during the first n hours?

(b) Use this formula to find the sum of the energy lost during the first 10 hours.

(c) If the energy loss continued indefinitely as assumed here, what would be the total sum of the energy lost?

4.7 Mathematical Induction

The purpose of this section is to present the principle of mathematical induction and to illustrate its use.

The *principle of mathematical induction* is used to prove various statements concerning integers, usually positive integers. Here are some examples of statements that can be proved by using this principle.

1. The sum of the squares of the first n positive integers is

$$\frac{n(n + 1)(2n + 1)}{6}.$$

$$1^2 + 2^2 + \cdots + n^2 = \frac{n(n + 1)(2n + 1)}{6} \qquad \text{for all integers } n \geq 1$$

2. The sum of the first n odd positive integers is n^2.

$$1 + 3 + 5 + \cdots + (2n - 1) = n^2 \qquad \text{for all integers } n \geq 1$$

3. For all integers $n \geq 5$, 2^n is greater than n^2.

$$2^n > n^2 \qquad \text{for all integers } n \geq 5$$

4. For real numbers x and y and for any positive integer n, $(xy)^n = x^n y^n$.

$$(xy)^n = x^n y^n \qquad \text{for all integers } n \geq 1$$

5. For real numbers x and y, and for any positive integer n, $x - y$ is a factor of $x^n - y^n$.

$$(x - y) \text{ is a factor of } (x^n - y^n) \qquad \text{for all integers } n \geq 1$$

We shall prove each of these statements after we have formally stated the principle of mathematical induction.

The *principle of mathematical induction* may be thought of as being the *domino theory*. If dominoes are lined up so that if any one of them falls, then the next one will fall, and if some one of the dominoes *does* fall, then it follows that all of the dominoes in line after that one will also fall. In politics, the domino theory refers to countries rather than dominoes. The theory is that certain countries are ''lined up'' so that if any one of them falls under unsavory influence, then the next one will also fall; it follows that if the first of these countries *does* fall, then all of the others after it will fall. In mathematics, statements concerning integers take the place of dominoes or countries. Roughly speaking, the principle of mathematical induction asserts that if a statement being true (falling) for any one integer makes the statement true (fall) for the next integer, and if the statement *is* true for some one integer, then it is true for all larger integers.

To prove a statement true for all positive integers n, we must prove two things.

1. If the statement is true for any positive integer k, then it is true for the next positive integer $k + 1$.

2. The statement *is* true for the first positive integer $n = 1$.

To prove a statement true for all integers n greater than or equal to some particular integer q, we must also prove two things.

1. If the statement is true for any integer $k \geq q$, then it is true for the next integer $k + 1$.

2. The statement *is* true for integer q.

These two parts of a "proof by induction" are usually proved in reverse order because it is usually easier to prove or disprove that a statement is true for a particular integer than it is to prove that truth for any one integer forces truth for the next integer.

The principle of mathematical induction is stated formally below. In the formal statement of the principle, the symbol P_n is used for a statement concerning integer n.

P_n means a statement concerning integer n.

Principle of Mathematical Induction

If a statement P_n concerning integer n satisfies the following two conditions, then P_n is true for all integers n greater than or equal to a particular integer q. The two conditions are

1. P_n is true for $n = q$.

2. If P_n is true for any integer $k \geq q$, then P_n is true for the next integer $k + 1$.

(Note that if $q = 1$, and if P_n satisfies the two conditions of the principle of mathematical induction, then P_n is true for all positive integers.)

The two conditions of the principle of mathematical induction can be stated symbolically as

1. P_q is true.

2. P_k true $\rightarrow P_{k+1}$ true, for integer $k \geq q$.

The first part of the symbolic form of the second condition is read, "If P_k is true, then P_{k+1} is true." Or it is read, "P_k true implies P_{k+1} true."

In using the principle of mathematical induction and letting q represent a particular integer, the proof that a statement P_n is true for all integers $n \geq q$ consists of two parts.

Part I   Prove: Statement P_n is true for $n = q$.

Part II   Prove: If P_n is true for $n = k$, then P_n is true for $n = k + 1$, where k is any integer greater than or equal to q.

Such a proof can be summarized as follows:

Part I   Prove: P_q is true.

Part II   Prove: P_k true $\rightarrow P_{k+1}$ true   for any integer $k \geq q$.

Conclusion   If P_q is true, and if P_k true $\rightarrow P_{k+1}$ true, for any integer $k \geq q$, then P_n is true for all integers $n \geq q$.

EXAMPLE 1   Prove that

$$1^2 + 2^2 + \cdots + n^2 = \frac{n(n+1)(2n+1)}{6}$$

for all positive integers n.

Solution   Let P_n be the statement

$$1^2 + 2^2 + \cdots + n^2 = \frac{n(n+1)(2n+1)}{6}. \qquad P_n$$

Part I   Prove: P_n is true for $n = 1$. (Prove: P_1 is true.)

When $n = 1$, the statement P_n becomes P_1.

$$1^2 = \frac{1(1+1)(2 \times 1 + 1)}{6} \qquad P_1$$

That is,

$$1 = \frac{(2)(3)}{6}. \qquad \text{This is true.}$$

Part II   Prove: If P_n is true for $n = k$ where k is any integer ≥ 1, then P_n is true for $n = k + 1$. (Prove: If P_k is true, then P_{k+1} is true, for any integer $k \geq 1$.)

Suppose P_n is true for $n = k$. That is, suppose

$$1^2 + 2^2 + \cdots + k^2 = \frac{k(k+1)(2k+1)}{6}. \qquad P_k$$

$$(P_n \text{ for } n = k)$$

On the basis of this supposition, we must prove that P_n is true for $n = k + 1$ and $k \geq 1$. That is, we must prove that P_k true implies that the following is true for $k \geq 1$.

$$1^2 + 2^2 + \cdots + (k+1)^2 = \frac{(k+1)(k+1+1)(2(k+1)+1)}{6}. \qquad P_{k+1}$$

$$(P_n \text{ for } n = k + 1)$$

This statement, P_{k+1}, was obtained from P_n by replacing n by $k + 1$. In this type of problem (a certain indicated sum is given by a certain formula), it is usually helpful to write in the next to the last term of the indicated sum on the left of P_{k+1}. If we do that here, and if we also collect terms in the factors on the right, P_{k+1} becomes

$$1^2 + 2^2 + \cdots + k^2 + (k + 1)^2 = \frac{(k + 1)(k + 2)(2k + 3)}{6}. \qquad P_{k+1}$$

Proceeding to prove that for integer $k \geq 1$, P_k true implies that P_{k+1} is true, we start with the left side of P_{k+1} and use the fact that if P_k is true, then $1^2 + 2^2 + \cdots + k^2$ can be replaced by

$$\frac{k(k + 1)(2k + 1)}{6}.$$

$$1^2 + 2^2 + \cdots + k^2 + (k + 1)^2 = \frac{k(k + 1)(2k + 1)}{6} + (k + 1)^2, \text{ if } P_k \text{ is true}$$

$$= \frac{k(k + 1)(2k + 1)}{6} + \frac{6(k + 1)^2}{6}$$

$$= \frac{(k + 1)[k(2k + 1) + 6(k + 1)]}{6}$$

$$= \frac{(k + 1)[2k^2 + k + 6k + 6]}{6}$$

$$= \frac{(k + 1)(2k^2 + 7k + 6)}{6}$$

$$= \frac{(k + 1)(k + 2)(2k + 3)}{6}$$

That is,

$$1^2 + 2^2 + \cdots + k^2 + (k + 1)^2 = \frac{(k + 1)(k + 2)(2k + 3)}{6}, \text{ if } P_k \text{ is true,}$$

which says P_{k+1} is true if P_k is true.

Conclusion Since P_1 is true, and since if P_k is true, then P_{k+1} is true for any integer $k \geq 1$, it follows that P_n is true for all positive integers n.

EXAMPLE 2 Use mathematical induction to prove that

$$1 + 3 + \cdots + (2n - 1) = n^2$$

for all integers $n \geq 1$.

Solution Let P_n be the statement

$$1 + 3 + \cdots + (2n - 1) = n^2. \qquad P_n$$

Part I Prove: P_1 is true.

When $n = 1$, P_n becomes P_1.

$$1 = 1^2. \qquad P_1$$

This is true.

Part II Prove: If P_k is true, then P_{k+1} is true, for any integer $k \geq 1$.

When $n = k$, P_n becomes P_k.

$$1 + 3 + \cdots + (2k - 1) = k^2 \qquad P_k$$

When $n = k + 1$, P_n becomes P_{k+1}.

$$1 + 3 + \cdots + [2(k + 1) - 1] = (k + 1)^2. \qquad P_{k+1}$$

If we simplify the last term on the left side of P_{k+1}, and if we write in the next to the last term, P_{k+1} becomes

$$1 + 3 + \cdots + (2k - 1) + (2k + 1) = (k + 1)^2. \qquad P_{k+1}$$

We must prove that for $k \geq 1$, if P_k is true, then P_{k+1} is true. To prove this, we start with the left side of P_{k+1} and note that if P_k is true, we can replace $1 + 3 + \cdots + (2k - 1)$ by k^2.

$$
\begin{aligned}
1 + 3 + \cdots + (2k - 1) + (2k + 1) &= k^2 + (2k + 1), \text{ if } P_k \text{ is true} \\
&= k^2 + 2k + 1 \\
&= (k + 1)^2
\end{aligned}
$$

That is,

$$1 + 3 + \cdots + (2k - 1) + (2k + 1) = (k + 1)^2, \text{ if } P_k \text{ is true.}$$

In other words, P_{k+1} is true if P_k is true.

Conclusion Since P_1 is true, and since P_k true $\rightarrow P_{k+1}$ true for any integer $k \geq 1$, it follows that P_n is true for all integers $n \geq 1$.

Note that in Example 2, the left side of P_n is the sum of the first n terms of an arithmetic sequence. In this case, we can therefore check the assertion that P_n is true by using the formula for the sum of the first n terms of an arithmetic sequence. Here the first term $a = 1$ and the last term $l = 2n - 1$, so according to that formula,

$$
\begin{aligned}
1 + 3 + \cdots + (2n - 1) &= \frac{n(1 + 2n - 1)}{2} \\
&= \frac{2n^2}{2} \\
&= n^2.
\end{aligned}
$$

Exercise 4.7, 1–20 This does check.

EXAMPLE 3

Use mathematical induction to prove that $2^n > n^2$ for all integers $n \geq 5$. (You may use the fact that $2^{k+1} = 2^k \times 2$ for any positive integer k.)

Solution

Let P_n be the statement

$$2^n > n^2. \qquad P_n$$

Part I   Prove: P_5 is true.

When $n = 5$, P_n becomes P_5.

$$2^5 > 5^2 \qquad P_5$$
$$32 > 25 \qquad \text{This is true.}$$

Part II   Prove: If P_k is true, then P_{k+1} is true, for any integer $k \geq 5$.

When $n = k$, P_n becomes P_k.

$$2^k > k^2 \qquad P_k$$

When $n = k + 1$, P_n becomes P_{k+1}.

$$2^{k+1} > (k + 1)^2 \qquad P_{k+1}$$

We must prove that for any integer $k \geq 5$, P_{k+1} is true if P_k is true. To prove this, we shall start with the left side of P_{k+1} and show that if P_k is true and $k \geq 5$, then the left side of P_{k+1} *is* greater than $(k + 1)^2$ so that P_{k+1} *is* true. In doing this, we shall use the fact that if P_k is true, $2^k > k^2$. We shall also use the fact that $k \geq 5$ so that $k > 3$ and $k > 1$. Here is the proof.

Left side
of P_{k+1}

$$
\begin{aligned}
2^{k+1} &= 2^k \times 2 \\
&> k^2 \times 2, &&\text{if } P_k \text{ is true} \\
&= k^2 + k^2 \\
&= k^2 + k \times k \\
&> k^2 + 3k &&k \geq 5 \text{ so } k > 3 \\
&= k^2 + 2k + k \\
&> k^2 + 2k + 1 &&k \geq 5 \text{ so } k > 1 \\
&= (k + 1)^2
\end{aligned}
$$

Right side
of P_{k+1}

This chain of inequalities and equalities says that $2^{k+1} > (k + 1)^2$, if P_k is true and integer $k \geq 5$. In other words, P_{k+1} is true if P_k is true and integer $k \geq 5$.

Since P_5 is true, and since P_k true $\rightarrow P_{k+1}$ true for any integer $k \geq 5$, then P_n is true for all integers $n \geq 5$. (Note that this P_n is *not* true for *all* integers n. In particular, it is not true for $n = 2$, 3, or 4.)

Exercise 4.7, 21–25

EXAMPLE 4

Use mathematical induction to prove that for real numbers x and y and for any positive integer n, $(xy)^n = x^n y^n$. (You may assume that for any real number a, $a^1 = a$, and also that $a^{k+1} = a^k \cdot a$ for any positive integer k.)

Solution

Let P_n be the statement

$$(xy)^n = x^n y^n. \qquad P_n$$

Part I Prove: P_1 is true.

When $n = 1$, P_n becomes P_1.

$$(xy)^1 = x^1 y^1 \qquad P_1$$

$$xy = xy \qquad \text{Since } a^1 = a \text{ for any real number } a.$$

This is true.

Part II Prove: If P_k is true, then P_{k+1} is true, for any integer $k \geq 1$.

When $n = k$, P_n becomes P_k.

$$(xy)^k = x^k y^k \qquad P_k$$

When $n = k + 1$, P_n becomes P_{k+1}.

$$(xy)^{k+1} = x^{k+1} y^{k+1} \qquad P_{k+1}$$

We must prove that for any integer $k \geq 1$, P_{k+1} is true if P_k is true. To prove this, we shall start with the left side of P_{k+1} and show that if k is a positive integer and P_k is true, then the left side of P_{k+1} *is* equal to $x^{k+1} y^{k+1}$ so that P_{k+1} *is* true. In doing this we shall use the fact that if P_k is true, then $(xy)^k = x^k y^k$.

We shall also use the fact that $(xy)^{k+1} = (xy)^k \cdot (xy)$. Here is the proof.

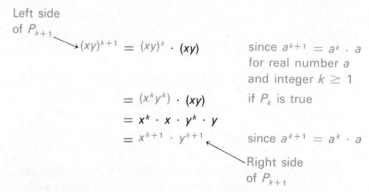

Left side
of P_{k+1}

$$(xy)^{k+1} = (xy)^k \cdot (xy) \qquad \text{since } a^{k+1} = a^k \cdot a \text{ for real number } a \text{ and integer } k \geq 1$$

$$= (x^k y^k) \cdot (xy) \qquad \text{if } P_k \text{ is true}$$

$$= x^k \cdot x \cdot y^k \cdot y$$

$$= x^{k+1} \cdot y^{k+1} \qquad \text{since } a^{k+1} = a^k \cdot a$$

Right side
of P_{k+1}

This chain of equalities says that

$$(xy)^{k+1} = x^{k+1} y^{k+1}, \text{ if } P_k \text{ is true and integer } k \geq 1$$

In other words, P_{k+1} is true if P_k is true and integer $k \geq 1$.

Conclusion

Exercise 4.7, 26–29 Since P_1 is true, and since P_k true $\rightarrow P_{k+1}$ true for any integer $k \geq 1$, then P_n is true for all positive integers n.

Some proofs by mathematical induction are quite straightforward; others are not. The next example demonstrates a proof that depends on the use of a bit of ingenuity.

EXAMPLE 5

Use mathematical induction to prove that for real numbers x and y and for any positive integer n, $(x - y)$ is a factor of $x^n - y^n$. (You may use the fact that for any real number a, $a^1 = a$, and that $a^{k+1} = a^k \cdot a$ for any positive integer k.)

Solution

Let P_n be the statement

$$(x - y) \text{ is a factor of } (x^n - y^n). \qquad P_n$$

Part I Prove: P_1 is true.

When $n = 1$, P_n becomes P_1.

$$(x - y) \text{ is a factor of } (x^1 - y^1). \qquad P_1$$

That is,

$$(x - y) \text{ is a factor of } (x - y). \qquad P_1$$

This is true because $(x - y) = 1 \cdot (x - y)$.

Part II Prove: If P_k is true, then P_{k+1} is true for any integer $k \geq 1$.

When $n = k$, P_n becomes P_k.

$$(x - y) \text{ is a factor of } (x^k - y^k). \qquad P_k$$

When $n = k + 1$, P_n becomes P_{k+1}.

$$(x - y) \text{ is a factor of } (x^{k+1} - y^{k+1}). \qquad P_{k+1}$$

We must prove that for any integer $k \geq 1$, P_{k+1} is true if P_k is true. To prove this, we shall consider $(x^{k+1} - y^{k+1})$. We shall add and subtract xy^k to $x^{k+1} - y^{k+1}$ so that we can group terms and factor. We shall then use the fact that $(x - y)$ is a factor of $(x^k - y^k)$ if P_k is true. Here is the proof.

$$x^{k+1} - y^{k+1} = x^{k+1} - xy^k + xy^k - y^{k+1}$$

$$= x^k \cdot x - xy^k + xy^k - y^k \cdot y \qquad \text{Since } a^{k+1} = a^k \cdot a$$

for real number a and positive integer k

$$= x(x^k - y^k) + y^k(x - y)$$

This says that

$$x^{k+1} - y^{k+1} = x(x^k - y^k) + y^k(x - y).$$

Note that $(x - y)$ is a factor of the second term on the right. Now if P_k is true, then $(x - y)$ is also a factor of $(x^k - y^k)$, and hence is a factor of both terms on the right of the expression for $x^{k+1} - y^{k+1}$. It follows that $(x - y)$ is a factor of $x^{k+1} - y^{k+1}$.

Conclusion

Exercise 4.7, 30–32

Since P_1 is true, and since P_k true $\rightarrow P_{k+1}$ is true for any integer $k \geq 1$, the statement P_n is true for all positive integers n.

The next example illustrates that it is not sufficient merely to prove Part II of an induction proof without proving Part I. There do exist false statements P_n for which Part II can be carried out successfully.

EXAMPLE 6

Suppose that we were trying to use induction to prove that $n > n + 2$ for all positive integers n. Show that Part II of the induction proof can be carried out successfully, but Part I cannot.

Solution

Let P_n be the statement

$$n > n + 2. \qquad P_n$$

To show that Part II of the induction proof can be carried out success-fully, we must show that if P_k is true, then P_{k+1} is true, for any integer $k \geq 1$. When $n = k$, P_n becomes P_k.

$$k > k + 2 \qquad P_k$$

When $n = k + 1$, P_n becomes P_{k+1}.

$$k + 1 > k + 1 + 2 \qquad P_{k+1}$$

We can show that P_{k+1} is true if P_k is true, by noting that if P_k is true, then the statement obtained by adding the number 1 to both sides of P_k is also true, and the statement so obtained is P_{k+1}. Indeed, if P_k is true, and if we add 1 to both sides of it, we obtain the true statement

$$k + 1 > k + 1 + 2,$$

which is P_{k+1}.

Part I of the induction proof cannot be done because P_1 is *not* true. The statement P_1 says that

$$1 > 1 + 2,$$

which is false. The statement P_n is not true for $n = 1$; in fact, it is not true for any positive integer.

Example 6 showed that for an induction proof, it is not sufficient merely to prove Part II. Likewise, merely proving Part I is not sufficient to guarantee that statement P_n is true for all integers $n \geq q$. Even verifying

PART FOUR | PRECALCULUS COLLEGE ALGEBRA

that P_n is true for several values of n is not sufficient. For example, the statement

$$1 + 2 + \cdots + n = \frac{n(n + 1)}{2} + (n - 1)(n - 2)(n - 3)$$

is true for $n = 1$, for $n = 2$, and for $n = 3$, but it is not true for all positive integers n. (Try it for $n = 4$.) In presenting a proof by mathematical induction, both Part I *and* Part II must be done.

Exercise 4.7, 33

Many theorems in mathematics are proved by using mathematical induction. The formula for the sum of the first n terms of a geometric or an arithmetic sequence can be established in this manner.

Exercise 4.7, 34–35

EXERCISE 4.7 MATHEMATICAL INDUCTION

▶ Use mathematical induction to prove that each of the following is true for all the prescribed positive integers n. If possible, check the truth of P_n by using the formula for the sum of the first n terms of an arithmetic sequence or a geometric sequence.

EXAMPLE $2 + 2^2 + \cdots + 2^n = 2(2^n - 1), \quad n \geq 1$

Solution   Let P_n be: $2 + 2^2 + \cdots + 2^n = 2(2^n - 1)$.

Part I P_1 is: $2 = 2(2 - 1)$. This is true.

Part II P_k is: $2 + 2^2 + \cdots + 2^k = 2(2^k - 1)$.

P_{k+1} is: $2 + 2^2 + \cdots + 2^k + 2^{k+1} = 2(2^{k+1} - 1)$.

Start with the left of P_{k+1}.

$$
\begin{aligned}
2 + 2^2 + \cdots + 2^k + 2^{k+1} &= 2(2^k - 1) + 2^{k+1}, \text{ if } P_k \text{ is true} \\
&= 2^{k+1} - 2 + 2^{k+1} \\
&= 2 \times 2^{k+1} - 2 \\
&= 2(2^{k+1} - 1)
\end{aligned}
$$

That is,

$$2 + 2^2 + \cdots + 2^k + 2^{k+1} = 2(2^{k+1} - 1), \text{ if } P_k \text{ is true.}$$

In other words,

$$P_{k+1} \text{ is true if } P_k \text{ is true.}$$

Conclusion P_n is true for all integers $n \geq 1$.

Check $2, 2^2, \ldots, 2^n$ is geometric with first term $a = 2$ and common ratio $r = 2$.

$$
\begin{aligned}
2 + 2^2 + \cdots + 2^n &= \frac{2(1 - 2^n)}{1 - 2} \\
&= -2(1 - 2^n) \\
&= 2(2^n - 1) \qquad \text{It checks.}
\end{aligned}
$$

1. $3 + 3^2 + \cdots + 3^n = \frac{3}{2}(3^n - 1)$, $n \geq 1$

2. $1 + 2 + \cdots + n = \dfrac{n(n + 1)}{2}$, $n \geq 1$

3. $1^3 + 2^3 + \cdots + n^3 = \dfrac{n^2(n + 1)^2}{4}$, $n \geq 1$

4. $2 + 4 + 6 + \cdots + 2n = n(n + 1)$, $n \geq 1$

5. $3 + 9 + 15 + \cdots + 3(2n - 1) = 3n^2$, $n \geq 1$

6. $4 + 8 + \cdots + 4n = 2n(n + 1)$, $n \geq 1$

7. $2 + 8 + 18 + \cdots + 2n^2 = \dfrac{n(n + 1)(2n + 1)}{3}$, $n \geq 1$

8. $\frac{1}{2} + \frac{1}{4} + \cdots + \dfrac{1}{2^n} = 1 - (\frac{1}{2})^n$, $n \geq 1$

9. $2 + 16 + \cdots + 2n^3 = \dfrac{n^2(n + 1)^2}{2}$, $n \geq 1$

10. $2 + 5 + 8 + \cdots + [2 + 3(n - 1)] = \dfrac{4n + 3n(n - 1)}{2}$, $n \geq 1$

11. $1 + 7 + 19 + \cdots + [n^3 - (n - 1)^3] = n^3$, $n \geq 1$

12. $0.8 + (0.8)^2 + (0.8)^3 + \cdots + (0.8)^n = 4[1 - (0.8)^n]$, $n \geq 1$

13. $1 + 15 + 66 + \cdots + [n^4 - (n - 1)^4] = n^4$, $n \geq 1$

14. $1 + \frac{1}{3} + \frac{1}{9} + \cdots + (\frac{1}{3})^{n-1} = \frac{3}{2}[1 - (\frac{1}{3})^n]$, $n \geq 1$

15. $1 + 5 + 5^2 + \cdots + 5^{n-1} = \dfrac{5^n - 1}{4}$, $n \geq 1$

16. $\dfrac{1}{1 \times 2} + \dfrac{1}{2 \times 3} + \dfrac{1}{3 \times 4} + \cdots + \dfrac{1}{n(n + 1)} = \dfrac{n}{n + 1}$, $n \geq 1$

17. $3 + 7 + 11 + \cdots + [3 + 4(n - 1)] = n(2n + 1)$, $n \geq 1$

18. $0.1 + (0.01) + (0.001) + \cdots + (0.1)^n = \dfrac{1 - (0.1)^n}{9}$, $n \geq 1$

19. $\dfrac{1}{6} + \dfrac{1}{6 \times 11} + \dfrac{1}{11 \times 16} + \cdots + \dfrac{1}{(5n - 4)(5n + 1)} = \dfrac{n}{5n + 1}$, $n \geq 1$

20. $5 + 10 + 15 + \cdots + [5 + 5(n - 1)] = \dfrac{5n(n + 1)}{2}$, $n \geq 1$

21. $n^3 > n^2$, $n \geq 2$

22. $2^n > 2n^2$, $n \geq 7$ (You may use the fact that $2^{k+1} = 2^k \times 2$ for any positive integer k.)

23. $n^3 > n$, $n \geq 2$

24. $n^2 > n$, $n \geq 2$

25. $3^n > 6n$, $n \geq 3$ (You may use the fact that $3^{k+1} = 3^k \times 3$ for any positive integer k.)

26. For real numbers x and y and for $n \geq 1$, $(x/y)^n = x^n/y^n$. (You may assume that for any real number a, $a^1 = a$, and also that $a^{k+1} = a^k \cdot a$ for any positive integer k.)

27. For real number x and $n \geq 1$, $x^2 \cdot x^n = x^{2+n}$. (You may assume that for any real number a, $a^1 = a$, and also that $a^{m+1} = a^m \cdot a$ for any positive integer m.)

28. For real number x and $n \geq 1$, $(x^2)^n = x^{2n}$. (You may assume that for any real number a, $a^1 = a$, $a^{m+1} = a^m \cdot a^1$ and also that $a^2 \cdot a^m = a^{2+m}$ for any positive integer m.)

29. For real number x and $n \geq 3$, $x^n/x^2 = x^{n-2}$. (You may assume that for any real number a, $a^1 = a$, and also that $a^{m+1} = a^m \cdot a$ for any positive integer m.)

30. For real numbers x and y and $n \geq 1$, $(x + y)$ is a factor of $x^{2n-1} + y^{2n-1}$. (You may use the fact that for any real number a and positive integer m, $a^{m+2} = a^m \cdot a^2$.)

31. For real numbers x and y and $n \geq 1$, $(x - y)$ is a factor of $(x^{2n} - y^{2n})$. (You may use the fact that for any real number a and positive integer m, $a^{m+2} = a^m \cdot a^2$.)

32. For $n \geq 1$, 2 is a factor of $n^2 + n$.

33. Suppose we were trying to use induction to prove that $n > n + 3$ for all positive integers n. Show that Part II of the induction proof can be carried out successfully, but Part I cannot.

34. For real number $r \neq 1$ and for $n \geq 1$, prove that

$$1 + r + r^2 + \cdots + r^{n-1} = \frac{1 - r^n}{1 - r}.$$

35. For real number d and for $n \geq 1$, prove that

$$1 + (1 + d) + (1 + 2d) + \cdots + [1 + (n - 1)d] = \frac{n[2 + (n - 1)d]}{2}.$$

CHECK LIST FOR THE STUDENT

If you have learned the material in Section 4.7, you should be able to do the following:

1. Tell what is meant by the domino theory.

2. Tell what two things must be proved in order to prove that a statement is true for all positive integers n or for all integers n greater than or equal to some particular integer q.

3. Complete the following statement of the principle of mathematical induction: If a statement P_n concerning integer n satisfies the following two conditions:

    1. _____

    2. _____

then P_n is true for all integers $n \geq q$.

4. Complete the following: The proof that a statement P_n is true for all integers $n \geq q$ consists of two parts.

Part I   Prove: _____

Part II   Prove: _____

5. Use mathematical induction to prove that certain given statements concerning integers are true; if possible, check the truth of a given statement by using the formula for the sum of the first n terms of an arithmetic or a geometric sequence.

6. Show by presenting examples that in a proof by mathematical induction, both Part I *and* Part II of the proof must be done.

SECTION QUIZ 4.7

Use mathematical induction to prove that each of the following is true for all the prescribed positive integers n. If possible, check the truth by using the formula for the sum of the first n terms of an arithmetic sequence or a geometric sequence.

1. $5 + 5^2 + 5^3 + \cdots + 5^n = \frac{5}{4}(5^n - 1)$, $n \geq 1$

2. $3 + 6 + 9 + \cdots + 3n = \frac{3}{2}(n + n^2)$, $n \geq 1$

3. $2^n > n + 2$, $n \geq 3$

4. $3 + 12 + \cdots + 3n^2 = \dfrac{n(n + 1)(2n + 1)}{2}$, $n \geq 1$

5. 2 is a factor of $n^2 + 3n$, $n \geq 1$

4.8 Binomial Theorem

The purpose of this section is to state and illustrate the binomial theorem and to present Pascal's triangle.

The binomial theorem provides us with a formula for expanding—that is, for writing out as a sum—the power of a binomial such as $(a + b)^3$, or $(a + b)^{25}$, or $(a + b)^n$ for any positive integer n. It enables us to write out any particular term of such an expansion without writing out the entire expansion. It can be extended to provide series representations for expressions such as $(1 + x)^r$ where r is any real number and $|x| < 1$.

As an introduction, let us write out $(a + b)^n$ for a few specific values of n so that we can see the pattern of the terms. We already know that $(a + b)^2 = a^2 + 2ab + b^2$. We can find successively higher powers of $(a + b)$ by successively multiplying by $a + b$.

$$(a + b)^3 = (a + b)^2(a + b),$$
$$(a + b)^4 = (a + b)^3(a + b),$$
$$(a + b)^5 = (a + b)^4(a + b), \text{ etc.}$$

Here are some powers of $(a + b)$ written out.

$(a + b)^1 = 1 \cdot a + 1 \cdot b$

$(a + b)^2 = 1 \cdot a^2 + 2ab + 1 \cdot b^2$

$(a + b)^3 = 1 \cdot a^3 + 3a^2b + 3ab^2 + 1 \cdot b^3$

$(a + b)^4 = 1 \cdot a^4 + 4a^3b + 6a^2b^2 + 4ab^3 + 1 \cdot b^4$

$(a + b)^5 = 1 \cdot a^5 + 5a^4b + 10a^3b^2 + 10a^2b^3 + 5ab^4 + 1 \cdot b^5$

To begin to see how the terms are formed, look at $(a + b)^4$ as an example. Note that the coefficient of a^4 is 1 and that the coefficient of b^4 is 1. The degree of each term is 4. We may think of the last term as being a^0b^4, and we may think of the first term as being a^4b^0. Thus proceeding left to right, the exponents of the a are successively 4, 3, 2, 1, 0, and the exponents of the b are successively 0, 1, 2, 3, 4.

Now look at $(a + b)^5$ as an example. The coefficient of a^5 is 1; the coefficient of b^5 is 1. The degree of each term is 5. Proceeding left to right, the exponents of the a are successively 5, 4, 3, 2, 1, 0; the exponents of the b are successively 0, 1, 2, 3, 4, 5.

The patterns that we observe in the expansions of $(a + b)^4$ and $(a + b)^5$ lead us to expect similar patterns in the expansion of $(a + b)^n$ for any positive integer n. As part of the binomial theorem, we can prove by mathematical induction that in the expansion of $(a + b)^n$,

1. The coefficient of a^n is 1; the coefficient of b^n is 1.

2. The degree of each term is n.

3. The exponents of the *a* are successively n, $(n - 1)$, $(n - 2)$, . . . , 2, 1, 0.

4. The exponents of the *b* are successively 0, 1, 2, . . . , $(n - 2)$, $(n - 1)$, n.

Knowing this, to write out $(a + b)^n$ for any positive integer n, we only need a formula for computing the coefficients.

EXAMPLE 1

Leaving blank parentheses for the coefficients of other than a^6 and b^6, write out $(a + b)^6$.

Solution

The coefficients of a^6 and b^6 are both 1. The degree of each term is 6; the exponents of *a* are successively 6, 5, 4, 3, 2, 1, 0; the exponents of *b* are successively 0, 1, 2, 3, 4, 5, 6.

$$(a + b)^6 = a^6 + (\ )a^5b + (\ )a^4b^2 + (\ )a^3b^3 + (\ )a^2b^4 + (\ )ab^5 + b^6$$

One way to compute the coefficients of $(a + b)^n$ for positive integer n was observed by the French mathematician Blaise Pascal in the seventeenth century and continues to be useful for small values of n. His method becomes apparent when the coefficients of $(a + b)^n$ for the first few positive integers n are written in an array called *Pascal's triangle.* Pascal's triangle for $n = 1, 2, 3, 4, 5$ is shown in Figure 4.12. Each row consists of the coefficients in the expansion of the expression written in the left margin. Except for the ones, each entry is the sum of the entry just to its left and the entry just to its right in the preceding row. For example, the 6 in the row of coefficients of $(a + b)^4$ is $3 + 3$, and the first 5 in the row of coefficients of $(a + b)^5$ is $1 + 4$. Notice that the array of coefficients in any row is symmetric in that the second one in from the right in each row is the same as the second one in from the left in that row, etc.

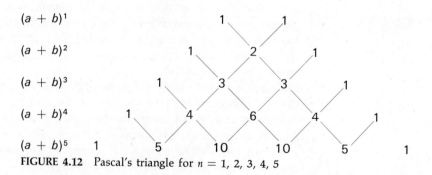

$(a + b)^1$

$(a + b)^2$

$(a + b)^3$

$(a + b)^4$

$(a + b)^5$

FIGURE 4.12 Pascal's triangle for $n = 1, 2, 3, 4, 5$

EXAMPLE 2

Use Pascal's triangle to find the coefficients and expand $(a + b)^6$.

Solution

Starting with the fifth row of Pascal's triangle as written in Figure 4.12, we can compute the sixth row.

$$1 \quad 5 \quad 10 \quad 10 \quad 5 \quad 1$$
$$6 \quad 15 \quad 20 \quad 15 \quad 6$$

Knowing the pattern of the terms, we can now write the expansion.

Exercise 4.8, 1–10

$$(a + b)^6 = a^6 + 6a^5b + 15a^4b^2 + 20a^3b^3 + 15a^2b^4 + 6ab^5 + b^6$$

Pascal's triangle is impractical to use for large values of n, so we are led to consider the binomial theorem. We shall see that according to the binomial theorem, we can find the coefficient of any term in the expansion of $(a + b)^n$ by considering the preceding term in *that* expansion rather than considering two coefficients in the expansion of $(a + b)^{n-1}$, as one has to do in Pascal's triangle. In fact, using the binomial theorem, we shall see that the following rule is true.

Binomial Coefficient Rule

To find the coefficient of any term after the first in the expansion $(a + b)^n$,

1. Consider the *preceding* term; multiply its coefficient by the exponent of a.

2. Divide this product by the number of the *preceding* term.

Let us check this rule in the expansion of $(a + b)^4$.

Number of term 1   2   3   4   5

$$1 \cdot a^4 + 4a^3b + 6a^2b^2 + 4a^1b^3 + 1 \cdot b^4$$

$$\frac{1 \times 4}{1} \quad \frac{4 \times 3}{2} \quad \frac{6 \times 2}{3} \quad \frac{4 \times 1}{4}$$

The rule checks. The student should verify that the rule also checks in the expansion of $(a + b)^n$ for 1, 2, 3, 5, and 6.

As a preliminary to stating the binomial theorem itself, Table 4.1 shows how the coefficients are formed when the binomial coefficient rule is applied to $(a + b)^n$ for any positive integer n. Note that in each term the exponent of b is one less than the number of the term. Where r is less than n, the chart gives a general formula for the term numbered $r + 1$. That general formula is based on the rth term and follows the patterns that may be observed in the formation of the first four terms. We shall see that the binomial theorem simply states that the formula shown in the chart is correct.

The entries in Table 4.1 can be written more concisely by employing *factorial* notation. For any positive integer n, the symbol $n!$, read "n factorial" represents the product $1 \times 2 \times 3 \times \cdots \times (n - 1) \times n$.

$$n! = 1 \times 2 \times 3 \times \cdots \times (n - 1) \times n$$
$$5! = 1 \times 2 \times 3 \times 4 \times 5$$

TABLE 4.1 FORMATION OF COEFFICIENTS OF $(a + b)^n$

Number of term	Coefficient	Term
1	1	$1 \times a^n$
2	$\dfrac{1 \times n}{1}$	$\dfrac{n}{1} a^{n-1} b$
3	$\dfrac{n}{1}(n - 1) \div 2$	$\dfrac{n(n - 1)}{1 \times 2} a^{n-2} b^2$
4	$\dfrac{n(n - 1)}{1 \times 2}(n - 2) \div 3$	$\dfrac{n(n - 1)(n - 2)}{1 \times 2 \times 3} a^{n-3} b^3$
$\vdots$	$\vdots$	$\vdots$
$r + 1$	$\dfrac{n(n - 1)(n - 2) \cdots (n - (r - 1))}{1 \times 2 \times 3 \cdots \times (r - 1)} \div r$	$\dfrac{n(n - 1)(n - 2) \cdots (n - (r - 1))}{1 \times 2 \times 3 \times \cdots \times (r - 1) \times r} a^{n-r} b^r$
$\vdots$	$\vdots$	$\vdots$
$n - 1$	n	nab^{n-1}
n	1	$1 \times b^n$

Note that $(n - 1)! \times n = n!$. Note also that

$$1! = 1$$

and

$$2! = 1 \times 2 = 2.$$

By definition,

$$0! = 1.$$

Using this notation in Table 4.1, we see that the denominator of the coefficient of term number 3 is 2!. The denominator of the coefficient of term number 4 is 3!, and the denominator of the coefficient of term number $r + 1$ is $r!$.

Let us now formally state the binomial theorem.

Binomial Theorem   If a and b represent real numbers, if n is a positive integer, and if positive integer r is less than n, then $(a + b)^n$ is the sum of $(n + 1)$ terms as follows:

$$(a + b)^n = a^n + na^{n-1}b + \frac{n(n - 1)}{2!} a^{n-2} b^2$$

$$+ \frac{n(n - 1)(n - 2)}{3!} a^{n-3} b^3 + \cdots$$

$$+ \frac{n(n - 1)(n - 2) \cdots (n - (r - 1))}{r!} a^{n-r} b^r + \cdots$$

$$+ \frac{n(n - 1)(n - 2) \cdots 1}{(n - 1)!} ab^{n-1} + b^n.$$

(The next to the last term can also be written nab^{n-1} because

$$\frac{n(n-1)(n-2)\cdots 1}{(n-1)!} = \frac{n \cdot (n-1)!}{(n-1)!} = n.)$$

Let us take a second look at the statement of this theorem. It does say that the coefficients are formed according to the rule used in Table 4.1. In fact, the binomial theorem states that in the expansion of $(a + b)^n$,

the coefficient of $a^{n-r}b^r$ is $\overbrace{\dfrac{n(n-1)(n-2)\cdots(n-(r-1))}{r!}}^{r \text{ factors}}$.

Note that this is the coefficient of term number $r + 1$; note that the exponent of the b is one less than the number of the term; note that there are r factors starting with the number n in the numerator of this coefficient, and r factorial is the denominator.

Exercise 4.8, 11–20

The binomial theorem is proved by mathematical induction. The proof will not be included here.

The symbol $\binom{n}{r}$ is commonly used for the coefficient of $a^{n-r}b^r$ in the expansion of $(a + b)^n$. The number $\binom{n}{r}$ is called the **binomial coefficient** of $a^{n-r}b^r$ in $(a + b)^n$. It is read "n over r."

$$\binom{n}{r} = \text{coefficient of } a^{n-r}b^r \text{ in } (a + b)^n$$

$$\binom{n}{r} = \frac{n(n-1)(n-2)\cdots(n-(r-1))}{r!} \quad \begin{array}{l} r \text{ factors} \\ \text{starting with } n \\ \\ r \text{ factorial} \end{array}$$

For example, the expansion of $(a + b)^7$, the coefficient of a^4b^3 is $\binom{7}{3}$.

$$\binom{7}{3} = \text{coefficient of } a^4b^3 \text{ in } (a + b)^7$$

$$\binom{7}{3} = \frac{7 \times 6 \times 5}{1 \times 2 \times 3} \quad \begin{array}{l} 3 \text{ factors} \\ \text{starting with } 7 \\ \\ 3 \text{ factorial} \end{array}$$

Using this notation, in $(a + b)^n$,

the coefficient of $a^{n-1}b$ is $\binom{n}{1} = \dfrac{n}{1}$, $\begin{array}{l} 1 \text{ factor} \\ \\ 1 \text{ factorial} \end{array}$

and   the coefficient of b^n is $\binom{n}{n} = \dfrac{\overset{\overset{\displaystyle n \text{ factors}}{\downarrow}}{n(n-1)(n-2)\cdots 1}}{\underset{\underset{\displaystyle n \text{ factorial}}{\uparrow}}{n!}}$

$$= \frac{n!}{n!}$$

$$= 1.$$

$$\binom{n}{n} = 1$$

Since the coefficient of $a^n = a^n b^0$ is also 1, for consistency we define $\binom{n}{0}$ to be 1.

$$\binom{n}{0} = 1$$

EXAMPLE 3

Find the value of the binomial coefficient $\binom{9}{6}$, and write the term having $\binom{9}{6}$ for its coefficient in the expansion of $(x + y)^9$.

Solution

$$\binom{9}{6} = \frac{9 \times 8 \times 7 \times 6 \times 5 \times 4}{1 \times 2 \times 3 \times 4 \times 5 \times 6}$$

$$= 84$$

$$\binom{9}{6} x^3 y^6 = 84 x^3 y^6$$

EXAMPLE 4

Use the notation $\binom{n}{r}$ for the coefficient of $a^{n-r} b^r$ in $(a + b)^n$, and write the first three rows of Pascal's triangle.

Solution

$n = 1$:
$$\binom{1}{0} \quad \binom{1}{1}$$

$n = 2$:
$$\binom{2}{0} \quad \binom{2}{1} \quad \binom{2}{2}$$

$n = 3$:
$$\binom{3}{0} \quad \binom{3}{1} \quad \binom{3}{2} \quad \binom{3}{3}$$

EXAMPLE 5   Show that the binomial coefficient $\binom{n}{r}$ can also be written in the form

$$\frac{n!}{r!\,(n-r)!}.$$

Solution $$\binom{n}{r} = \frac{n(n-1)\cdots(n-(r-1))}{r!}$$

Multiplying numerator and denominator by $(n-r)!$, we have

$$\binom{n}{r} = \frac{n(n-1)\cdots(n-r+1)(n-r)!}{r!\,(n-r)!}.$$

Since $(n-r)! = (n-r)(n-r-1)\cdots 1$, the numerator is simply $n!$. It follows that

Exercise 4.8, 21–55

$$\binom{n}{r} = \frac{n!}{r!\,(n-r)!}.$$

In calculus it is shown that for *any* number n (not just for *positive integer n*), and for real number x such that $|x| < 1$, the expression $(1 + x)^n$ can be represented by an infinite series. It is shown that the terms of the series representing $(1 + x)^n$ are formed according to the same rule as that given by the binomial theorem. This is illustrated for $(1 + x)^{1/2}$ in the next example.

EXAMPLE 6   Assume that there is an infinite series representing $(1 + x)^{1/2}$ and that the terms of the series are formed according to the rule given in the binomial theorem. Find the first four terms.

Solution   According to the binomial theorem,

$$(a + b)^n = a^n + na^{n-1}b + \frac{n(n-1)}{2!}a^{n-2}b^2$$

$$+ \frac{n(n-1)(n-2)}{3!}a^{n-3}b^3 + \cdots$$

If we take $a = 1$, $b = x$, and $n = \frac{1}{2}$, we have

$$(1 + x)^{1/2} = 1^{1/2} + \tfrac{1}{2}(1)^{(1/2)-1}x + \frac{\frac{1}{2}(\frac{1}{2} - 1)}{2!}(1)^{(1/2)-2}x^2$$

$$+ \frac{\frac{1}{2}(\frac{1}{2} - 1)(\frac{1}{2} - 2)}{3!}(1)^{(1/2)-3}x^3 + \cdots$$

Simplifying, we have

$$(1 + x)^{1/2} = 1 + \frac{1}{2}x + \frac{1}{2}\left(\frac{-1}{2}\right)\left(\frac{1}{2}\right)x^2 + \left(\frac{1}{2}\right)\left(\frac{-1}{2}\right)\left(\frac{-3}{2}\right)\left(\frac{1}{2 \times 3}\right)x^3 + \cdots$$

$$= 1 + \frac{x}{2} - \frac{x^2}{2^3} + \frac{x^3}{2^4} + \cdots .$$

Exercise 4.8, 56–60

EXERCISE 4.8 BINOMIAL THEOREM

▶ Use Pascal's triangle to find the coefficients and expand each of the following.

EXAMPLES

(a) $(1 + x)^7$ (b) $(x - y)^4$

Solutions

(a) Sixth row: 1 6 15 20 15 6 1

Seventh row: 1 7 21 35 35 21 7 1

Since 1^7, 1^6, 1^5, 1^4, 1^3, and 1^2 are all equal to 1,

$(1 + x)^7 = 1 + 7x + 21x^2 + 35x^3 + 35x^4 + 21x^5 + 7x^6 + x^7.$

(b) Fourth row: 1 4 6 4 1

$(x - y)^4 = x^4 + 4x^3(-y) + 6x^2(-y)^2 + 4x(-y)^3 + (-y)^4$

$= x^4 - 4x^3y + 6x^2y^2 - 4xy^3 + y^4$

1. $(1 + x)^8$ **2.** $(x - y)^6$ **3.** $(x + 2y)^4$ **4.** $(a + b)^9$

5. $(1 + x)^6$ **6.** $(a - b)^{10}$ **7.** $(x - y)^5$ **8.** $(x - 1)^7$

9. $(2a + 3b)^4$ **10.** $(a - 2b)^5$

▶ Use the binomial theorem and write out the first four terms in the expansion of each of the following.

EXAMPLES

(a) $(a + b)^{12}$ (b) $(x - 2y)^5$

Solutions

(a) $(a + b)^{12} = a^{12} + 12a^{11}b + \dfrac{12 \cdot 11}{2!}a^{10}b^2 + \dfrac{12 \cdot 11 \cdot 10}{3!}a^9b^3 + \cdots$

$= a^{12} + 12a^{11}b + 66a^{10}b^2 + 220a^9b^3 + \cdots$

(b) $(x - 2y)^5 = x^5 + 5x^4(-2y) + \dfrac{5 \cdot 4}{2!}x^3(-2y)^2$

$+ \dfrac{5 \cdot 4 \cdot 3}{3!}x^2(-2y)^3 + \cdots$

$= x^5 - 10x^4y + 40x^3y^2 - 80x^2y^3 + \cdots$

11. $(a + b)^{20}$ **12.** $(a - b)^{20}$ **13.** $(1 + x)^{12}$ **14.** $(1 - x)^{12}$

15. $(3x - y)^6$ **16.** $(2x + y)^5$ **17.** $(x - y)^{10}$ **18.** $(1 + 2x)^7$

19. $(x + y)^{30}$ **20.** $(x - y)^{18}$

▶ Find the numerical value of each of the given binomial coefficients, and write the term having that binomial coefficient in the expansion indicated.

EXAMPLE $\binom{8}{3}$, $(x + y)^8$

Solution $\binom{8}{3} = \dfrac{8 \times 7 \times 6}{1 \times 2 \times 3} = 56$, $\binom{8}{3}x^5y^3 = 56x^5y^3$

21. $\binom{8}{5}$, $(x + y)^8$ 22. $\binom{9}{4}$, $(a + b)^9$

23. $\binom{10}{6}$, $(x + y)^{10}$ 24. $\binom{11}{4}$, $(x - y)^{11}$

25. $\binom{30}{2}$, $(x + 2y)^{30}$ 26. $\binom{15}{15}$, $(a + b)^{15}$

27. $\binom{12}{10}$, $(a - b)^{12}$ 28. $\binom{23}{1}$, $(x + y)^{23}$

29. $\binom{16}{4}$, $(x - y)^{16}$ 30. $\binom{25}{24}$, $(2x + y)^{25}$

▶ Find the coefficient of each of the following in the expansion indicated.

EXAMPLE x^7y^3, $(x - 2y)^{10}$

Solution $\binom{10}{3}(x)^7(-2y)^3 = \dfrac{10 \times 9 \times 8}{1 \times 2 \times 3}x^7(-8y^3)$

$\qquad\qquad\qquad = -960x^7y^3$

31. x^2y^5, $(3x + 2y)^7$ 32. x^5y^5, $(2x - y)^{10}$

33. a^9b^2, $(a - 5b)^{11}$ 34. a^3b^9, $(3a + b)^{12}$

35. x^7y^8, $(2x - y)^{15}$

▶ Use the notation $\binom{n}{r}$ for the coefficient of $a^{n-r}b^r$ in $(a + b)^n$, and write the indicated row of Pascal's triangle.

EXAMPLE Row 4

Solution $\binom{4}{0}$ $\binom{4}{1}$ $\binom{4}{2}$ $\binom{4}{3}$ $\binom{4}{4}$

36. Row 5   37. Row 6   38. Row 7   39. Row 8   40. Row 9

▶ Verify each of the following by finding the values of the binomial coefficients.

EXAMPLE

$$\binom{3}{1} = \binom{2}{0} + \binom{2}{1}$$

Solution

$$\binom{3}{1} = \frac{3}{1} = 3, \quad \binom{2}{0} = 1, \quad \binom{2}{1} = \frac{2}{1} = 2$$

$$3 = 1 + 2$$

41. $\binom{4}{1} = \binom{3}{0} + \binom{3}{1}$ 42. $\binom{6}{1} = \binom{5}{0} + \binom{5}{1}$ 43. $\binom{6}{2} = \binom{5}{1} + \binom{5}{2}$

44. $\binom{6}{4} = \binom{5}{3} + \binom{5}{4}$ 45. $\binom{7}{1} = \binom{6}{0} + \binom{6}{1}$ 46. $\binom{7}{6} = \binom{6}{5} + \binom{6}{6}$

47. $\binom{4}{1} = \binom{4}{3}$ 48. $\binom{5}{2} = \binom{5}{3}$ 49. $\binom{6}{1} = \binom{6}{5}$

50. $\binom{8}{3} = \binom{8}{5}$

▶ For the given n and r, show that $n(n-1)\cdots(n-r+1)(n-r)!$ is $n!$.

EXAMPLE

$n = 8, r = 3$

Solution

$8 \times 7 \times 6 \times 5! = 8 \times 7 \times 6 \times 5 \times 4 \times 3 \times 2 \times 1 = 8!$

51. $n = 9, r = 3$ 52. $n = 7, r = 2$ 53. $n = 10, r = 4$

54. $n = 8, r = 4$ 55. $n = 11, r = 4$

▶ Assume that there is an infinite series representing each of the following and that the terms of the series are formed according to the rule given in the binomial theorem. Find the first four terms.

EXAMPLE

$(1 + x)^{-2}$

Solution

$$(1 + x)^{-2} = 1 + (-2)(1)^{-3}x + \frac{(-2)(-3)}{2!}(1)^{-4}x^2$$

$$+ \frac{(-2)(-3)(-4)}{3!}(1)^{-5}x^3 + \cdots$$

$$= 1 - 2x + 3x^2 - 4x^3 + \cdots$$

56. $(1 + x)^{-1}$ 57. $(1 + x)^{1/3}$ 58. $(1 + x)^{-1/2}$

59. $(1 + x)^{3/2}$ 60. $(1 + x)^{-3}$

CHECK LIST FOR THE STUDENT

If you have learned the material in Section 4.8, you should be able to do the following:

1. Give the coefficients of a^n and b^n, and give the degree of each term in the expansion $(a + b)^n$, for any positive integer n.

2. Proceeding left to right, tell what the exponents of the a and of the b are in successive terms of the expansion of $(a + b)^n$, for any positive integer n.

3. Show how to form the rows of Pascal's triangle.

4. Use Pascal's triangle and expand $(a + b)^n$ for some given positive integers n.

5. Use factorial notation. Give the values of 1! and 0!. Show that

$$\frac{n!}{(n - 1)!} = n.$$

6. State the binomial theorem.

7. Use the binomial theorem and expand $(a + b)^n$ for some given positive integers n.

8. Give the coefficient of $a^{n-r}b^r$ in the expansion of $(a + b)^n$.

9. Give the number of the term having $a^{n-r}b^r$ as a factor in the expansion of $(a + b)^n$.

10. Tell what coefficient it is that $\binom{n}{r}$ represents in $(a + b)^n$.

11. Find the value of $\binom{n}{r}$ for given n and r.

12. Give the values of $\binom{n}{1}$, $\binom{n}{n}$, and $\binom{n}{0}$, for any positive integer n.

13. Use the notation $\binom{n}{r}$ and write the first few rows of Pascal's triangle.

14. Show that

$$\binom{n}{r} = \frac{n(n - 1) \cdots (n - r + 1)}{r!}$$

can also be written

$$\frac{n!}{r!(n - r)!}.$$

15. Assume there is an infinite series representing $(1 + x)^r$ for a given real number r, and use the rule given in the binomial theorem to find the first few terms.

1. Leave blank parentheses for the coefficients of other than a^7 and b^7, and write out $(a + b)^7$.

2. Use Pascal's triangle to find the coefficients and expand $(x + y)^5$.

3. Use the binomial theorem and write out the first four terms in the expansion of $(a + b)^{10}$.

4. Find the numerical value of the binomial coefficient $\binom{12}{4}$, and write the term having that binomial coefficient in the expansion of $(x + y)^{12}$.

5. Find the coefficient of $x^2 y^4$ in the expansion of $(2x + y)^6$.

Part 4 Final Test

1. Use the remainder theorem to determine whether or not 4 is a root of the equation $x^3 - 4x^2 - 2x + 8 = 0$.

2. Use the factor theorem to construct an equation with the lowest possible degree having the roots 2, -3, 0, and -1.

3. Consider the equation $12x^4 - 5x^3 - 26x^2 + 10x + 4 = 0$.
   (a) What is the greatest number of distinct complex-number roots that this equation may have?
   (b) Find all the rational roots.

4. If $2 + 3i$ is a root of polynomial equation $P(x) = 0$, what is an irreducible quadratic factor of $P(x)$?

5. Use the location-of-roots theorem to show that there is at least one root of $x^3 - 6x^2 + 6x + 8 = 0$ between 2 and 3.

If possible, solve each of the following systems with the Gaussian elimination method. If there is not exactly one solution, say so.

6. $3x - 2y + 5z = -7$
   $2x + y - z = 6$
   $x - 4y + z = 3$

7. $5x - 3y = 10$
   $10x - 6y = 5$

8. Find the value of $\begin{vmatrix} 2 & 5 \\ -3 & 4 \end{vmatrix}$.

9. Use expansion by minors of elements in the first row and find the value of
$$\begin{vmatrix} 2 & 1 & 0 \\ 4 & 2 & -6 \\ -1 & 3 & 0 \end{vmatrix}.$$

10. Use Cramer's rule to find the solution of the following system.
$$x + y - z = -1$$
$$-x - y + 3z = 3$$
$$2x + y + z = 1$$

11. Express the following as a sum of partial fractions.
$$\frac{x^2 - 9x + 20}{x(x - 1)(x + 5)}$$

12. Write out the form of the partial-fraction decomposition of the following.
$$\frac{3x^2 - 7x + 2}{(x + 1)^2(x^2 + 5)^2}$$

Sketch the graph of each of the following.

13. $x^2 + y^2 \leq 9$

14. $2x + y + 1 \leq 0$

Sketch the graph of the solution of each of the following systems.

15. $y \geq x^2$ **16.** $y \leq x + 2$

      $y < 2$ $x + y - 5 \geq 0$

                                  $x \leq 4$

17. Write out the first four terms and the general term a_n of $\{(\tfrac{1}{5})^n\}$, and give $\lim\limits_{n\to\infty} a_n$.

Classify each of the following as either geometric or arithmetic and write out a formula for the sum of the first n terms.

18. $7, 7^2, 7^3, \ldots, 7^n, \ldots$

19. $7, 14, 21, \ldots, 7n, \ldots$

20. (a) Write out $\displaystyle\sum_{n=1}^{5} \frac{n}{n + 3}$.

     (b) Write out the first four terms and write the nth term of $\displaystyle\sum_{n=1}^{\infty} n^3$.

21. (a) A substance decays at such a rate that the amount of decay during the nth hour is $(\tfrac{1}{2})^n$ for any positive integer n. Write an expression for the sum of the total amount decayed during the first n hours.

     (b) State whether or not the series $1 + \tfrac{1}{5} + \tfrac{1}{25} + \cdots + (\tfrac{1}{5})^{n-1} + \cdots$ converges. If it converges, find its sum.

Use mathematical induction to prove that each of the following is true for all the prescribed positive integers n.

22. $4 + 32 + \cdots + 4n^3 = n^2(n + 1)^2, \quad n \geq 1$

23. $5^n > n + 5, \quad n \geq 2$

24. (a) Leave blank parentheses for the coefficients of other than a^8 and b^8, and write out $(a + b)^8$.

     (b) Use Pascal's triangle in determining the coefficients and expand $(x + 2y)^4$.

25. (a) Use the binomial theorem and write out the first four terms in the expansion $(a + b)^9$.

     (b) Find the numerical value of the binomial coefficient $\dbinom{15}{3}$, and write the term having that coefficient in the expansion of $(x + y)^{15}$.

Appendixes

PROPERTIES OF THE REAL NUMBERS

The following properties are true for all real numbers a, b, and c.

1. $a + b$ is a real number   Closure law for addition
   ab is a real number   Closure law for multiplication

2. $a + b = b + a$ Commutative law for addition
   $ab = ba$ Commutative law for multiplication

3. $(a + b) + c = a + (b + c) = a + b + c$ Associative law for addition
   $(ab)c = a(bc) = abc$ Associative law for multiplication

4. $a(b + c) = ab + ac$ Distributive law

5. $\left. \begin{array}{l} a + 0 = a \\ 0 + a = a \end{array} \right\}$ 0 is the additive identity

6. $\left. \begin{array}{l} a \cdot 0 = 0 \\ 0 \cdot a = 0 \end{array} \right\}$

7. $\left. \begin{array}{l} a \cdot 1 = a \\ 1 \cdot a = a \end{array} \right\}$ 1 is the multiplicative identity

8. $\left. \begin{array}{l} a + (-a) = 0 \\ (-a) + a = 0 \end{array} \right\}$ $-a$ is the additive inverse of a

9. $\left. \begin{array}{l} a \cdot \left(\dfrac{1}{a}\right) = 1 \\[2ex] \left(\dfrac{1}{a}\right) \cdot a = 1 \end{array} \right\}$ provided $a \neq 0$ $\dfrac{1}{a}$ is the multiplicative inverse of a

TABLE 1 | COMMON LOGARITHMS

x	0	1	2	3	4	5	6	7	8	9
1.0	.0000	.0043	.0086	.0128	.0170	.0212	.0253	.0294	.0334	.0374
1.1	.0414	.0453	.0492	.0531	.0569	.0607	.0645	.0682	.0719	.0755
1.2	.0792	.0828	.0864	.0899	.0934	.0969	.1004	.1038	.1072	.1106
1.3	.1139	.1173	.1206	.1239	.1271	.1303	.1335	.1367	.1399	.1430
1.4	.1461	.1492	.1523	.1553	.1584	.1614	.1644	.1673	.1703	.1732
1.5	.1761	.1790	.1818	.1847	.1875	.1903	.1931	.1959	.1987	.2014
1.6	.2041	.2068	.2095	.2122	.2148	.2175	.2201	.2227	.2253	.2279
1.7	.2304	.2330	.2355	.2380	.2405	.2430	.2455	.2480	.2504	.2529
1.8	.2553	.2577	.2601	.2625	.2648	.2672	.2695	.2718	.2742	.2765
1.9	.2788	.2810	.2833	.2856	.2878	.2900	.2923	.2945	.2967	.2989
2.0	.3010	.3032	.3054	.3075	.3096	.3118	.3139	.3160	.3181	.3201
2.1	.3222	.3243	.3263	.3284	.3304	.3324	.3345	.3365	.3385	.3404
2.2	.3424	.3444	.3464	.3483	.3502	.3522	.3541	.3560	.3579	.3598
2.3	.3617	.3636	.3655	.3674	.3692	.3711	.3729	.3747	.3766	.3784
2.4	.3802	.3820	.3838	.3856	.3874	.3892	.3909	.3927	.3945	.3962
2.5	.3979	.3997	.4014	.4031	.4048	.4065	.4082	.4099	.4116	.4133
2.6	.4150	.4166	.4183	.4200	.4216	.4232	.4249	.4265	.4281	.4298
2.7	.4314	.4330	.4346	.4362	.4378	.4393	.4409	.4425	.4440	.4456
2.8	.4472	.4487	.4502	.4518	.4533	.4548	.4564	.4579	.4594	.4609
2.9	.4624	.4639	.4654	.4669	.4683	.4698	.4713	.4728	.4742	.4757
3.0	.4771	.4786	.4800	.4814	.4829	.4843	.4857	.4871	.4886	.4900
3.1	.4914	.4928	.4942	.4955	.4969	.4983	.4997	.5011	.5024	.5038
3.2	.5051	.5065	.5079	.5092	.5105	.5119	.5132	.5145	.5159	.5172
3.3	.5185	.5198	.5211	.5224	.5237	.5250	.5263	.5276	.5289	.5302
3.4	.5315	.5328	.5340	.5353	.5366	.5378	.5391	.5403	.5416	.5428
3.5	.5441	.5453	.5465	.5478	.5490	.5502	.5514	.5527	.5539	.5551
3.6	.5563	.5575	.5587	.5599	.5611	.5623	.5635	.5647	.5658	.5670
3.7	.5682	.5694	.5705	.5717	.5729	.5740	.5752	.5763	.5775	.5786
3.8	.5798	.5809	.5821	.5832	.5843	.5855	.5866	.5877	.5888	.5899
3.9	.5911	.5922	.5933	.5944	.5955	.5966	.5977	.5988	.5999	.6010
4.0	.6021	.6031	.6042	.6053	.6064	.6075	.6085	.6096	.6107	.6117
4.1	.6128	.6138	.6149	.6160	.6170	.6180	.6191	.6201	.6212	.6222
4.2	.6232	.6243	.6253	.6263	.6274	.6284	.6294	.6304	.6314	.6325
4.3	.6335	.6345	.6355	.6365	.6375	.6385	.6395	.6405	.6415	.6425
4.4	.6435	.6444	.6454	.6464	.6474	.6484	.6493	.6503	.6513	.6522
4.5	.6532	.6542	.6551	.6561	.6571	.6580	.6590	.6599	.6609	.6618
4.6	.6628	.6637	.6646	.6656	.6665	.6675	.6684	.6693	.6702	.6712
4.7	.6721	.6730	.6739	.6749	.6758	.6767	.6776	.6785	.6794	.6803
4.8	.6812	.6821	.6830	.6839	.6848	.6857	.6866	.6875	.6884	.6893
4.9	.6902	.6911	.6920	.6928	.6937	.6946	.6955	.6964	.6972	.6981
5.0	.6990	.6998	.7007	.7016	.7024	.7033	.7042	.7050	.7059	.7067
5.1	.7076	.7084	.7093	.7101	.7110	.7118	.7126	.7135	.7143	.7152
5.2	.7160	.7168	.7177	.7185	.7193	.7202	.7210	.7218	.7226	.7235
5.3	.7243	.7251	.7259	.7267	.7275	.7284	.7292	.7300	.7308	.7316
5.4	.7324	.7332	.7340	.7348	.7356	.7364	.7372	.7380	.7388	.7396
x	0	1	2	3	4	5	6	7	8	9

TABLE 1 (*continued*)

x	0	1	2	3	4	5	6	7	8	9
5.5	.7404	.7412	.7419	.7427	.7435	.7443	.7451	.7459	.7466	.7474
5.6	.7482	.7490	.7497	.7505	.7513	.7520	.7528	.7536	.7543	.7551
5.7	.7559	.7566	.7574	.7582	.7589	.7597	.7604	.7612	.7619	.7627
5.8	.7634	.7642	.7649	.7657	.7664	.7672	.7679	.7686	.7694	.7701
5.9	.7709	.7716	.7723	.7731	.7738	.7745	.7752	.7760	.7767	.7774
6.0	.7782	.7789	.7796	.7803	.7810	.7818	.7825	.7832	.7839	.7846
6.1	.7853	.7860	.7868	.7875	.7882	.7889	.7896	.7903	.7910	.7917
6.2	.7924	.7931	.7938	.7945	.7952	.7959	.7966	.7973	.7980	.7987
6.3	.7993	.8000	.8007	.8014	.8021	.8028	.8035	.8041	.8048	.8055
6.4	.8062	.8069	.8075	.8082	.8089	.8096	.8102	.8109	.8116	.8122
6.5	.8129	.8136	.8142	.8149	.8156	.8162	.8169	.8176	.8182	.8189
6.6	.8195	.8202	.8209	.8215	.8222	.8228	.8235	.8241	.8248	.8254
6.7	.8261	.8267	.8274	.8280	.8287	.8293	.8299	.8306	.8312	.8319
6.8	.8325	.8331	.8338	.8344	.8351	.8357	.8363	.8370	.8376	.8382
6.9	.8388	.8395	.8401	.8407	.8414	.8420	.8426	.8432	.8439	.8445
7.0	.8451	.8457	.8463	.8470	.8476	.8482	.8488	.8494	.8500	.8506
7.1	.8513	.8519	.8525	.8531	.8537	.8543	.8549	.8555	.8561	.8567
7.2	.8573	.8579	.8585	.8591	.8597	.8603	.8609	.8615	.8621	.8627
7.3	.8633	.8639	.8645	.8651	.8657	.8663	.8669	.8675	.8681	.8686
7.4	.8692	.8698	.8704	.8710	.8716	.8722	.8727	.8733	.8739	.8745
7.5	.8751	.8756	.8762	.8768	.8774	.8779	.8785	.8791	.8797	.8802
7.6	.8808	.8814	.8820	.8825	.8831	.8837	.8842	.8848	.8854	.8859
7.7	.8865	.8871	.8876	.8882	.8887	.8893	.8899	.8904	.8910	.8915
7.8	.8921	.8927	.8932	.8938	.8943	.8949	.8954	.8960	.8965	.8971
7.9	.8976	.8982	.8987	.8993	.8998	.9004	.9009	.9015	.9020	.9025
8.0	.9031	.9036	.9042	.9047	.9053	.9058	.9063	.9069	.9074	.9079
8.1	.9085	.9090	.9096	.9101	.9106	.9112	.9117	.9122	.9128	.9133
8.2	.9138	.9143	.9149	.9154	.9159	.9165	.9170	.9175	.9180	.9186
8.3	.9191	.9196	.9201	.9206	.9212	.9217	.9222	.9227	.9232	.9238
8.4	.9243	.9248	.9253	.9258	.9263	.9269	.9274	.9279	.9284	.9289
8.5	.9294	.9299	.9304	.9309	.9315	.9320	.9325	.9330	.9335	.9340
8.6	.9345	.9350	.9355	.9360	.9365	.9370	.9375	.9380	.9385	.9390
8.7	.9395	.9400	.9405	.9410	.9415	.9420	.9425	.9430	.9435	.9440
8.8	.9445	.9450	.9455	.9460	.9465	.9469	.9474	.9479	.9484	.9489
8.9	.9494	.9499	.9504	.9509	.9513	.9518	.9523	.9528	.9533	.9538
9.0	.9542	.9547	.9552	.9557	.9562	.9566	.9571	.9576	.9581	.9586
9.1	.9590	.9595	.9600	.9605	.9609	.9614	.9619	.9624	.9628	.9633
9.2	.9638	.9643	.9647	.9652	.9657	.9661	.9666	.9671	.9675	.9680
9.3	.9685	.9689	.9694	.9699	.9703	.9708	.9713	.9717	.9722	.9727
9.4	.9731	.9736	.9741	.9745	.9750	.9754	.9759	.9763	.9768	.9773
9.5	.9777	.9782	.9786	.9791	.9795	.9800	.9805	.9809	.9814	.9818
9.6	.9823	.9827	.9832	.9836	.9841	.9845	.9850	.9854	.9859	.9863
9.7	.9868	.9872	.9877	.9881	.9886	.9890	.9894	.9899	.9903	.9908
9.8	.9912	.9917	.9921	.9926	.9930	.9934	.9939	.9943	.9948	.9952
9.9	.9956	.9961	.9965	.9969	.9974	.9978	.9983	.9987	.9991	.9996
x	0	1	2	3	4	5	6	7	8	9

TABLE 1 COMMON LOGARITHMS

A3

TABLE 2 | NATURAL LOGARITHMS

n	$\log_e n$	n	$\log_e n$	n	$\log_e n$
	*	4.5	1.5041	9.0	2.1972
0.1	7.6974	4.6	1.5261	9.1	2.2083
0.2	8.3906	4.7	1.5476	9.2	2.2192
0.3	8.7960	4.8	1.5686	9.3	2.2300
0.4	9.0837	4.9	1.5892	9.4	2.2407
0.5	9.3069	5.0	1.6094	9.5	2.2513
0.6	9.4892	5.1	1.6292	9.6	2.2618
0.7	9.6433	5.2	1.6487	9.7	2.2721
0.8	9.7769	5.3	1.6677	9.8	2.2824
0.9	9.8946	5.4	1.6864	9.9	2.2925
1.0	0.0000	5.5	1.7047	10	2.3026
1.1	0.0953	5.6	1.7228	11	2.3979
1.2	0.1823	5.7	1.7405	12	2.4849
1.3	0.2624	5.8	1.7579	13	2.5649
1.4	0.3365	5.9	1.7750	14	2.6391
1.5	0.4055	6.0	1.7918	15	2.7081
1.6	0.4700	6.1	1.8083	16	2.7726
1.7	0.5306	6.2	1.8245	17	2.8332
1.8	0.5878	6.3	1.8405	18	2.8904
1.9	0.6419	6.4	1.8563	19	2.9444
2.0	0.6931	6.5	1.8718	20	2.9957
2.1	0.7419	6.6	1.8871	25	3.2189
2.2	0.7885	6.7	1.9021	30	3.4012
2.3	0.8329	6.8	1.9169	35	3.5553
2.4	0.8755	6.9	1.9315	40	3.6889
2.5	0.9163	7.0	1.9459	45	3.8067
2.6	0.9555	7.1	1.9601	50	3.9120
2.7	0.9933	7.2	1.9741	55	4.0073
2.8	1.0296	7.3	1.9879	60	4.0943
2.9	1.0647	7.4	2.0015	65	4.1744
3.0	1.0986	7.5	2.0149	70	4.2485
3.1	1.1314	7.6	2.0281	75	4.3175
3.2	1.1632	7.7	2.0412	80	4.3820
3.3	1.1939	7.8	2.0541	85	4.4427
3.4	1.2238	7.9	2.0669	90	4.4998
3.5	1.2528	8.0	2.0794	100	4.6052
3.6	1.2809	8.1	2.0919	110	4.7005
3.7	1.3083	8.2	2.1041	120	4.7875
3.8	1.3350	8.3	2.1163	130	4.8676
3.9	1.3610	8.4	2.1282	140	4.9416
4.0	1.3863	8.5	2.1401	150	5.0106
4.1	1.4110	8.6	2.1518	160	5.0752
4.2	1.4351	8.7	2.1633	170	5.1358
4.3	1.4586	8.8	2.1748	180	5.1930
4.4	1.4816	8.9	2.1861	190	5.2470

*Subtract 10 for $n < 1$. Thus $\log_e 0.1 = 7.6974 - 10 = -2.3026$.

TABLE 3 | VALUES OF THE TRIGONOMETRIC FUNCTIONS TO FOUR PLACES

Real number x or θ radians	θ degrees	sin x or sin θ	tan x or tan θ	cot x or cot θ	cos x or cos θ		
.0000	0°00′	.0000	.0000	—	1.0000	90°00′	1.5708
.0029	10′	.0029	.0029	343.8	1.0000	89°50′	1.5679
.0058	20′	.0058	.0058	171.9	1.0000	40′	1.5650
.0087	30′	.0087	.0087	114.6	1.0000	30′	1.5621
.0116	40′	.0116	.0116	85.94	.9999	20′	1.5592
.0145	50′	.0145	.0145	68.75	.9999	10′	1.5563
.0175	1°00′	.0175	.0175	57.29	.9998	89°00′	1.5533
.0204	10′	.0204	.0204	49.10	.9998	88°50′	1.5504
.0233	20′	.0233	.0233	42.96	.9997	40′	1.5475
.0262	30′	.0262	.0262	38.19	.9997	30′	1.5446
.0291	40′	.0291	.0291	34.37	.9996	20′	1.5417
.0320	50′	.0320	.0320	31.24	.9995	10′	1.5388
.0349	2°00′	.0349	.0349	28.64	.9994	88°00′	1.5359
.0378	10′	.0378	.0378	26.43	.9993	87°50′	1.5330
.0407	20′	.0407	.0407	25.54	.9992	40′	1.5301
.0436	30′	.0436	.0437	22.90	.9990	30′	1.5272
.0465	40′	.0465	.0466	21.47	.9989	20′	1.5243
.0495	50′	.0494	.0495	20.21	.9988	10′	1.5213
.0524	3°00′	.0523	.0524	19.08	.9986	87°00′	1.5184
.0553	10′	.0552	.0553	18.07	.9985	86°50′	1.5155
.0582	20′	.0581	.0582	17.17	.9983	40′	1.5126
.0611	30′	.0610	.0612	16.35	.9981	30′	1.5097
.0640	40′	.0640	.0641	15.60	.9980	20′	1.5068
.0669	50′	.0669	.0670	14.92	.9978	10′	1.5039
.0698	4°00′	.0698	.0699	14.30	.9976	86°00′	1.5010
.0727	10′	.0727	.0729	13.73	.9974	85°50′	1.4981
.0756	20′	.0756	.0758	13.20	.9971	40′	1.4952
.0785	30′	.0785	.0787	12.71	.9969	30′	1.4923
.0814	40′	.0814	.0816	12.25	.9967	20′	1.4893
.0844	50′	.0843	.0846	11.83	.9964	10′	1.4864
.0873	5°00′	.0872	.0875	11.43	.9962	85°00′	1.4835
.0902	10′	.0901	.0904	11.06	.9959	84°50′	1.4806
.0931	20′	.0929	.0934	10.71	.9957	40′	1.4777
.0960	30′	.0958	.0963	10.39	.9954	30′	1.4748
.0989	40′	.0987	.0992	10.08	.9951	20′	1.4719
.1018	50′	.1016	.1022	9.788	.9948	10′	1.4690
.1047	6°00′	.1045	.1051	9.514	.9945	84°00′	1.4661
.1076	10′	.1074	.1080	9.255	.9942	83°50′	1.4632
.1105	20′	.1103	.1110	9.010	.9939	40′	1.4603
.1134	30′	.1132	.1139	8.777	.9936	30′	1.4573
.1164	40′	.1161	.1169	8.556	.9932	20′	1.4544
.1193	50′	.1190	.1198	8.345	.9929	10′	1.4515
.1222	7°00′	.1219	.1228	8.144	.9925	83°00′	1.4486
		cos x or cos θ	cot x or cot θ	tan x or tan θ	sin x or sin θ	θ degrees	Real number x or θ radians

TABLE 3 VALUES OF THE TRIGONOMETRIC FUNCTIONS TO FOUR PLACES   A5

TABLE 3 (*continued*)

Real number x or θ radians	θ degrees	$\sin x$ or $\sin \theta$	$\tan x$ or $\tan \theta$	$\cot x$ or $\cot \theta$	$\cos x$ or $\cos \theta$		
.1222	7°00′	.1219	.1228	8.144	.9925	83°00′	1.4486
.1251	10′	.1248	.1257	7.953	.9922	82°50′	1.4457
.1280	20′	.1276	.1287	7.770	.9918	40′	1.4428
.1309	30′	.1305	.1317	7.596	.9914	30′	1.4399
.1338	40′	.1334	.1346	7.429	.9911	20′	1.4370
.1376	50′	.1363	.1376	7.269	.9907	10′	1.4341
.1396	8°00′	.1392	.1405	7.115	.9903	82°00′	1.4312
.1425	10′	.1421	.1435	6.968	.9899	81°50′	1.4283
.1454	20′	.1449	.1465	6.827	.9894	40′	1.4254
.1484	30′	.1478	.1495	6.691	.9890	30′	1.4224
.1513	40′	.1507	.1524	6.561	.9886	20′	1.4195
.1542	50′	.1536	.1554	6.435	.9881	10′	1.4166
.1571	9°00′	.1564	.1584	6.314	.9877	81°00′	1.4137
.1600	10′	.1583	.1614	6.197	.9872	80°50′	1.4108
.1629	20′	.1622	.1644	6.084	.9868	40′	1.4079
.1658	30′	.1650	.1673	5.976	.9863	30′	1.4050
.1687	40′	.1679	.1703	5.871	.9858	20′	1.4021
.1716	50′	.1708	.1733	5.769	.9853	10′	1.3992
.1745	10°00′	.1736	.1763	5.671	.9848	80°00′	1.3963
.1774	10′	.1765	.1793	5.576	.9843	79°50′	1.3934
.1804	20′	.1794	.1823	5.485	.9838	40′	1.3904
.1833	30′	.1822	.1853	5.396	.9833	30′	1.3875
.1862	40′	.1851	.1883	5.309	.9827	20′	1.3846
.1891	50′	.1880	.1914	5.226	.9822	10′	1.3817
.1920	11°00′	.1908	.1944	5.145	.9816	79°00′	1.3788
.1949	10′	.1937	.1974	5.066	.9811	78°50′	1.3759
.1978	20′	.1965	.2004	4.989	.9805	40′	1.3730
.2007	30′	.1994	.2035	4.915	.9799	30′	1.3701
.2036	40′	.2022	.2065	4.843	.9793	20′	1.3672
.2065	50′	.2051	.2095	4.773	.9787	10′	1.3643
.2094	12°00′	.2079	.2126	4.705	.9781	78°00′	1.3614
.2123	10′	.2108	.2156	4.638	.9775	77°50′	1.3584
.2153	20′	.2136	.2186	4.574	.9769	40′	1.3555
.2182	30′	.2164	.2217	4.511	.9763	30′	1.3526
.2211	40′	.2193	.2247	4.449	.9757	20′	1.3497
.2240	50′	.2221	.2278	4.390	.9750	10′	1.3468
.2269	13°00′	.2250	.2309	4.331	.9744	77°00′	1.3439
.2298	10′	.2278	.2339	4.275	.9737	76°50′	1.3410
.2327	20′	.2306	.2370	4.219	.9730	40′	1.3381
.2356	30′	.2334	.2401	4.165	.9724	30′	1.3352
.2385	40′	.2363	.2432	4.113	.9717	20′	1.3323
.2414	50′	.2391	.2462	4.061	.9710	10′	1.3294
.2443	14°00′	.2419	.2493	4.011	.9703	76°00′	1.3265
		$\cos x$ or $\cos \theta$	$\cot x$ or $\cot \theta$	$\tan x$ or $\tan \theta$	$\sin x$ or $\sin \theta$	θ degrees	Real number x or θ radians

TABLE 3 (*continued*)

Real number x or θ radians	θ degrees	sin x or sin θ	tan x or tan θ	cot x or cot θ	cos x or cos θ		Real number x or θ radians
.2443	14°00′	.2419	.2493	4.011	.9703	76°00′	1.3265
.2473	10′	.2447	.2524	3.962	.9696	75°50′	1.3235
.2502	20′	.2476	.2555	3.914	.9689	40′	1.3206
.2531	30′	.2504	.2586	3.867	.9681	30′	1.3177
.2560	40′	.2532	.2617	3.821	.9674	20′	1.3148
.2589	50′	.2560	.2648	3.776	.9667	10′	1.3119
.2618	15°00′	.2588	.2679	3.732	.9659	75°00′	1.3090
.2647	10′	.2616	.2711	3.689	.9652	74°50′	1.3061
.2676	20′	.2644	.2742	3.647	.9644	40′	1.3032
.2705	30′	.2672	.2773	3.606	.9636	30′	1.3003
.2734	40′	.2700	.2805	3.566	.9628	20′	1.2974
.2763	50′	.2728	.2836	3.526	.9621	10′	1.2945
.2793	16°00′	.2756	.2867	3.487	.9613	74°00′	1.2915
.2822	10′	.2784	.2899	3.450	.9605	73°50′	1.2886
.2851	20′	.2812	.2931	3.412	.9596	40′	1.2857
.2880	30′	.2840	.2962	3.376	.9588	30′	1.2828
.2909	40′	.2868	.2994	3.340	.9580	20′	1.2799
.2938	50′	.2896	.3026	3.305	.9572	10′	1.2770
.2967	17°00′	.2824	.3057	3.271	.9563	73°00′	1.2741
.2996	10′	.2952	.3089	3.237	.9555	72°50′	1.2712
.3025	20′	.2979	.3121	3.204	.9546	40′	1.2683
.3054	30′	.3007	.3153	3.172	.9537	30′	1.2654
.3083	40′	.3035	.3185	3.140	.9528	20′	1.2625
.3113	50′	.3062	.3217	3.108	.9520	10′	1.2595
.3142	18°00′	.3090	.3249	3.078	.9511	72°00′	1.2566
.3171	10′	.3118	.3281	3.047	.9502	71°50′	1.2537
.3200	20′	.3145	.3314	3.018	.9492	40′	1.2508
.3229	30′	.3173	.3346	2.989	.9483	30′	1.2479
.3258	40′	.3201	.3378	2.960	.9474	20′	1.2450
.3287	50′	.3228	.3411	2.932	.9465	10′	1.2421
.3316	19°00′	.3256	.3443	2.904	.9455	71°00′	1.2392
.3345	10′	.3283	.3476	2.877	.9446	70°50′	1.2363
.3374	20′	.3311	.3508	2.850	.9436	40′	1.2334
.3403	30′	.3338	.3541	2.824	.9426	30′	1.2305
.3432	40′	.3365	.3574	2.798	.9417	20′	1.2275
.3462	50′	.3393	.3607	2.773	.9407	10′	1.2246
.3491	20°00′	.3420	.3640	2.747	.9397	70°00′	1.2217
.3520	10′	.3448	.3673	2.723	.9387	69°50′	1.2188
.3549	20′	.3475	.3706	2.699	.9377	40′	1.2159
.3578	30′	.3502	.3739	2.675	.9367	30′	1.2130
.3607	40′	.3529	.3772	2.651	.9356	20′	1.2101
.3636	50′	.3557	.3805	2.628	.9346	10′	1.2072
.3665	21°00′	.3584	.3839	2.605	.9336	69°00′	1.2043
		cos x or cos θ	cot x or cot θ	tan x or tan θ	sin x or sin θ	θ degrees	Real number x or θ radians

TABLE 3 (*continued*)

Real number x or θ radians	θ degrees	$\sin x$ or $\sin \theta$	$\tan x$ or $\tan \theta$	$\cot x$ or $\cot \theta$	$\cos x$ or $\cos \theta$		
.3665	21°00′	.3584	.3839	2.605	.9336	69°00′	1.2043
.3694	10′	.3611	.3872	2.583	.9325	68°50′	1.2014
.3723	20′	.3638	.3906	2.560	.9315	40′	1.1985
.3752	30′	.3665	.3939	2.539	.9304	30′	1.1956
.3782	40′	.3692	.3873	2.517	.9293	20′	1.1926
.3811	50′	.3719	.4006	2.496	.9283	10′	1.1897
.3840	22°00′	.3746	.4040	2.475	.9272	68°00′	1.1868
.3869	10′	.3773	.4074	2.455	.9264	67°50′	1.1839
.3898	20′	.3800	.4108	2.434	.9250	40′	1.1810
.3927	30′	.3827	.4142	2.414	.9239	30′	1.1781
.3956	40′	.3854	.4176	2.394	.9228	20′	1.1752
.3985	50′	.3881	.4210	2.375	.9216	10′	1.1723
.4014	23°00′	.3907	.4245	2.356	.9205	67°00′	1.1694
.4043	10′	.3934	.4279	2.337	.9194	66°50′	1.1665
.4072	20′	.3961	.4314	2.318	.9182	40′	1.1636
.4102	30′	.3987	.4348	2.300	.9171	30′	1.1606
.4131	40′	.4014	.4383	3.282	.9159	20′	1.1577
.4160	50′	.4041	.4417	2.264	.9147	10′	1.1548
.4189	24°00′	.4067	.4452	2.246	.9135	66°00′	1.1519
.4218	10′	.4094	.4487	2.229	.9124	65°50′	1.1490
.4247	20′	.4120	.4522	2.211	.9112	40′	1.1461
.4276	30′	.4147	.4557	2.194	.9100	30′	1.1432
.4305	40′	.4173	.4592	2.177	.9088	20′	1.1403
.4334	50′	.4200	.4628	2.161	.9075	10′	1.1374
.4363	25°00′	.4226	.4663	2.145	.9063	65°00′	1.1345
.4392	10′	.4253	.4699	2.128	.9051	64°50′	1.1316
.4422	20′	.4279	.4734	2.112	.9038	40′	1.1286
.4451	30′	.4305	.4770	2.097	.9026	30′	1.1257
.4480	40′	.4331	.4806	2.081	.9013	20′	1.1228
.4509	50′	.4358	.4841	2.066	.9001	10′	1.1199
.4538	26°00′	.4384	.4877	2.050	.8988	64°00′	1.1170
.4567	10′	.4410	.4913	2.035	.8975	63°50′	1.1141
.4596	20′	.4436	.4950	2.020	.8962	40′	1.1112
.4625	30′	.4462	.4986	2.006	.8949	30′	1.1083
.4654	40′	.4488	.5022	1.991	.8936	20′	1.1054
.4683	50′	.4514	.5059	1.977	.8923	10′	1.1025
.4712	27°00′	.4540	.5095	1.963	.8910	63°00′	1.0996
.4741	10′	.4566	.5132	1.949	.8897	62°50′	1.0966
.4771	20′	.4592	.5169	1.935	.8884	40′	1.0937
.4800	30′	.4617	.5206	1.921	.8870	30′	1.0908
.4829	40′	.4643	.5243	1.907	.8857	20′	1.0879
.4858	50′	.4669	.5280	1.894	.8843	10′	1.0850
.4887	28°00′	.4695	.5317	1.881	.8829	62°00′	1.0821
		$\cos x$ or $\cos \theta$	$\cot x$ or $\cot \theta$	$\tan x$ or $\tan \theta$	$\sin x$ or $\sin \theta$	θ degrees	Real number x or θ radians

TABLE 3 (*continued*)

Real number x or θ radians	θ degrees	sin x or sin θ	tan x or tan θ	cot x or cot θ	cos x or cos θ		
.4887	28°00'	.4695	.5317	1.881	.8829	62°00'	1.0821
.4916	10'	.4720	.5354	1.868	.8816	61°50'	1.0792
.4945	20'	.4746	.5392	1.855	.8802	40'	1.0763
.4874	30'	.4772	.5430	1.842	.8788	30'	1.0734
.5003	40'	.4797	.5467	1.829	.8774	20'	1.0705
.5032	50'	.4823	.5505	1.816	.8760	10'	1.0676
.5061	29°00'	.4848	.5543	1.804	.8746	61°00'	1.0647
.5091	10'	.4874	.5581	1.792	.8732	60°50'	1.0617
.5120	20'	.4899	.5619	1.780	.8718	40'	1.0588
.5149	30'	.4924	.5658	1.767	.8704	30'	1.0559
.5178	40'	.4950	.5696	1.756	.8689	20'	1.0530
.5207	50'	.4975	.5735	1.744	.8675	10'	1.0501
.5236	30°00'	.5000	.5774	1.732	.8660	60°00'	1.0472
.5265	10'	.5025	.5812	1.720	.8646	59°50'	1.0443
.5294	20'	.5050	.5851	1.709	.8631	40'	1.0414
.5323	30'	.5075	.5890	1.698	.8616	30'	1.0385
.5352	40'	.5100	.5930	1.686	.8601	20'	1.0356
.5381	50'	.5125	.5969	1.675	.8587	10'	1.0327
.5411	31°00'	.5150	.6009	1.664	.8572	59°00'	1.0297
.5440	10'	.5175	.6048	1.653	.8557	58°50'	1.0268
.5469	20'	.5200	.6088	1.643	.8542	40'	1.0239
.5498	30'	.5225	.6128	1.632	.8526	30'	1.0210
.5527	40'	.5250	.6168	1.621	.8511	20'	1.0181
.5556	50'	.5275	.6208	1.611	.8496	10'	1.0152
.5585	32°00'	.5299	.6249	1.600	.8480	58°00'	1.0123
.5614	10'	.5324	.6289	1.590	.8465	57°50'	1.0094
.5643	20'	.5348	.6330	1.580	.8450	40'	1.0065
.5672	30'	.5373	.6371	1.570	.8434	30'	1.0036
.5701	40'	.5398	.6412	1.560	.8418	20'	1.0007
.5730	50'	.5422	.6453	1.550	.8403	10'	.9977
.5760	33°00'	.5446	.6494	1.540	.8387	57°00'	.9948
.5789	10'	.5471	.6536	1.530	.8371	56°50'	.9919
.5818	20'	.5495	.6577	1.520	.8355	40'	.9890
.5847	30'	.5519	.6619	1.511	.8339	30'	.9861
.5876	40'	.5544	.6661	1.501	.8323	20'	.9832
.5905	50'	.5568	.6703	1.492	.8307	10'	.9803
.5934	34°00'	.5592	.6745	1.483	.8290	56°00'	.9774
.5963	10'	.5616	.6787	1.473	.8274	55°50'	.9745
.5992	20'	.5640	.6830	1.464	.8258	40'	.9716
.6021	30'	.5664	.6873	1.455	.8241	30'	.9687
.6050	40'	.5688	.6916	1.446	.8225	20'	.9657
.6080	50'	.5712	.6959	1.437	.8208	10'	.9628
.6109	35°00'	.5736	.7002	1.428	.8192	55°00'	.9599
		cos x or cos θ	cot x or cot θ	tan x or tan θ	sin x or sin θ	θ degrees	Real number x or θ radians

TABLE 3 VALUES OF THE TRIGONOMETRIC FUNCTIONS TO FOUR PLACES   A9

TABLE 3 (*continued*)

Real number x or θ radians	θ degrees	sin x or sin θ	tan x or tan θ	cot x or cot θ	cos x or cos θ		
.6109	35°00′	.5736	.7002	1.428	.8192	55°00′	.9599
.6138	10′	.5760	.7046	1.419	.8175	54°50′	.9570
.6167	20′	.5783	.7089	1.411	.8158	40′	.9541
.6196	30′	.5807	.7133	1.402	.8141	30′	.9512
.6225	40′	.5831	.7177	1.393	.8124	20′	.9483
.6254	50′	.5854	.7221	1.385	.8107	10′	.9454
.6283	36°00′	.5878	.7265	1.376	.8090	54°00′	.9425
.6312	10′	.5901	.7310	1.368	.8073	53°50′	.9396
.6341	20′	.5925	.7355	1.360	.8056	40′	.9367
.6370	30′	.5948	.7400	1.351	.8039	30′	.9338
.6440	40′	.5972	.7445	1.343	.8021	20′	.9308
.6429	50′	.5995	.7490	1.335	.8004	10′	.9279
.6458	37°00′	.6018	.7536	1.327	.7986	53°00′	.9250
.6487	10′	.6041	.7581	1.319	.7969	52°50′	.9221
.6516	20′	.6065	.7627	1.311	.7951	40′	.9192
.6545	30′	.6088	.7673	1.303	.7934	30′	.9163
.6574	40′	.6111	.7720	1.295	.7916	20′	.9134
.6603	50′	.6134	.7766	1.288	.7898	10′	.9105
.6632	38°00′	.6157	.7813	1.280	.7880	52°00′	.9076
.6661	10′	.6180	.7860	1.272	.7862	51°50′	.9047
.6690	20′	.6202	.7907	1.265	.7844	40′	.9018
.6720	30′	.6225	.7954	1.257	.7826	30′	.8988
.6749	40′	.6248	.8002	1.250	.7808	20′	.8959
.6778	50′	.6271	.8050	1.242	.7790	10′	.8930
.6807	39°00′	.6293	.8098	1.235	.7771	51°00′	.8901
.6836	10′	.6316	.8146	1.228	.7753	50°50′	.8872
.6865	20′	.6338	.8195	1.220	.7735	40′	.8843
.6894	30′	.6361	.8243	1.213	.7716	30′	.8814
.6923	40′	.6383	.8292	1.206	.7698	20′	.8785
.6952	50′	.6406	.8342	1.199	.7679	10′	.8756
.6981	40°00′	.6428	.8391	1.192	.7660	50°00′	.8727
.7010	10′	.6450	.8441	1.185	.7642	49°50′	.8698
.7039	20′	.6472	.8491	1.178	.7623	40′	.8668
.7069	30′	.6494	.8541	1.171	.7604	30′	.8639
.7098	40′	.6517	.8591	1.164	.7585	20′	.8610
.7127	50′	.6539	.8642	1.157	.7566	10′	.8581
.7156	41°00′	.6561	.8693	1.150	.7547	49°00′	.8552
.7185	10′	.6583	.8744	1.144	.7528	48°50′	.8523
.7214	20′	.6604	.8796	1.137	.7509	40′	.8494
.7243	30′	.6626	.8847	1.130	.7490	30′	.8465
.7272	40′	.6648	.8899	1.124	.7470	20′	.8436
.7301	50′	.6670	.8952	1.117	.7451	10′	.8407
.7330	42°00′	.6691	.9004	1.111	.7431	48°00′	.8378
		cos x or cos θ	cot x or cot θ	tan x or tan θ	sin x or sin θ	θ degrees	Real number x or θ radians

TABLE 3 (*continued*)

Real number x or θ radians	θ degrees	sin x or sin θ	tan x or tan θ	cot x or cot θ	cos x or cos θ		
.7330	42°00′	.6691	.9004	1.111	.7431	48°00′	.8378
.7359	10′	.6713	.9057	1.104	.7412	47°50′	.8348
.7389	20′	.6734	.9110	1.098	.7392	40′	.8319
.7418	30′	.6756	.9163	1.091	.7373	30′	.8290
.7447	40′	.6777	.9217	1.085	.7353	20′	.8261
.7476	50′	.6799	.9271	1.079	.7333	10′	.8232
.7505	43°00′	.6820	.9325	1.072	.7314	47°00′	.8203
.7534	10′	.6841	.9380	1.066	.7294	46°50′	.8174
.7563	20′	.6862	.9435	1.060	.7274	40′	.8145
.7592	30′	.6884	.9490	1.054	.7254	30′	.8116
.7621	40′	.6905	.9545	1.048	.7234	20′	.8087
.7650	50′	.6926	.9601	1.042	.7214	10′	.8058
.7679	44°00′	.6947	.9657	1.036	.7193	46°00′	.8029
.7709	10′	.6967	.9713	1.030	.7173	45°50′	.7999
.7738	20′	.6988	.9770	1.024	.7153	40′	.7970
.7767	30′	.7009	.9827	1.018	.7133	30′	.7941
.7796	40′	.7030	.9884	1.012	.7112	20′	.7912
.7825	50′	.7050	.9942	1.006	.7092	10′	.7883
.7854	45°00′	.7071	1.0000	1.000	.7071	45°00′	.7854
		cos x or cos θ	cot x or cot θ	tan x or tan θ	sin x or sin θ	θ degrees	Real number x or θ radians

Answers to Quizzes and Tests

Section Quiz 1.1

1. (a) Positive integers, integers, rational numbers, real numbers, complex numbers (b) Integers, rational numbers, real numbers, complex numbers   **2.** (a) Negative integers, integers, rational numbers, real numbers, complex numbers (b) Integers, rational, real, and complex numbers   **3.** (a) Positive integers, integers, rational numbers, real numbers, complex numbers (b) Rational numbers, real numbers, complex numbers
4. (a) Irrational numbers, real numbers, complex numbers (b) Complex numbers, imaginary numbers
5. -3, -2, 0, 2, $\frac{3}{5}$, 2.3, $2.\overline{3}$, 9, $\frac{22}{7}$, 3.14, $\frac{-2}{3}$, 1.732 **6.** $\sqrt{3}$, π **7.** (a) Irrational, nonterminating, and nonrepeating (b) Rational, repeating (c) Rational, terminating   **8.** (a) $\{x \mid x \geq 0\}$ ———————•———
0

(b) $\{x \mid x < 0\}$ ————————○———
0   **9.** (a) $\{x \mid x \leq 0\}$ ————————•———
0   (b) $\{x \mid x > 0\}$ ————————○———
0
10. -4

Section Quiz 1.2

1. (a) 125 (b) 5 (c) $\frac{1}{125}$ (d) 1 (e) 0 **2.** (a) 4 (b) -4 (c) 3 (d) -3 (e) -5 **3.** (a) 5 (b) 0 (c) 3
(d) 3 (e) $\frac{25}{9}$ **4.** (a) 5 (b) 0 (c) Not defined (d) Not defined (e) Not real   **5.** (a) 7 (b) 3 (c) 10
(d) $5, 5$ (e) $5, 5$ **6.** $\frac{y}{x^2}$, $x \neq 0$, $y \neq 0$ **7.** $\frac{1}{x^2 y^5}$, $x \neq 0$, $y \neq 0$ **8.** $\frac{y^2}{x^3}$, $x \neq 0$ **9.** $\frac{9x^6}{16y^8}$, $y \neq 0$
10. (a) 1 (b) $-i$

Section Quiz 1.3

1. (a) $\sqrt[4]{5^3}$, $(\sqrt[4]{5})^3$ (b) $5^{2/3}$ **2.** (a) 16 (b) $\frac{1}{16}$ (c) 16 (d) 8 (e) 0 **3.** (a) $\frac{y^2}{x^{2/3}}$ (b) $\frac{1}{x^{1/4} y^2}$ **4.** (a) 12
(b) $\frac{-3}{5}$ **5.** (a) $2xy^2$ (b) $15xy$ (c) $2y$

Section Quiz 1.4

1. (a) No (b) No (c) Yes (d) Yes (e) Yes   **2.** (a) 4 (b) $-5x^3 - 2x^2 - 3x + 7$ **3.** (a) -9 (b) 4
(c) -9 **4.** $2x^3 + 4x^2 - 9x - 13$ **5.** $2x^3 + 2x^2 - x + 1$ **6.** $2x^5 - 5x^4 - 31x^3 - 7x^2 + 59x + 42$
7. $2x + 11 + \frac{53x + 71}{x^2 - 4x - 7}$ **8.** $3x^2 + 6xh + 3h^2 - 2x - 2h - 5$ **9.** $2x^2 + 5x - \frac{6}{x - 1}$
10. (a) $5 - 2i$ (b) $-1 + 12i$ (c) $41 + i$

Section Quiz 1.5

1. (a) $x^2 - y^2$ (b) $x^2 + 2xy + y^2$ (c) $x^3 - y^3$ (d) $3x^2 - 19x - 14$ (e) $-6x^3 + 2x^2 - 2x$
2. $(3x - 2)^2$ **3.** $(3x + 2)(3x - 2)$ **4.** $(3x + 2)(9x^2 - 6x + 4)$ **5.** $(x + 3)^2$ **6.** Not factorable
7. $(3x + 4)(3x - 2)$ **8.** $(a + b)(2 - 3a)$ **9.** $2a^2 b(a^3 - 3ab + 7)$ **10.** 36, $(x - 6)^2$

Section Quiz 1.6

1. (a) -1 (b) $a^2 - 4a + 3$ **2.** $\dfrac{x + 3}{2(x + 2)}$ **3.** $(x - 5)^3(x + 5)^2(x + 1)$ **4.** $\dfrac{5x + 10}{7x^2}$ **5.** $\dfrac{x - 3}{7x + 7}$

6. $\dfrac{5x^3 + 12x^2 - 2x}{x^3 + 4x^2 + x - 6}$ **7.** $\dfrac{3x + 7}{x^2 + 2x - 3}$ **8.** $3x^4 - 2x^3$ **9.** $\dfrac{3x^2 - 5x}{7x^3 - 1}$ **10.** (a) $\dfrac{3(\sqrt{5} + 1)}{4}$

(b) $\dfrac{-29}{53} + \dfrac{31}{53}i$

Section Quiz 1.7

1. $\dfrac{9}{8}$ **2.** $\dfrac{-2}{13}$ **3.** $0, 1, -3$ **4.** $B = \dfrac{A - CA}{C}$ **5.** (a) 2 (b) 0   **6.** (a) 1 (b) 2   **7.** (a) 0, 3

(b) $5, -2$ **8.** $\dfrac{2 \pm \sqrt{5}}{3}$ **9.** $2, \dfrac{-1}{5}$ **10.** $\sqrt{9}\ \sqrt{(x - 1)^2 - 2^2},\ h = 1,\ k = 2$

Section Quiz 1.8

1. -3 **2.** $\dfrac{x^2}{2^2} + \dfrac{y^2}{1^2} = 1;\ a = 2,\ b = 1$ **3.** 3   **4.** $5, \dfrac{-1}{3}$
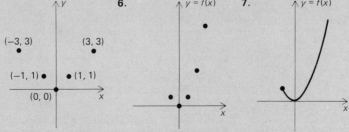

5. (a) $\pm \sqrt{5}$ (b) $-4, 3$

Section Quiz 1.9

1. $x < 9$ **2.** $x \leq -2$ **3.** $x \geq -2$

4. $-1 < x < 6$ **5.** $x \geq 3$ or $x \leq \dfrac{5}{3}$

Section Quiz 1.10

1. $x = 0, y = 2; x = \dfrac{-4}{3}, y = 0; x = 1, y = \dfrac{7}{2}; x = \dfrac{-2}{3}, y = 1$ **2.** $x = -2, y = -1$ **3.** $x = 3, y = 4$

4. $x = -1, y = 2, z = 3$ **5.** $A = 3, B = 4, C = 0$

Section Quiz 2.1

1. $\{12, 3, 4, 7\}$ **2.** $5, 3, a^2 - 3a + 5$ **3.** It is a function; domain of $f = \{\frac{1}{3}, 1, 0, 2\}; \frac{1}{3} \to -4, 1 \to -2,$
$0 \to -5, 2 \to -2$; range $= \{-4, -2, -5\}$ **4.** (a) $\{x \mid x \neq -5\}$ (b) $\{x \mid x \geq -5\}$

5. (a) $-, -$ (b)   **6.**   **7.**

8. (a) Function 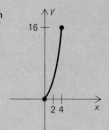 (b) Not a function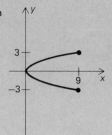

9. (a) Yes (b) No **10.** $2x + h - 5$

Section Quiz 2.2

1. 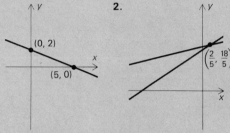 **2.** 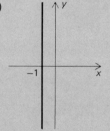 **3.** (a) 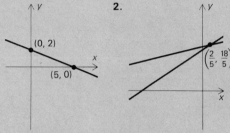 (b)

4. (a) $\dfrac{-7}{4}$ (b) $\dfrac{f(a + h) - f(a)}{h}$ **5.**  **6.** $2x - 5y - 11 = 0$ **7.** $y = \dfrac{3x}{2} + 5, \dfrac{3}{2}, 5$

8. (a) $\dfrac{2}{9}$ (b) $\dfrac{-9}{2}$ **9.** $4x - y - 7 = 0$ **10.** (a) (b) $\dfrac{x}{2} + \dfrac{5}{6}, \dfrac{5}{6}, \dfrac{4}{3}$

Section Quiz 2.3

1. $(0, 2)$, $(0, -2)$, $(5, 0)$, $(-5, 0)$ **2.** (a) Yes (b) No (c) Yes (d) No   **3.** (a) No (b) Yes (c) Yes (d) No
4. (a) No (b) No (c) Yes (d) Yes   **5.** (a) $x = 3$, $x = -3$ (b) No vertical asymptotes (c) $x = 9$
6. (a) $y = 0$ (b) $y = \frac{2}{5}$ (c) No horizontal asymptotes   **7.** $x > 4$ and $y < 0$, $-1 < x < 4$ and $y > 0$,
$-1 > x$ and $y < 0$ **8.** $y < -2$, $y > 2$, $x < -5$, $x > 5$ **9.** Intercepts: $(0, 5)$, $(0, -5)$, $(-2, 0)$, $(2, 0)$;
Symmetric with respect to the x-axis, the y-axis, and the origin; No asymptotes; Excluded areas are $y > 5$,
$y < -5$, $x > 2$, $x < -2$.

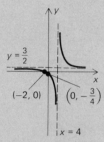

10. Intercepts: $(0, \frac{-3}{4})$, $(-2, 0)$; Asymptotes: $x = 4$, $y = \frac{3}{2}$; Excluded areas: $x > 4$ and $y \leq 0$, $-2 < x < 4$ and
$y \geq 0$, $-2 > x$ and $y \leq 0$, $y > \frac{3}{2}$ and $x \leq 0$, $\frac{-3}{4} < y < \frac{3}{2}$ and $x \geq 0$, $\frac{-3}{4} > y$ and $x \leq 0$; Not symmetric with
respect to the origin, the x-axis, or the y-axis.

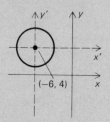

Section Quiz 2.4

1. (a) 5 (b) $(4, 3)$ **2.** $(x - 2)^2 + (y + 5)^2 = 49$ **3.** Center $(-4, 1)$, radius 6
4. (a) $(x - 7)^2 + (y + 2)^2 = 106$ (b) $(x - 1)^2 + (y + 2)^2 = -9$; no graph

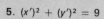

5. $(x')^2 + (y')^2 = 9$

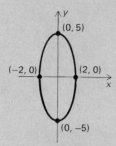

Section Quiz 2.5

1. $(0, 0)$, $x = 0$, $(0, 5)$, $y = -5$

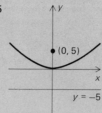
(0, 5)
$y = -5$

2. $(0, 0)$, $y = 0$, $(-5, 0)$, $x = 5$

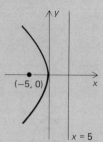
$(-5, 0)$
$x = 5$

3.

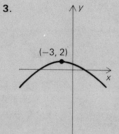
$(-3, 2)$

4.

$(2, -3)$

5. $(y - 3)^2 = 20x$

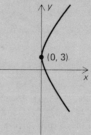
$(0, 3)$

6. $(x - 1)^2 = 8(y + 1)$

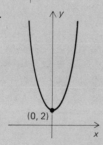
$(1, -1)$

7. (a) Parabola, up   (b) Parabola, down

8.

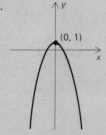
$(0, 1)$

9.

$(0, 2)$

10. 7 seconds

Section Quiz 2.6

1. Center (0, 0)

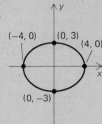

2. $\dfrac{x'^2}{9} + \dfrac{y'^2}{25} = 1$

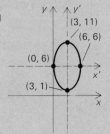

3. $\dfrac{(x-3)^2}{9} + \dfrac{(y+4)^2}{4} = 1$; ellipse   **4.** $\dfrac{(x-3)^2}{9} - \dfrac{(y+4)^2}{9} = 1$; hyperbola   **5.** Center (0, 0); vertices:

$(-5, 0)$, $(5, 0)$; asymptotes: $y = \pm \dfrac{3x}{5}$

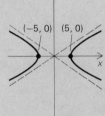

6. $\dfrac{(y')^2}{25} - \dfrac{(x')^2}{9} = 1$; asymptotes: $y' = \dfrac{-5}{3}x'$, $y' = \dfrac{5}{3}x'$; asymptotes: $y - 6 = \dfrac{5}{3}(x-3)$, $y - 6 = \dfrac{-5}{3}(x-3)$

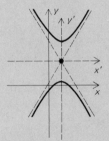

7. (a) Ellipse (b) Circle   **8.** (a) Hyperbola (b) Two intersecting lines   **9.** (a) No graph (b) Parabola
10. (a) Hyperbola (b) Parabola

Section Quiz 2.7

1. (a) Not a function   (b) Function

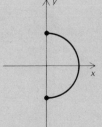
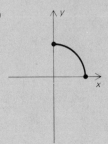

2. (a) $\{(x, y) \mid x = 2y + 7\}$, function   (b) $\{(x, y) \mid x = 8y^2\}$, not a function   **3.** (a) Yes   (b) No   **4.** (a) No

(b) Yes   (c) No   **5.** Domain of $f^{-1} = \{x \mid x \geq 3\}$, range of $f^{-1} = \{y \mid y \geq 0\}$, $f^{-1}(x) = \dfrac{x}{2} - \dfrac{3}{2}$

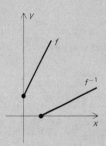

Section Quiz 2.8

1.

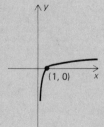

2.

3. Domain $= \{x \mid x > 0\}$, range $=$ set of all real numbers, $x = 3^y$

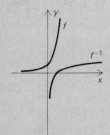

4. (a) 6   (b) 0   (c) 1   (d) -1 (e) -3

5.

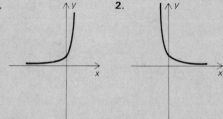

Section Quiz 2.9

1. (a) $2^b = a$ (b) $\log_3 y = 5$ (c) $3 = e^{\ln 3}$ (d) $\dfrac{\ln 37}{\ln 10}$ **2.** (a) 0 (b) 1 (c) 5 (e) 2 **3.** (a) 2.8451

(b) $0.8451 - 1$ **4.** (a) 1.5441 (b) 0.1461 **5.** (a) 2.5353 (b) 0.2330

Section Quiz 2.10

1. (a) $\{x | 0 \le x \le 3\}$ (b) $\{x | 0 \le x < 3\}$ **2.** (a) 33 (b) -29 (c) 62 **3.** (a) $\dfrac{2}{31}$ (b) $2\sqrt[3]{4}$ (c) 6

4. (a) $(f + g)(x) = \sqrt{x} + x^2 + 9$, $\{x | x \ge 0\}$ (b) $(f - g)(x) = \sqrt{x} - x^2 - 9$, $\{x | x \ge 0\}$

(c) $fg(x) = x^2\sqrt{x} + 9\sqrt{x}$, $\{x | x \ge 0\}$ **5.** (a) $\dfrac{f}{g}(x) = \dfrac{\sqrt{x}}{x^2 + 9}$, $\{x | x \ge 0\}$ (b) $f \circ g(x) = \sqrt{x^2 + 9}$, set of all real

numbers (c) $g \circ f(x) = x + 9$, set of all real numbers

Section Quiz 3.1

1. (a) $\dfrac{20\pi}{9}$ (b) $300°$ (c) $\dfrac{360°}{\pi}$

2. (a) $25°$, $\dfrac{5\pi}{36}$ radians

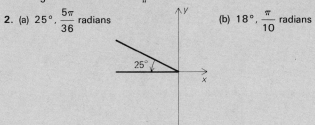

(b) $18°$, $\dfrac{\pi}{10}$ radians

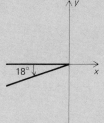

(c) $68°$, $\dfrac{17\pi}{45}$ radians

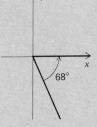

(d) $68°$, $\dfrac{17\pi}{45}$ radians

(e) $17°$, $\dfrac{17\pi}{180}$ radians

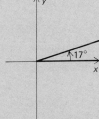

3. $\dfrac{28\pi}{9}$ cm **4.** 2 radians **5.** (a) $\dfrac{\pi}{2}$, $(0, 1)$ (b) $-540°$, $(-1, 0)$

Section Quiz 3.2

1. $\cos A = \dfrac{1}{3}$, $\sin A = \dfrac{-\sqrt{8}}{3}$, $\tan A = -\sqrt{8}$, $\cot A = \dfrac{-1}{\sqrt{8}}$, $\sec A = 3$, $\csc A = \dfrac{-3}{\sqrt{8}}$

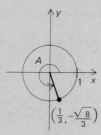

$\left(\dfrac{1}{3}, -\dfrac{\sqrt{8}}{3}\right)$

2. $\cos \alpha = \dfrac{-1}{\sqrt{5}}$, $\sin \alpha = \dfrac{2}{\sqrt{5}}$, $\tan \alpha = -2$, $\cot \alpha = \dfrac{-1}{2}$, $\sec \alpha = -\sqrt{5}$, $\csc \alpha = \dfrac{\sqrt{5}}{2}$

3. (a) $(0, 1)$ (b) $(-1, 0)$ (c) $(0, -1)$ (d) $(1, 0)$ (e) $(1, 0)$ (f) $(-1, 0)$ (g) $(0, 1)$ (h) $(0, -1)$

4. (a) 0 (b) 1 (c) -1 (d) 0 (e) 0 (f) -1 (g) -1 (h) 1 (i) 1

5. $\sin A = \dfrac{3}{5}$, $\cos A = \dfrac{4}{5}$, $\tan A = \dfrac{3}{4}$, $\cot A = \dfrac{4}{3}$, $\sec A = \dfrac{5}{4}$, $\csc A = \dfrac{5}{3}$, $\sin B = \dfrac{4}{5}$, $\cos B = \dfrac{3}{5}$

Section Quiz 3.3

1. (a) $\dfrac{\sqrt{3}}{2}$ (b) $\dfrac{\sqrt{3}}{2}$ (c) 1 (d) $\sqrt{2}$ (e) $\sqrt{3}$ (f) $\dfrac{2}{\sqrt{3}}$ 2. (a) $\dfrac{-\sqrt{3}}{2}$ (b) $\dfrac{\sqrt{2}}{2}$ (c) -1 (d) $\dfrac{\sqrt{3}}{3}$ (e) $-\sqrt{2}$
(f) -2 3. (a) 0.9397 (b) -0.6428 (c) 0.3420 (d) -0.8718 (e) 0.2532 (f) 0.9674 4. (a) $-40°$
(b) $\dfrac{8\pi}{9}$ radians 5. $14°$

Section Quiz 3.4

1. (a) $-\sin \theta$ (b) $\cos \theta$ (c) $\sin \theta$ (d) $\cos \theta$ (e) $\cos \theta$ (f) $\sin \theta$
2. Amplitude $= 5$, period $= \pi$

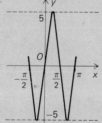

3. Amplitude $= 1$, period $= 4\pi$

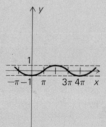

4. Amplitude $= 5$, period $= \pi$

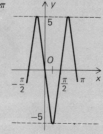

5. Amplitude $= 3$, period $= \dfrac{\pi}{2}$

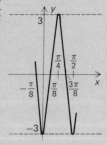

Section Quiz 3.5

1. (a) $-\tan \theta$ (b) $\tan \theta$ (c) $\cot \theta$; range is the set of all real numbers.

2.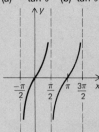

3. Range is the set of all real numbers.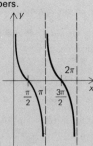

4. Range is $\{y \mid |y| \geq 1\}$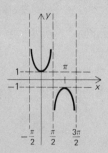

5. Range is $\{y \mid |y| \geq 1\}$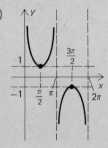

Section Quiz 3.6

1. $\sin \theta = \dfrac{5}{13}$, $\tan \theta = \dfrac{-5}{12}$, $\cot \theta = \dfrac{-12}{5}$, $\cos \theta = \dfrac{-12}{13}$, $\csc \theta = \dfrac{13}{5}$ **2.** $\cos \theta = \dfrac{-2\sqrt{6}}{5}$, $\tan \theta = \dfrac{-1}{2\sqrt{6}}$,

$\cot \theta = -2\sqrt{6}$, $\sec \theta = \dfrac{-5}{2\sqrt{6}}$, $\csc \theta = 5$ **3.** $\dfrac{\sqrt{2}(\sqrt{3} - 1)}{4}$ **4.** $\dfrac{-120}{169}$ **5.** $-\sqrt{\dfrac{3}{10}}$

6. $-\sqrt{\dfrac{3 - \sqrt{8}}{6}}$ **7.** $\dfrac{1}{2}\sqrt{2 - \sqrt{3}}$ **8.** $\dfrac{1}{\sin^2 \theta} + \dfrac{1}{\cos \theta}$

9. $\cos^2 \theta - \cot^2 \theta = \cos^2 \theta - \dfrac{\cos^2 \theta}{\sin^2 \theta}$

$= \dfrac{\sin^2 \theta \cos^2 \theta - \cos^2 \theta}{\sin^2 \theta}$

$= \dfrac{\cos^2 \theta(\sin^2 \theta - 1)}{\sin^2 \theta}$

$= \dfrac{-\cos^2 \theta(1 - \sin^2 \theta)}{\sin^2 \theta}$

$= \dfrac{-\cos^2 \theta \cdot \cos^2 \theta}{\sin^2 \theta} = -\cot^2 \theta \cos^2 \theta$

10.

$$\frac{\sin 2\theta \cos 2\theta}{\tan^2 \theta} = \frac{\sin 2\theta(\cos^2 \theta - \sin^2 \theta)}{\tan^2 \theta}$$

$$= \frac{\cos^2 \theta}{\sin^2 \theta} \cdot \sin 2\theta(\cos^2 \theta - \sin^2 \theta)$$

$$= \frac{\cos^4 \theta \sin 2\theta}{\sin^2 \theta} - \frac{\cos^2 \theta \sin 2\theta \sin^2 \theta}{\sin^2 \theta}$$

$$= \frac{\cos^4 \theta(2 \sin \theta \cos \theta)}{\sin^2 \theta} - \cos^2 \theta \sin 2\theta$$

$$= 2 \cos^4 \theta \frac{\cos \theta}{\sin \theta} - \cos^2 \theta \sin 2\theta$$

$$= 2 \cot \theta \cos^4 \theta - \sin 2\theta \cos^2 \theta$$

Section Quiz 3.7

1. $5\sqrt{6}$ **2.** (a) 0   (b) 1   (c) 2   (d) 1   (e) 1   **3.** $60°$ **4.** $2\sqrt{34 - 15\sqrt{3}}$ **5.** $\frac{3}{5}$

Section Quiz 3.8

1. (a) Range $= \{y | -1 \leq y \leq 1\}$ (b) Domain $= \{x | -1 \leq x \leq 1\}$, range $= \left\{y \left| \frac{-\pi}{2} \leq y \leq \frac{\pi}{2}\right.\right\}$

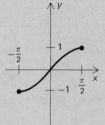
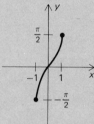

2. (a) Range $= \{y | -1 \leq y \leq 1\}$ (b) Domain $= \{x | -1 \leq x \leq 1\}$, range $= \{y | 0 \leq y \leq \pi\}$

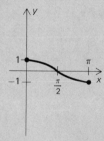
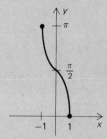

3. (a) Range is the set of all real numbers. (b) Domain is the set of all real numbers. Range $= \left\{ y \left| \dfrac{-\pi}{2} < y < \dfrac{\pi}{2} \right. \right\}$

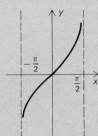

4. (a) $\dfrac{\pi}{3}$ (b) $\dfrac{-\pi}{2}$ (c) $\dfrac{\pi}{2}$ (d) 0 (e) $\dfrac{\pi}{4}$ (f) $\dfrac{-\pi}{4}$ **5.** (a) 0.78 (b) $\dfrac{4}{5}$

Section Quiz 3.9

1. (a) $\theta = \dfrac{2\pi}{3}$ (b) $\theta = \dfrac{2\pi}{3}, \dfrac{4\pi}{3}$; $\left\{ \theta \left| \theta = \dfrac{2\pi}{3} + 2k\pi \text{ or } \dfrac{4\pi}{3} + 2k\pi \right. \right\}$ **2.** $\theta = \dfrac{-\pi}{4}$; $\left\{ \theta \left| \theta = \dfrac{3\pi}{4} + k\pi \right. \right\}$

3. $\left\{ \theta \left| \theta = \dfrac{\pi}{2} + k\pi, \text{ or } \theta = \dfrac{\pi}{3} + 2k\pi, \text{ or } \theta = \dfrac{5\pi}{3} + 2k\pi \right. \right\}$ **4.** $\left\{ \theta \left| \theta = k\pi \text{ or } \theta = \dfrac{\pi}{4} + k\pi \right. \right\}$

5. $\left\{ \theta \left| \theta = \dfrac{\pi}{6} + 2k\pi, \text{ or } \theta = \dfrac{7\pi}{6} + 2k\pi, \text{ or } \theta = \dfrac{5\pi}{6} + 2k\pi, \text{ or } \theta = \dfrac{11\pi}{6} + 2k\pi \right. \right\}$

Section Quiz 3.10

1. $(-5, 220°), (-5, -140°), (5, -320°)$ **2.** $\left(\dfrac{3\sqrt{2}}{2}, \dfrac{-3\sqrt{2}}{2} \right)$ **3.** $(5\sqrt{2}, 135°)$

4. $r^2 - 6r \sin^2 \theta - 16 = 0$ **5.** $x^2 + y^2 = 4x - 6y$

6.   **7.**   **8.**

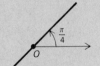

9.   **10.**

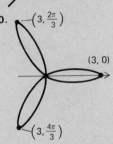

Section Quiz 3.11

1. $3y = x + 14$, Line **2.** $(x - 2)^2 + (y + 1)^2 = 49$, Circle with center at $(2, -1)$ and radius 7

3. $\dfrac{x^2}{4} + \dfrac{y^2}{25} = 1$, Ellipse **4.** $\dfrac{x^2}{9} - \dfrac{y^2}{4} = 1$, Hyperbola **5.** $\dfrac{(y - 1)^2}{4} - \dfrac{(x + 3)^2}{4} = 1$, Hyperbola

6. $4y - 8x - x^2 = 0$, Parabola **7.** They each represent the line $y = 2x - 3$.

8. Since t is squared, $x \geq 0$.   **9.** $y = \sqrt{3}x - \dfrac{gx^2}{1800}$, Parabola

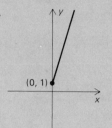

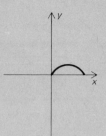

10. $(2\pi, 4)$

Section Quiz 4.1

1. 6   **2.** $x = \dfrac{-b \pm \sqrt{b^2 - 4ac}}{2a}$ **3.** $4x^3 - 3x^2 + 5x - 30 = (x - 2)(4x^2 + 5x + 15) + 0$, so 2 *is* a root.

4. $x(x - 1)(x - 3)(x + 4) = 0$ **5.** 0, multiplicity 2; 3, multiplicity 4; -1, multiplicity 1

6. $x^3(x + 4)^2(x - 6) = 0$ **7.** 4   **8.** 5   **9.** $(2 - i)^2 - 4(2 - i) + 5 = 4 - 4i - 1 - 8 + 4i + 5 = 0$

$(2 + i)^2 - 4(2 + i) + 5 = 4 + 4i - 1 - 8 - 4i + 5 = 0$ **10.** $(x - (3 - 4i))(x - (3 + 4i)) = x^2 - 6x + 25$

11. $1^3 - 5(1)^2 + 11(1) - 7 = 1 - 5 + 11 - 7 = 0$; $x^3 - 5x^2 + 11x - 7 = (x - 1)(x^2 - 4x + 7)$, so the

roots are 1, $2 + \sqrt{3}i$, $2 - \sqrt{3}i$.   **12.** ± 1, $\pm\tfrac{1}{7}$, $\pm\tfrac{3}{7}$, $\pm\tfrac{5}{7}$, $\pm\tfrac{15}{7}$, ± 3, ± 5, ± 15 **13.** 3, -2, $\tfrac{1}{2}$

14. 2, $3i$, $-3i$ **15.** Let $P(x) = 5x^3 + 2x^2 - 16x + 4 = 0$. $P(0) = 4$ (positive), and $P(1) = -5$ (negative),

so there is at least one root between 0 and 1.

Section Quiz 4.2

1. 3 rows; 4 columns   **2.** 9   **3.** 2, 7, 5

4. $\begin{pmatrix} 1 & 0 & 6 \\ 0 & 1 & 4 \\ 0 & 0 & 5 \end{pmatrix}$ **5.** $\begin{pmatrix} 4 & -3 & 7 \\ -4 & 4 & -6 \\ 5 & 9 & -8 \end{pmatrix}$ **6.** $\begin{pmatrix} 1 & -1 & 2 \\ 0 & 7 & 9 \\ 0 & 1 & 8 \end{pmatrix}$

7. $x = 2$, $y = -3$ **8.** $x = 1$, $y = -2$, $z = -3$ **9.** Not exactly one solution (no solution)

10. Not exactly one solution (unlimited number of solutions)

Section Quiz 4.3

1. Row 3, column 2   **2.** Row 2, column 3; minus position   **3.** $\begin{vmatrix} 3 & 4 \\ 0 & 8 \end{vmatrix}$ **4.** 31

5. $3(-7 - 30) - (-4 - 10) - 2(24 - 14) = 105$ **6.** (a) 0   (b) 0   **7.** (a) 1   (b) 2   **8.** (a) 1   (b) 1

9. $x = -2, y = -4$ **10.** $x = \dfrac{8}{7}, y = \dfrac{19}{7}, z = \dfrac{4}{7}$

Section Quiz 4.4

1. $\dfrac{A}{x - 4} + \dfrac{B}{x + 5} + \dfrac{C}{(x + 5)^2}$ **2.** $\dfrac{A}{x - 3} + \dfrac{Bx + C}{x^2 - 4x + 29}$ **3.** $\dfrac{A}{x - 3} + \dfrac{Bx + C}{x^2 - 4x + 29} + \dfrac{Dx + E}{(x^2 - 4x + 29)^2}$

4. $\dfrac{2}{x - 3} + \dfrac{3}{(x - 3)^2}$ **5.** $\dfrac{-3}{x} - \dfrac{1}{(x + 1)} + \dfrac{4}{x - 5}$

Section Quiz 4.5

1. 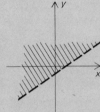   **2.** 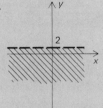   **3.** 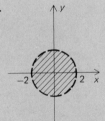   **4.**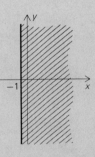

5. 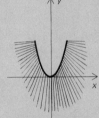   **6.**    **7.** 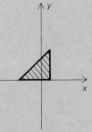   **8.**

9. 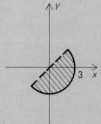   **10.** (a) $x \geq 2, y \geq 1, x + y \leq 20$ (b)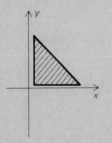

(c) $69, $443, $494   (d) 2 rows of lettuce, 18 rows of beans

Section Quiz 4.6

1. (a) $\{2n + 3\}$ (b) $1, 4, 7, \ldots, 3n - 2, \ldots$ **2.** $\dfrac{1}{3}, \dfrac{1}{9}, \dfrac{1}{27}, \dfrac{1}{81}; \dfrac{1}{3^n}; 0$ **3.** Arithmetic, 4

4. Geometric, 3 **5.** $\dfrac{5^{n+1} - 5}{4}$ **6.** $\dfrac{3n(n + 1)}{2}$ **7.** Converges, $\frac{3}{2}$ **8.** Does not converge

9. (a) $\dfrac{1}{3}, \dfrac{1}{2}, \dfrac{3}{5}, \dfrac{2}{3}; \dfrac{n}{n + 2}$ (b) $\dfrac{1}{3}, \dfrac{1}{3} + \dfrac{1}{2}, \dfrac{1}{3} + \dfrac{1}{2} + \dfrac{3}{5}$ **10.** (a) $\dfrac{1 - (0.1)^n}{9}$ (b) $\dfrac{1 - (0.1)^{10}}{9}$ (c) $\dfrac{1}{9}$

Section Quiz 4.7

1. Part 1: $5 = \frac{5}{4}(5^1 - 1)$ since $5 = \dfrac{5 \cdot 4}{4}$

Part 2: Suppose $5 + 5^2 + 5^3 + \cdots + 5^k = \frac{5}{4}(5^k - 1)$. Then

$$5 + 5^2 + 5^3 + \cdots + 5^k + 5^{k+1} = \tfrac{5}{4}(5^k - 1) + 5^{k+1}$$
$$= \dfrac{5(5^k - 1) + 4 \cdot 5^{k+1}}{4}$$
$$= \dfrac{5(5^k - 1 + 4 \cdot 5^k)}{4}$$
$$= \dfrac{5(5 \cdot 5^k - 1)}{4}$$
$$= \tfrac{5}{4}(5^{k+1} - 1)$$

Check: $S_n = \dfrac{5 - 5 \cdot 5^n}{1 - 5} = \dfrac{5(1 - 5^n)}{-4} = \frac{5}{4}(5^n - 1)$

2. Part 1: $3 = \frac{3}{2}(1 + 1^2)$ since $3 = \frac{3}{2} \cdot 2$

Part 2: Suppose $3 + 6 + 9 + \cdots + 3k = \frac{3}{2}(k + k^2)$. Then

$$3 + 6 + 9 + \cdots + 3k + 3(k + 1) = \tfrac{3}{2}(k + k^2) + 3(k + 1)$$
$$= \dfrac{3(k + k^2) + 6(k + 1)}{2}$$
$$= \dfrac{3k + 3k^2 + 6k + 6}{2}$$
$$= \dfrac{3(k^2 + 3k + 2)}{2}$$
$$= \dfrac{3((k^2 + 2k + 1) + (k + 1))}{2}$$
$$= \dfrac{3[(k + 1)^2 + (k + 1)]}{2}$$
$$= \dfrac{3[(k + 1) + (k + 1)^2]}{2}$$

Check: $l = 3 + (n - 1)3$

$$S_n = \dfrac{n(3 + 3 + (n - 1)3)}{2} = \dfrac{3(n + n^2)}{2}$$

3. Part 1: $2^3 > 3 + 2$ since $8 > 5$.

Part 2: $2^{k+1} = 2^k \cdot 2$
$$> (k + 2) \cdot 2$$
$$= 2k + 4$$
$$= k + k + 4$$
$$\geq k + 3 + 4$$
$$> k + 3$$
$$= (k + 1) + 2$$
$$2^{k+1} > (k + 1) + 2$$

4. Part 1: $3 = \dfrac{1(1 + 1)(2 \cdot 1 + 1)}{2}$ since $3 = \dfrac{1(2)(3)}{3}$

Part 2: If $3 + 12 + \cdots + 3k^2 = \dfrac{k(k + 1)(2k + 1)}{2}$, then

$$3 + 12 + \cdots + 3k^2 + 3(k + 1)^2 = \frac{k(k + 1)(2k + 1)}{2} + 3(k + 1)^2$$

$$= \frac{(k + 1)[k(2k + 1) + 6(k + 1)]}{2} = \frac{(k + 1)(2k^2 + 7k + 6)}{2}$$

$$= \frac{(k + 1)(k + 2)(2k + 3)}{2} = \frac{(k + 1)(k + 2)[2(k + 1) + 1]}{2}.$$

5. Part 1: 2 is a factor of $1^2 + 3 \cdot 1$ since 2 is a factor of 4

Part 2: $(k + 1)^2 + 3(k + 1) = k^2 + 5k + 4$
$$= k^2 + 5k + 4 + [k^2 + 3k - (k^2 + 3k)]$$
$$= 2k^2 + 8k + 4 - (k^2 + 3k)$$
$$= 2(k^2 + 4k + 2) - (k^2 + 3k)$$

2 is a factor of $2(k^2 + 4k + 2)$, so if 2 is a factor of $k^2 + 3k$, then 2 is a factor of $(k + 1)^2 + 3(k + 1)$.

Section Quiz 4.8

1. $a^7 + (\quad)a^6b + (\quad)a^5b^2 + (\quad)a^4b^3 + (\quad)a^3b^4 + (\quad)a^2b^5 + (\quad)ab^6 + b^7$
2. $x^5 + 5x^4y + 10x^3y^2 + 10x^2y^3 + 5xy^4 + y^5$ **3.** $a^{10} + 10a^9b + 45a^8b^2 + 120a^7b^3$ **4.** 495, $495x^8y^4$ **5.** 60

Part 1 Final Test

1. (a) $\{0, 1, 2, 3, \ldots\}$ (b) $\{x \mid x > 0\}$ ——$\circ$———————
   0

2. (a) $-5, 0, |-5|, 0.5, \dfrac{1}{5}, 2.\overline{51}, \dfrac{22}{7}, 3.14, \dfrac{-4}{5}$ (b) $\sqrt{5}, \pi, \sqrt{2}$ **3.** (a) 243   (b) 3   (c) 1   (d) $\dfrac{1}{243}$ (e) 0

4. (a) 3   (b) Not defined   (c) Not real   (d) 5   (e) $\dfrac{3}{2}$ **5.** (a) $\dfrac{x^6}{y^7}, y \neq 0$ (b) $\dfrac{9x^4}{y^4}, y \neq 0, x \neq 0$

6. (a) 16   (b) 5   (c) $\frac{1}{2}$ (d) 9   (e) 0   **7.** (a) y (b) xy **8.** (a) $2x^3y$ (b) $\dfrac{x^2}{2y^2}$

9. (a) $-2x^3 - 4x^2 + 3x - 4$ (b) $3x^3 - 23x^2 + 15x - 7$ **10.** $4x^2 - 14x + 43 - \dfrac{134}{x + 3}$

11. $5x^2 + 10xh + 5h^2 - 3x - 3h - 7$ **12.** (a) $(5x - 6)(5x + 6)$ (b) $(5x + 2)^2$
(c) $(3x - 2)(9x^2 + 6x + 4)$ (d) $(5x + 2)(5x - 4)$ **13.** 169; $(x - 13)^2$ **14.** (a) $(x - 2)(x + 3)$
(b) $\dfrac{3x - 6}{x^2 - 3x}$ **15.** (a) $\dfrac{-2x^2 - 2x - 3}{x(x - 1)(x + 3)}$ (b) $\dfrac{x^2 - 5x + 4}{x + 4}$ **16.** $\dfrac{3x^2 - 2x}{2 + x^3}$ **17.** (a) $x = \dfrac{7}{9}$ (b) $x = 0, -2, 5$

18. (a) $x = 0, \dfrac{5}{3}$ (b) $x = \dfrac{5 \pm \sqrt{2}}{3}$ **19.** $x = -2$ **20.** $x = 5$ **21.** (a) $x = 3, \dfrac{-3}{5}$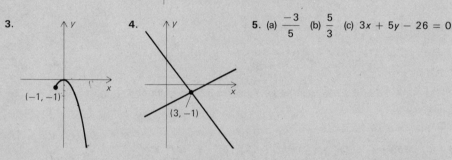

(b) $x = 0, 3, 1, 2$ **22.** $x \le \dfrac{-8}{9}$ **23.** $x > 6$

24. $\dfrac{-1}{8} \le x \le \dfrac{7}{8}$ **25.** $x = \dfrac{-1}{5}, y = 2$

Part 2 Final Test

1. (a) $\{1, 3, 4, 6\}$ (b)   **2.** (a) $\{x \mid x \ge 4\}$ (b) $\{x \mid x \ne 0\}$

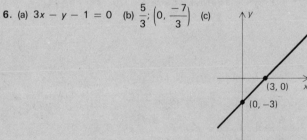

3.   **4.**   **5.** (a) $\dfrac{-3}{5}$ (b) $\dfrac{5}{3}$ (c) $3x + 5y - 26 = 0$

6. (a) $3x - y - 1 = 0$ (b) $\dfrac{5}{3}; \left(0, \dfrac{-7}{3}\right)$ (c)

7. (a) $(1, 0), (-1, 0)$; Symmetric with respect to the x-axis, the y-axis, and the origin; No vertical or horizontal asymptotes   (b) $(0, 0)$; Not symmetric with respect to the x-axis or the y-axis, symmetric with respect to the origin; Vertical asymptotes: $x = 2, x = -2$, horizontal asymptote: $y = 0$

8.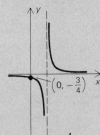

$x = 4$

9. (a) 5 (b) $\left(\dfrac{-1}{2}, 5\right)$ **10.** (a) $(x - 3)^2 + (y + 7)^2 = 36$ (b) Center $(5, -2)$; radius 9

11. $x'^2 + y'^2 = 4$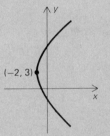

$(-4, -2)$

12. (a)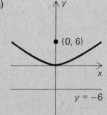

$(0, 6)$

$y = -6$

(b)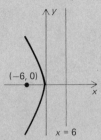

$(-6, 0)$

$x = 6$

13. $(y - 3)^2 = 24(x + 2)$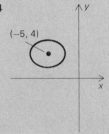

$(-2, 3)$

14.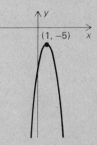

$(1, -5)$

15. Major axis 6, minor axis 4

$(-5, 4)$

16. Asymptotes $y = \pm \dfrac{3}{2}(x - 5)$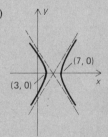

$(7, 0)$

$(3, 0)$

17. (a) Line (b) Parabola (c) No graph (d) Circle (e) Point   **18.** (a) Not one-to-one (b) One-to-one

19. Domain of $f^{-1} = \{x \mid -1 \le x \le 5\}$; $f^{-1}(x) = \dfrac{x + 1}{3}$

20.

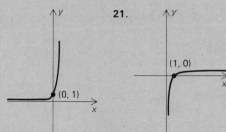

21.

22. (a) 0   (b) 1   (c) 9   (d) -4 (e) -2 **23.** (a) 2.4771   (b) $0.6021 - 3$ (c) 1.0792   (d) 0.1250
(e) 0.2007   (f) 1.9084   **24.** (a) $\sqrt{x-3}$ (b) $\sqrt{x}-3$ **25.** (a) $\{x\,|\,x \geq 0\}$ (b) $\{x\,|\,x \geq 0 \text{ and } x \neq 3\}$
(c) $\{x\,|\,x \geq 3\}$ (d) $\{x\,|\,x \geq 0\}$

Part 3 Final Test

1. $\dfrac{\pi}{8}$, $22.5°$ **2.** (a) 4π cm   (b) $\dfrac{720°}{\pi}$ **3.** (a) 5π, $(-1, 0)$ (b) $450°$, $(0, 1)$ **4.** $\cos \alpha = \dfrac{1}{\sqrt{10}}$,

$\sin \alpha = \dfrac{-3}{\sqrt{10}}$, $\tan \alpha = -3$, $\cot \alpha = \dfrac{-1}{3}$, $\sec \alpha = \sqrt{10}$, $\csc \alpha = \dfrac{-\sqrt{10}}{3}$ **5.** $\sin A = \dfrac{2}{\sqrt{13}}$, $\cos A = \dfrac{3}{\sqrt{13}}$,

$\tan A = \dfrac{2}{3}$, $\cot A = \dfrac{3}{2}$, $\sec A = \dfrac{\sqrt{13}}{3}$, $\csc A = \dfrac{\sqrt{13}}{2}$, $\sin B = \dfrac{3}{\sqrt{13}}$, $\tan B = \dfrac{3}{2}$ **6.** (a) $\dfrac{1}{2}$ (b) $-\sqrt{2}$ (c) $\dfrac{1}{\sqrt{3}}$

(d) 0   (e) 0   (f) $-\sqrt{2}$ **7.** $30°$ **8.** (a) Domain $= \{x\,|\,x \text{ is a real number}\}$, range $= \{y\,|\,{-3} \leq y \leq 3\}$,
period $= 2\pi$ (b) Domain $= \{x\,|\,x \neq k\pi,\ k \text{ an integer}\}$, range $= \{y\,|\,y \text{ is a real number}\}$, period $= \pi$
(c) Domain $= \{x\,|\,x \neq k\pi,\ k \text{ an integer}\}$, range $= \{y\,|\,y \geq 3 \text{ or } y \leq -3\}$, period $= 2\pi$

9. Amplitude $= \dfrac{1}{2}$, period $= \dfrac{\pi}{2}$

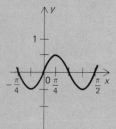

10. Amplitude $= 3$, period $= 4\pi$

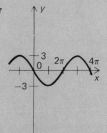

11. Range $= \{y\,|\,y \text{ is a real number}\}$

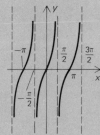

12. Range $= \{y\,|\,y \geq 1 \text{ or } y \leq -1\}$

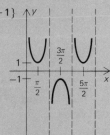

13. $\sin \theta = \dfrac{-12}{13}$, $\cos \theta = \dfrac{-15}{13}$, $\cot \theta = \dfrac{5}{12}$, $\sec \theta = \dfrac{-13}{5}$, $\csc \theta = \dfrac{-13}{12}$ **14.** (a) $\dfrac{\sqrt{6}-\sqrt{2}}{4}$ (b) $\dfrac{-24}{25}$

15. $\sin \theta = -\sqrt{\dfrac{5}{8}}$

16. $\cos^2 \theta - \sin^2 \theta \tan^2 \theta = (1 - \sin^2 \theta) - \sin^2 \theta \tan^2 \theta$

$$= 1 - \sin^2 \theta(1 + \tan^2 \theta)$$

$$= 1 - \sin^2 \theta(\sec^2 \theta)$$

$$= 1 - \frac{\sin^2 \theta}{\cos^2 \theta} = 1 - \tan^2 \theta$$

17. (a) $50\sqrt{2}n$ (b) $\alpha = 30°$ **18.** $a = 7$

19.

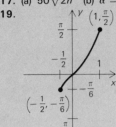

20. (a) $\frac{\pi}{3}$ (b) $\frac{\pi}{6}$ (c) $\frac{-\pi}{3}$ (d) 2   (e) $\frac{1}{2\sqrt{6}}$ **21.** (a) No solution   (b) $\theta = \frac{5\pi}{4}$ or $\frac{7\pi}{4}$;

$\left\{\theta \mid \theta = \frac{5\pi}{4} + 2k\pi \text{ or } \theta = \frac{7\pi}{4} + 2k\pi\right\}$ **22.** $\left\{\theta \mid \theta = \frac{\pi}{3} + k\pi \text{ or } \theta = \frac{2\pi}{3} + k\pi\right\}$ **23.** (a) $(7, 310°)$, $(-7, 130°)$,

$(-7, -230°)$ (b) $\left(\frac{-3}{2}, \frac{-3\sqrt{3}}{2}\right)$ (c) $(5, 120°)$

24. (a)   (b)

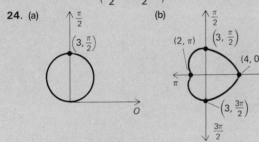

25. $\frac{x^2}{25} + \frac{y^2}{4} = 1$; ellipse

Part 4 Final Test

1. Remainder $= 0$; 4 is a root   **2.** $x(x - 2)(x + 3)(x + 1) = 0$ **3.** (a) 4   (b) $\frac{2}{3}, \frac{-1}{4}$ **4.** $x^2 - 4x + 13$

5. $P(2) = 4 > 0$, $P(3) = -1 < 0$ **6.** $x = 2$, $y = -1$, $z = -3$ **7.** Not exactly one solution (no solution)

8. 23   **9.** 42   **10.** $x = 0$, $y = 0$, $z = 1$ **11.** $\frac{-4}{x} + \frac{2}{x - 1} + \frac{3}{x + 5}$

12. $\frac{A}{x + 1} + \frac{B}{(x + 1)^2} + \frac{Cx + D}{x^2 + 5} + \frac{Ex + F}{(x^2 + 5)^2}$

13.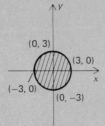

14.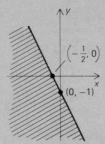

15.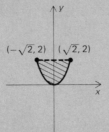

16.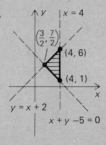

17. $\dfrac{1}{5}, \dfrac{1}{25}, \dfrac{1}{125}, \dfrac{1}{625}, \cdots, \left(\dfrac{1}{5}\right)^n, \ldots; 0$ **18.** Geometric, $S_n = \dfrac{7 - 7^{n+1}}{-6}$ **19.** Arithmetic, $S_n = \dfrac{7n + 7n^2}{2}$

20. (a) $\dfrac{1}{4} + \dfrac{2}{5} + \dfrac{1}{2} + \dfrac{4}{7} + \dfrac{5}{8}$ (b) $1 + 8 + 27 + 64 + \cdots + n^3 + \cdots$ **21.** (a) $1 - \dfrac{1}{2^n}$ (b) Converges; $\dfrac{5}{4}$

22. *Part I* $4 = 1^2(1 + 1)^2$ since $4 = 1 \times 4$.

Part II If $4 + 32 + \cdots + 4k^3 = k^2(k + 1)^2$, then

$$4 + 32 + \cdots + 4k^3 + 4(k + 1)^3 = k^2(k + 1)^2 + 4(k + 1)^3$$
$$= (k + 1)^2(k^2 + 4k + 4)$$
$$= (k + 1)^2(k + 2)^2$$
$$= (k + 1)^2(k + 1 + 1)^2$$

23. *Part I* $5^2 > 2 + 7$ since $25 > 9$.

Part II If $5^k > k + 5$, then

$$5 \times 5^k > 5(k + 5)$$
$$5^{k+1} > 5k + 25$$
$$= k + 1 + 5 + 4k + 19$$
$$> k + 1 + 5$$

24. (a) $a^8 + (\quad)a^7b + (\quad)a^6b^2 + (\quad)a^5b^3 + (\quad)a^4b^4 + (\quad)a^3b^5 + (\quad)a^2b^6 + (\quad)ab^7 + b^8$

(b) $x^4 + 8x^3y + 24x^2y^2 + 32xy^3 + 16y^4$ **25.** (a) $a^9 + 9a^8b + 36a^7b^2 + 84a^6b^3$

(b) $455, 455x^{12}y^3$

Odd-Numbered Exercise Answers;
Selected Graphs and Proofs

Exercise 1.1

1. 0, 1, 4, 7, 11   **3.** $-5, -1$ **5.** 0, 4   **7.** $-5, -1, 0, 1, 4, \dfrac{3}{4}, 7, 11, \dfrac{22}{7}, -1.5, 1.\overline{5}$

9. $-5, -1, 0, 1, 4, \dfrac{3}{4}, \pi, 7, 11, \sqrt{3}, e, \dfrac{22}{7}, -1.5, 1.\overline{5}, 1.020020002\ldots$ **11.** $0, 1, 4, \dfrac{3}{4}, \pi, 7, 11, \sqrt{3},$
$e, \dfrac{22}{7}, 1.\overline{5}, 1.020020002\ldots$ **13.** $-3i, 4 + 3i$ **15.** $-5, -1, 0, -1.5$

17. Real numbers, positive real numbers, rational numbers, nonnegative real numbers, complex numbers
19. Real numbers, positive real numbers, irrational numbers, nonnegative real numbers, complex numbers
21. Real numbers, negative integers, integers, odd integers, rational numbers, negative real numbers, nonpositive real numbers, complex numbers
23. Real numbers, positive real numbers, rational numbers, nonnegative real numbers, complex numbers
25. Real numbers, positive real numbers, irrational numbers, complex numbers, nonnegative real numbers
27. Real numbers, negative integers, integers, odd integers, rational numbers, negative real numbers, nonpositive real numbers, complex numbers
29. Real numbers, positive real numbers, irrational numbers, nonnegative real numbers, complex numbers
31. Real numbers, positive real numbers, irrational numbers, nonnegative real numbers, complex numbers
33. Complex numbers
35. Imaginary numbers, complex numbers
37. Real numbers, irrational numbers, negative real numbers, nonpositive real numbers, complex numbers
39. Real numbers, negative real numbers, nonpositive real numbers, irrational numbers, complex numbers
41. Complex numbers
43. Imaginary numbers, complex numbers
45. Real numbers, positive real numbers, rational numbers, nonnegative real numbers, complex numbers
47. Repeating; rational   **49.** Repeating; rational   **51.** Terminating; rational   **53.** Terminating; rational
55. Nonterminating, nonrepeating; irrational   **57.** Terminating; rational   **59.** Terminating; rational
61. $\{x \mid x > 2\}$ **63.** $\{x \mid 2 < x < 5\}$
65. $\{x \mid -2 \le x \le 2\}$ **67.** $\{x \mid 0 \le x \le 9 \text{ and } x \text{ is an integer}\}$
69. $\{x \mid |x| < 5 \text{ and } x \text{ is an integer}\}$
71. $\{x \mid |x| \ge 5\}$ **73.** $\{x \mid 0 \le x < 7 \text{ and } x \text{ is an integer}\}$
75. $\{x \mid x \ge -3\}$ **77.** 0   **79.** -8 **81.** -2 **83.** 6   **85.** -10

Exercise 1.2

1. $\dfrac{9}{25}$ **3.** $\dfrac{25}{9}$ **5.** 0   **7.** -3 **9.** $\dfrac{1}{5}$ **11.** 17   **13.** -8 **15.** 2   **17.** 2   **19.** $\dfrac{1}{2}$ **21.** 0   **23.** -2
25. 11   **27.** 7   **29.** 7   **31.** 4   **33.** -4 **35.** 9   **37.** 7   **39.** 7   **41.** 0   **43.** Not defined   **45.** 0

47. Not defined **49.** 0 **51.** Not defined **53.** 0 **55.** Not defined **57.** Not defined **59.** 0
61. Not defined **63.** Not defined **65.** 0 **67.** 36 **69.** Not real **71.** Not real **73.** 1 **75.** -1
77. x^8y^9 **79.** $2x^2y^2$ **81.** $2x$ **83.** $\dfrac{1}{x^3y^3}$; $x \neq 0, y \neq 0$ **85.** $\dfrac{1}{y}$; $y \neq 0$ **87.** x^4y^{10}
89. $\dfrac{25x^8}{2y^{12}}$; $x \neq 0, y \neq 0$ **91.** $\dfrac{y}{x}$; $x \neq 0, y \neq 0$ **93.** $\dfrac{2}{x^2y^4}$; $x \neq 0, y \neq 0$ **95.** $\dfrac{x^2}{3}$ **97.** $+i$ **99.** $+1$
101. $-i$ **103.** -1 **105.** $+1$ **107.** -1 **109.** $+i$ **111.** $x^2 + xx + x^2 = 3x^2$
113. $x^5 + x^4x + x^3x^2 + x^2x^3 + xx^4 + x^5 = 6x^5$

Exercise 1.3

1. $\sqrt[5]{8^4}$, $(\sqrt[5]{8})^4$ **3.** $\sqrt[3]{25}$, $(\sqrt[3]{5})^2$ **5.** $\sqrt{\pi^{-1}}$, $(\sqrt{\pi})^{-1}$ **7.** $\sqrt{8}$, $(\sqrt{2})^3$ **9.** $\sqrt[8]{6^7}$, $(\sqrt[8]{6})^7$ **11.** $7^{2/3}$
13. $3^{1/2}$ **15.** $7^{3/2}$ **17.** $5^{1/2}$ **19.** $2^{3/2}$ **21.** 343 **23.** 9 **25.** $\frac{1}{9}$ **27.** $\frac{1}{64}$ **29.** 10 **31.** $\frac{2}{5}$
33. 100 **35.** -2 **37.** $\frac{1}{12}$ **39.** 9 **41.** 0.3 **43.** 0.2 **45.** 0.1 **47.** 2 **49.** 16
51. $\dfrac{y^{7/2}}{x}$ **53.** $\dfrac{x^2}{y^{1/2}}$ **55.** $x^{10}y^2$ **57.** $x^{3/2}y^2$ **59.** $a^{7/3}b^{7/2}$ **61.** $\dfrac{21x^{1/2}}{2\sqrt{7}}$ **63.** $\dfrac{8x^{1/3}}{3\sqrt[3]{4}}$ **65.** $\dfrac{-21}{2\sqrt{7^3}}x^{-5/2}$
67. $\dfrac{-15}{\sqrt{6^3}}x^{-7/2}$ **69.** $\dfrac{\sqrt{2}}{6}x^{-5/6}$ **71.** 50 **73.** 6 **75.** $\frac{1}{45}$ **77.** 4×10^5 **79.** 20 **81.** $2xy\sqrt[3]{y}$
83. $9x^3y\sqrt{y}$ **85.** $-2xy^2\sqrt[3]{2y}$ **87.** $2x^2y^2\sqrt[5]{2y}$ **89.** $xz^2\sqrt[5]{-81x^4y^3z^2}$ **91.** $\dfrac{\sqrt{6x}}{3x}$ **93.** $\dfrac{\sqrt[3]{20x^2y}}{2x}$
95. $\dfrac{\sqrt{5}}{5}$ **97.** $\dfrac{\sqrt[4]{3xy^2}}{y}$ **99.** $\dfrac{\sqrt[3]{4}}{2}$

Exercise 1.4

1. $2x^2y$, $5x^2y$; $3x$, $-7x$; $7y$, $8y$ **3.** x^5y^2, $-x^5y^2$; $2x^2y^5$, $-x^2y^5$ **5.** $-x^5y$, $7x^5y$; $7x^4y^2$, $-x^4y^2$; $3y$, $19y$
7. x^3y, $2x^3y$; xy^3, $2xy^3$; $2xy$, xy **9.** $\sqrt{2}x^2y$, $3x^2y$; $\sqrt{2}xy$, $3xy$ **11.** Polynomial, but not over the integers
13. Not polynomial **15.** Polynomial over the integers **17.** Polynomial over the integers
19. Polynomial, but not over the integers **21.** Polynomial, but not over the integers **23.** Not polynomial
25. Polynomial over the integers **27.** Polynomial, but not over the integers **29.** Not polynomial
31. Polynomial **33.** Not polynomial **35.** Polynomial **37.** Polynomial **39.** Not polynomial
41. Tenth; fourth in x; seventh in y **43.** Fifth; fifth in x; second in y **45.** First; first in x; first in y
47. Second; second in x; first in y **49.** First; first in x; first in y **51.** $-5x^4 + 11x^3 - 7$
53. $x^2 - 5xy + y^2 + 7$ **55.** $3h$ **57.** $5h$ **59.** $3x^2h + 3xh^2 + h^3$ **61.** $-3, -2, -7, 4, -36, 90$
63. 0, 21, 18, 33, 990, 936 **65.** $3a^2 - 2a$, $2a + 7$, $12a^2 + 80a + 133$, $6a^2 - 4a + 7$
67. $x^3 + x^2 - 1$ **69.** $4x^4 - 2x^3 + 10x^2 - x + 8$ **71.** 0 **73.** $-5x^5 + 5x^4 + 7x^3 + 6x^2 - 16$
75. $3x^3 + 5x^2 - 12x - 1$ **77.** $48x^5 - 40x^4 - 12x^3 + 18x^2 - 2$ **79.** $5x^5 - 17x^4 + 16x^3 + x^2 - 2x$
81. $9x^2 - 3x - 56$ **83.** $x^2 - 3x - 28$ **85.** $x^2 - 4$ **87.** $x^3 + 8$ **89.** $x^3 - y^3$ **91.** $x^3 + y^3$
93. $x^2 - 10x + 25$ **95.** $a^2 - 2ab + b^2$ **97.** 7 **99.** -1 **101.** $x - 6\sqrt{x} + 9$
103. $x - 2\sqrt{xy} + y$ **105.** $x + 4\sqrt{x - 2} + 2$ **107.** $5x + 4\sqrt{x^2 - 3x} - 3$
109. $53 - 14\sqrt{x + 4} + x$ **111.** $2a^2 + 4ah + 2h^2 - 3a - 3h + 5$, $2x^2 + 4xh + 2h^2 - 3x - 3h + 5$
113. $a^2 + 2ah + h^2 + 5a + 5h - 3$, $x^2 + 2xh + h^2 + 5x + 5h - 3$
115. $a^3 + 3a^2h + 3ah^2 + h^3 - a^2 - 2ah - h^2 + 3a + 3h + 2$,
$x^3 + 3x^2h + 3xh^2 + h^3 - x^2 - 2xh - h^2 + 3x + 3h + 2$ **117.** $7a + 7h + 15$, $7x + 7h + 15$
119. $9a^2 + 18ah + 9h^2 + 1$, $9x^2 + 18xh + 9h^2 + 1$ **121.** $6xh + 3h^2 - 7h$
123. $6x^2h + 6xh^2 + 2h^3 - 2xh - h^2$ **125.** $12x^2h + 12xh^2 + 4h^3$ **127.** $-2xh - h^2 + h$
129. $6x^2h + 6xh^2 + 2h^3 - 5h$ **131.** $45x^2 + 69x + 26$ **133.** $x^3 + 6x^2 + 12x + 8$
135. $9x^4 - 21x^2 + 10$ **137.** 3 **139.** $2x^2 + x$ **141.** $5x + 13 + \dfrac{10x - 80}{x^2 - 3x + 6}$
143. $2x - 4 + \dfrac{11}{2x^2 + x + 3}$ **145.** $2x + 9 + \dfrac{48x^2 - 12x - 59}{x^3 - 5x^2 + 6}$ **147.** $2x + 9$, Remainder 0

149. $x + 3$, Remainder 0   **151.** $6; 2x + 5 + \dfrac{6}{x - 3}$ **153.** $-8; 3x^2 + 2x - 1 + \dfrac{-8}{x - 1}$

155. $-6; 2x^2 + 3x - 1 + \dfrac{-6}{x - 2}$ **157.** $-10; 5x^2 - 3x + \dfrac{-10}{x + 4}$ **159.** $-4; 7x^2 - 2x + 3 + \dfrac{-4}{x + 5}$

161. $x^4 + 3x^2 + 5x + 12 + \dfrac{25}{x - 2}$ **163.** $3x^3 - 11x^2 + 33x - 92 + \dfrac{272}{x + 3}$

165. $5x + 18 + \dfrac{79}{x - 4}$ **167.** $2x^4 + 2x^3 - x^2 + 4x + 4 + \dfrac{2}{x - 1}$

169. $5x^3 + 35x^2 + 242x + 1690 + \dfrac{11836}{x - 7}$ **171.** 28   **173.** -23 **175.** -237 **177.** 304

179. -234 **181.** $7 - i$ **183.** $54 - 10i$ **185.** $15 - 35i$ **187.** $-2 - 7i$ **189.** $59 - 26i$
191. $5 + 0i$ **193.** $0 + 0i$ **195.** $\frac{34}{25} + 0i$ **197.** 0, 0   **199.** 2, 2

Exercise 1.5

1. $5^2 \times 3^2$ **3.** $2^3 \times 5^2 \times 7$ **5.** 2×13^2 **7.** $2 \times 3 \times 7 \times 17$ **9.** $2^2 \times 5 \times 11 \times 41$
11. $a^2 - b^2$ **13.** $a^2 - 2ab + b^2$ **15.** $a^3 - b^3$ **17.** $a^5 - b^5$ **19.** $x^2 - 2x - 15$
21. $x^2 + 2x - 24$ **23.** $6x^2 - 11x - 7$ **25.** $12x^2 + 10x - 12$ **27.** $21x^2 + 22x - 8$
29. $21x^7 - 6x^6 + 27x^5$ **31.** $5x^5 - 4x^3 + 2x^2 - 7x$ **33.** $-5x^6 - 15x^4 + 10x^3 + 45x^2$
35. $3x^6 - 24x^5 - 2x^4 + 17x^3 - 8x^2 - x + 8$ **37.** $3x^8 + 7x^7 - 6x^6 - 14x^5 + 18x + 42$
39. $x^2 - 4y^2$ **41.** $9x^2 - 24x + 16$ **43.** $x^2 + 4xy + 4y^2$ **45.** $x^6 - y^6$ **47.** $x^6 + 2x^3y^3 + y^6$
49. $x^2 - 4x + 4$ **51.** $9x^2 - 4$ **53.** $x^6 - 4$ **55.** $x^2 + 2xh + h^2$ **57.** $x^2 - 2x + 1$ **59.** $9x^2 - 49$
61. $9x^2 - 18x - 7$ **63.** $x^3 - 1$ **65.** $27x^3 + 343$ **67.** $x^3 + 8y^3$ **69.** $16 - 81x^4$
71. $(5x + 4)(5x - 4)$ **73.** $x^2y(3z - 2x^3y + xyz)$ **75.** $(2w + 3z)(4w^2 - 6wz + 9z^2)$
77. $(3x + 7y)(3x - 7y)$ **79.** $xy(3x - 4y)(9x^2 + 12xy + 16y^2)$ **81.** $(x^2 + y^2)^2(x + y)^2(x - y)^2$
83. $(x - y)(x^2 + xy + y^2)(x^6 + x^3y^3 + y^6)$ **85.** $(x - 11)(x + 4)$
87. $(2x - y)(2x + y)(4x^2 + 2xy + y^2)(4x^2 - 2xy + y^2)$ **89.** $(2x + 5)(x - 4)$ **91.** $4(3x + y)(3x - y)$
93. $2a^2b^2(a + b)^2$ **95.** $(11a + 13y)(11a - 13y)$ **97.** $(3w + 8z)^2$ **99.** Not factorable
101. $b(3a + 2b)(9a^2 - 6ab + 4b^2)$ **103.** $x(4xy + 5z)(4xy - 5z)$ **105.** $x^2y(x - 8)(x + 7)$
107. Not factorable   **109.** $6(x - 2)(x + 1)$ **111.** $(x^2 + 4y^2)(x + 2y)(x - 2y)$
113. $(a - b)(a^4 + a^3b + a^2b^2 + ab^3 + b^4)(a + b)(a^4 - a^3b + a^2b^2 - ab^3 + b^4)$
115. $(x - y)(x^2 + xy + y^2)(x + y)(x^2 - xy + y^2)(x^2 + y^2)(x^4 - x^2y^2 + y^4)$

117. $\dfrac{49}{4}; \left(x + \dfrac{7}{2}\right)^2$ **119.** $\dfrac{81}{4}; \left(x - \dfrac{9}{2}\right)^2$ **121.** $1; (x - 1)^2$ **123.** $1; (x + 1)^2$

125. $\dfrac{b^2}{4a^2}; \left(x + \dfrac{b}{2a}\right)^2$ **127.** $(y^7 - 3)(x + 2)$ **129.** $a^3(3a^2 - 2)(a - 1)$ **131.** $2x + h$

133. $(x + h)^5 + (x + h)^4x + (x + h)^3x^2 + (x + h)^2x^3 + (x + h)x^4 + x^5$
135. $(x + h)^7 + (x + h)^6x + (x + h)^5x^2 + (x + h)^4x^3 + (x + h)^3x^4 + (x + h)^2x^5 + (x + h)x^6 + x^7$

Exercise 1.6

1. Minus   **3.** Plus   **5.** Plus   **7.** Plus   **9.** Minus   **11.** Minus   **13.** Minus   **15.** Plus   **17.** $\dfrac{2}{9}$ **19.** $\dfrac{7}{4}$

21. In lowest terms   **23.** In lowest terms   **25.** In lowest terms   **27.** In lowest terms   **29.** $\dfrac{6(x + 1)}{x + 4}$

31. $\dfrac{x + 7}{x + 4}$ **33.** $\dfrac{2x^2 - x + 5}{x(x - 5)}$ **35.** In lowest terms   **37.** $\dfrac{x - 5}{x - 7}$ **39.** $\dfrac{x^2 + x + 1}{x - 1}$ **41.** 110   **43.** 75

45. $19(x + 2)$ **47.** $5(x + 2)$ **49.** $x^2(x^2 + 2x + 4)$ **51.** 180   **53.** $567x^5$ **55.** $(x - 1)^3(x^2 + x + 1)$

57. $x^3(x - 2)^7(x + 1)^3(x - 1)$ **59.** $(x - y)^2(x + y)$ **61.** $\dfrac{35}{88}$ **63.** $\dfrac{111}{88}$ **65.** $\dfrac{24}{5}$ **67.** $\dfrac{-101}{630}$

69. $\dfrac{269}{420}$ **71.** $\dfrac{153}{40}$ **73.** $\dfrac{1427}{1302}$ **75.** $\dfrac{253}{2808}$ **77.** $\dfrac{1}{3}$ **79.** $\dfrac{1}{\sqrt{3}}$ **81.** $\dfrac{10x^3}{(x^2-9)(x-3)}$

83. $\dfrac{(x+2)(x+1)^2}{(x-2)(x+3)(x-3)}$ **85.** $\dfrac{16a^2b^2y}{3x}$ **87.** $\dfrac{4x^2a^3b^3c^3}{9yz^3}$ **89.** $\dfrac{7(x+2)}{x-3}$ **91.** $\dfrac{5x^2+3x}{x^2-9}$

93. $\dfrac{7x^2+6x}{x^2-9}$ **95.** $\dfrac{2x^2-27x+64}{(x-2)(x^2-10x+25)}$ **97.** $\dfrac{2x^2+4}{(x-1)(x+2)}$ **99.** $\dfrac{7x^2+3x+9}{5}$

101. $\dfrac{2}{3x-1}+\dfrac{5}{x+2}=\dfrac{2(x+2)+5(3x-1)}{(3x-1)(x+2)}=\dfrac{17x-1}{3x^2+5x-2}$

103. $\dfrac{3}{x-5}+\dfrac{6}{x+5}+\dfrac{2}{x}=\dfrac{3x(x+5)+6x(x-5)+2(x-5)(x+5)}{x(x+5)(x-5)}$

$$=\dfrac{11x^2-15x-50}{x^3-25x}$$

105. $\dfrac{7}{x+3}+\dfrac{1}{(x+3)^2}=\dfrac{7(x+3)+1}{(x+3)^2}=\dfrac{7x+22}{x^2+6x+9}$

107. $\dfrac{3x+2}{x^2+x+1}+\dfrac{1}{x-1}=\dfrac{(3x+2)(x-1)+(x^2+x+1)}{(x-1)(x^2+x+1)}=\dfrac{4x^2-1}{x^3-1}$

109. $\dfrac{7}{x}+\dfrac{1}{x+3}+\dfrac{4}{(x+3)^2}=\dfrac{7(x+3)^2+x(x+3)+4x}{x(x+3)^2}$

$$=\dfrac{8x^2+49x+63}{x^3+6x^2+9x}$$

111. $\dfrac{10x^3-6}{6x^6+x^4}$ **113.** $\dfrac{3x^2-14x}{4x^2-6x-4}$ **115.** $\dfrac{-x^2+2x+1}{4x^2-3}$ **117.** $\dfrac{20-4\sqrt{2}}{23}$ **119.** $\sqrt{5}-\sqrt{3}$

121. $\dfrac{22}{41}+\dfrac{7}{41}i$ **123.** $\dfrac{17}{10}+\dfrac{1}{10}i$ **125.** $9-7i$

Exercise 1.7

1. Any solution of (1) is a solution of (2), but (2) may have additional solutions.
3. The solutions of (1) and (2) are the same
5. Any solution of (1) is a solution of (2), but (2) may have additional solutions.
7. Any solution of (1) is a solution of (2), but (2) may have additional solutions.
9. Any solution of (1) is a solution of (2), but (2) may have additional solutions.

11. $x=3$ **13.** $x=8$ **15.** $x=\dfrac{7}{11}$ **17.** $x=\dfrac{-4}{3}$ **19.** $x=-19$ **21.** $h=\dfrac{V}{2\pi r}$ **23.** $h=\dfrac{3V}{\pi r^2}$

25. $P=\dfrac{A}{1+ni}$ **27.** $y'=\dfrac{-2x-y}{x+2},\ x\neq-2$ **29.** $y'=\dfrac{-x}{y},\ y\neq0$ **31.** 2 real solutions

33. 2 real solutions **35.** 2 imaginary solutions **37.** 1 real solution **39.** 2 imaginary solutions

41. $x=8,-4,7$ **43.** $x=-4,-3,2$ **45.** $x=-4,3$ **47.** $x=0,3$ **49.** $x=4,\pm2$

51. $x=9,-2$ **53.** $x=7,-5$ **55.** $x=\dfrac{1}{3},\dfrac{-1}{2}$ **57.** $x=\dfrac{-1}{8},\dfrac{1}{7}$ **59.** $x=\dfrac{-5}{3}$

61. $x=\dfrac{1}{3},-2$ **63.** $x=1\pm\sqrt{11}$ **65.** $x=\dfrac{8\pm6i}{5}$ **67.** $x=\pm3$ **69.** $x=\dfrac{-1}{2}\pm\dfrac{5}{2}i$

71. $x=\dfrac{-1}{2},\dfrac{5}{3}$ **73.** $x=\dfrac{-1}{3},\dfrac{1}{5}$ **75.** $x=5$ **77.** $x=7,\dfrac{1}{3}$ **79.** $x=1,\dfrac{-5}{2}$ **81.** $x=\pm1,\pm4i$

83. $x=\pm3i,\pm2$ **85.** $x=3,1,\dfrac{-3\pm3\sqrt{3}i}{2},\dfrac{-1\pm\sqrt{3}i}{2}$

87. $\dfrac{(x-1)^2}{9}-\dfrac{(y+2)^2}{4}=1;\ h=1,a=3,k=-2,b=2$

89. $(y - 3)^2 = 8(x - (-1))$; $k = 3$, $p = 4$, $h = -1$

91. $\dfrac{(y - 3)^2}{4} - \dfrac{(x + 1)^2}{9} = 1$; $k = 3$, $a = 2$, $h = -1$, $b = 3$

93. $\dfrac{(x + 5)^2}{4} + (y - 4)^2 = 1$; $h = -5$, $a = 2$, $k = 4$, $b = 1$

95. $\dfrac{(x - 3)^2}{16} + \dfrac{(y + 1)^2}{4} = 1$; $h = 3$, $a = 4$, $k = -1$, $b = 2$ **97.** $3\sqrt{1 - (x - 2)^2}$; $h = 2$, $k = 1$

97. $3\sqrt{1 - (x - 2)^2}$; $h = 2$, $k = 1$

99. $2\sqrt{\left(x - \dfrac{3}{2}\right)^2 - 36}$; $h = \dfrac{3}{2}$, $k = 6$ **101.** $\sqrt{(x + 5)^2 + 9}$; $h = -5$, $k = 3$

103. $2\sqrt{\left(x - \dfrac{5}{2}\right)^2 - 9}$; $h = \dfrac{5}{2}$, $k = 3$ **105.** $2\sqrt{\left(x + \dfrac{1}{2}\right)^2 + 4}$; $h = \dfrac{-1}{2}$, $k = 2$

Exercise 1.8

1. 3   **3.** No solution   **5.** No solution   **7.** $x = 20$ **9.** $x = 12$ **11.** $\dfrac{x^2}{25} + \dfrac{y^2}{9} = 1$; $a = 5$, $b = 3$

13. $\dfrac{x^2}{169} + \dfrac{y^2}{25} = 1$; $a = 13$, $b = 5$ **15.** $\dfrac{x^2}{100} + \dfrac{y^2}{64} = 1$; $a = 10$, $b = 8$

17. $\dfrac{x^2}{144} - \dfrac{y^2}{25} = 1$; $a = 12$, $b = 5$ **19.** $\dfrac{x^2}{4} - \dfrac{y^2}{9} = 1$; $a = 2$, $b = 3$ **21.** $y^2 = 20x$; $\dfrac{p}{2} = 5$

23. $x^2 = -20y$; $\dfrac{p}{2} = -5$ **25.** $y^2 = -8x$; $\dfrac{p}{2} = -2$ **27.** No solution   **29.** $x = 4$ **31.** $x = 1, -5$

33. $x = 2$ **35.** $x = 4$ **37.** $x = \dfrac{3}{2}, \dfrac{-7}{2}$ **39.** $x = \dfrac{10}{3}, \dfrac{-2}{3}$

41. $x = 1, -10$ **43.** $x = -1, -7$

45. $x = 2, -2$

47. $x = -2, -1, \dfrac{-3 \pm \sqrt{97}}{2}$

49. $x = \pm 3, \pm\sqrt{3}$ **51.** $\dfrac{(x - 2)^2}{25} + \dfrac{(y - 1)^2}{16} = 1$; $h = 2$, $a = 5$, $k = 1$, $b = 4$

53. $\dfrac{(x + 1)^2}{16} - \dfrac{(y + 2)^2}{9} = 1$; $h = -1$, $a = 4$, $k = -2$, $b = 3$

55. $(y - 2)^2 = 16x$, $\dfrac{p}{2} = 4$

Exercise 1.9

1. $x \geq \dfrac{-8}{5}$ **3.** $x \leq 0$ **5.** $x \leq 5$

7. $x > -1$ ———○|●———
                    -1 0

9. $x \geq 3$ ———|●———
                    0 3

11. $x \leq -3$ or $x \geq -\dfrac{5}{3}$
    $-\frac{5}{3}$
    ———●|●|———
    -3 0

13. $x \leq -1$ or $x \geq 2$ ———●|●———
                                -1 0 2

15. $-2 < x < 2$ ———○|○———
                        -2 0 2

17. $-1 \leq x \leq \dfrac{-1}{2}$
    ———●●|———
    -1 $-\frac{1}{2}$ 0

19. $x \leq 1$ or $x \geq 4$ ———|● ●———
                                0 1 4

21. $|(2x - 5) - 1| = |2x - 6| = 2|x - 3| < 2\delta$ whenever $|x - 3| < \delta$. But if $\delta \leq \dfrac{\epsilon}{2}$, then $2\delta \leq \epsilon$ so that $|(2x - 5) - 1| < \epsilon$ whenever $|x - 3| < \delta$.

23. $|(3x + 7) - 10| = |3x - 3| = 3|x - 1| < 3\delta$ whenever $|x - 1| < \delta$. But if $\delta \leq \dfrac{\epsilon}{3}$, then $3\delta \leq \epsilon$ so that $|(3x + 7) - 10| < \epsilon$ whenever $|x - 1| < \delta$.

25. $|(4x + 1) + 7| = |4x + 8| = 4|x + 2| < 4\delta$ whenever $|x + 2| < \delta$. But if $\delta \leq \dfrac{\epsilon}{4}$, then $4\delta \leq \epsilon$ so $|(4x + 1) + 7| < \epsilon$ whenever $|x + 2| < \delta$.

27. $\delta \leq \dfrac{\epsilon}{3}$ **29.** $\delta \leq \dfrac{\epsilon}{2}$

Exercise 1.10

1. $x = 1, y = -2$ **3.** $A = \dfrac{-72}{25}, B = \dfrac{102}{25}$ **5.** $A = -1, B = 1$ **7.** $x = 8, y = -9$ **9.** $x = \dfrac{3}{2}, y = -1$

11. $A = -4, B = -1$ **13.** $x = \dfrac{29}{107}, y = \dfrac{268}{107}, z = \dfrac{-300}{107}$ **15.** $A = -1, B = -1, C = 3$

17. $x = -2, y = 3, z = 4$ **19.** $x = 1, y = -1, z = 3, w = 2$ **21.** $A = 7, B = 4, C = 5, D = 2$

23. $A = 1, B = -1, C = 2, D = -2, E = 3$ **25.** $x = 1, y = 0, z = -1, w = 2, q = -2$

27. $A = -3, B = -5$ **29.** $A = 3, B = -5, C = 7$ **31.** $A = 3, B = -3, C = 4$

33. $A = 1, B = 2, C = 1, D = 2$ **35.** $A = 1, B = 4, C = 5, D = -1, E = -2$ **37.** Corn, 5; rice, 8

39. Carberry, 0; Smiley, 50   **41.** 12   **43.** Tuba, 33; piccolo, 6   **45.** Silver, 7; new pennies, 8; wheat-ears, 6

Exercise 2.1

1. $-3 \to 10, \dfrac{-1}{2} \to \dfrac{5}{4}, 0 \to 1, \dfrac{1}{2} \to \dfrac{5}{4}, 3 \to 10; \left\{10, \dfrac{5}{4}, 1\right\}$

3. $0 \to 0, 1 \to 1, \dfrac{9}{4} \to \dfrac{3}{2}, 4 \to 2, 9 \to 3; \left\{0, 1, \dfrac{3}{2}, 2, 3\right\}$

5. $\dfrac{-1}{8} \to \dfrac{-1}{2}, -1 \to -1, 0 \to 0, 1 \to 1, 8 \to 2; \left\{\dfrac{-1}{2}, -1, 0, 1, 2\right\}$

7. $1, -1, 1, a^3 - a^2 + 1, a^3 + 3a^2h + 3ah^2 + h^3 - a^2 - 2ah - h^2 + 1$

9. $\dfrac{1}{2}, \dfrac{7}{6}, 2, \dfrac{1}{2a} + 1, \dfrac{1}{2(a + h)} + 1$ **11.** $x_1^2 + 2x_1(\Delta x) + (\Delta x)^2 - 5x_1 - 5(\Delta x) + 3$

13. $4x_1^2 + 8x_1(\Delta x) + 4(\Delta x)^2 - 2x_1 - 2(\Delta x) + 1$ **15.** $2x_1^3 + 6x_1^2(\Delta x) + 6x_1(\Delta x)^2 + 2(\Delta x)^3$

17. $2x^2 + 4xh + 2h^2 + 3x + 3h - 4$ **19.** $5x^2 + 10xh + 5h^2 - x - h + 7$

21. Function; domain $= \left\{\dfrac{-1}{2}, \dfrac{1}{2}, \dfrac{1}{3}, \dfrac{3}{2}\right\}$; $\dfrac{-1}{2} \to -1, \dfrac{1}{2} \to 0, \dfrac{1}{3} \to 0, \dfrac{3}{2} \to 1$; range $= \{-1, 0, 1\}$

23. Function; domain $= \{-2, 0, -1, 2\}$; $-2 \to -8, 0 \to 0, -1 \to -1, 2 \to 8$; range $= \{-8, 0, -1, 8\}$

25. Not a function   **27.** $\{x \,|\, x \geq 1\}$ **29.** $\{x \,|\, x \geq -2\}$

31.

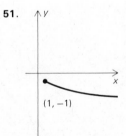

33.

35.

37. $(-5, -2)$ **39.** $(-2, 0)$

41.

43.

45.

47.

49.

51.

53.

55.

57.

x	y
-3	2
-2	1
-1	0
0	1
1	2

59.

x	y
-3	-1
-2	0
-1	1
0	2
1	3
2	4
3	5

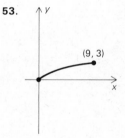
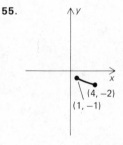

61. Not a function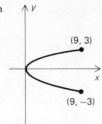

63. Function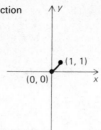

65. Not a function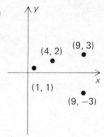

67. Not a function   **69.** Not a function   **71.** Function   **73.** Not a function   **75.** Function
77. Not a function   **79.** Function   **81.** $3h + 6x - 2$ **83.** $4h + 8x + 3$ **85.** $7h + 14x - 4$
87. $3x_1^2 + 3x_1(\Delta x) + (\Delta x)^2$ **89.** $3x_1^2 + 3x_1(\Delta x) + (\Delta x)^2 - 2$ **91.** $3x + 3a - 2$ **93.** $4x + 4a + 3$
95. $7x + 7a - 4$ **97.** $x_1^2 + x_1 x + x^2$ **99.** $x_1^2 + x_1 x + x^2 - 2$ **101.** $6x_1 + 3h - 2$
103. $8x_1 + 4h + 3$ **105.** $14x_1 + 7h - 4$ **107.** $r^2 + rx + x^2$ **109.** $r^2 + rx + x^2 - 2$

Exercise 2.2

1.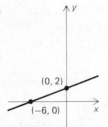

7.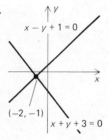

11.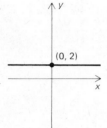

13.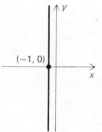

17. $\dfrac{-1}{5}$ **19.** $\dfrac{-4}{3}$ **21.** $(1) - (2) + 1 = 0$; $(3) - (4) + 1 = 0$; 1

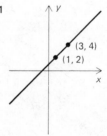

23. $3(2) + 5(1) - 11 = 0$; $3(-3) + 5(4) - 11 = 0$; $\dfrac{-3}{5}$ **25.** $(-3) + 4(2) - 5 = 0$; $1 + 4(1) - 5 = 0$; $\dfrac{-1}{4}$

27.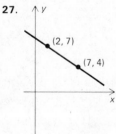

31. $2x + 5y - 26 = 0$ **33.** $x + y + 1 = 0$ **35.** $3x - 8y + 43 = 0$ **37.** $3x + 4y + 20 = 0$

39. $x - y + 3 = 0$ **41.** $y = \frac{2}{7}x + 4; \frac{2}{7}; (0, 4)$ **43.** $y = -5x + 11; -5; (0, 11)$

45. $y = \frac{5}{8}x + \frac{3}{2}; \frac{5}{8}; \left(0, \frac{3}{2}\right)$ **47.** $\frac{-1}{2}; 2$ **49.** $-2; \frac{1}{2}$ **51.** $5x + y - 11 = 0$ **53.** $x - 5y - 14 = 0$

55. $5x - 9y + 64 = 0$ **57.** $x + y - 1 = 0$ **59.** $3x + 2y - 6 = 0$ **61.** $3x - 4y = 0$

63. $x - 2y = 0$ **65.** $x - y + 2 = 0$

67.

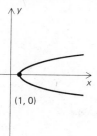

77. $2x + 18; 20; 18$ **79.** $-x + 6; 5; 6$ **81.** $\dfrac{f(x) - f(a)}{x - a}$ **83.** $\dfrac{f(x) - f(x_1)}{x - x_1}$ **85.** $\dfrac{f(x_1 + h) - f(x_1)}{h}$

87. $6x_1 + 3\Delta x - 5$ **89.** $2x + 2a + 4$ **91.** $7x + 7r + 3$ **93.** $8c + 4h - 3$

95. $3x_1^2 + 3x_1(\Delta x) + (\Delta x)^2$

Exercise 2.3

1. Intercept: $(1, 0)$; symmetric with respect to the x-axis; excluded area: $x < 1$

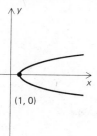

3. Intercept: $(0, 0)$; symmetric with respect to the x-axis; excluded area: $x < 0$

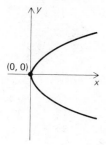

5. Intercept: (1, 0); symmetric with respect to the x-axis; excluded area: $x < 1$ **7.** Intercept: $(0, -1)$; symmetric with respect to the y-axis; excluded area: $y < -1$ **9.** Intercept: (0, 0); symmetric with respect to the y-axis; excluded area: $y < 0$ **11.** Intercept: (0, 0); symmetric with respect to the origin   **13.** Intercept: (0, 0); symmetric with respect to the origin   **15.** Intercept: (0, 0); symmetric with respect to the origin
17. Intercept: (0, 0); symmetric with respect to the origin   **19.** Intercept: (0, 1); asymptotes: $y = 0$, $x = -1$; excluded areas: $x > -1$ and $y < 0$, $x < -1$ and $y > 0$

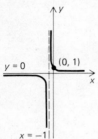

21. Intercept: $(0, \frac{1}{3})$; asymptotes: $x = -3$, $y = 0$; excluded areas: $x > -3$ and $y < 0$, $x < -3$ and $y > 0$
23. Intercept: $(0, -3)$; asymptotes: $x = 1$, $y = 0$; excluded areas: $x > 1$ and $y < 0$, $x < 1$ and $y > 0$
25. Symmetric with respect to the origin; asymptotes: $x = 0$, $y = 0$; excluded areas: $x > 0$ and $y > 0$, $x < 0$ and $y < 0$

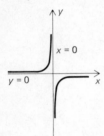

27. Intercepts: $(0, -6)$, $(2, 0)$; asymptotes: $x = \dfrac{-1}{2}$, $y = \dfrac{3}{2}$; excluded areas: $x > 2$ and $y < 0$, $\dfrac{-1}{2} < x < 2$ and $y > 0$, $y < \dfrac{-1}{2}$ and $y < 0$

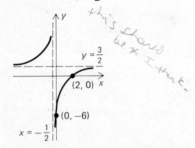

29. Intercepts: $\left(0, \dfrac{-10}{3}\right)$, $(-2, 0)$; asymptotes: $x = 3$, $y = 5$; excluded areas: $x > 3$ and $y < 0$, $-2 < x < 3$ and $y > 0$, $x < -2$ and $y < 0$ **31.** Intercepts: $\left(0, \dfrac{-4}{3}\right)$, $(4, 0)$; asymptotes: $x = -2$, $y = \dfrac{2}{3}$; excluded areas: $x > 4$ and $y < 0$, $-2 < x < 4$ and $y > 0$, $x < -2$ and $y < 0$

33. Intercept: (0, 0); asymptotes: $x = 2$, $x = -2$, $y = \frac{1}{3}$; excluded areas: $x > 2$ and $y < 0$, $-2 < x < 2$ and $y > 0$, $x < -2$ and $y < 0$

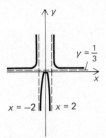

35. Intercept: (0, 0); asymptotes: $x = 2$, $x = -2$, $y = 3$; excluded areas: $x > 2$ and $y < 0$, $-2 < x < 2$ and $y > 0$, $x < -2$ and $y < 0$ **37.** Intercept: (0, 0); asymptote: $x = 1$; excluded areas: $x > 1$ and $y < 0$, $x < 1$ and $y > 0$ **39.** Intercept: (0, 0); asymptote: $x = -1$; excluded areas: $x > -1$ and $y < 0$, $x < -1$ and $y > 0$
41. Intercepts: (0, 3), (0, −3), (2, 0), (−2, 0); excluded areas: $x > 2$, $x < -2$, $y > 3$, $y < -3$

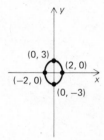

43. Intercepts: (0, 3), (0, −3) (1, 0), (−1, 0); excluded areas: $y > 3$, $y < -3$, $x > 1$, $x < -1$
45. Intercepts: (0, 1), (0, −1), (3, 0), (−3, 0); excluded areas: $x > 3$, $x < -3$, $y > 1$, $y < -1$
47. Intercepts: (3, 0), (−3, 0); excluded area: $-3 < x < 3$

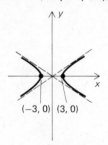

49. Intercepts: (0, 4), (0, −4), (5, 0), (−5, 0); excluded areas: $y > 4$, $y < -4$, $x > 5$, $x < -5$

Exercise 2.4

1. 8   **3.** $\sqrt{85}$ **5.** $4\sqrt{2}$ **7.** $\frac{4\sqrt{226}}{5}$ **9.** $\frac{\sqrt{1145}}{3}$ **11.** (6, 4)   **13.** $\left(\frac{1}{2}, \frac{7}{2}\right)$ **15.** $\left(\frac{-7}{2}, \frac{1}{2}\right)$
17. $(x + 4)^2 + (y + 5)^2 = \frac{9}{16}$ **19.** $(x - 5)^2 + y^2 = \frac{49}{16}$ **21.** (−3, 5); 6   **23.** (6, −2); $4\sqrt{5}$

25. $(7, -8); \sqrt{2}$

27. $(x + 1)^2 + (y - 3)^2 = 0$

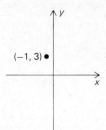

$(-1, 3)$ •

29. $(x - 6)^2 + (y - 1)^2 = 4$

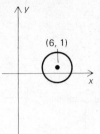

$(6, 1)$

31. $\left(x - \dfrac{1}{2}\right)^2 + (y + 3)^2 = \dfrac{-19}{4}$; No graph   **33.** $(x - 4)^2 + (y - 3)^2 = 0$ **35.** $x^2 + y^2 = 4$

37. $x'^2 + y'^2 = 9$

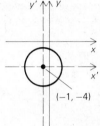

$(-1, -4)$

39. $x'^2 + y'^2 = 1$ **41.** $x'^2 + y'^2 = 49$ **43.** $x'^2 + y'^2 = 36$

45. $x'^2 + y'^2 = 64$ **47.** $y' = x'^2$

$(0, 3)$

49. $y' = x'^2$ **51.** $y' = x'^2$

53. $y' = |x'|$

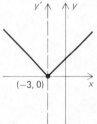

$(-3, 0)$

55. $y' = x'^3$

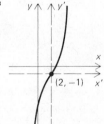

$(2, -1)$

57. $y' = 3x'$

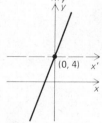

$(0, 4)$

59. $y' = -|x'|$

61. $y' = \dfrac{1}{x'}$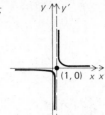

63. $y' = \dfrac{1}{x'}$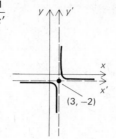

65. $y' = \dfrac{3x'^2}{x' - 1}$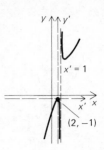

Exercise 2.5

1. Vertex $(0, 0)$; axis: $y = 0$; focus $(-2, 0)$; directrix: $x = 2$

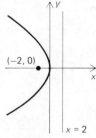

3. Vertex $(0, 0)$; axis: $y = 0$; focus $(2, 0)$; directrix: $x = -2$ **5.** Vertex $(0, 0)$; axis: $y = 0$; focus $(-1, 0)$; directrix: $x = 1$ **7.** Vertex $(0, 0)$; axis: $x = 0$; focus $(0, -1)$; directrix: $y = 1$ **9.** Vertex $(0, 0)$; axis: $y = 0$; focus $\left(\dfrac{-1}{4}, 0\right)$; directrix: $x = \dfrac{1}{4}$

11. Vertex $(5, 0)$; axis: $y = 0$; focus $(2, 0)$; directrix: $x = 8$; $y'^2 = -12x'$

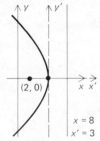

13. Vertex $(0, 5)$; axis: $x = 0$; focus $(0, 1)$; directrix: $y = 9$; $x'^2 = -16y'$ **15.** Vertex $(-1, 3)$; axis: $y = 3$; focus $(1, 3)$; directrix: $x = -3$; $y'^2 = 8x'$ **17.** Vertex $(-3, 2)$; axis: $y = 2$; focus $\left(\dfrac{-5}{2}, 2\right)$; directrix: $x = \dfrac{-7}{2}$; $y'^2 = 2x'$ **19.** Vertex $(-4, -1)$; axis: $y = -1$; focus $\left(\dfrac{-15}{4}, -1\right)$; directrix: $x = \dfrac{-17}{4}$; $y'^2 = x'$

21. $(x - 1)^2 = 4(y + 2)$; vertex $(1, -2)$; axis: $x = 1$

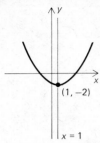

23. $(y + 1)^2 = \dfrac{3}{4}(x - 2)$; vertex $(2, -1)$; axis: $y = -1$ **25.** $(y - 5)^2 = \dfrac{7}{4}(x - 6)$; vertex $(6, 5)$; axis: $y = 5$

27. $(y - 5)^2 = \dfrac{5}{2}(x + 4)$; vertex $(-4, 5)$; axis: $y = 5$ **29.** $(x - 2)^2 = \dfrac{5}{2}(y + 1)$; vertex $(2, -1)$; axis: $x = 2$

31. Vertex $(5, 1)$; axis: $x = 5$

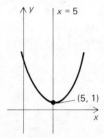

33. Vertex: $(-4, 3)$; axis: $x = -4$ **35.** Vertex $(-2, 1)$; axis: $x = -2$

37. Maximum height, $t = 5$; returns to ground, $t = 10$

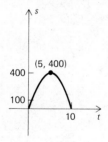

39. Maximum height, $t = \dfrac{3}{2}$; returns to ground, $t = 3$

Exercise 2.6

1. Center (0, 0)

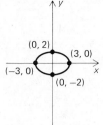

3. Center (0, 0)   **5.** Center (0, 0)   **7.** Center (0, 0)

9. Center (0, 0)   **11.** Center $(-3, 4)$; $\dfrac{x'^2}{9} + \dfrac{y'^2}{16} = 1$

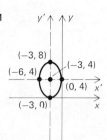

13. Center $(-5, 2)$; $\dfrac{x'^2}{16} + \dfrac{y'^2}{4} = 1$ **15.** Center $(1, -2)$; $\dfrac{x'^2}{9} + \dfrac{y'^2}{1} = 1$ **17.** Center $(-3, 1)$; $\dfrac{x'^2}{\frac{25}{16}} + \dfrac{y'^2}{\frac{9}{4}} = 1$

19. Center $(-3, 5)$; $\dfrac{x'^2}{4} + \dfrac{y'^2}{16} = 1$ **21.** $\dfrac{(x-2)^2}{\frac{1}{4}} + \dfrac{(y+5)^2}{\frac{1}{9}} = 1$

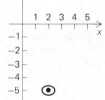

23. $\dfrac{(x-3)^2}{9} + \dfrac{(y+1)^2}{4} = 1$ **25.** $\dfrac{(x-3)^2}{\frac{1}{4}} + \dfrac{(y+1)^2}{\frac{1}{9}} = 0$ **27.** $\dfrac{(x-5)^2}{81} + \dfrac{y^2}{9} = 1$

29. $\dfrac{x^2}{81} + \dfrac{(y-4)^2}{9} = 1$

31. Center (0, 0); vertices (3, 0), $(-3, 0)$; asymptotes $y = \dfrac{2x}{3}$, $y = \dfrac{-2x}{3}$

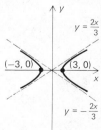

33. Center $(0, 0)$; vertices $\left(0, \dfrac{1}{3}\right)$, $\left(0, \dfrac{-1}{3}\right)$; asymptotes $y = \dfrac{2x}{3}$, $y = \dfrac{-2x}{3}$

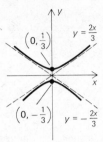

35. Center $(0, 0)$; vertices $\left(\dfrac{9}{4}, 0\right)$, $\left(\dfrac{-9}{4}, 0\right)$; asymptotes $y = \dfrac{2x}{3}$, $y = \dfrac{-2x}{3}$ **37.** Center $(0, 0)$; vertices $(0, 10)$, $(0, -10)$; asymptotes $y = 5x$, $y = -5x$ **39.** Center $(0, 0)$; vertices $(3, 0)$, $(-3, 0)$; asymptotes $y = \dfrac{x}{3}$, $y = \dfrac{-x}{3}$

41. Center $(-3, 4)$; vertices $(0, 4)$, $(-6, 4)$; $\dfrac{x'^2}{9} - \dfrac{y'^2}{16} = 1$; asymptotes $y' = \pm\dfrac{4x'}{3}$, $(y - 4) = \pm\dfrac{4(x + 3)}{3}$

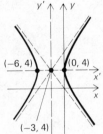

43. Center $(-5, -2)$; vertices $(-1, -2)$, $(-9, -2)$; $\dfrac{x'^2}{16} - \dfrac{y'^2}{4} = 1$; asymptotes $y' = \pm\dfrac{x'}{2}$, $y + 2 = \pm\dfrac{1}{2}(x + 5)$ **45.** Center $(6, -3)$; vertices $(7, -3)$, $(5, -3)$; $\dfrac{x'^2}{1} - \dfrac{y'^2}{9} = 1$; asymptotes $y' = \pm 3x'$, $y + 3 = \pm 3(x - 6)$ **47.** Center $(1, -4)$; vertices $(4, -4)$, $(-2, -4)$; $\dfrac{x'^2}{9} - \dfrac{y'^2}{4} = 1$; asymptotes $y' = \pm\dfrac{2x'}{3}$, $y + 4 = \pm\dfrac{2}{3}(x - 1)$ **49.** Center $(5, 4)$; vertices $(5, 14)$, $(5, -6)$; $\dfrac{y'^2}{100} - \dfrac{x'^2}{4} = 1$; asymptotes $y' = \pm 5x'$, $y - 4 = \pm 5(x - 5)$

51. $\dfrac{(x - 2)^2}{\frac{1}{4}} - \dfrac{(y + 5)^2}{\frac{1}{9}} = 1$

53. $\dfrac{(y+1)^2}{\frac{1}{9}} - \dfrac{(x-3)^2}{\frac{1}{4}} = 1$ **55.** $(x+1)^2 - y^2 = 0$ **57.** $(x+1)^2 - y^2 = 1$

59. $\dfrac{(y+1)^2}{4} - \dfrac{(x-3)^2}{9} = 1$ **61.** Ellipse   **63.** Circle   **65.** Hyperbola   **67.** Two intersecting lines

69. Point   **71.** $\pi\left(9 - \dfrac{9}{4}x^2\right);\ 0 < x < 2$ **73.** $\pi(4 - x^2);\ 0 < x < 2$ **75.** $\pi\left(\dfrac{1}{9} - x^2\right);\ 0 < x < \dfrac{1}{3}$

77. $2\pi\sqrt{9 - \dfrac{9x^2}{25}};\ 1 \le x \le 5$ **79.** $\pi\dfrac{x^2}{2};\ 0 \le x \le 3$

Exercise 2.7

1. Not a function

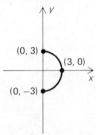

3. Not a function   **5.** Function   **7.** Function   **9.** Not a function   **11.** $\{(x, y)\,|\,7y - 8x + 9 = 0\}$
13. $\{(x, y)\,|\,x^2 + y^2 = 9\}$ **15.** $\{(x, y)\,|\,x = y^2 - 6y + 9\}$ **17.** $\{(x, y)\,|\,x = 10^y\}$ **19.** $\{(x, y)\,|\,x = e^y\}$
21. $\{(x, y)\,|\,x = 3y - 7\}$; function   **23.** $\{(x, y)\,|\,3y - 7x + 10 = 0\}$; function   **25.** $\{(x, y)\,|\,x = y^4\}$; not a
function   **27.** $\{(x, y)\,|\,x = \sqrt{y}\}$; function   **29.** $\{(x, y)\,|\,x = -y\}$; function   **31.** One-to-one
33. Not one-to-one   **35.** One-to-one   **37.** Not one-to-one   **39.** One-to-one   **41.** One-to-one
43. Not one-to-one   **45.** Not one-to-one   **47.** One-to-one   **49.** Not one-to-one

51. Domain $= \{x\,|\,x \ge 2\}$; range $= \{y\,|\,y \ge -1\}$; $f^{-1}(x) = \dfrac{x - 5}{3}$

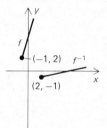

53. Domain $= \{x\,|\,x \ge 4\}$; range $= \{y\,|\,y \ge 2\}$; $f^{-1}(x) = \sqrt{x}$

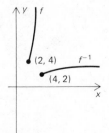

55. Domain $= \{x|x \geq -5\}$; range $= \{y|y \geq 0\}$; $f^{-1}(x) = \dfrac{x+5}{2}$ **57.** Domain $= \{x|x \geq 0\}$;

range $= \{y|y \leq 0\}$; $f^{-1}(x) = -\sqrt{x}$ **59.** Domain $= \{x|x \text{ is a real number}\}$; range $= \{y|y \text{ is a real number}\}$;

$f^{-1}(x) = \dfrac{x+4}{3}$

Exercise 2.8

1.

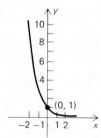

7.

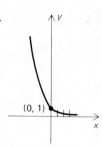

9.

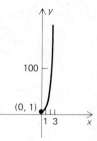

11. Range $= \{y|y \text{ is a real number}\}$; domain $= \{x|x > 0\}$; rule of f^{-1}: $y = \log_2 x$

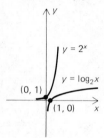

13. Range $= \{y|y \text{ is a real number}\}$; domain $= \{x|x > 0\}$; rule of f^{-1}: $y = \log_{1/3} x$

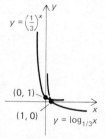

17. 0   **19.** 4   **21.** -2 **23.** -1 **25.** 1   **27.** 2   **29.** 4

31.

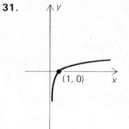

33.

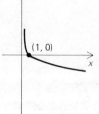

35.

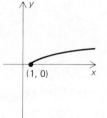

EXERCISE ANSWERS; SELECTED GRAPHS AND PROOFS

Exercise 2.9

1. $3^2 = 9$ **3.** $5^2 = 25$ **5.** $10^5 = x$ **7.** $e^5 = x$ **9.** $e^3 = x^2$ **11.** $\log_2 y = 6$ **13.** $\ln y = x^2$
15. $\log_2 8 = 3$ **17.** $\ln y = x + h$ **19.** $\ln y = m + n$ **21.** u **23.** 1 **25.** 0 **27.** x **29.** x **31.** x
33. x **35.** $2x$ **37.** -5 **39.** -5 **41.** $e^{x \ln 5}$ **43.** $e^{x^2 \ln 2}$ **45.** $e^{x \ln \sqrt{2}}$ **47.** $e^{5 \ln x}$ **49.** $e^{1/3 \ln x}$
51. Yes **53.** Yes **55.** No **57.** Let $x = \ln M$. Then $M = e^x$. Let $y = \ln N$. Then $N = e^y$. So
$M/N = e^x/e^y = e^{x-y}$ or $x - y = \ln M/N$. That is, $\ln M - \ln N = \ln M/N$. **59.** Let $x = \log M$. Then
$M = 10^x$. So $M^k = 10^{kx}$ and hence $M^k = 10^{k \log M}$ or $k \log M = \log M^k$. **61.** 2.9652 **63.** $0.7619 - 4$
65. 4.7959 **67.** 5.8492 **69.** 3.2959 **71.** -2.0099 **73.** 3.3223 **75.** 2.7713
77. $(\log e)(\ln(3x + 1))$ **79.** $(\log e)(\ln 38)$ **81.** 8250 **83.** 35,700 **85.** 187,000 **87.** 256,000,000
89. 5.27 **91.** 185 **93.** 16.7 **95.** 2.59×10^{-10}
97. $\ln y = \frac{2}{3}\ln(x + 2) - \frac{1}{2}\ln(1 + x^2) - \frac{1}{2}\ln(1 + x) - \frac{1}{2}\ln(1 - x)$
99. $\ln y = \ln(x - 2) + \frac{1}{2}\ln(x - 1) - \frac{1}{2}\ln(x + 4)$

Exercise 2.10

1. $(f + g)(x) = e^{3x} + \sqrt{-x}$, domain $= \{x | x \le 0\}$; $(f - g)(x) = e^{3x} - \sqrt{-x}$, domain $= \{x | x \le 0\}$;
$(fg)(x) = e^{3x}\sqrt{-x}$, domain $= \{x | x \le 0\}$; $\left(\dfrac{f}{g}\right)(x) = \dfrac{e^{3x}}{\sqrt{-x}}$, domain $= \{x | x < 0\}$; $(f \circ g)(x) = e^{3\sqrt{-x}}$,
domain $= \{x \le 0\}$; $(g \circ f)(x) = \sqrt{-e^{3x}}$, domain is empty **3.** $(f + g)(x) = \ln|x| + x^3$, domain $= \{x | x \ne 0\}$;
$(f - g)(x) = \ln|x| - x^3$, domain $= \{x | x \ne 0\}$; $(fg)(x) = x^3 \ln|x|$, domain $= \{x | x \ne 0\}$;
$\left(\dfrac{f}{g}\right)(x) = \dfrac{\ln|x|}{x^3}$, domain $= \{x | x \ne 0\}$; $(f \circ g)(x) = \ln|x^3|$, domain $= \{x | x \ne 0\}$; $(g \circ f)(x) = (\ln|x|)^3$,
domain $= \{x | x \ne 0\}$ **5.** $(f + g)(x) = \ln x + 2x$, domain $= \{x | 0 < x \le 1\}$; $(f - g)(x) = \ln x - 2x$,
domain $= \{x | 0 < x \le 1\}$; $(fg)(x) = 2x \ln x$, domain $= \{x | 0 < x \le 1\}$; $\left(\dfrac{f}{g}\right)(x) = \dfrac{\ln x}{2x}$,
domain $= \{x | 0 < x \le 1\}$; $(f \circ g)(x) = \ln 2x$, domain $= \{x | 0 < x \le 1\}$; $(g \circ f)(x) = 2 \ln x$,
domain $= \{x | e^{-1} \le x \le e\}$ **7.** $(f + g)(x) = \sqrt{x + 3} + x^2$, domain $= \{x | x \ge -3\}$;
$(f - g)(x) = \sqrt{x + 3} - x^2$, domain $= \{x | x \ge -3\}$; $(fg)(x) = x^2\sqrt{x + 3}$, domain $= \{x | x \ge -3\}$;
$\left(\dfrac{f}{g}\right)(x) = \dfrac{\sqrt{x + 3}}{x^2}$, domain $= \{x | x \ge -3 \text{ and } x \ne 0\}$; $(f \circ g)(x) = \sqrt{x^2 + 3}$, domain $= \{x | x \text{ is a real number}\}$;
$(g \circ f)(x) = x + 3$, domain $= \{x | x \text{ is a real number}\}$ **9.** $(f + g)(x) = x^2 + x + 5$, domain $= \{x | x \le 0\}$;
$(f - g)(x) = x^2 - x - 5$, domain $= \{x | x \le 0\}$; $(fg)(x) = x^3 + 5x^2$, domain $= \{x | x \le 0\}$;
$\left(\dfrac{f}{g}\right)(x) = \dfrac{x^2}{x + 5}$, domain $= \{x | x \le 0 \text{ and } x \ne -5\}$; $(f \circ g)(x) = x^2 + 10x + 25$, domain $= \{x | x \le -5\}$;
$(g \circ f)(x) = x^2 + 5$, domain $= \{x | x \le 0\}$ **11.** 11 **13.** 18 **15.** $\sqrt{10}$ **17.** $\sqrt{65}$ **19.** 1
21. $u = g(x) = x^2 - 5$, $f(u) = u^{1/3}$ **23.** $u = g(x) = x^3$, $f(u) = e^u$ **25.** $u = g(x) = x^3 - 2x + 1$,
$f(u) = u^{1/2}$ **27.** $u = g(x) = -x$, $f(u) = 10^u$ **29.** $u = g(x) = x^2 + 3x - 5$, $f(u) = u^{-4/5}$

Exercise 3.1

1. $-120°$, $\dfrac{-2\pi}{3}$ **3.** $180°$, π **5.** $-270°$, $\dfrac{-3\pi}{2}$ **7.** $-240°$, $\dfrac{-4\pi}{3}$ **9.** $-540°$, -3π **11.** $\dfrac{7\pi}{6}$

13. $\dfrac{5\pi}{3}$ **15.** $\dfrac{7\pi}{3}$ **17.** $315°$ **19.** $\dfrac{720°}{\pi}$ **21.** $\dfrac{5\pi}{18}$ **23.** $\dfrac{5\pi}{18}$

25. $\dfrac{11\pi}{36}$ **27.** $\dfrac{\pi}{6}$ **29.** $\dfrac{\pi}{3}$ **31.** $\dfrac{4\pi}{3}$ ft **33.** $\dfrac{92\pi}{9}$ cm **35.** $\dfrac{40\pi}{9}$ in. **37.** $\dfrac{63\pi}{2}$ cm **39.** $\dfrac{3\pi}{4}$ cm

41. $229.3°$ **43.** $57.3°$ **45.** $258.0°$ **47.** $143.3°$ **49.** $298.1°$ **51.** $3\pi, (-1, 0)$ **53.** $5\pi, (-1, 0)$

55. $\dfrac{9\pi}{2}, (0, 1)$ **57.** $-\pi, (-1, 0)$ **59.** $\dfrac{7\pi}{2}, (0, -1)$ **61.** $0°, (1, 0)$ **63.** $180°, (-1, 0)$

65. $360°, (1, 0)$ **67.** $630°, (0, -1)$ **69.** $1260°, (-1, 0)$ **71.** $-360°, (1, 0)$ **73.** $900°, (-1, 0)$

75. $-1620°, (-1, 0)$

Exercise 3.2

1. $\cos A = \dfrac{\sqrt{2}}{2}$, $\sin A = \dfrac{\sqrt{2}}{2}$, $\tan A = 1$, $\sec A = \sqrt{2}$, $\csc A = \sqrt{2}$, $\cot A = 1$ **3.** $\cos A = \frac{1}{2}$, $\sin A = \dfrac{\sqrt{3}}{2}$,
$\tan A = \sqrt{3}$, $\sec A = 2$, $\csc A = \dfrac{2\sqrt{3}}{3}$, $\cot A = \dfrac{\sqrt{3}}{3}$ **5.** $\cos A = \dfrac{-1}{2}$, $\sin A = \dfrac{-\sqrt{3}}{2}$, $\tan A = \sqrt{3}$,
$\sec A = -2$, $\csc A = \dfrac{-2\sqrt{3}}{3}$, $\cot A = \dfrac{\sqrt{3}}{3}$ **7.** $\cos \alpha = \dfrac{-\sqrt{3}}{2}$, $\sin \alpha = \dfrac{-1}{2}$, $\tan \alpha = \dfrac{\sqrt{3}}{3}$,
$\sec \alpha = \dfrac{-2\sqrt{3}}{3}$, $\csc \alpha = -2$, $\cot \alpha = \sqrt{3}$ **9.** $\cos \alpha = \frac{1}{5}$, $\sin \alpha = \dfrac{2\sqrt{6}}{5}$, $\tan \alpha = 2\sqrt{6}$, $\sec \alpha = 5$,
$\csc \alpha = \dfrac{5\sqrt{6}}{12}$, $\cot \alpha = \dfrac{\sqrt{6}}{12}$ **11.** $\cos 450° = 0$, $\sin 450° = 1$, $\tan 450°$ undefined, $\sec 450°$ undefined,
$\csc 450° = 1$, $\cot 450° = 0$ **13.** $\cos \dfrac{5\pi}{2} = 0$, $\sin \dfrac{5\pi}{2} = 1$, $\tan \dfrac{5\pi}{2}$ undefined, $\sec \dfrac{5\pi}{2}$ undefined,
$\csc \dfrac{5\pi}{2} = 1$, $\cot \dfrac{5\pi}{2} = 0$ **15.** $\cos 720° = 1$, $\sin 720° = 0$, $\tan 720° = 0$, $\sec 720° = 1$,
$\csc 720°$ undefined, $\cot 720°$ undefined **17.** $\cos\left(\dfrac{-9\pi}{2}\right) = 0$, $\sin\left(\dfrac{-9\pi}{2}\right) = -1$, $\tan\left(\dfrac{-9\pi}{2}\right)$ undefined,
$\sec\left(\dfrac{-9\pi}{2}\right)$ undefined, $\csc\left(\dfrac{-9\pi}{2}\right) = -1$, $\cot\left(\dfrac{-9\pi}{2}\right) = 0$ **19.** $\cos 810° = 0$, $\sin 810° = 1$,
$\tan 810°$ undefined, $\sec 810°$ undefined, $\csc 810° = 1$, $\cot 810° = 0$ **21.** $\cos 5\pi = -1$, $\sin 5\pi = 0$,
$\tan 5\pi = 0$, $\sec 5\pi = -1$, $\csc 5\pi$ undefined, $\cot 5\pi$ undefined **23.** $\cos \dfrac{7\pi}{2} = 0$, $\sin \dfrac{7\pi}{2} = -1$, $\tan \dfrac{7\pi}{2}$
undefined, $\sec \dfrac{7\pi}{2}$ undefined, $\csc \dfrac{7\pi}{2} = -1$, $\cot \dfrac{7\pi}{2} = 0$ **25.** $\cos \dfrac{9\pi}{2} = 0$, $\sin \dfrac{9\pi}{2} = 1$,
$\tan \dfrac{9\pi}{2}$ undefined, $\sec \dfrac{9\pi}{2}$ undefined, $\csc \dfrac{9\pi}{2} = 1$, $\cot \dfrac{9\pi}{2} = 0$ **27.** $\cos 6\pi = 1$, $\sin 6\pi = 0$, $\tan 6\pi = 0$,
$\sec 6\pi = 1$, $\csc 6\pi$ undefined, $\cot 6\pi$ undefined **29.** $\cos \dfrac{13\pi}{2} = 0$, $\sin \dfrac{13\pi}{2} = 1$, $\tan \dfrac{13\pi}{2}$ undefined,
$\sec \dfrac{13\pi}{2}$ undefined, $\csc \dfrac{13\pi}{2} = 1$, $\cot \dfrac{13\pi}{2} = 0$

31. $\cos A = \frac{4}{5}$, $\sin A = \frac{3}{5}$, $\tan A = \frac{3}{4}$, $\sec A = \frac{5}{4}$, $\csc A = \frac{5}{3}$, $\cot A = \frac{4}{3}$, $\cos B = \frac{3}{5}$, $\sin B = \frac{4}{5}$, $\tan B = \frac{4}{3}$,
$\sec B = \frac{5}{3}$, $\csc B = \frac{5}{4}$, $\cot B = \frac{3}{4}$

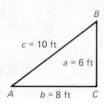

33. $\cos A = \frac{4}{5}$, $\sin A = \frac{3}{5}$, $\tan A = \frac{3}{4}$, $\sec A = \frac{5}{4}$, $\csc A = \frac{5}{3}$, $\cot A = \frac{4}{3}$, $\cos B = \frac{3}{5}$, $\sin B = \frac{4}{5}$, $\tan B = \frac{4}{3}$,
$\sec B = \frac{5}{3}$, $\csc B = \frac{5}{4}$, $\cot B = \frac{3}{4}$ **35.** $\cos A = \frac{4}{5}$, $\sin A = \frac{3}{5}$, $\tan A = \frac{3}{4}$, $\sec A = \frac{5}{4}$, $\csc A = \frac{5}{3}$, $\cot A = \frac{4}{3}$,
$\cos B = \frac{3}{5}$, $\sin B = \frac{4}{5}$, $\tan B = \frac{4}{3}$, $\sec B = \frac{5}{3}$, $\csc B = \frac{5}{4}$, $\cot B = \frac{3}{4}$ **37.** 250 ft   **39.** 44 ft

41. $\dfrac{\cos A}{\sin A} = \dfrac{\frac{4}{5}}{\frac{3}{5}} = \dfrac{4}{3} = \cot A$ **43.** $\dfrac{1}{\sin A} = \dfrac{1}{\frac{3}{5}} = \dfrac{5}{3} = \csc A$ **45.** $\dfrac{1}{\cos B} = \dfrac{1}{\frac{3}{5}} = \dfrac{5}{3} = \sec B$

47. $\cos\theta = \dfrac{3}{\sqrt{9+x^2}}$, $\sin\theta = \dfrac{x}{\sqrt{9+x^2}}$, $\tan\theta = \dfrac{x}{3}$, $\sec\theta = \dfrac{\sqrt{9+x^2}}{3}$, $\csc\theta = \dfrac{\sqrt{9+x^2}}{x}$, $\cot\theta = \dfrac{3}{x}$

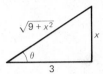

49. $\cos\theta = \dfrac{\sqrt{4-x^2}}{2}$, $\sin\theta = \dfrac{x}{2}$, $\tan\theta = \dfrac{x}{\sqrt{4-x^2}}$, $\sec\theta = \dfrac{2}{\sqrt{4-x^2}}$, $\csc\theta = \dfrac{2}{x}$, $\cot\theta = \dfrac{\sqrt{4-x^2}}{x}$

Exercise 3.3

1. $\cos 120° = \dfrac{-1}{2}$, $\sin 120° = \dfrac{\sqrt{3}}{2}$, $\tan 120° = -\sqrt{3}$, $\sec 120° = -2$, $\csc 120° = \dfrac{2\sqrt{3}}{3}$,
$\cot 120° = \dfrac{-\sqrt{3}}{3}$ **3.** $\cos 150° = \dfrac{-\sqrt{3}}{2}$, $\sin 150° = \frac{1}{2}$, $\tan 150° = \dfrac{-\sqrt{3}}{3}$, $\sec 150° = \dfrac{-2\sqrt{3}}{3}$,
$\csc 150° = 2$, $\cot 150° = -\sqrt{3}$ **5.** $\cos 240° = \dfrac{-1}{2}$, $\sin 240° = \dfrac{-\sqrt{3}}{3}$, $\tan 240° = \sqrt{3}$,
$\sec 240° = -2$, $\csc 240° = \dfrac{-2\sqrt{3}}{3}$, $\cot 240° = \dfrac{\sqrt{3}}{3}$ **7.** $\cos 315° = \dfrac{\sqrt{2}}{2}$, $\sin 315° = \dfrac{-\sqrt{2}}{2}$,
$\tan 315° = -1$, $\sec 315° = \sqrt{2}$, $\csc 315° = -\sqrt{2}$, $\cot 315° = -1$ **9.** $\cos(-45°) = \dfrac{\sqrt{2}}{2}$,
$\sin(-45°) = \dfrac{-\sqrt{2}}{2}$, $\tan(-45°) = -1$, $\sec(-45°) = \sqrt{2}$, $\csc(-45°) = -\sqrt{2}$, $\cot(-45°) = -1$

11. $\cos(-390°) = \dfrac{\sqrt{3}}{2}$, $\sin(-390°) = \dfrac{-1}{2}$, $\tan(-390°) = \dfrac{-\sqrt{3}}{3}$, $\sec(-390°) = \dfrac{2\sqrt{3}}{3}$, $\csc(-390°) = -2$,
$\cot(-390°) = -\sqrt{3}$ **13.** $\cos 1020° = \frac{1}{2}$, $\sin 1020° = \dfrac{-\sqrt{3}}{2}$, $\tan 1020° = -\sqrt{3}$, $\sec 1020° = 2$,
$\csc 1020° = \dfrac{-2\sqrt{3}}{3}$, $\cot 1020° = \dfrac{-\sqrt{3}}{3}$ **15.** $\cos\dfrac{3\pi}{4} = \dfrac{-\sqrt{2}}{2}$, $\sin\dfrac{3\pi}{4} = \dfrac{\sqrt{2}}{2}$, $\tan\dfrac{3\pi}{4} = -1$,
$\sec\dfrac{3\pi}{4} = -\sqrt{2}$, $\csc\dfrac{3\pi}{4} = \sqrt{2}$, $\cot\dfrac{3\pi}{4} = -1$ **17.** $\cos\dfrac{7\pi}{6} = \dfrac{-\sqrt{3}}{2}$, $\sin\dfrac{7\pi}{6} = \dfrac{-1}{2}$, $\tan\dfrac{7\pi}{6} = \dfrac{\sqrt{3}}{3}$,
$\sec\dfrac{7\pi}{6} = \dfrac{-2\sqrt{3}}{3}$, $\csc\dfrac{7\pi}{6} = -2$, $\cot\dfrac{7\pi}{6} = \sqrt{3}$ **19.** $\cos\dfrac{4\pi}{3} = \dfrac{-1}{2}$, $\sin\dfrac{4\pi}{3} = \dfrac{-\sqrt{3}}{2}$, $\tan\dfrac{4\pi}{3} = \sqrt{3}$,
$\sec\dfrac{4\pi}{3} = -2$, $\csc\dfrac{4\pi}{3} = \dfrac{-2\sqrt{3}}{3}$, $\cot\dfrac{4\pi}{3} = \dfrac{\sqrt{3}}{3}$ **21.** $\cos\dfrac{7\pi}{4} = \dfrac{\sqrt{2}}{2}$, $\sin\dfrac{7\pi}{4} = \dfrac{-\sqrt{2}}{2}$, $\tan\dfrac{7\pi}{4} = -1$,
$\sec\dfrac{7\pi}{4} = \sqrt{2}$, $\csc\dfrac{7\pi}{4} = -\sqrt{2}$, $\cot\dfrac{7\pi}{4} = -1$ **23.** $\cos\left(\dfrac{-11\pi}{4}\right) = \dfrac{-\sqrt{2}}{2}$, $\sin\left(\dfrac{-11\pi}{4}\right) = \dfrac{-\sqrt{2}}{2}$,
$\tan\left(\dfrac{-11\pi}{4}\right) = 1$, $\sec\left(\dfrac{-11\pi}{4}\right) = -\sqrt{2}$, $\csc\left(\dfrac{-11\pi}{4}\right) = -\sqrt{2}$, $\cot\left(\dfrac{-11\pi}{4}\right) = 1$ **25.** $\cos\left(\dfrac{-7\pi}{3}\right) = \dfrac{1}{2}$,
$\sin\left(\dfrac{-7\pi}{3}\right) = \dfrac{-\sqrt{3}}{2}$, $\tan\left(\dfrac{-7\pi}{3}\right) = -\sqrt{3}$, $\sec\left(\dfrac{-7\pi}{3}\right) = 2$, $\csc\left(\dfrac{-7\pi}{3}\right) = \dfrac{-2\sqrt{3}}{3}$, $\cot\left(\dfrac{-7\pi}{3}\right) = \dfrac{-\sqrt{3}}{3}$ **27.** Second

29. First   **31.** Fourth   **33.** Third   **35.** Second   **37.** −1.483   **39.** 2.246   **41.** Table, −0.7585 (Calculator, −0.7594)   **43.** Table, −2.651 (Calculator; −2.663)   **45.** Table, 0.1593 (Calculator, 0.1632) **47.** Table, −0.9903 (Calculator, −0.9900)   **49.** Table, −0.1392 (Calculator, −0.1411) **51.** Table, 68° (Calculator, 68.00°)   **53.** Table, 17° (Calculator, 17.06°)   **55.** Table, 83° (Calculator, 82.99°)   **57.** Table, 0.2618 (Calculator, 0.2618)   **59.** Table, −0.1047 (Calculator, −0.1047) **61.** Table, 133° (Calculator, 133.0°)   **63.** Table, 162° (Calculator, 162.0°)   **65.** Table, 2.2499 (Calculator, 2.2514)   **67.** $B = 30°$ **69.** 2 ft   **71.** 7 mm   **73.** Table, $A = 36°50'$ (Calculator: $A = 36°52'$); $c = 10$ ft   **75.** $a = 6$ m, $B = 70°$ **77.** $\dfrac{\pi}{4}, \dfrac{\pi}{3}$ **79.** $\dfrac{\pi}{4}, \dfrac{\pi}{3}$

Exercise 3.4

1. Amplitude 1; period 2π 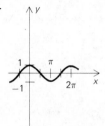   **3.** Amplitude 1; period 2π

5. Amplitude 1; period 2π 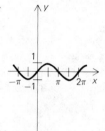   **7.** Amplitude 1; period 2π **9.** Amplitude 1; period 2π

11. Amplitude 4; period 2π 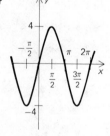   **13.** Amplitude $\frac{1}{4}$; period 2π **15.** Amplitude 4; period 2π

17. Amplitude 2, period $\dfrac{\pi}{2}$

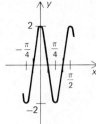

19. Amplitude 3; period 2   **21.** Amplitude 3; period 4π

23. Amplitude 1; period 4   **25.** Amplitude 1; period 16π **27.** Amplitude 3; period $\dfrac{\pi}{2}$

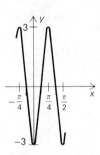

29. Amplitude $\frac{1}{4}$; period 2π **31.** Amplitude 2; period π **33.** Amplitude 1; period 2π
35. Amplitude $\frac{1}{2}$; period 4π **37.** $\pi\cos^2 x$ **39.** $\pi\cos^2 2x$

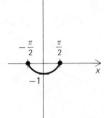

Exercise 3.5

1. Range $= \{y \mid -1 \le y \le 1\}$ **7.** Range $= \{y \mid y \text{ is a real number}\}$

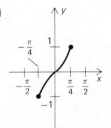

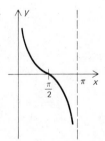

9. Range $= \{y \mid y \le 0\}$

11. Range $= \{y \mid 1 \le y \le \sqrt{2}\}$

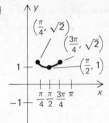

13. Range $= \{y \mid y \le -1\}$ **15.** Range $= \{y \mid |y| \ge 1\}$

17. Range $= \{y \mid y \ge 1\}$

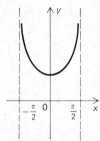

19. Range $= \{y \mid |y| \ge 1\}$

21. $\cot(-\theta) = \dfrac{\cos(-\theta)}{\sin(-\theta)} = \dfrac{\cos\theta}{-\sin\theta} = -\cot\theta$ **23.** $\csc(-\theta) = \dfrac{1}{\sin(-\theta)} = \dfrac{1}{-\sin\theta} = -\csc\theta$

Exercise 3.6

1. $\sin\theta = \frac{-4}{5}$, $\tan\theta = \frac{4}{3}$, $\sec\theta = \frac{-5}{3}$, $\csc\theta = \frac{-5}{4}$, $\cot\theta = \frac{3}{4}$ **3.** $\sin\theta = \frac{4}{5}$, $\cos\theta = \frac{-3}{5}$, $\tan\theta = \frac{-4}{3}$, $\csc\theta = \frac{5}{4}$, $\cot\theta = \frac{-3}{4}$ **5.** $\sin\theta = \frac{8}{17}$, $\cos\theta = \frac{-15}{17}$, $\csc\theta = \frac{17}{8}$, $\sec\theta = \frac{-17}{15}$, $\cot\theta = \frac{-15}{8}$ **7.** $\cos\theta = \frac{8}{17}$, $\tan\theta = \frac{-15}{8}$, $\sec\theta = \frac{17}{8}$, $\csc\theta = \frac{-17}{15}$, $\cot\theta = \frac{-8}{15}$ **9.** $\sin\theta = \frac{12}{13}$, $\cos\theta = \frac{-5}{13}$, $\tan\theta = \frac{-12}{5}$, $\csc\theta = \frac{13}{12}$, $\sec\theta = \frac{-13}{5}$ **11.** $\dfrac{\sqrt{2} + \sqrt{6}}{4}$ **13.** $\dfrac{\sqrt{2} + \sqrt{6}}{4}$ **15.** $\dfrac{3 - \sqrt{3}}{3 + \sqrt{3}}$ **17.** $\dfrac{3 + \sqrt{3}}{3 - \sqrt{3}}$ **19.** $\dfrac{\sqrt{2} + \sqrt{6}}{4}$

21. $\sin 2\theta = \frac{-24}{25}$, $\cos 2\theta = \frac{7}{25}$, $\tan 2\theta = \frac{-24}{7}$ **23.** $\sin 2\theta = \frac{-240}{289}$, $\cos 2\theta = \frac{161}{289}$, $\tan 2\theta = \frac{-240}{161}$

25. $\sin 2\theta = \frac{120}{169}$, $\cos 2\theta = \frac{119}{169}$, $\tan 2\theta = \frac{120}{119}$ **27.** $\sin\theta = \pm\dfrac{2}{\sqrt{5}}$, $\cos\theta = -\dfrac{1}{\sqrt{5}}$ **29.** $\sin\theta = \frac{1}{2}$, $\cos\theta = -\dfrac{\sqrt{3}}{2}$ **31.** $\sin\dfrac{\theta}{2} = \dfrac{2}{\sqrt{5}}$, $\cos\dfrac{\theta}{2} = \dfrac{1}{\sqrt{5}}$, $\tan\dfrac{\theta}{2} = 2$ **33.** $\sin\dfrac{\theta}{2} = \dfrac{1}{\sqrt{17}}$, $\cos\dfrac{\theta}{2} = \dfrac{-4}{\sqrt{17}}$, $\tan\dfrac{\theta}{2} = \dfrac{-1}{4}$

35. $\sin\dfrac{\theta}{2} = \dfrac{2}{\sqrt{5}}$, $\cos\dfrac{\theta}{2} = \dfrac{-1}{\sqrt{5}}$, $\tan\dfrac{\theta}{2} = -2$ **37.** $\dfrac{\sqrt{2 + \sqrt{2}}}{2}$ **39.** $\dfrac{-\sqrt{2 + \sqrt{2}}}{2}$ **41.** $\dfrac{\sqrt{2 + \sqrt{2}}}{2}$

43. $\dfrac{\sqrt{2 - \sqrt{2}}}{\sqrt{2 + \sqrt{2}}}$ **45.** $\dfrac{\sqrt{3}}{3}$

47. $4\sin\theta\cos^3\theta - 4\sin^3\theta\cos\theta$ **49.** $6\sin\theta\cos^5\theta - 20\sin^3\theta\cos^3\theta + 6\cos\theta\sin^5\theta$

51. $\sec\theta + \tan\theta = \dfrac{1}{\cos\theta} + \dfrac{\sin\theta}{\cos\theta}$

$= \dfrac{1 + \sin\theta}{\cos\theta}$

67. Let $x = 3\sec\theta$ where $0 < \theta < \dfrac{\pi}{2}$; $\sqrt{x^2 - 9} = 3\tan\theta$ **71.** Let $u = \tan\dfrac{\theta}{2}$; $\dfrac{1}{\cos\theta} = \dfrac{1 + u^2}{1 - u^2}$

Exercise 3.7

1. $\beta = 100°$, $c = 3.47$ m, $a = 8.79$ m   **3.** $\gamma = 55°$, $c = 106.95$ m, $b = 126.10$ m   **5.** $\beta = 50°$,
$a = 1.81$ m, $c = 9.04$ m   **7.** $\gamma = 90°$, $\beta = 30°$ **9.** $\gamma = 74°10'$, $\beta = 45°50'$ or $\gamma = 105°50'$,
$\beta = 14°10'$ **11.** $\alpha = 45°$, $\gamma = 90°$ **13.** $\gamma = 36°10'$, $\alpha = 98°50'$ **15.** No triangle   **17.** $b = 11.56$ m
19. $c = 15.72$ m   **21.** $b = 9.62$ m   **23.** $c = 8.42$ m   **25.** $b = 16.08$ m   **27.** $\alpha = 114°$, $\beta = 42°$,
$\gamma = 25°$ **29.** $\alpha = 47°$, $\beta = 39°$, $\gamma = 94°$ **31.** 33 m   **33.** 80 m   **35.** Either approximately 124 m or
approximately 29 m

Exercise 3.8

1. $\dfrac{\pi}{3}$ **3.** 0   **5.** $\dfrac{-\pi}{2}$ **7.** $\dfrac{2\pi}{3}$ **9.** $\dfrac{\pi}{4}$ **11.** $\dfrac{\pi}{6}$ **13.** 0   **15.** $\dfrac{5\pi}{6}$ **17.** $\dfrac{2\pi}{3}$ **19.** $\dfrac{\pi}{6}$ **21.** $\dfrac{\pi}{4}$ **23.** π

25. $\dfrac{\pi}{4}$ **27.** $\dfrac{\pi}{3}$ **29.** $\dfrac{-\pi}{2}$ **31.**

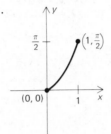   **35.** 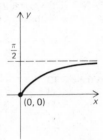   **37.**

41. 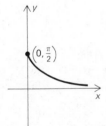   **43.** 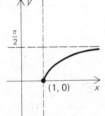   **47.**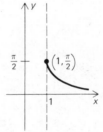

51. 0.1542   **53.** 0.0698   **55.** 0.2298   **57.** 0.4043   **59.** 2.8782   **61.** $\frac{4}{5}$ **63.** $\frac{5}{3}$ **65.** $\frac{12}{5}$

67. $\dfrac{\sqrt{1 - u^2}}{u}$, where $u = \cos\theta$ **69.** $\sqrt{1 - u^2}$, where $u = \cos\theta$ **71.** $\dfrac{u}{\sqrt{1 - u^2}}$, where $u = \sin\theta$

73. $\dfrac{1}{u}$, where $u = \sin\theta$ **75.** $\dfrac{1}{u}$, where $u = \sec\theta$

Exercise 3.9

1. $\theta = \dfrac{\pi}{6}$ **3.** $\theta = \dfrac{\pi}{3}$ **5.** $\theta = \dfrac{\pi}{3}$ **7.** $\theta = \dfrac{5\pi}{6}$ **9.** $\theta = \dfrac{-\pi}{3}$

11. $\theta = \dfrac{\pi}{6}, \dfrac{11\pi}{6}$; $\left\{\theta \mid \theta = \dfrac{\pi}{6} + 2k\pi \text{ or } \theta = \dfrac{11\pi}{6} + 2k\pi\right\}$ **13.** $\theta = \dfrac{\pi}{3}$; $\left\{\theta \mid \theta = \dfrac{\pi}{3} + k\pi\right\}$ **15.** $\theta = 0$; $\{\theta \mid \theta = 2k\pi\}$

17. $\theta = \dfrac{5\pi}{6}, \dfrac{7\pi}{6}$; $\left\{\theta \mid \theta = \dfrac{5\pi}{6} + 2k\pi \text{ or } \theta = \dfrac{7\pi}{6} + 2k\pi\right\}$ **19.** $\theta = \dfrac{2\pi}{3}$, $\left\{\theta \mid \theta = \dfrac{2\pi}{3} + k\pi\right\}$

21. $\left\{\theta \mid \theta = \dfrac{\pi}{6} + k\pi\right\}$ **23.** $\left\{\theta \mid \theta = \dfrac{\pi}{3} + 2k\pi \text{ or } \theta = \dfrac{5\pi}{3} + 2k\pi\right\}$ **25.** $\left\{\theta \mid \theta = \dfrac{2\pi}{3} + 2k\pi \text{ or } \theta = \dfrac{4\pi}{3} + 2k\pi\right\}$

27. $\left\{\theta \middle| \theta = \dfrac{\pi}{2} + k\pi \text{ or } \theta = \dfrac{2\pi}{3} + 2k\pi \text{ or } \theta = \dfrac{4\pi}{3} + 2k\pi\right\}$ **29.** $\{\theta \middle| \theta = (2k + 1)\pi\}$

31. $\left\{\theta \middle| \theta = k\pi \text{ or } \theta = \dfrac{\pi}{4} + k\pi\right\}$ **33.** $\left\{\theta \middle| \theta = \dfrac{2\pi}{3} + k\pi\right\}$

35. $\left\{\theta \middle| \theta = \dfrac{\pi}{6} + 2k\pi \text{ or } \theta = \dfrac{5\pi}{6} + 2k\pi \text{ or } \theta = \dfrac{7\pi}{6} + 2k\pi \text{ or } \theta = \dfrac{11\pi}{6} + 2k\pi\right\}$

37. $\{\theta \middle| \theta = \text{Arctan } \sqrt{2} + k\pi \text{ or } \theta = \text{Arctan } (-\sqrt{2}) + k\pi\}$ **39.** $\left\{\theta \middle| \theta = \dfrac{\pi}{3} + k\pi \text{ or } \theta = \dfrac{2\pi}{3} + k\pi\right\}$

41. $\left\{\theta \middle| \theta = \dfrac{\pi}{4} + 2k\pi, \theta = \dfrac{3\pi}{4} + 2k\pi, \theta = \dfrac{5\pi}{4} + 2k\pi, \text{ or } \theta = \dfrac{7\pi}{4} + 2k\pi\right\}$

43. $\left\{\theta \middle| \theta = \dfrac{\pi}{12} + 2k\pi \text{ or } \theta = \dfrac{11\pi}{12} + 2k\pi\right\}$ **45.** $\left\{\theta \middle| \theta = \dfrac{k\pi}{2}\right\}$

47. $\{\theta \middle| \theta = 1.57 + 6.28k \text{ or } \theta = 1.0647 + 6.28k \text{ or } \theta = 2.0753 + 6.28k\}$

49. $\{\theta \middle| \theta = 1.05 + 6.28k \text{ or } \theta = 5.23 + 6.28k, \text{ or } \theta = 1.3177 + 6.28k \text{ or } \theta = 4.9623 + 6.28k\}$

51. $x = 0, \theta = \dfrac{\pi}{2}; x = 3, \theta = 0$ **53.** $x = \sqrt{3}, \theta = \text{Arcsin } \dfrac{\sqrt{3}}{3}; x = 3, \theta = \dfrac{\pi}{2}$ **55.** $x = \frac{7}{2}, \theta = \dfrac{\pi}{6}; x = 7, \theta = \dfrac{\pi}{2}$

Exercise 3.10

1. $(-3, -45°)$

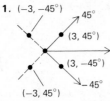

11. $(-3, 225°), (-3, -135°), (3, -315°)$ **13.** $(2, -225°), (-2, -45°), (-2, 315°)$
15. $(3, 90°), (3, -270°), (-3, -90°)$ **17.** $(-7, -280°), (7, 260°), (7, -100°)$ **19.** $(3, 105°),$
$(3, -255°), (-3, 285°)$ **21.** $\left(\dfrac{3}{2}, \dfrac{3\sqrt{3}}{2}\right)$ **23.** $(0, -3)$ **25.** $\left(\dfrac{-3\sqrt{2}}{2}, \dfrac{3\sqrt{2}}{2}\right)$ **27.** $\left(\dfrac{-5\sqrt{3}}{2}, \dfrac{-5}{2}\right)$
29. $(-1, \sqrt{3})$ **31.** $(3\sqrt{2}, 225°)$ **33.** $(4, 30°)$ **35.** $(4, 330°)$ **37.** $(5, 90°)$ **39.** $(6, 120°)$
41. $r^2 - 4r \cos \theta - 5 = 0$ **43.** $r^2 - 6r \cos \theta - 2r \sin \theta + 6 = 0$ **45.** $2r \cos \theta - 3r \sin \theta + 5 = 0$
47. $r^2 \sin^2 \theta = 4r \cos \theta$ **49.** $4r^2 \cos^2 \theta + 9r^2 \sin^2 \theta = 36$ **51.** $x^2 = 4(y + 1)$
53. $x^4 + y^4 + 2x^2y^2 - 4x^3 - 4xy^2 = 5x^2 + 9y^2$ **55.** $3x^2 + 4y^2 = 16 - 8x$ **57.** $y^2 - 3x^2 + 32x = 64$
59. $x^2 + y^2 = 2x + 3y$ **61.** **63.** **67.**

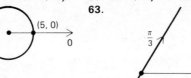

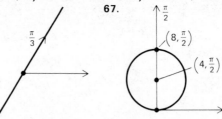

71.

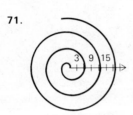

77.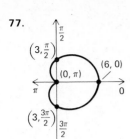

83.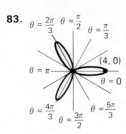

85.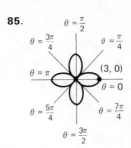

89.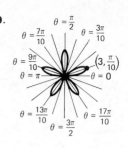

Exercise 3.11

1. $(x + 2)^2 + (y - 1)^2 = 16$; Circle   **3.** $\dfrac{x^2}{25} + \dfrac{y^2}{144} = 1$; Ellipse   **5.** $x - 3y + 11 = 0$; Line

7. $y = 4x$; Line   **9.** $\dfrac{y^2}{16} - \dfrac{x^2}{9} = 1$; Hyperbola   **11.** $\dfrac{(x + 1)^2}{9} + \dfrac{(y - 2)^2}{16} = 1$; Ellipse

13. $\dfrac{y^2}{9} - \dfrac{x^2}{4} = 1$; Hyperbola   **15.** $\dfrac{(y + 1)^2}{9} - \dfrac{(x - 5)^2}{4} = 1$; Hyperbola   **17.** $x = y^2$; Parabola

19. $y = 3x^2 + 1$; Parabola   **21.** $y = \dfrac{\sqrt{3}x}{3} - \dfrac{4.9x^2}{1200}$; Parabola   **23.** $y = x - \dfrac{4.9x^2}{50}$; Parabola

25. $x = 3\pi$, $y = 6$ **27.** $x = 10\pi$, $y = 0$ **29.** $x = 0$, $y = 0$

Exercise 4.1

1. 7   **3.** 2   **5.** 4   **7.** Remainder 0, -8 is root; remainder $120 \neq 0$, -2 not root   **9.** Remainder 0,
$\frac{3}{2}$ is root; remainder $42 \neq 0$, 3 not root   **11.** $(x - 3)(x + 5)(x - 4) = 0$ **13.** $(x^2 - 9)(x^2 - 16) = 0$
15. $(x^2 + 4)(x - 3) = 0$ **17.** 0, 2, -4 **19.** 0, -3, 2   **21.** 0, multiplicity 2; -1, multiplicity 3;
4, multiplicity 5   **23.** 3, multiplicity 2; i, multiplicity 3; $-i$, multiplicity 3   **25.** 0, multiplicity 4;
2, multiplicity 3; -9, multiplicity 1   **27.** $(x + 5)^2(x + 4)^3(x + 1)^4 = 0$ **29.** $(x - 4)^3(x - 3i)^5(x + 3i)^5 = 0$
31. 4, 4   **33.** 7, 7   **35.** 5, 5   **37.** 3 distinct roots: -1, 3, -4; 10   **39.** 4 distinct roots: -5, 2, i, $-i$; 7
41. $(-5i)^2 + 25 = 0$ **43.** $(2 - 4i)^2 - 4(2 - 4i) + 20 = 0$ **45.** $(1 + 2i)^2 - 2(1 + 2i) + 5 = 0$
47. $x^2 - 4x + 29$ **49.** $x^2 - 4x + 40$ **51.** $1, \dfrac{-3 \pm \sqrt{11}i}{2}$ **53.** $-2, \dfrac{-1 \pm \sqrt{7}i}{2}$ **55.** $4, -1 \pm 2i$

57. $\pm 1, \pm 2, \pm 4, \pm 8$ **59.** $\pm 1, \pm 3, \pm 5, \pm 15, \pm\dfrac{1}{7}, \pm\dfrac{3}{7}, \pm\dfrac{5}{7}, \pm\dfrac{15}{7}$ **61.** $1, \dfrac{3}{2}, -2$ **63.** $-4, \dfrac{1}{5}, \dfrac{-5}{2}$

65. 2, 3   **67.** $3, -5, \dfrac{1}{4}$ **69.** 0, 4, 5   **71.** $\dfrac{2}{11}, \dfrac{-1 \pm \sqrt{11}i}{2}$ **73.** $-6, \dfrac{-5 \pm \sqrt{21}}{2}$

75. $\dfrac{3}{8}, \dfrac{1 \pm \sqrt{2}i}{3}$ **77.** $P(0) = 12 > 0$, $P(1) = -36 < 0$ **79.** $P(0) = 9 > 0$, $P(2) = -3 < 0$
81. $P(0) = 15 > 0$, $P(2) = -21 < 0$ **83.** $P(-3) = 10 > 0$, $P(-4) = -36 < 0$
85. $P(0) = 60 > 0$, $P(-2) = -80 < 0$

Exercise 4.2

1. $x = 3$, $y = -1$ **3.** $x = -2$, $y = 3$ **5.** $x = -1$, $y = -4$ **7.** $x = 1$, $y = -1$ **9.** $x = -5$, $y = -2$
11. $x = -22$, $y = 3$, $z = -1$ **13.** $x = -1$, $y = -2$, $z = -3$ **15.** $A = 1$, $B = -2$, $C = 3$,
$D = -1$, $E = 4$ **17.** $x = 7$, $y = -5$, $z = 0$ **19.** $A = 2$, $B = 3$, $C = 4$ **21.** Not exactly one solution
23. $x = 2$, $y = 1$ **25.** No solution   **27.** No solution   **29.** No solution

Exercise 4.3

1. -1 **3.** 4 **5.** 8 **7.** Second row, third column, minus **9.** Fourth row, first column, minus

11. $\begin{vmatrix} 3 & 0 \\ 6 & -7 \end{vmatrix}$ **13.** $\begin{vmatrix} -1 & 5 \\ 0 & 4 \end{vmatrix}$ **15.** $\begin{vmatrix} -1 & 5 \\ -7 & 8 \end{vmatrix}$ **17.** -42 **19.** 9 **21.** -25 **23.** -292 **25.** -27

27. 686 **29.** 454 **31.** $(-)$ **33.** 2 **35.** 0

Exercise 4.4

1. $\dfrac{A}{x} + \dfrac{B}{x-3} + \dfrac{C}{x+7}$ **3.** $\dfrac{A}{x-3} + \dfrac{B}{(x-3)^2}$ **5.** $\dfrac{A}{x-1} + \dfrac{B}{(x-1)^2} + \dfrac{Cx+D}{3x^2+x+1}$

7. $\dfrac{1}{x} + \dfrac{3}{x-1} - \dfrac{2}{x-3}$ **9.** $\dfrac{-1}{x-2} + \dfrac{x+3}{x^2+x+1} + \dfrac{2x}{(x^2+x+1)^2}$ **11.** $x+5 + \dfrac{3}{x+4} + \dfrac{1}{x-2}$

13. $\dfrac{3}{x} + \dfrac{3}{x+5} - \dfrac{5}{x-2}$ **15.** $\dfrac{-1}{x-2} + \dfrac{2x+1}{x^2+4} + \dfrac{3x-1}{(x^2+4)^2}$

Exercise 4.5

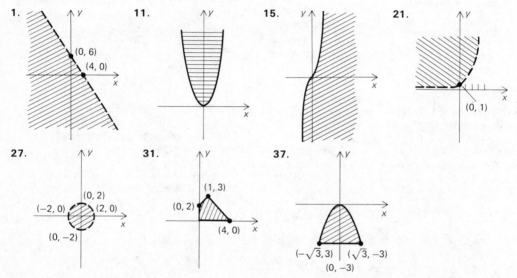

1.

(0, 6)
(4, 0)

11.

15.

21.

(0, 1)

27.

(0, 2)
(−2, 0)   (2, 0)
(0, −2)

31.

(1, 3)
(0, 2)
(4, 0)

37.

(−$\sqrt{3}$, 3)   ($\sqrt{3}$, −3)
(0, −3)

41. Corn, 300 acres; rye, 200 acres **43.** A, 5000 boxes; B, 0 boxes **45.** Emergencies, 6; regular check-ups, 18; 9 hours

Exercise 4.6

1. $\{3n\}$; $f(n) = 3n$ **3.** $\{n^2+1\}$; $f(n) = n^2+1$ **5.** $\{2n^2+1\}$; $f(n) = 2n^2+1$ **7.** 3, 17, 55, . . . ,

$2n^3+1$, . . . **9.** 2, 6, 12, . . . , $n(n+1)$, . . . **11.** $\dfrac{3}{4}$, $\left(\dfrac{3}{4}\right)^2$, $\left(\dfrac{3}{4}\right)^3$, . . . , $\left(\dfrac{3}{4}\right)^n$, . . . **13.** 0.75, $(0.75)^2$,

$(0.75)^3$, . . . , $(0.75)^n$, . . . **15.** 5, 8, 11, . . . , $5+3(n-1)$, . . .

17. $3+\dfrac{1}{5}$, $3+\dfrac{1}{25}$, $3+\dfrac{1}{125}$, $3+\dfrac{1}{625}$, . . . , $3+\left(\dfrac{1}{5}\right)^n$, . . . ; $\displaystyle\lim_{n\to\infty} 3 + \left(\dfrac{1}{5}\right)^n = 3$

19. $1 + \dfrac{1}{6}, 1 + \dfrac{1}{36}, 1 + \dfrac{1}{216}, 1 + \dfrac{1}{1296}, \ldots, 1 + \left(\dfrac{1}{6}\right)^n, \ldots; \lim\limits_{n \to \infty} 1 + \left(\dfrac{1}{6}\right)^n = 1$

21. $r = -2$; diverges **23.** $r = -1$; diverges **25.** $r = \dfrac{1}{2}$; converges to 0 **27.** $r = -1$; diverges

29. $r = 2$; diverges **31.** $S_n = \dfrac{1 - 4^n}{-3}$; $S_5 = 341$ **33.** $S_n = \dfrac{1}{2} - \dfrac{1}{2}\left(\dfrac{1}{3}\right)^n$; $S_5 = \dfrac{121}{243}$

35. $S_n = \dfrac{-2 + 2(-2)^n}{3}$; $S_5 = -22$ **37.** $S_n = \dfrac{1 - (-1)^n}{2}$; $S_6 = 0$ **39.** $S_n = 8 - 8\left(\dfrac{1}{2}\right)^n$; $S_3 = 7$

41. $d = 2$; $S_n = n^2 + n$; $S_5 = 30$ **43.** $d = -2$; $S_n = -n^2 + 2n$; $S_5 = -15$

45. $d = -3$; $S_n = \dfrac{9n - 3n^2}{2}$; $S_5 = -15$ **47.** Geometric; $r = 2$ **49.** Geometric; $r = 4$

51. $\sum\limits_{n=1}^{\infty} 3 \times 2^{n-1}$ **53.** $\sum\limits_{n=1}^{\infty} n^2$ **55.** $\sum\limits_{n=1}^{\infty} n(n + 3)$ **57.** $3^4 + 3^5 + 3^6 + 3^7 + 3^8 + 3^9$

59. $\dfrac{1}{2} + \dfrac{1}{4} + \dfrac{1}{6} + \cdots + \dfrac{1}{2n} + \cdots$ **61.** $\dfrac{3}{10} + \dfrac{4}{17} + \dfrac{5}{26} + \dfrac{6}{37} + \dfrac{7}{50}$ **63.** $\dfrac{1}{3} + \dfrac{8}{9} + 1 + \cdots + \dfrac{n^3}{3^n} + \cdots$

65. $1 + \sqrt{2} + \sqrt{3} + 2 + \sqrt{5} + \sqrt{6} + \sqrt{7} + 2\sqrt{2} + 3$

67. $S_1 = 1, S_2 = 1 - \dfrac{1}{3}, S_3 = 1 - \dfrac{1}{3} + \dfrac{1}{9}; S_n = 1 - \dfrac{1}{3} + \dfrac{1}{9} + \cdots + \left(\dfrac{-1}{3}\right)^{n-1}, S_n = \dfrac{3}{4}\left[1 - \left(\dfrac{-1}{3}\right)^n\right]; \dfrac{3}{4}$

69. $S_1 = 1, S_2 = 1 + \dfrac{1}{6}, S_3 = 1 + \dfrac{1}{6} + \dfrac{1}{36}; S_n = 1 + \dfrac{1}{6} + \dfrac{1}{36} + \cdots + \left(\dfrac{1}{6}\right)^{n-1}, S_n = \dfrac{6}{5}\left[1 - \left(\dfrac{1}{6}\right)^n\right]; \dfrac{6}{5}$

71. Converges; $\dfrac{4}{3}$ **73.** Converges; $\dfrac{3}{4}$ **75.** Diverges **77.** 2046· **79.** $100(1.06)^{10}$ dollars

81. 8 days; 1 day **83.** $\dfrac{1}{2}(x_1^2 + x_2^2 + x_3^2 + x_4^2 + x_5^2 + x_6^2); \dfrac{1}{2}\sum\limits_{i=1}^{6} x_i^2$ **85.** 2.7

Exercise 4.7

1. *Part I* $3 = \tfrac{3}{2}(3^1 - 1)$ since $3 = \dfrac{3 \times 2}{2}$.

Part II If $3 + 3^2 + \cdots + 3^k = \tfrac{3}{2}(3^k - 1)$, then $3 + 3^2 + \cdots + 3^k + 3^{k+1} = \tfrac{3}{2}(3^k - 1) + 3^{k+1}$

$$= \dfrac{3(3^k - 1) + 2 \times 3^{k+1}}{2}$$

$$= \dfrac{3(3^k - 1 + 2 \times 3^k)}{2}$$

$$= \dfrac{3(3 \times 3^k - 1)}{2}$$

$$= \tfrac{3}{2}(3^{k+1} - 1)$$

Check $S_n = \dfrac{3 - 3 \times 3^n}{1 - 3} = \dfrac{3(1 - 3^n)}{-2} = \tfrac{3}{2}(3^n - 1)$

21. *Part I* $2^3 > 2^2$ since $8 > 4$.

Part II $(k + 1)^3 = k^3 + 3k^2 + 3k + 1$
$> k^2 + 3k^2 + 3k + 1$, if $k^3 > k^2$
$= 4k^2 + 3k + 1$
$> k^2 + 2k + 1$, for $k \geq 2$
$= (k + 1)^2$
$(k + 1)^3 > (k + 1)^2$, if $k^3 > k^2, k \geq 2$

27. *Part I* $x^2 \cdot x^1 = x^{2+1}$ since $x^2 \cdot x^1 = x^3$.
Part II $x^2 \cdot x^{k+1} = x^2 \cdot x^k \cdot x$
$$= x^{2+k} \cdot x, \text{ if } x^2 \cdot x^k = x^{2+k}$$
$$= x^{2+k+1}$$

31. *Part I* $(x - y)$ is a factor of $x^{2(1)} - y^{2(1)} = x^2 - y^2$ since $x^2 - y^2 = (x - y)(x + y)$.
Part II $x^{2(k+1)} - y^{2(k+1)} = x^{2k+2} - y^{2k+2}$
$$= x^{2k}x^2 - y^{2k}y^2 \text{ since } a^{m+2} = a^m \cdot a^2$$
$$= x^{2k}x^2 - x^2y^{2k} + x^2y^{2k} - y^{2k}y^2$$
$$= x^2(x^{2k} - y^{2k}) + y^{2k}(x^2 - y^2)$$
Note that $(x - y)$ is a factor of $x^2 - y^2$, so if $(x - y)$ is a factor of $x^{2k} - y^{2k}$, then $(x - y)$ is a factor of $x^{2(k+1)} - y^{2(k+1)}$.

33. *Part I* 1 is not greater than $1 + 3$. *Part II* If we assume that $k > k + 3$, then adding 1 to both sides we have $k + 1 > (k + 1) + 3$.

Exercise 4.8

1. $x^8 + 8x^7 + 28x^6 + 56x^5 + 70x^4 + 56x^3 + 28x^2 + 8x + 1$ **3.** $x^4 + 8x^3y + 24x^2y^2 + 32xy^3 + 16y^4$
5. $1 + 6x + 15x^2 + 20x^3 + 15x^4 + 6x^5 + x^6$ **7.** $x^5 - 5x^4y + 10x^3y^2 - 10x^2y^3 + 5xy^4 - y^5$
9. $16a^4 + 96a^3b + 216a^2b^2 + 216ab^3 + 81b^4$ **11.** $a^{20} + 20a^{19}b + 190a^{18}b^2 + 1140a^{17}b^3 + \cdots$
13. $1 + 12x + 66x^2 + 220x^3 + \cdots$ **15.** $729x^6 - 1458x^5y + 1215x^4y^2 - 540x^3y^3 + \cdots$
17. $x^{10} - 10x^9y + 45x^8y^2 - 120x^7y^3 + \cdots$ **19.** $x^{30} + 30x^{29}y + 435x^{28}y^2 + 4060x^{27}y^3 + \cdots$
21. $56; 56x^3y^5$ **23.** $210; 210x^4y^6$ **25.** $435; 1740x^{28}y^2$ **27.** $66; 66a^2b^{10}$ **29.** $1820; 1820x^{12}y^4$
31. 6048 **33.** 1375 **35.** 823680 **37.** $\binom{6}{0}\binom{6}{1}\binom{6}{2}\binom{6}{3}\binom{6}{4}\binom{6}{5}\binom{6}{6}$ **39.** $\binom{8}{0}\binom{8}{1}\binom{8}{2}\binom{8}{3}\binom{8}{4}\binom{8}{5}\binom{8}{6}\binom{8}{7}\binom{8}{8}$

41. $\binom{4}{1} = 4, \binom{3}{0} = 1, \binom{3}{1} = 3; 4 = 1 + 3$ **43.** $\binom{6}{2} = 15, \binom{5}{1} = 5, \binom{5}{2} = 10; 15 = 5 + 10$

45. $\binom{7}{1} = 7, \binom{6}{0} = 1, \binom{6}{1} = 6; 7 = 1 + 6$ **47.** $\binom{4}{1} = 4, \binom{4}{3} = 4$ **49.** $\binom{6}{1} = 6, \binom{6}{5} = 6$

51. $9 \times 8 \times 7 \times 6! = 9 \times 8 \times 7 \times 6 \times 5 \times 4 \times 3 \times 2 \times 1 = 9!$
53. $10 \times 9 \times 8 \times 7 \times 6! = 10 \times 9 \times 8 \times 7 \times 6 \times 5 \times 4 \times 3 \times 2 \times 1 = 10!$
55. $11 \times 10 \times 9 \times 8 \times 7! = 11 \times 10 \times 9 \times 8 \times 7 \times 6 \times 5 \times 4 \times 3 \times 2 \times 1 = 11!$
57. $1 + \frac{1}{3}x - \frac{1}{9}x^2 + \frac{5}{81}x^3 + \cdots$ **59.** $1 + \frac{3}{2}x + \frac{3}{8}x^2 - \frac{1}{16}x^3 + \cdots$

Index